REVIEWS in MINERALOGY
and GEOCHEMISTRY

Volume 39 2000

Transformation Processes in Minerals

Editors:

Simon A.T. Redfern & Michael A. Carpenter

Department of Earth Sciences
University of Cambridge
Cambridge, U.K.

COVER PHOTOGRAPH:

Computer simulation of the tweed microstructure induced during a ferroelastic phase transition. Green indicates ares of the crystal in which the order parameter is positive, and red indicates areas in which it is negative. Yellow areas are domain boundaries in which the order parameter is close to zero.

Series Editor: ***Paul H. Ribbe***

Virginia Polytechnic Institute and State University
Blacksburg, Virginia

MINERALOGICAL SOCIETY of AMERICA

Washington, DC

REVIEWS IN MINERALOGY AND GEOCHEMISTRY

(Formerly: REVIEWS IN MINERALOGY)

ISSN 1529-6466

Volume 39

Transformation Processes in Minerals

ISBN 0-939950-51-0

** This volume is the first of a series of review volumes published jointly under the banner of the Mineralogical Society of America and the Geochemical Society. The newly titled *Reviews in Mineralogy and Geochemistry* has been numbered contiguously with the previous series, *Reviews in Mineralogy*.

Additional copies of this volume as well as others in this series may be obtained at moderate cost from:

THE MINERALOGICAL SOCIETY OF AMERICA
1015 EIGHTEENTH STREET, NW, SUITE 601
WASHINGTON, DC 20036 U.S.A.

Transformation Processes in Minerals

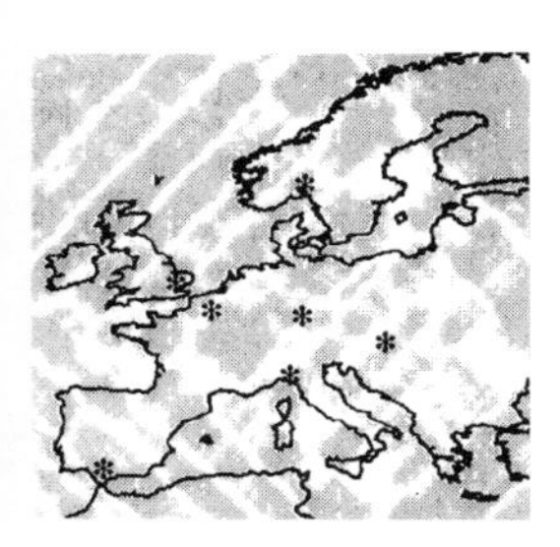

EU NETWORK ON MINERAL TRANSFORMATIONS
(*ERB-FMRX-CT97-0108*)

EUROPEAN MINERALOGICAL UNION

PREFACE

Phase transformations occur in most types of materials, including ceramics, metals, polymers, diverse organic and inorganic compounds, minerals, and even crystalline viruses. They have been studied in almost all branches of science, but particularly in physics, chemistry, engineering, materials science and earth sciences. In some cases the objective has been to produce materials in which phase transformations are suppressed, to preserve the structural integrity of some engineering product, for example, while in other cases the objective is to maximise the effects of a transformation, so as to enhance properties such as superconductivity, for example. A long tradition of studying transformation processes in minerals has evolved from the need to understand the physical and thermodynamic properties of minerals in the bulk earth and in the natural environment at its surface. The processes of interest have included magnetism, ferroelasticity, ferroelectricity, atomic ordering, radiation damage, polymorphism, amorphisation and many others—in fact there are very few minerals which show no influence of transformation processes in the critical range of pressures and temperatures relevant to the earth. As in all other areas of science, an intense effort has been made to turn qualitative understanding into quantitative description and prediction via the simultaneous development of theory, experiments and simulations. In the last few years rather fast progress has been made in this context, largely through an interdisciplinary effort, and it seemed to us to be timely to produce a review volume for the benefit of the wider scientific community which summarises the current state of the art. The selection of transformation processes covered here is by no means comprehensive, but represents a coherent view of some of the most important processes which occur specifically in minerals. A number of the contributors have been involved in a European Union funded research network with the same theme, under the Training and Mobility of Researchers programme, which has stimulated much of the most recent progress in some of the areas covered. This support is gratefully acknowledged.

The organisers of this volume, and the short course held in Cambridge, UK, to go with it, are particularly grateful to the Mineralogical Society of Great Britain and Ireland, the German Mineralogical Society, the European Mineralogical Union and the Natural Environment Research Council of Great Britain for their moral and financial support of the short course. The society logos are reproduced here, along with the logo for the Mineralogical

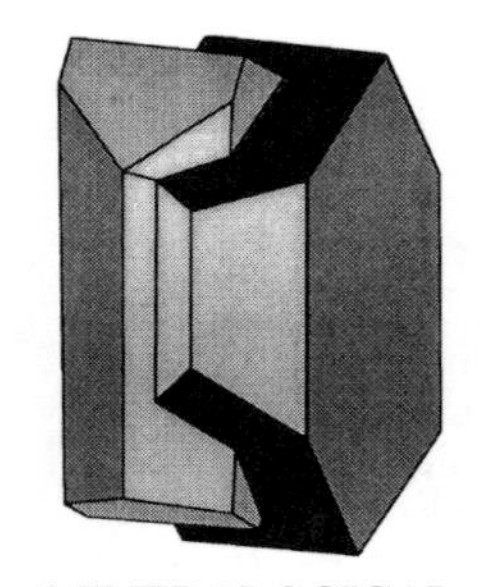

MINERALOGICAL
SOCIETY OF GREAT
BRITAIN AND IRELAND

Society of America, both in acknowledgement of this support and also to emphasise the coherency of the mineralogical communities in Europe and North America. The short course is the first MSA- sponsored short course to be held in Europe, and it is our hope that it will promote, further, the strong ties of scientific collaboration and personal friendship that draw us all together.

Simon Redfern, Michael Carpenter
Cambridge, July 2000

TABLE OF CONTENTS

1 Rigid Unit Modes in Framework Structures: Theory, Experiment and Applications

Martin T. Dove, Kostya O. Trachenko, Matthew G. Tucker, David A. Keen

2 Strain and Elasticity at Structural Phase Transitions in Minerals

Michael A. Carpenter

3 Mesoscopic Twin Patterns in Ferroelastic and Co-Elastic Minerals

Ekhard K. H. Salje

4 High-Pressure Structural Phase Transitions

R.J. Angel

5 Order-Disorder Phase Transitions

Simon A. T. Redfern

6 Phase Transformations Induced by Solid Solution

Peter J. Heaney

7 Magnetic Transitions in Minerals

Richard J. Harrison

8 NMR Spectroscopy of Phase Transitions in Minerals

Brian L. Phillips

9 Insights into Phase Transformations from Mössbauer Spectroscopy

Catherine A. McCammon

10 Hard Mode Spectroscopy of Phase Transitions

Ulli Bismayer

11 Synchrotron Studies of Phase Transformations

John B. Parise

12 Radiation-Induced Amorphization

Rodney C. Ewing, Alkiviathes Meldrum
LuMin Wang, ShiXin Wang

1 Rigid Unit Modes in Framework Structures: Theory, Experiment and Applications

Martin T. Dove, Kostya O. Trachenko, Matthew G. Tucker

Department of Earth Sciences
University of Cambridge
Downing Street
Cambridge CB2 3EQ, UK

David A. Keen

ISIS Facility, Rutherford Appleton Laboratory
Chilton, Didcot, Oxfordshire OX11 0QX, UK

INTRODUCTION

The theoretical construction of the "Rigid Unit Mode" model arose from asking a few simple questions about displacive phase transitions in silicates (Dove 1997a,b). The simplest of these was why displacive phase transitions are so common? Another was why Landau theory should seemingly be so successful in describing these phase transitions? As the construction developed, it became clear that the model is able to describe a wide range of properties of silicates, in spite of the gross over-simplifications that appear so early in the development of the approach. In this review, we will outline the basic principles of the Rigid Unit Mode model, the experimental evidence in support of the model, and some of the applications of the model.

The Rigid Unit Mode (RUM) model was developed to describe the behavior of materials with crystal structures that can be described as frameworks of linked polyhedra. In aluminosilicates, the polyhedra are the SiO_4 or AlO_4 tetrahedra, and in perovskites, these may be the TiO_6 octahedra. At the heart of the RUM model is the observation that the SiO_4 or AlO_4 tetrahedra are very stiff. One measure of the stiffness of the tetrahedra might be that the vibrations involving significant distortions of the tetrahedra have squared frequency values above 1000 THz^2 (Strauch and Dorner 1993), whereas vibrations in which there are only minimal distortions of the tetrahedra and a buckling of the framework have squared frequency values of around 1 THz^2 (see later). The values of the squared frequency directly reflect the force constants associated with the motions of a vibration, so clearly the stiffness of a tetrahedron is 2–3 orders of magnitude larger than the stiffness of the framework against motions in which the tetrahedra can move without distorting. This large range of stiffness constants invites the simple approximation of assigning a value for the stiffness of the tetrahedra that is in Figure 1. Any modes of motion that do not involve any distortions of the tetrahedra will have zero restoring force in this approximation, and hence low energy. The question of whether such modes can exist had already been answered in papers such as that of Grimm and Dorner (1975), in which the phase transition in quartz was described using a simple mechanical model involving rotations of rigid tetrahedra. The rotations associated with the phase transition cause the large changes in the lattice parameters that are measured experimentally. The model of Grimm and Dorner (1975) did not give exact agreement with experiment, but we will show later that this is because a simple static model will automatically overestimate the volume of a dynamically disordered high-temperature phase.

1529-6466/00/0039-0001$05.00

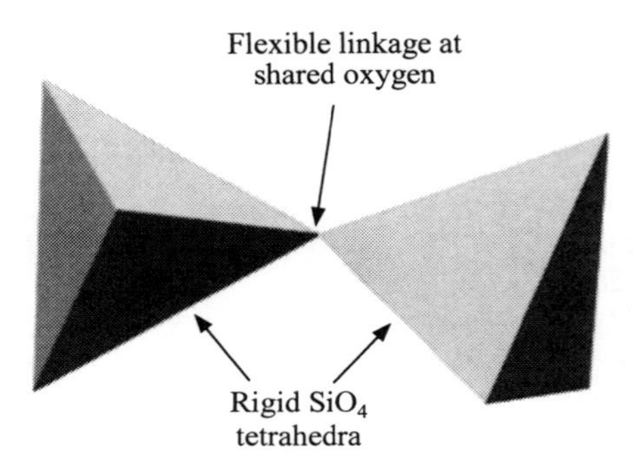

Figure 1. Two linked tetrahedra, showing stiff and floppy forces.

The significant step forward was due to Vallade and co-workers in Grenoble (Berge et al. 1986, Bethke et al. 1987, Dolino et al. 1989, 1992, Vallade et al. 1992). They were interested in the incommensurate phase transition in quartz (Dolino 1992). The model they proposed is very simple. The important point is that there is a soft optic mode of zero wave vector whose frequency decreases on cooling towards the phase transition. This mode is the RUM identified by Grimm and Dorner (1975). We now consider wave vectors along $\mathbf{a}^*$. If the soft mode is part of a dispersion branch that has little variation in frequency across the range of wave vectors, the whole branch may soften on cooling. Now suppose (as is the case) that there is an acoustic mode that has a different symmetry at $\mathbf{k} = 0$, but the same symmetry for $\mathbf{k} \neq 0$. This will lead to an energy that involves the product of the amplitude of the pure acoustic mode and the pure optic mode on increasing **k**, which in turn leads to hybridisation the motions. Because of the symmetry is different at $\mathbf{k} = 0$, this interaction will vanish as $\mathbf{k} \rightarrow 0$, and therefore varies as k^2. The overall effect is demonstrated in Figure 2. The softening optic mode causes the acoustic mode frequency to be depressed, but the effect only sets in on increasing **k**. As a result, when the optic mode has softened sufficiently, it causes the acoustic branch to dip to zero frequency at a wave vector that is close to $\mathbf{k} = 0$ but which has to be greater than zero. This gives rise to the incommensurate phase transition (Tautz et al. 1991). So far we have said nothing about RUMs, but it is not difficult to appreciate that for a whole phonon branch to have a low frequency each phonon along the branch from $\mathbf{k} = 0$ to $\mathbf{k} = \mathbf{a}^*/2$ must be a RUM. Vallade set about establishing this essential ingredient in the picture of the incommensurate phase transition, and developed an analytical method to compute the whole spectrum of RUMs in β-quartz (Berge et al. 1986, Vallade et al. 1992). The important finding was that for **k** along $\mathbf{a}^*$ there is a whole branch of RUMs, not just at $\mathbf{k} = 0$, together with other lines or planes in reciprocal space that have RUMs.

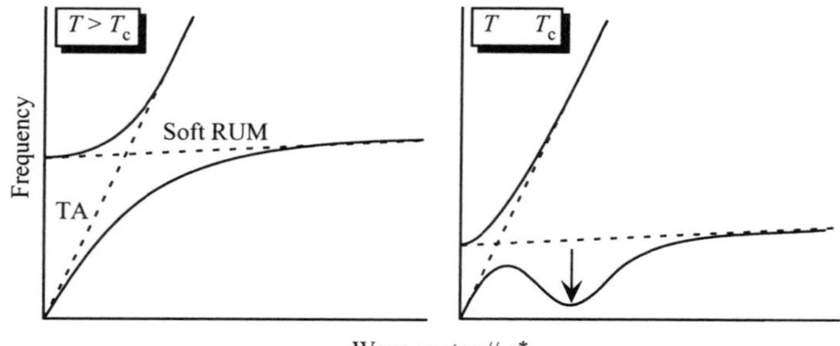

Figure 2. Phonon picture of the origin of the incommensurate phase transition in quartz. The two plots show the $\mathbf{a}^*$ dispersion curves for the transverse acoustic mode (TA) and the soft optic RUM, at temperatures above (left) and close to (right) the incommensurate phase transition. The RUM has a frequency that is almost constant with **k**, and as it softens it drives the TA mode soft at an incommensurate wave vector owing to the fact that the strength of the coupling between the RUM and the acoustic mode varies as k^2.

This description of the mechanism of the phase transition in quartz raised the question of whether RUMs could act as soft modes for other displacive phase transitions in silicates (Dove 1997a,b, Hammonds et al. 1996). To address this issue a practical computational method was essential, and the easiest idea to implement was the "split-atom" method (Giddy et al. 1993), which we will describe later. This has enabled us to search for RUMs in any framework structure with little effort (Hammonds et al. 1996). From these calculations we have been able to rationalise a number of phase transitions in important silicate minerals (Dove et al. 1995). By analysing the complete flexibility of specific framework structures we have also been able to interpret the nature of the crystal structures of high-temperature phases, and the theory has had a close link to a number of experiments, as will be discussed later in this article. The RUM model has also given a number of other new insights. One of these is that we have been able to link the value of the transition temperature to the stiffness of the tetrahedra (Dove et al. 1993, 1995, 1999a), which is a particularly nice insight. The RUM model has also been used to explain the origin of negative thermal expansion in framework structures (Pryde et al. 1996, Welche et al. 1998, Heine et al. 1999), and to understand the catalytic behavior of zeolites (Hammonds et al. 1997a,b; 1998a). More recently, the RUM model has given new insights into the links between crystalline and amorphous materials at an atomic level, from both the structure and dynamic perspectives (Keen and Dove 1999, 2000; Dove et al. 2000a, Harris et al. 2000).

We need to stress that the simple model of stiff tetrahedra and no forces between tetrahedra will give results that are necessarily modified when forces between tetrahedra are taken into account. One extension to the model is to include a harmonic force operating between the centres of neighboring tetrahedra (Hammonds et al. 1994, 1996; Dove et al. 1995). This has the effect of increasing the frequencies of all RUMs that do not involve simple torsional motions of neighboring tetrahedra. When realistic inter-tetrahedral forces are taken into account, all frequencies are modified, and the RUMs will no longer have zero frequency. They will, however, retain their basic RUM character, and in a harmonic lattice dynamics calculation on a high-temperature phase the RUMs may have imaginary values for their frequencies (Dove et al. 1992, 1993), reflecting the fact that RUM distortions can drive displacive phase transitions with the RUMs playing the role of the classic soft mode (Dove1997a,b; Dove et al. 1995, Hammonds et al. 1996).

FLEXIBILITY OF NETWORK STRUCTURES: SOME BASIC PRINCIPLES

Engineering principles

Before we plunge into the science that has developed from the RUM model, we should take note of some insights from the engineering perspectives. In fact, it was the great 19th Century physicist, James Clerk Maxwell (1864), who laid the foundation for the study of the rigidity of frameworks by a simple consideration of the requirements of engineering structures, and the basic engineering principles are still named after him. Any structure will be built from many individual parts, which each have their own degrees of freedom, and these individual parts will be held together by a set of engineering constraints. For example, a structure built from rigid rods will have 5 degrees of freedom associated with each rod. The rods are linked by corner pins, which mean that the ends of two rods are constrained to lie at the same point in three-dimensional space. This gives three constraints per joint. An alternative approach is to consider the joints to be the objects with three degrees of freedom, and to treat each rod as giving one constraint that acts to ensure that two joints lie at a fixed separation. Both methods of assigning and counting degrees of freedom and constraints should lead to equivalent conclusions about the degree of flexibility of a network of objects.

To evaluate the rigidity of a structure, the total numbers of degrees of freedom and constraints are calculated. If the number of degrees of freedom exceed the number of constraints by 6, the structure has some degree of internal flexibility, and it is not completely rigid. The structure is underconstrained. On the other hand, if the number of constraints exceeds the number of degrees of freedom minus 6, the structure is overconstrained and cannot easily be deformed. This is the key to designing engineering structures such as girder bridges, but the ideas can be applied to understanding the flexibility of crystal structures.

The second engineering approach reviewed above is equivalent to treating each atom as an object with three degrees of freedom, and each bond as giving one constraint. This is a natural and often-used perspective, and is particularly appropriate for atomic systems such as the chalcogenide glasses (He and Thorpe 1985, Cai and Thorpe 1989, Thorpe et al. 1997). However, for the study of silicates we prefer an alternative viewpoint, which is to treat each SiO_4 tetrahedron as a rigid object with six degrees of freedom, and each corner linked to another as having three constraints. In a network structure such as quartz, where all tetrahedra are linked to four others, the number of constraints per tetrahedron is equal to $3 \times$ the number of corners $\div 2$ (to account for the sharing of constraints by two tetrahedra). Thus the number of constraints is equal to the number of degrees of freedom, and this implies that a framework silicate has a fine balance between the number of degrees of freedom and the number of constraints. The alternative atomic approach is to note that there are 9 degrees of freedom associated with the three atoms assigned to each tetrahedron (i.e. one SiO_2 formula unit), and the rigidity of the tetrahedron is assured by 9 bond constraints (this is one less than the number of bonds). This analysis leads to the same conclusion as before, namely that there is an exact balance between the numbers of degrees of freedom and constraints.

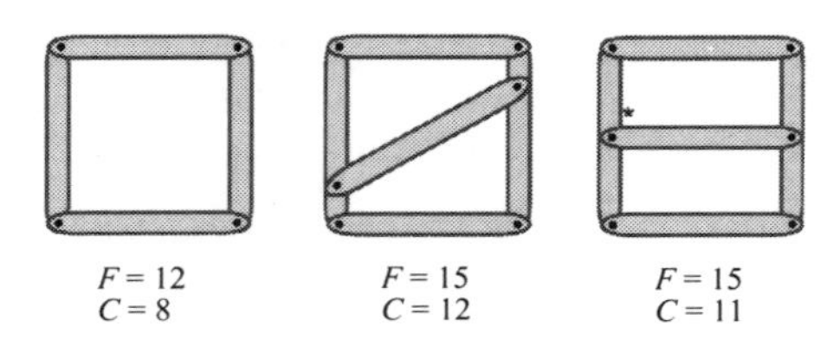

Figure 3. Three structures of jointed bars in two dimensions, with a count of the numbers of constraints, C, and degrees of freedom, F, based on there being three degrees of freedom for each rod and 2 constraints for each joint. The structures on the left and in the centre are trivial to analyse, and the existence of an internal shear mode that is equivalent to a RUM only in the under-constrained structure is intuitive and in accord with the standard method of Maxwell counting. The structure on the right shows the effect of symmetry, By replacing the two constraints at the joint marked with an asterisk by a single constraint that keeps the centre bar parallel to the top bar we see how the balance between the numbers of degrees of freedom and constraints allows the existence of the internal shear RUM, even though the structure has the same topology as the overconstrained structure in the middle.

The role of symmetry

This simple analysis assumes that none of the constraints is dependent on each other. One simple example of the role of symmetry is shown in Figure 3, in which we compare the rigidity of three two-dimensional structures consisting of hinged rods (Giddy et al. 1993, Dove et al. 1995, Dove 1997b). A square of four hinged rods has an internal shear instability. This can be removed by the addition of a fifth cross-bracing rod. In both cases, the simple counting of the number of degrees of freedom (3 per rod) and the number of constraints (2 per linkage) is consistent with our intuitive expectations. However, if the fifth rod is set to be parallel to the top and bottom rods, the shear instability is now enabled. This can be reconciled with the constraint count by noting that one of the pair of constraints at one end of the fifth rod can be replaced by a single constraint that ensures that the fifth bar remains parallel to the top bar. In this sense, symmetry has forced two of the constraints to become dependent on each other, or "degenerate", reducing the number of independent constraints by 1.

Thus we might expect that symmetry could reduce the number of independent constraints in a framework silicate, allowing the total number of degrees of freedom to exceed the number of independent constraints, so that the structure will have some internal degrees of flexibility. We might also expect that as symmetry is lowered, perhaps as a result of a displacive phase transition, the number of independent constraints will be increased and the flexibility of the structure will be reduced. We will see that this is found in practice (Hammonds et al. 1996).

THE SPECTRUM OF RIGID UNIT MODES IN SILICATES

The "split-atom" method

Our approach to calculating the spectrum of RUMs in a network structure recognises the existence of two types of constraint. One is where the SiO_4 tetrahedra are rigid, and the other is where the corners of two linked tetrahedra should remain joined. There may be any number of ways of investigating the flexibility of a structure subject to these constraints, but we have developed a pragmatic solution that lends itself to the interpretation of the flexibility in terms of low-energy phonon modes. This approach uses the formalism of molecular lattice dynamics (Pawley 1992). The tetrahedra are treated as rigid molecules. The oxygen atoms at the corners of the tetrahedra are treated as two atoms, called the "split atoms", which are assigned to each of the two tetrahedra sharing a common corner (Giddy et al. 1993). This is illustrated in Figure 4. The split atoms are linked by a stiff spring of zero equilibrium length, so that if a pair of tetrahedra move relative to each other in such a way as to open up the pair of split atoms there will be a significant energy penalty. Conversely, if the structure can flex in such a way that none of the split atoms are separated, there will be no energy costs associated with the deformation. Putting this into the language of lattice dynamics, if a vibration of a given wave vector **k** can propagate with the tetrahedra rotating or being displaced but without any of the split atoms opening up, it is equivalent to a vibration in which the tetrahedra can move without distorting. These are the modes of motion that are the RUMs, and our task is to enumerate the RUMs associated with any wave vector. We have argued that the stiffness of the spring is related to the stiffness of the tetrahedra (Dove 1997, Dove et al. 1999).

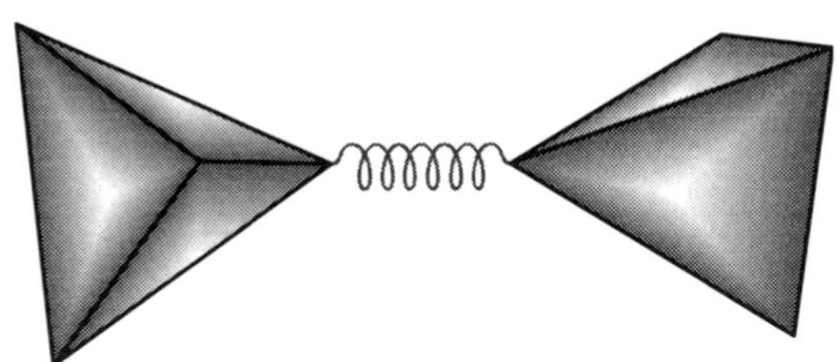

Figure 4. Representation of the "split-atom" method. The spring only opens up for motions that would otherwise distort the tetrahedra.

We have developed a molecular lattice dynamics program, based on the split-atom method, for the calculation of the RUMs in any framework structure (Giddy et al. 1993, Hammonds et al. 1994). This solves the dynamical matrix for any given wave vector, and the modes with zero frequency are the RUMs associated with that wave vector. The

Table 1. Numbers of rigid unit modes for symmetry points in the Brillouin zones of some aluminosilicates, excluding the trivial acoustic modes at $k = 0$. The "—" indicates that the wave vector is not of special symmetry in the particular structure. The numbers in brackets denote the numbers of RUMs that remain in any lower-symmetry low-temperature phases. (Taken from Hammonds et al. 1996.)

k	*Quartz*	*Cristobalite*	*Tridymite*	*Sanidine*	*Leucite*	*Cordierite*
	$P6_222$	$Fd3m$	$P6_3/mmc$	$C2/m$	$Ia3d$	$Cccm$
$0,0,0$	1 (0)	3 (1)	6	0	5 (0)	6
$0,0,\frac{1}{2}$	3 (1)	—	6	1	—	6
$\frac{1}{2},0,0$	2 (1)	—	3	—	—	6
$\frac{1}{3},\frac{1}{3},0$	1 (1)	—	1	—	—	6
$\frac{1}{3},\frac{1}{3},\frac{1}{2}$	1 (1)	—	2	—	—	0
$\frac{1}{2},0,\frac{1}{2}$	1 (1)	—	2	—	4 (0)	2
$0,1,0$	—	2	—	1	—	—
$\frac{1}{2},\frac{1}{2},\frac{1}{2}$	—	3 (0)	—	0	0	—
$0,1,\frac{1}{2}$	—	—	—	1	—	—
$0,0,\xi$	3 (0)	2 (0)	6	—	0	6
$0,\xi,0$	2 (0)	2 (2)	3	1	0	6
$\xi,\xi,0$	1 (1)	1 (0)	1	—	4 (0)	6
ξ,ξ,ξ	—	3 (0)	—	—	0	—
$\frac{1}{2},0,\xi$	1 (0)	—	2	—	0	2
$\xi,\xi,\frac{1}{2}$	1 (1)	—	0	—	0	0
$\frac{1}{2}-\xi,2\xi,0$	1 (1)	—	1	—	—	6
$\frac{1}{2}-\xi,2\xi,\frac{1}{2}$	1 (1)	—	0	—	—	0
$0,\xi,\frac{1}{2}$	0 (0)	—	1	1	—	—
$\xi,1,\xi$	—	1 (0)	—	—	0	—
$\xi,\zeta,0$	1 (0)	0	1	—	0	6
$\xi,0,\zeta$	0 (0)	0	2	1	0	0
$\xi,1,\zeta$	—	0	—	1	0	—
ξ,ξ,ζ	0 (0)	1 (0)	0	—	0	0

program, which we call CRUSH, is straightforward to run for any periodic structure (Hammonds et al. 1994). Some sample results are given in Table 1. In this table we show the numbers of RUMs for wave vectors lying on special points, on special lines or on special planes in reciprocal space.

Three-dimensional distribution of RUMs

Table 1 actually only presents part of the story. We have also found that there are cases where RUMs are found for wave vectors lying on exotic curved surfaces in reciprocal space (Dove et al. 1996, Dove et al. 1999b, Hayward et al. 2000). Examples are given in Figure 5. Initially we thought that since the existence of RUMs may be due

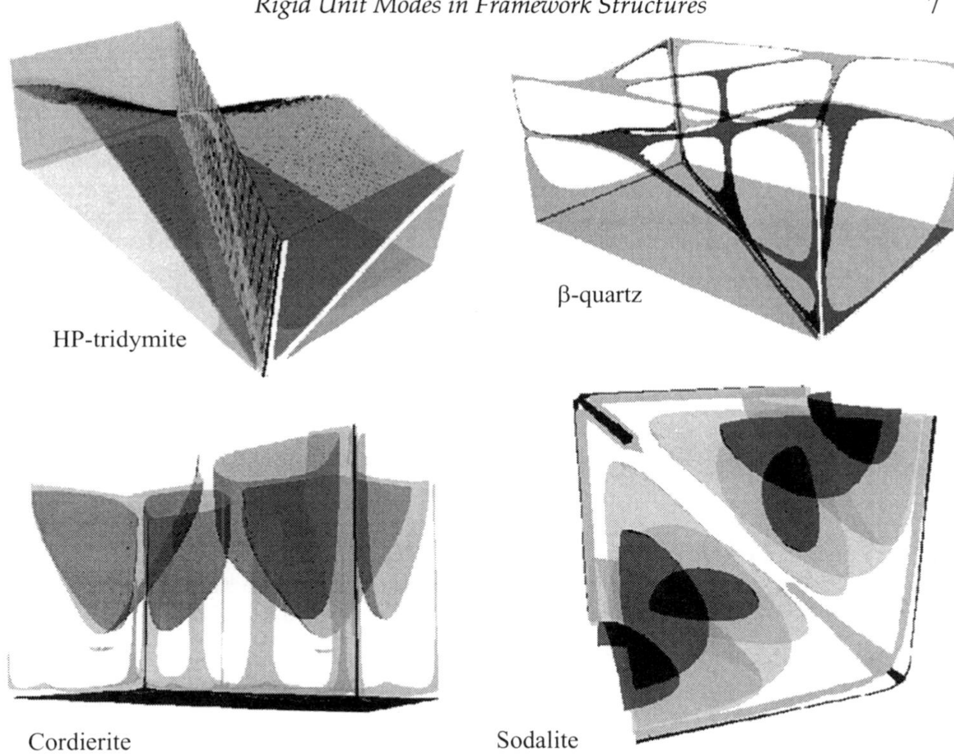

Figure 5. Four examples of three-dimensional loci in reciprocal space of wave vectors containing one or more RUMs.

to the presence of symmetry, the RUMs would only exist for wave vectors of special symmetry. However, it is now clear that this is too simplistic, but we do not yet understand how these curved surfaces arise. We would stress, though, that at least in the case of tridymite there is extremely good experimental data in support of the curved surfaces we have calculated (Withers et al. 1994, Dove et al. 1996).

In the examples given in Table 1 and Figure 5, the RUMs lie on planes of wave vectors. The number of RUMs lying on a plane is but a tiny fraction of the total number of vibrations, although we will see later that this in no way diminishes the importance of these RUMs. However, another great surprise was to discover that for some high-symmetry zeolites, there are one or more RUMs for each wave vector (Hammonds et al. 1997a,b; 1998a). In this case, the fraction of all vibrations that are also RUMs is now not vanishingly small. This discovery has been exploited to give a new understanding of the methods by which zeolite structures can flex to give localised distortions. This will be briefly discussed in the last section.

"Density of states" approach

The existence of RUMs on complex surfaces in reciprocal space makes it hard to measure the actual flexibility of a structure when using calculations of RUMs only for a few representative wave vectors. We have developed the approach of using a density of states calculation to characterise the RUM flexibility (Hammonds et al. 1998b). In a material that has no RUMs, the density of states at low frequency ω has the usual form

$g(\omega) \propto \omega^2$. But if there are RUMs, $g(\omega)$ as calculated by CRUSH will have a non-zero value at $\omega = 0$. This is not to say that when all the real interactions are taken into account the RUMs will have zero frequency. For example, when we modify the simple model by including a Si...Si harmonic interaction, many of the RUMs no longer have zero frequencies (Hammonds et al. 1996). And when the full set of interatomic interactions are included, as in a full lattice dynamics calculation, none of the RUMs will have zero frequency. In fact, if the calculation is performed on a high-temperature phase, some of the RUMs will have calculated frequencies that are imaginary, representing the fact that

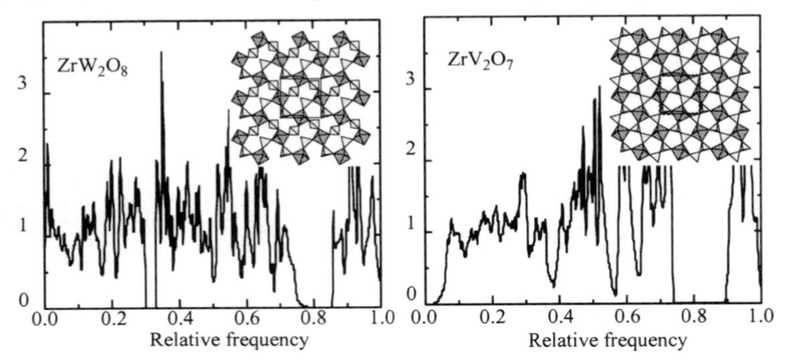

Figure 6. RUM density of states of two network structures, ZrW_2O_8 and ZrV_2O_7, with the crystal structures shown as inserts.

they could act as soft modes for displacive phase transitions. This was seen in a lattice dynamics calculation for β-cristobalite, where all the RUMs for wave vectors along symmetry directions in reciprocal space had imaginary frequencies (Dove et al. 1993). In Figure 6, we show the CRUSH density of states for two similar network materials, ZrW_2O_8 and ZrV_2O_7 (Pryde et al. 1996). Both structures have frameworks of corner-linked octahedra and tetrahedra, but in the first material, there is one non-bridging bond on each tetrahedron whereas these bonds are linked in the latter material. The structures are shown in Figure 6. There are RUMs in ZrW_2O_8, and this is shown as the non-zero value in the density of states at small frequencies. On the other hand, there are no RUMs in ZrV_2O_7, and as a result, the density of states has the normal ω^2 dependence at low frequency. The value of $g(\omega)$ at low ω gives a measure of the RUM flexibility of a structure.

Framework structures containing octahedra

The existence of RUMs in the cubic perovskite structure (Giddy et al. 1993) suggests that we may consider framework structures that contain octahedral units as well as the tetrahedral units we have concentrated on so far. However, the simple Maxwell counting suggests that the existence of octahedral polyhedra may cause a structure to be overconstrained. Each octahedron has the same 6 degrees of freedom as a tetrahedron, but with 6 corners there will be 9 constraints per octahedron rather than the 6 constraints for a tetrahedron. The solution to this problem has been discussed in more detail for the case of perovskite (Dove et al. 1999b). However, unless the symmetry causes a significant number of constraints to be degenerate, the greater number of constraints for an octahedron does mean that there will be no RUMs in a network containing octahedra. This has been demonstrated by a number of calculations by Hammonds et al. (1998b). The case of ZrW_2O_8, which contains ZrO_6 octahedra and WO_4 tetrahedra, can have RUMs because some of the W–O bonds are non-bridging and the simple count of

RUMs because some of the W–O bonds are non-bridging and the simple count of constraints and degrees of freedom gives an exact balance (Pryde et al. 1996). The related material ZrP_2O_7 differs in that the non-bridging bonds are now joined up, and this material is overconstrained and has no RUMs (Pryde et al. 1996).

EXPERIMENTAL OBSERVATIONS 1: MEASUREMENTS OF DIFFUSE SCATTERING IN ELECTRON DIFFRACTION

The intensity of one-phonon scattering in the high-temperature limit is given by the standard result from the theory of lattice dynamics (Dove 1993)

$$S(\mathbf{Q},\omega) \propto \frac{k_B T}{\omega^2} \sum_k \delta(\omega \pm \omega_k) \left| \sum_j b_j \exp(i\mathbf{Q}\cdot\mathbf{r}_j) \times (\mathbf{Q}\cdot\mathbf{e}_{j,k}) \right|^2$$

The first sum is over all phonons for a given scattering vector $\mathbf{Q}$, the second sum is over all atoms in the unit cell, and $\mathbf{e}_{j,k}$ represents the components of the eigenvector of the phonon. Since the intensity is inversely proportional to ω^2, we anticipate that low-frequency RUMs will be seen in measurements of diffuse scattering, which will be proportional to

$$S(Q) = \int S(\mathbf{Q},\omega)\,\mathrm{d}\omega \propto \frac{k_B T}{\omega^2}$$

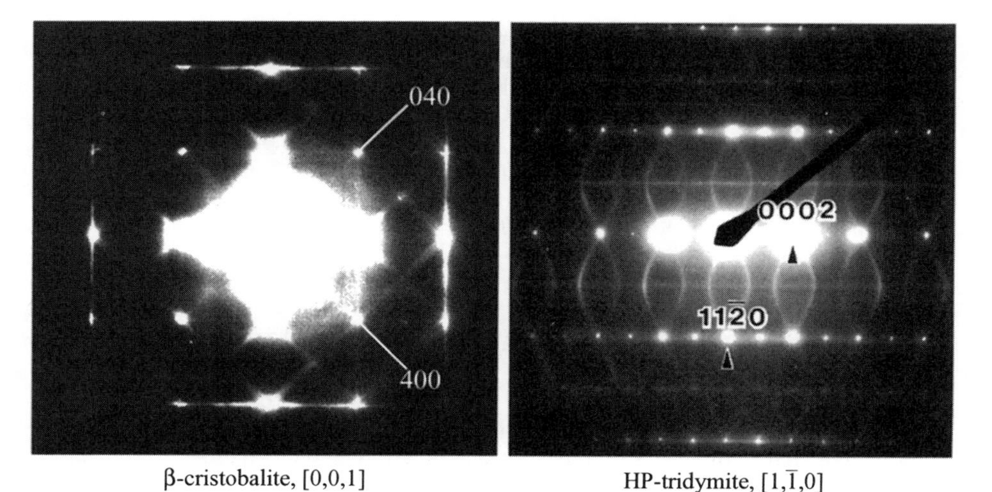

Figure 7. Electron diffraction measurements of diffuse scattering in the high-temperature phases of cristobalite (left, Hua et al. 1988) and tridymite (right, Withers et al. 1994), showing RUMs as streaks and curves.

In Figure 7, we show the diffuse scattering from the high-temperature phases of cristobalite (Hua et al. 1988, Welberry et al. 1989, Withers et al. 1989) and tridymite (Withers et al. 1994, Dove et al. 1996) measured by electron diffraction. The streaks of diffuse scattering correspond exactly to the planes of wave vectors containing the RUMs as given by our CRUSH calculations (Hammonds et al. 1996). The case of the high-temperature phase of tridymite is particularly interesting, because the second diffraction pattern also shows curves of diffuse scattering. We have shown that these curves of diffuse scattering correspond exactly to the exotic three-dimensional curved surface of

RUMs found in our CRUSH calculations as shown in Figure 5 (Dove et al. 1996).

The observation of the diffuse scattering is a direct confirmation of the RUM calculations. Moreover, there is important information in the diffuse scattering about the properties of the RUMs in real systems. First, the lines and curves of diffuse scattering are very sharp. If we consider a phonon dispersion curve around a wave vector $\mathbf{k}_{\text{RUM}}$ that contains a RUM:

$$\omega^2(\mathbf{k}-\mathbf{k}_{\text{RUM}}) = \omega_{\text{RUM}}^2 + \frac{1}{2}\alpha(\mathbf{k})\left|\mathbf{k}-\mathbf{k}_{\text{RUM}}\right|^2$$

where ω_{RUM} is the frequency of the RUM, which in a CRUSH calculation would be zero. The sharpness of the lines of diffuse scattering show that ω_{RUM} is relatively low, and that the dispersion coefficient α is relatively large. Second, the fact that the lines and curves of diffuse scattering remain reasonably strong across the region of reciprocal space shows that the RUM frequencies do not change very much with wave vector.

The important point is that none of the RUMs appear to dominate the diffuse scattering. The CRUSH calculations give the RUMs in the limit of infinitesimal atomic displacements. Detailed analysis shows that when finite displacements are taken into account, many of the RUMs will develop a non-zero frequency within the split-atom model and hence will no longer be RUMs. The main demonstration of this is when a static RUM distortion is imposed on a structure as in a displacive phase transition (Hammonds et al. 1996). It might be argued, therefore, that when all RUMs are excited with finite amplitude (note that the square of the amplitude increases linearly with temperature in the classical limit), they will interfere with each other and act to destroy the basic RUM character of the vibrations. On the other hand, it could also be argued that the amplitude associated with each individual RUM will remain small even at high temperatures, so the effect may not be significant. The observation of continuous lines and curves of diffuse scattering effectively show that the RUMs are not destroyed by the interactions between them. The same conclusion has been reached by Molecular Dynamics simulation studies (Gambhir et al. 1999).

EXPERIMENTAL OBSERVATIONS 2: INELASTIC NEUTRON SCATTERING MEASUREMENTS

Single crystal measurements

Inelastic neutron scattering measurements are not trivial, and therefore the number of experimental studies of RUMs by this approach is few. The only single-crystal inelastic neutron scattering measurement has been on quartz (Dolino et al. 1992) and leucite, $KAlSi_2O_6$ (Boysen 1990). The RUMs for these two materials are shown in Table 1.

There have been several measurements of the lattice dynamics of quartz by inelastic neutron scattering. Early results showed that the soft mode in the high-temperature phase is overdamped (Axe 1971). Other work on RUMs at wave vectors not directly associated with the phase transition showed that on cooling through the phase transition the RUMs rapidly increase in frequency since they are no longer RUMs in the low-temperature phase (Boysen et al. 1980). The most definitive study of the RUMs associated with the phase transition was that of Dolino et al. (1992).

There has been one lattice dynamics study of leucite by inelastic neutron scattering (Boysen 1990). The low-energy dispersion curves were measured for the high-temperature cubic phase along a few symmetry directions in reciprocal space. The results were consistent with the predictions of a CRUSH calculation (Dove et al. 1995, 1997b). In addition to the importance of confirming the CRUSH calculations, the experimental data also showed that the RUM excitations are heavily damped and are within the energy range 0–5 meV (0–1 THz).

Measurements on polycrystalline samples

We have performed a number of inelastic neutron scattering measurements from polycrystalline samples. The first measurements involved an average of the inelastic scattering over all values of the scattering vector **Q**, i.e.

$$S(E) = \int S(\mathbf{Q}, E)\, d\mathbf{Q}$$

This is equivalent to a phonon density of states weighted by the atomic scattering lengths. Data were obtained on the crystal analyser spectrometer TFXA at the ISIS pulsed neutron source for the low-temperature and high-temperature phases of cristobalite, and the results are shown in Figure 8 (Swainson and Dove 1993). The data in the higher-temperature phase show a significant increase in the number of low-energy vibrations, and the energy scale of 0–1 THz is consistent with the energy scale probed in the single-crystal measurements on quartz and leucite. This was the first direct dynamic experimental evidence for the existence of RUMs.

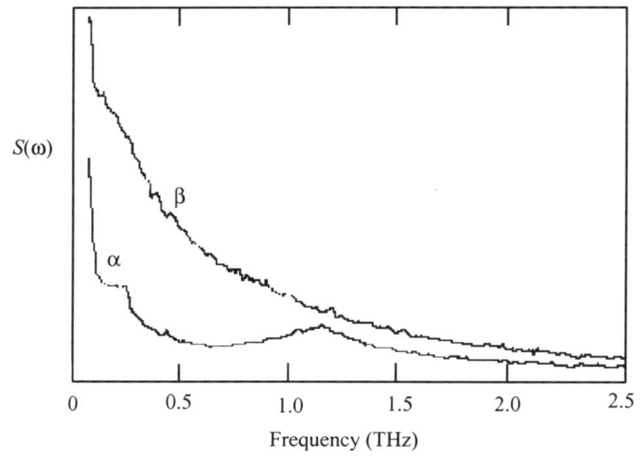

Figure 8. Inelastic neutron scattering data for the two phases of cristobalite integrated over the scattering vector (Swainson and Dove 1993).

More recently, we have obtained inelastic neutron scattering measurements of $S(Q,E)$ for a range of crystalline and amorphous silicates which only involve an average over the orientations of **Q** (unpublished). These data are collected on the chopper spectrometer MARI at ISIS. In Figure 9 we show the $S(Q,E)$ measurements for the high-temperature and room-temperature phases of cristobalite and tridymite. What is clear from the data is that on heating into the high-temperature phase there is a sudden increase in the number of RUMs which are seen as a considerable growth in the low-energy intensity across the range of Q. Again, the energy scale for the RUMs is 0–5 meV.

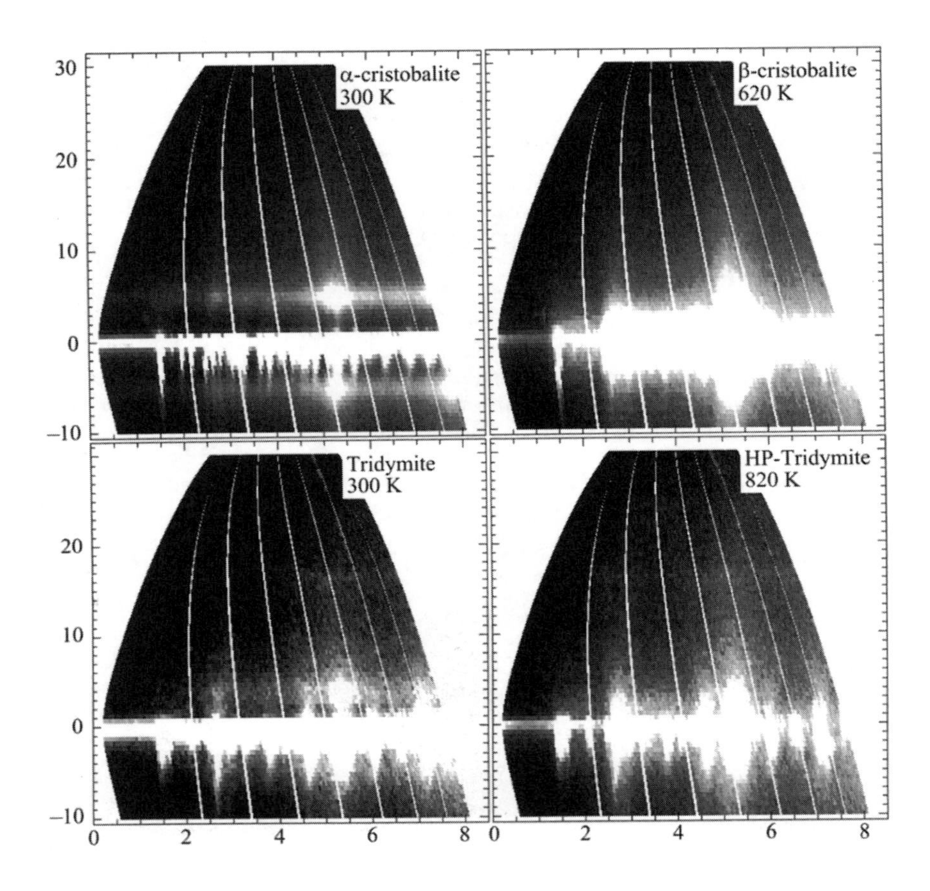

Figure 9. Map of the Q–E inelastic neutron spectra for cristobalite and tridymite at room temperature and temperatures corresponding to their high-temperature phases. Lighter regions correspond to larger values of the inelastic scattering function S(Q,E).

EXPERIMENTAL OBSERVATIONS 3: STRUCTURE MODELLING USING NEUTRON DIFFUSE SCATTERING DATA FROM POLYCRYSTALLINE SAMPLES

Total scattering measurements

There has been a lot of effort in recent years to use the diffuse scattering component of powder diffraction measurements. In an ordered crystal the diffuse scattering can appear to be little more than a flat background, but in a disordered crystal the background in a powder diffraction measurement can have structure in the form of oscillations or even broad peaks. The total diffraction pattern, which includes both the Bragg and diffuse scattering, contains information about both the long-range and short-range structure. To make full use of this information, it is necessary to collect data to a large value of Q (of the order of 20–50 Å^{-1}) in order to obtain satisfactory spatial resolution. Moreover, it is also necessary to take proper account of corrections due to instrument background and beam attenuation by the sample and sample environment, and to then make calibration measurements (usually from a sample of vanadium) to obtain a properly normalised set of data. The final result will be the total scattering function, $S(Q)$. Formally, S(Q) is a measurement of the total scattering, elastic and inelastic, integrated over all energy transfers:

$$S(Q) = \int S(Q,E)\,\mathrm{d}E$$

The Fourier transform gives the pair distribution function. This procedure is exactly that followed in diffraction studies of glasses and fluids (Wright 1993, 1997; see also Dove and Keen 1999).

The total scattering function $S(Q)$ is related to the pair distribution functions by

$$S(Q) = \rho\int_0^\infty 4\pi r^2 G(r)\frac{\sin Qr}{Qr}\mathrm{d}r + \sum_j c_j \overline{b_j^2}$$

where

$$G(r) = \sum_{i,j} c_i c_j b_i b_j \left(g_{ij}(r) - 1\right)$$

c_i and c_j are the concentrations of atomic species i and j, and b_i and b_j are the corresponding neutron scattering lengths. $g_{ij}(r)$ is the pair distribution function, defined such that the probability of finding an atom of species j within the range of distances r to $r + \mathrm{d}r$ from an atom of atomic species i is equal to $c_j 4\pi r^2 g_{ij}(r)\mathrm{d}r$. It is common to define the quantity $T(r)$ as

$$T(r) = rG(r) + r\left(\sum_j c_j b_j\right)^2$$

The importance of this function is that the component $rG(r)$ is the significant quantity in the Fourier transform that gives $S(Q)$ above (Wright 1993, 1997).

To illustrate the main points, in Figure 10 we show the $S(Q)$ functions for the two phases of cristobalite (Dove et al. 1997) and quartz (Tucker et al. 2000a,b). In $S(Q)$ the Bragg peaks are seen as the sharp peaks. However, there is a substantial background of diffuse scattering, particularly in the high-temperature phases. At higher values of Q, there are significant oscillations in $S(Q)$. These correspond to the Fourier transform of the SiO_4 tetrahedra. In Figure 11 we show the $T(r)$ functions for the two phases of cristobalite (Dove et al. 1997). The first two large peaks correspond to the Si–O and O–O bonds within the SiO_4 tetrahedra. The positions of the peaks correspond to the average instantaneous interatomic distances, and the heights of the peaks correspond to the numbers of neighbors weighted by the scattering lengths of the atoms. The weak third peak corresponds to the nearest-neighbor Si–Si distance. In cristobalite, the internal tetrahedral interatomic distances are similar in both phases, but the structures of the two phases shown by the $T(r)$ functions are clearly different for distances greater than 5 Å, as highlighted in Figure 11.

The Reverse Monte Carlo method

The main advance in recent years has been the development of methods to obtain models of structures that are consistent with the total diffraction pattern. One method is the Reverse Monte Carlo (RMC) method (McGreevy and Pusztai 1988, McGreevy 1995, Keen 1997, 1998). In this method, the Monte Carlo technique is used to modify a configuration of atoms in order to give the best agreement with the data. This can be carried out using either $S(Q)$ or $T(r)$ data, or both simultaneously. We also impose a

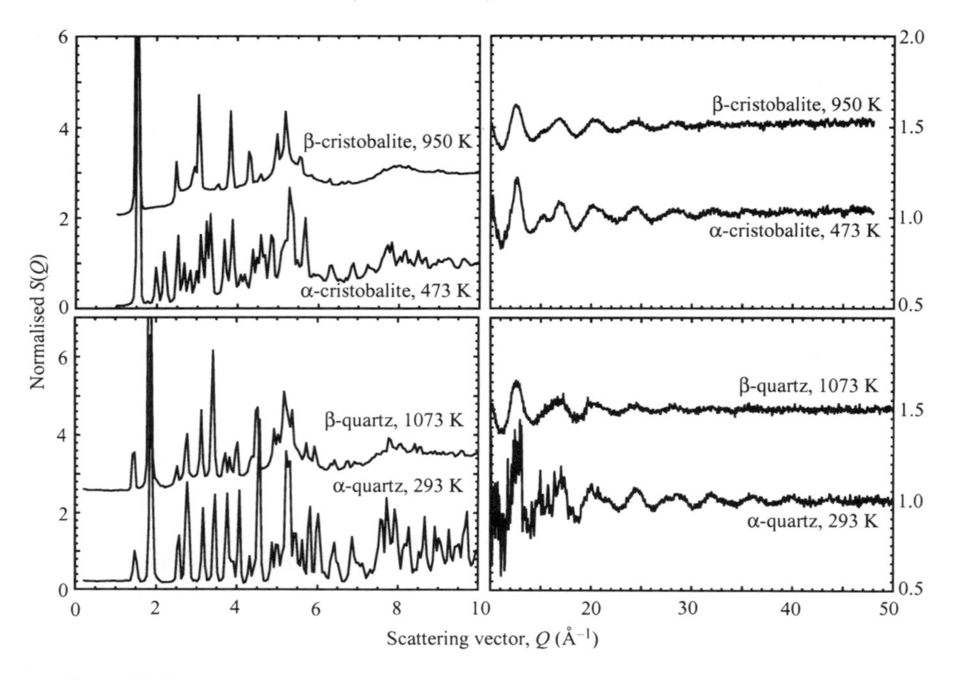

Figure 10. Neutron total scattering S(Q) functions from powdered samples of the high and low temperature phases of cristobalite and quartz, shown over two ranges of Q. The scattering functions show well-defined Bragg peaks at lower values of Q superimposed above structured diffuse scattering, and at high Q the scattering shows well-defined oscillations.

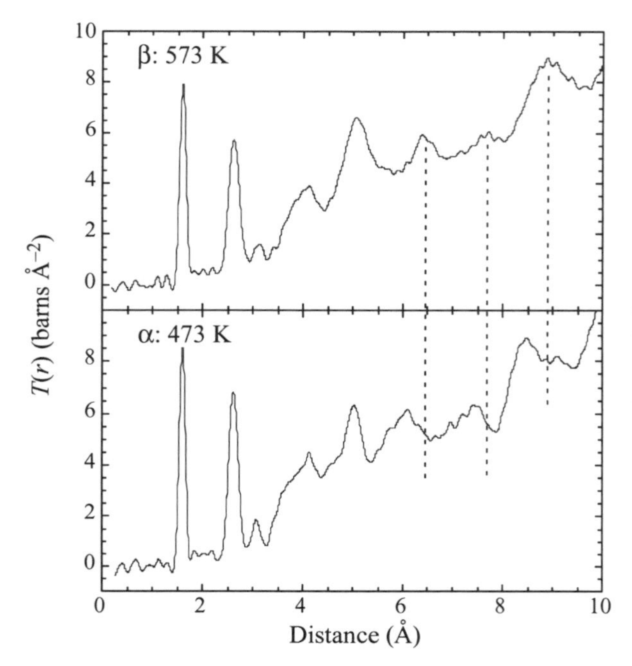

Figure 11. T(r) functions for the high and low temperatures phases of cristobalite obtained from neutron total scattering measurements. The two peaks at low-r correspond to the Si–O and O–O bonds. Dashed lines trace features in T(r) for β-cristobalite that are not seen in the T(r) for α-cristobalite.

constraint that the nearest-neighbor Si–O distances and O–Si–O angles should not deviate from specified distributions (Keen 1997, 1998). These constraints are actually fixed by the experimental data for $T(r)$ and so do not represent external constraints, but they are useful because there is nothing otherwise to ensure that the atomic coordinations do not change. The real external constraint is that each Si atom should always only be bonded to 4 O atoms. This is not a trivial point because the diffraction data only contain information about the average coordination. Thus the diffraction data tell us that silicon atoms have an average of 4 neighboring oxygen atoms, but the data do not tell you that the coordination number is *always* 4, and a mixture of coordination numbers 3 and 5 can be made to be consistent with the data. We now include a new constraint in the RMC method, which is to include the intensities of the Bragg peaks obtained by a separate fit to the diffraction data (Tucker 2000a,c). By including the Bragg peaks we are effectively adding a new constraint on the long-range order and on the distribution of positions of individual atoms. The Bragg peaks also provide more information about the three-dimensional aspects of the structure, since $S(Q)$ and $T(r)$ can only provide one-dimensional information.

The configurations generated by the RMC method can be analysed to give information about the fluctuations in short-range order within the constraints of the long-range symmetry. This approach is particularly useful for the study of disordered crystalline materials. We have applied this to a study of the structural disorder in the high-temperature phases of cristobalite (Tucker et al. 2000d) and quartz (Tucker et al. 2000a,b). In the latter case it has been possible to use a number of internal checks to show that the RMC is giving reasonable results. In particular, we are confident that the procedure is properly capturing the extent of the growth of disorder on heating without any exaggeration, with the correct balance between long-range and short-range order.

Application of RMC modelling to the phase transition in cristobalite

The atomic configurations of the two phases of cristobalite obtained by RMC analysis are shown in Figure 12, where we use a polyhedral representation. Both configurations are viewed down a common axis, which is the [1,1,1] direction of the high-temperature cubic phase. In this representation the structure has rings of six tetrahedra, and in the idealised structure of β-cristobalite these rings would be perfect hexagons. The effects of thermal fluctuations are clearly seen in the large distortions of these rings. The effect of the phase transition is seen in the configuration of α-cristobalite. The rings are distorted in a uniform manner, but there are still significant effects of thermal fluctuations causing localised distortions of these rings.

The driving force for the disorder in β-cristobalite is the fact that the average structure, namely the structure that is given by the analysis of the Bragg diffraction, appears to have straight Si–O–Si bonds (Schmahl et al. 1992). Because there is a high-energy cost associated with straight bonds, it is unlikely that these bonds are really straight, but instead are bent, but in this case the symmetry of the whole structure implies that there must be disorder in the orientations of the SiO_4 tetrahedra. That this must be so can be deduced from the position of the nearest-neighbor Si–O peak in the $T(r)$ function, which has a distance corresponding to the typical length of an Si–O bond, and which is longer than the distance between the mean Si and O positions in the cubic structure (Dove et al. 1997). The disorder is clearly seen in the configurations in Figure 12. There has been a lot of discussion about the nature of this disorder, including whether there is a prominent distribution of domains with the structure of the α-phase (Hatch and Ghose 1991), but the impression from Figure 12 is that the distortions of the rings of tetrahedra are random. We noted that the $T(r)$ functions for the two phases shown in Figure 11 are different for distances above 5 Å, which is evidence against the presence of domains with

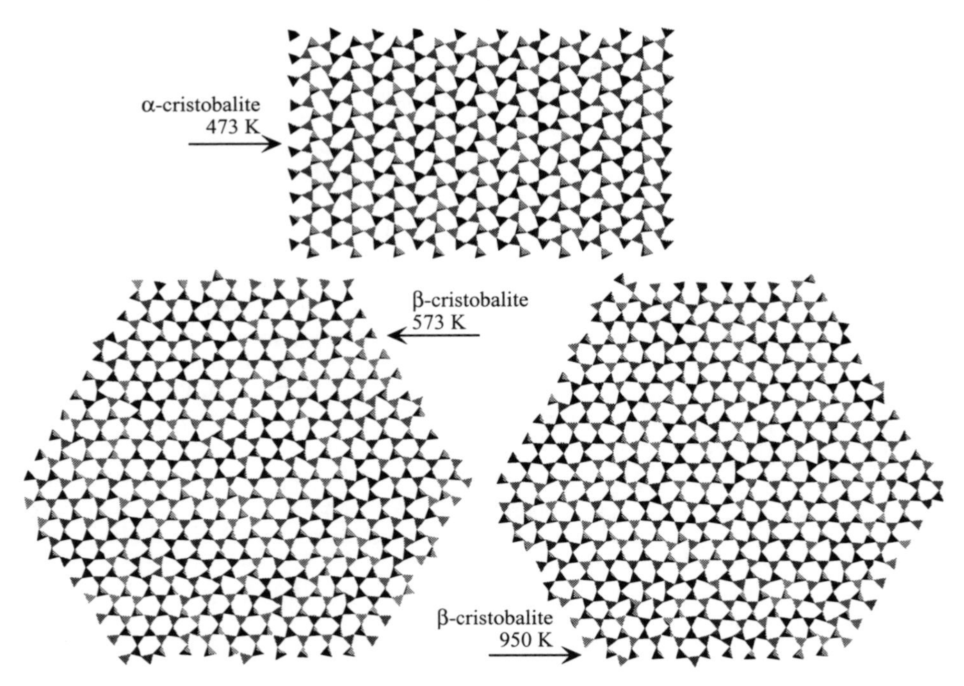

Figure 12. Polyhedral representations of the atomic configurations (from RMC) of the high and low temperature phases of cristobalite, viewed down a common direction that corresponds to [111] in β-cristobalite.

the structure of the α-phase. The absence of any clear correlations that would be associated with small domains of the α-temperature phase is highlighted in calculations of the Si–O–Si and Si–Si–Si angle distribution functions shown in Figure 13 (Tucker et al. 2000). The Si–O–Si angle distribution function for β-cristobalite is a broad single peak centred on $\cos\theta = -1$, but for α-cristobalite the distribution function has a maximum at $\cos\theta = -0.85$, which corresponds to a most probable Si–O–Si angle of 148°. Note that the corresponding distribution function in terms of angle, $P(\theta)$, is related to the distribution function in terms of $\cos\theta$ by $P(\theta) = P(\cos\theta)\sin\theta$, which means that the peak in $P(\cos\theta)$ at $\cos\theta = -1$ will correspond to a value of zero in $P(\theta)$. The Si–Si–Si angle distribution function is a broad single-peaked function centred on $\cos^{-1}(-1/3)$ (most probable angle of 109.47°) for β-cristobalite at all temperatures. However, for α-cristobalite the angle distribution function has three peaks, with the two outside peaks extended beyond the range of the distribution function of β-cristobalite. The significant differences of both angle distribution functions between the two phases show that the short-range structures of both phases are quite different, consistent with the qualitative impressions give by the configurations in Figure 12. The widths of the distribution functions for α-cristobalite show that there is still considerable structural disorder in this phase, consistent with the structural distortions seen in the atomic configuration of Figure 12.

The nature of the structural disorder in β-cristobalite has been discussed from the perspective of the RUM model (Swainson and Dove 1993, 1995a; Hammonds et al. 1996, Dove et al. 1997, 1998, 1999b). The basic idea is that there are whole planes of wave

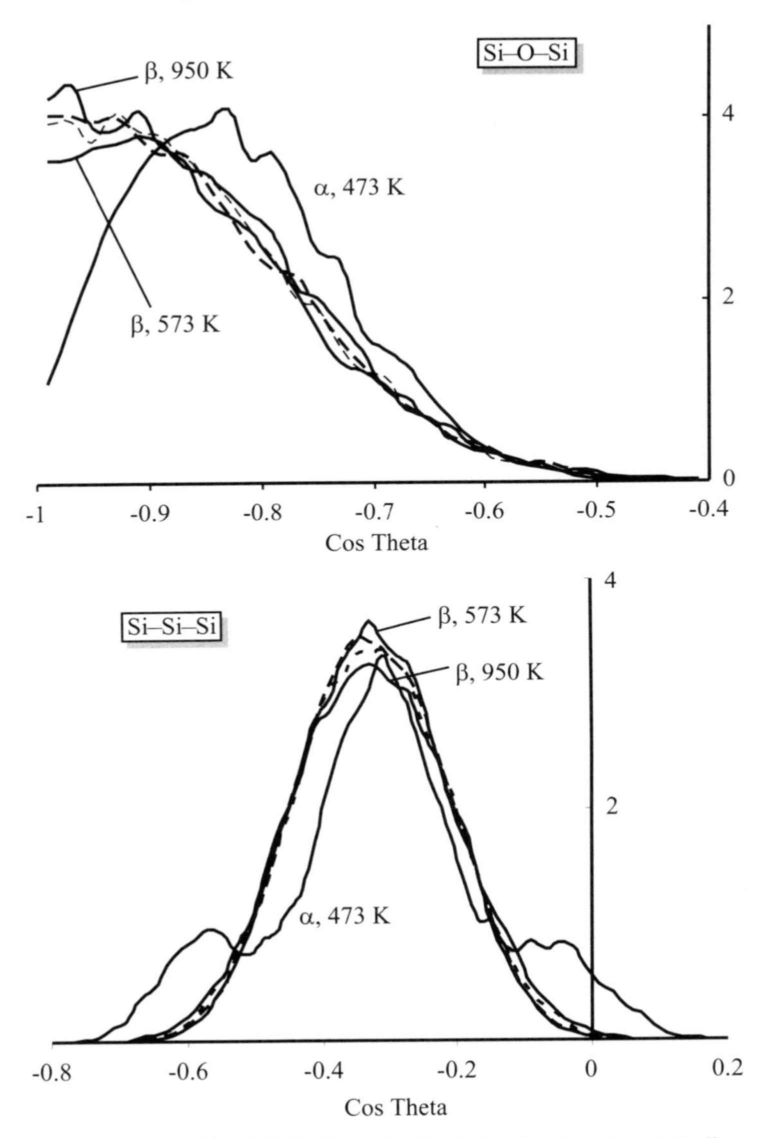

Figure 13. Si–O–Si and Si–Si–Si angular distribution functions for cristobalite at various temperatures, obtained from analysis of the RMC configurations. The two unmarked plots lying between the two labelled plots for β-cristobalite correspond to intermediate temperatures. The plots for α-cristobalite show features not seen in any of the plots for β-cristobalite.

vectors that contain RUMs. All of these RUMs can distort the structure by different rotations of the tetrahedra with different phases, and at any instance the structure will be distorted by a linear combination of all these RUMs acting together. This will give rise to a dynamically disordered structure, with constant reorientations of tetrahedra (Swainson and Dove 1995a). The same picture has been found to apply to the high-temperature

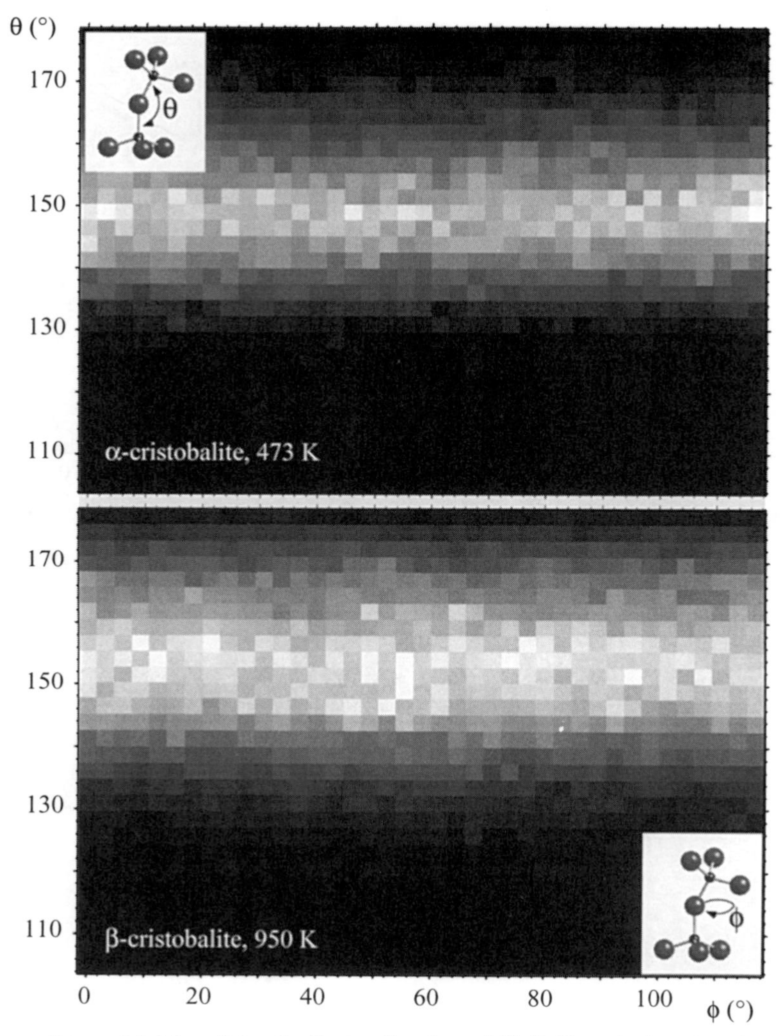

Figure 14. Map of the distribution functions of Si–O–Si angle θ against the orientation of the tilt orientation of the participating Si–O bond ϕ for the two phases of cristobalite. Lighter areas correspond to larger values of the distribution functions. The data were obtained from the configurations produced by the RMC simulations. There is a complete spread of values of ϕ in both phases, and the spread of values of θ is around 20° in the β-phase and 15° in the α-phase.

phase of tridymite also (Dove et al. 1998, 1999b, 2000b).

The extent of the disorder can also be seen in the distribution of orientations of the Si–O bonds (Tucker et al. 2000d). Figure 14 shows a two-dimensional map of the orientational distribution functions for the two phases of cristobalite. The two angles are defined as an insert to the figure: θ is the Si–O–Si angle, and ϕ gives the orientation of the displaced central O atom with respect to the base of one of the SiO_4 tetrahedra. The map for β-cristobalite shows that the positions of the oxygen atoms, which are reproduced in the values of ϕ, are equally distributed around the Si–Si vector on an annulus with no preferred positions. This annulus is broad, as seen in the spread of values of θ across a range of around 20°. The distribution for α-cristobalite also shows a high degree of disorder. The average crystal structure has a single value of θ but several values of ϕ, but the large thermal motion seen in Figures 12 and 13 causes a broad spread in the range of values of both θ and ϕ. The range of values of θ is only slightly tighter than that of β-cristobalite, and the range of values of the angle ϕ still seems unrestricted. The existence of disorder in the low-temperature α-phase anticipates the behavior found in quartz that will be discussed below.

One interesting aspect of the detailed analysis of the RMC configurations is that the three-dimensional diffuse scattering calculated from the β-cristobalite configurations is consistent with the experimental data shown in Figure 7 (Dove et al. 1998, Keen 1998). The calculated diffuse scattering is greatly diminished in the α-phase, consistent with the experimental data (Hua et al. 1988, Welberry et al. 1989, Withers et al. 1989). The ability to reproduce three-dimensional diffraction data from models based on one-dimensional powder diffraction data is one indication that the RMC method is giving a realistic simulation of the real crystals.

Application of RMC modelling to the phase transition in quartz

The temperature evolution of the short-range structure of quartz is seen in the behavior of the $T(r)$ functions shown in Figure 15 (Tucker et al. 2000a,b). At the lowest temperatures the peaks in $T(r)$ are very sharp, reflecting the small thermal motion, and on heating the peaks broaden. It is noticeable that there are no dramatic changes in the $T(r)$ functions associated with heating through the phase transition, although the positions of the higher-r peaks change in line with the changes in the volume of the unit cell (Carpenter et al. 1998). The atomic configurations of quartz at three temperatures obtained by RMC analysis are shown in Figure 16, where again we use a polyhedral representation. These configurations are viewed down the [110] direction, and Figure 16 also shows the average structure as inserts. As in β-cristobalite, the structure of the high-temperature β-phase shows a considerable amount of orientational disorder without any obvious formation of domains with the structure of the low-temperature α-phase. This nature of the structure of β-quartz has been the subject of some debate over many years (Heaney 1994). Our view is that the RMC models show that the structure of β-quartz is dynamically disordered as a result of the presence of RUMs, following exactly the same argument as for the structure of β-cristobalite and the high-temperature phase of tridymite (Dove et al. 1999b, 2000a—note also that the maximum in the lattice parameters, which has been cited as evidence for the existence of domains, has now been shown to be consistent with the complete RUM picture. Welche et al. 1998). What is new here is that the RMC results show that the disorder grows on heating within the temperature range of α-quartz, as seen in the $T(r)$ functions in Figure 15 and in the configurations shown in Figure 16. This arises because, on heating towards the transition temperature, the frequencies of the phonon modes that will become RUMs in β-quartz rapidly decrease (Boysen et al. 1980, Hammonds et al. 1996), leading to an increase in their amplitudes. This increase in the amplitudes of the phonons that correspond to quasi-RUM motions will give increased orientational disorder, exactly as seen in Figure 16.

We can sharpen the argument by more detailed analysis of the RMC configurations (Tucker et al. 2000a,b). Figure 17 shows the temperature-dependence of the mean Si–O bond, which we denote as ⟨Si–O⟩, and this is compared with the distance between the average positions of the Si and O atoms, which we denote as ⟨Si⟩–⟨O⟩. At low temperatures we expect ⟨Si–O⟩ → ⟨Si⟩–⟨O⟩, as seen in Figure 17. On heating ⟨Si–O⟩

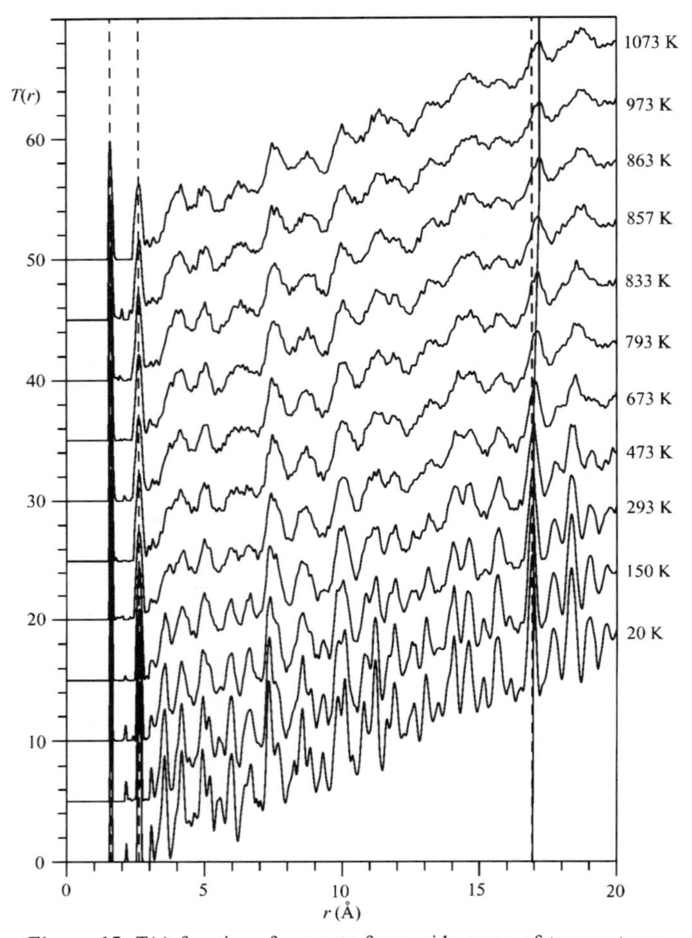

Figure 15. T(r) functions for quartz for a wide range of temperatures obtained by total neutron scattering measurements. The peaks associated with the intratetrahedral Si–O and O–O distances are marked with a vertical dashed curve to show that these distances do not change much with temperature. A peak at around 17 Å is marked with a continuous curve which, when compared with the vertical dashed line, shows the effects of the increase in volume on heating through the phase transition.

increases due to normal thermal expansion of the Si–O bond (Tucker et al. 2000e), but at the same time ⟨Si⟩–⟨O⟩ is seen to decrease. This is completely due to the growth of the orientational disorder of the SiO_4 tetrahedra, which brings the average positions of the Si and O atoms closer together, and is in spite of the fact that the volume of the unit cell expands on heating towards the phase transition. In Figure 18 we show the distribution of Si–Si–Si angles, similar to that shown for cristobalite in Figure 13. The distribution function includes two peaks at low temperature that broaden and coalesce on heating, and in Figure 13 we also show the variation of the positions and widths of these peaks with temperature. The phase transition, which reflects the long-range order, is clearly seen in this distribution function, but the growth in short-range disorder is also clear.

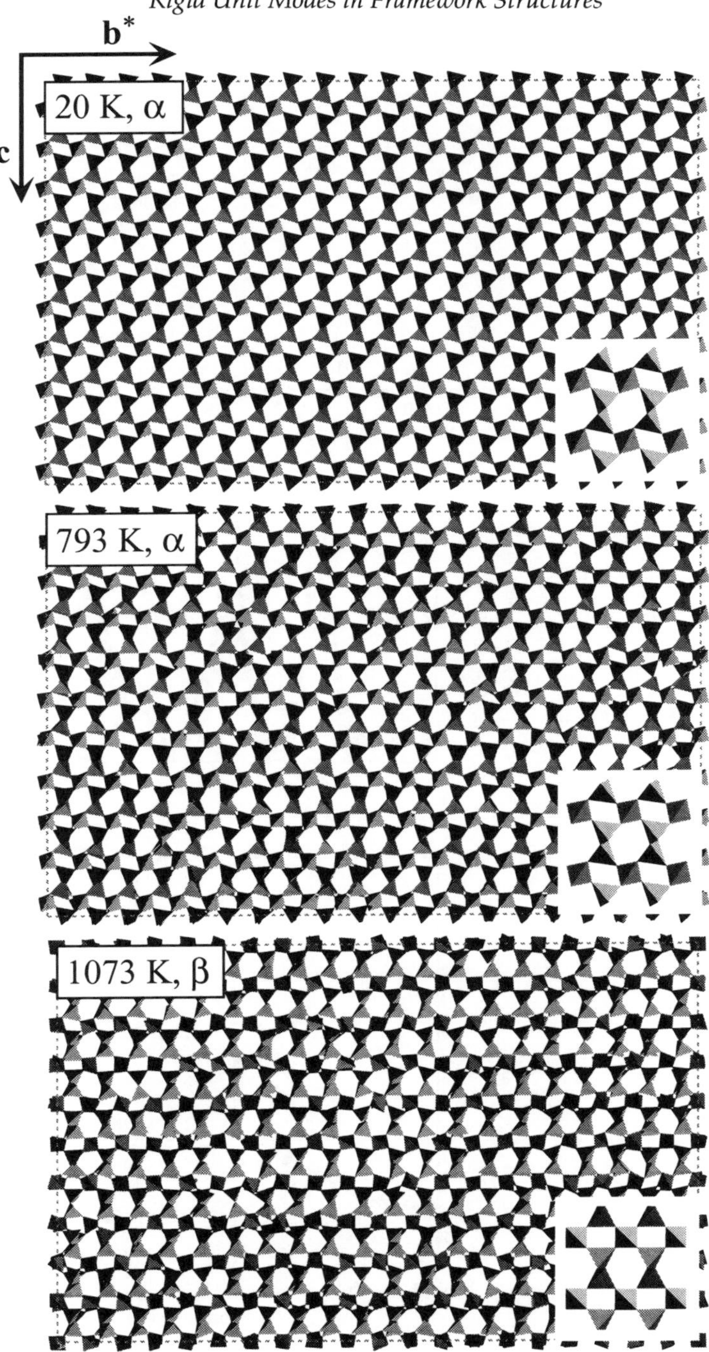

Figure 16. Polyhedral representations of the atomic configurations (RMC) of quartz at three temperatures, viewed down [100].

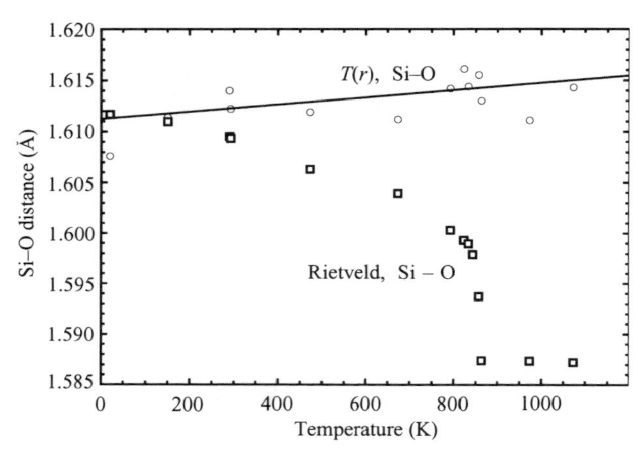

Figure 17. Temperature dependence of the mean Si–O bond length, determined from the T(r) functions, ⟨Si–O⟩, compared with the distance between the mean positions of the Si and O atoms, ⟨Si⟩–⟨O⟩, determined by Rietveld refinement methods.

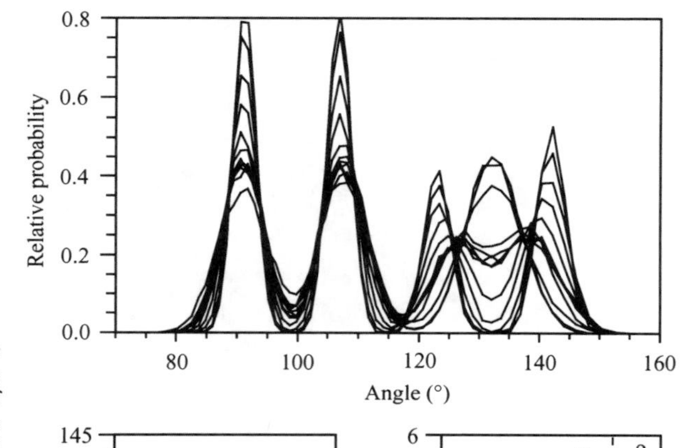

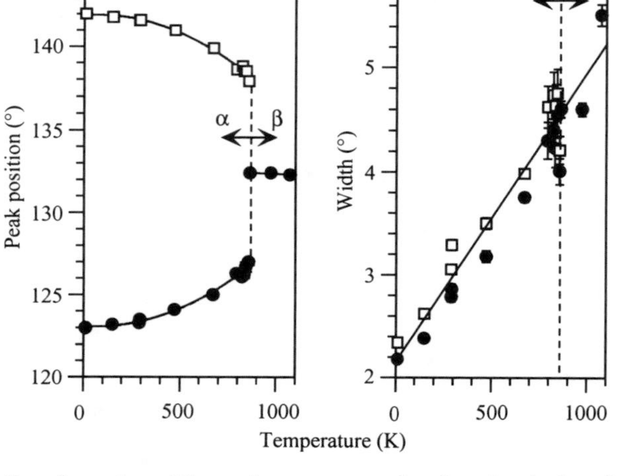

Figure 18. Si–Si–Si angle distribution function for quartz for the same range of temperatures as shown in Figure 15 (top). The positions and widths of the two peaks that coalesce are shown as a function of temperature in the lower plots. The distribution functions were obtained from analysis of the RMC configurations.

We argue that in quartz, the phase transition arises as a result of a classical soft-mode instability (Dolino et al 1992, Tezuka et al. 1991), but unlike in the classical soft-mode model the phase transition into the high-temperature phase also allows the excitation of a large number of low-frequency modes that contribute to the orientational disorder (Tucker et al. 2000a). These all have the effect of allowing the SiO_4 tetrahedra to rotate, but the different modes have different patterns of rotations and different phases, and the growth in the amplitudes of these modes acting together gives the appearance of the growth of considerable orientational disorder. This way of looking at the structures of high-temperature phases, based on the RUM model but now backed up by experimental data, gives a new understanding of the structures of high-temperature phases.

APPLICATIONS OF THE RIGID UNIT MODE (RUM) MODEL

Displacive phase transitions

Because RUMs are low-energy deformations of a framework structure, they are natural candidates for soft modes associated with displacive phase transitions (Dove 1997a,b, Dove et al. 1992, 1993, 1995, Hammonds et al. 1996). Indeed, we started by noting that the soft mode that gives the displacive α–β phase transition in quartz is a RUM, and we summarised the model by which the existence of a line of RUMs gives rise to the intermediate incommensurate phase transition. We have used the RUM analysis to explain the displacive phase transitions in a number of silicates (Dove et al. 1995, Hammonds et al. 1996).

The simplest example to understand the role of RUMs in phase transitions is the octahedral-rotation phase transition in the perovskite structure. The RUMs in the perovskite only exist for wave vectors along the edges of the cubic Brillouin zone, namely for wave vectors between $(\frac{1}{2},\frac{1}{2},0)$ and $(\frac{1}{2},\frac{1}{2},\frac{1}{2})$, and the symmetrically-related sets of wave vectors (Giddy et al. 1993). Experimental measurements of phonon dispersion curves for $SrTiO_3$ show that the whole branch of phonons between these wave vectors has low frequency (Stirling 1972). All the RUMs along this branch have the rotational motions shown in Figure 19. This shows a single layer of octahedra, and the effect of changing the wave vector from $(\frac{1}{2},\frac{1}{2},0)$ to $(\frac{1}{2},\frac{1}{2},\frac{1}{2})$ is to change the relative signs of the rotations of neighboring layers along [001]. The mechanism of the phase transition in $SrTiO_3$ involves the softening of the RUM of wave vector $(\frac{1}{2},\frac{1}{2},\frac{1}{2})$.

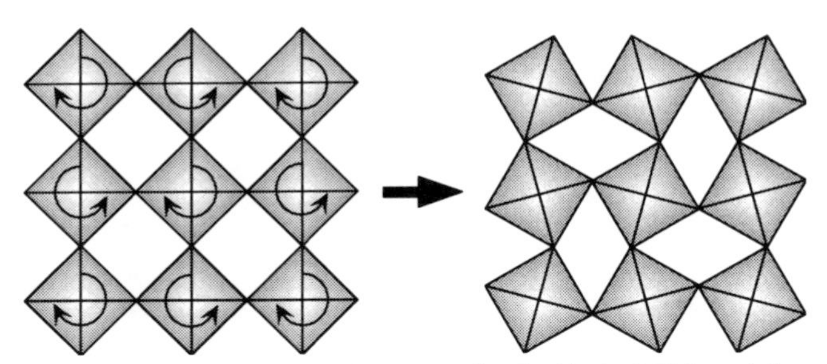

Figure 19. Rotations of octahedra associated with the RUMs and the displacive phase transition in the perovskite structure.

Cristobalite undergoes a first-order displacive phase transition at around 500°K at ambient pressure (Schmahl et al. 1992, Swainson and Dove 1993, 1995a). The distortion of the structure can be associated with a RUM with wave vector (1,0,0), which is at the corner of the Brillouin zone (Dove et al. 1993, Swainson and Dove 1993, 1995a; Hammonds et al. 1996). At ambient temperature there is another first-order displacive phase transition to a monoclinic phase on increasing pressure (Palmer and Finger 1994). A recent solution of the crystal structure of the monoclinic phase (Dove et al. 2000c) has

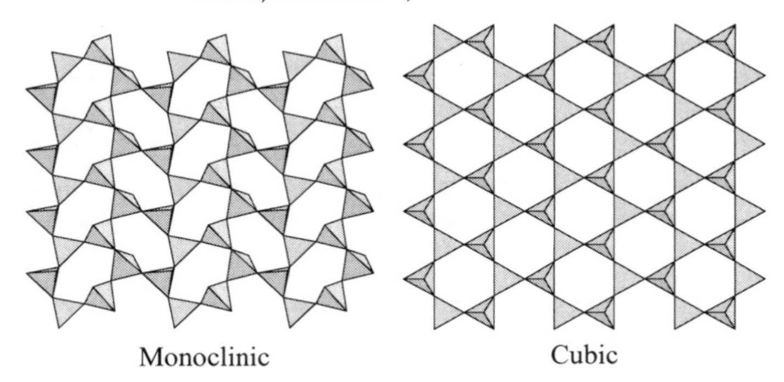

Figure 20. Crystal structures of the monoclinic (left) and cubic (right) phases of cristobalite (Dove et al., 2000), viewed down a common direction that corresponds to [1,1,1] in the cubic phase. The comparison of the two structures show the RUM distortions associated with the high-pressure phase.

shown that the phase transition cannot involve a RUM distortion of the ambient-pressure tetragonal phase, as anticipated in an earlier theoretical study (Hammonds et al. 1996). However, the high-pressure phase can be described as involving a RUM distortion of the high-symmetry cubic phase, as shown in Figure 20. The transition between the tetragonal and monoclinic phases involves a change from one minimum of the free-energy surface to another, rather than a continuous evolution of the free-energy surface as usually envisaged in treatments based on a Landau free-energy function.

The case of tridymite is particularly interesting, since there are many displacive phase transitions. Recently the RUM model was used to provide a consistent overall view of the various sequences of phase transitions (Pryde and Dove 1998, Dove et al. 2000b). Many of the phase transitions could be described as involving RUM deformations of parent structures, but it is clear that there are several sequences starting from the parent high-temperature hexagonal phase. Some phase transitions involve jumping from one RUM sequence to another. This is shown schematically in Figure 21. In this sense, the behavior of the phase transitions in tridymite follows the way that the high-pressure phase transition in cristobalite involves distinctly different distortions of the parent cubic phase.

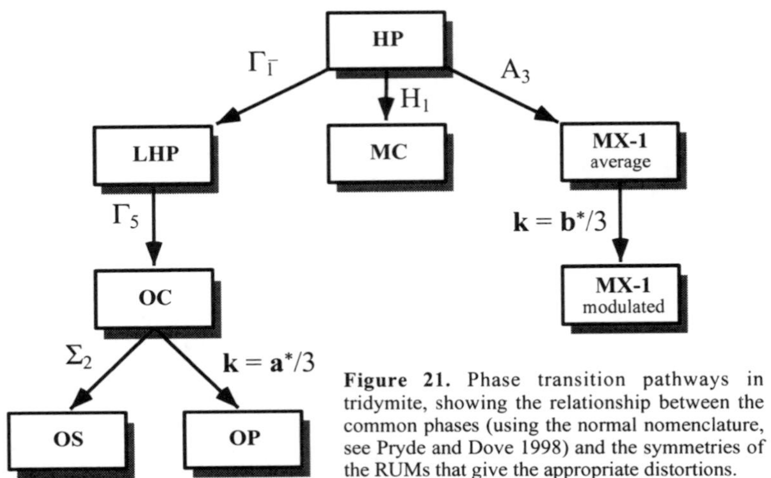

Figure 21. Phase transition pathways in tridymite, showing the relationship between the common phases (using the normal nomenclature, see Pryde and Dove 1998) and the symmetries of the RUMs that give the appropriate distortions.

The work that has been carried out so far shows that it is possible to interpret many displacive phase transitions in terms of RUM distortions, with the RUMs acting as the classic soft modes (Dove et al. 1995, Dove 1997b, Hammonds et al. 1996). One advantage of the RUM model is that it gives an intuitive understanding of the mechanism of the phase transition. Moreover, we have also shown that by including an Si–Si interaction in the RUM calculations, which has the effect of highlighting the RUMs that involve torsional rotations of neighboring tetrahedra, it is possible to understand why one RUM is preferred as the soft mode over the other RUMs (Dove et al. 1995, Hammonds et al. 1996).

Theory of the transition temperature

A lot of theoretical work on displacive phase transitions has focussed on a simple model in which atoms are connected by harmonic forces to their nearest neighbors, and each neighbor also sees the effect of the rest of the crystal by vibrating independently in a local potential energy well (Bruce and Cowley 1980). For a phase transition to occur, this double well must have two minima, and can be described by the following function:

$$E(u) = -\frac{\kappa_2}{2}u^2 + \frac{\kappa_4}{4}u^4$$

where u is a one-dimensional displacement of the atom. A one-dimensional version of this model is illustrated in Figure 22. The interactions between neighboring atoms j and $j + 1$ are described by the simple harmonic function

$$E(u_j, u_{j+1}) = \frac{1}{2}J(u_j - u_{j+1})^2$$

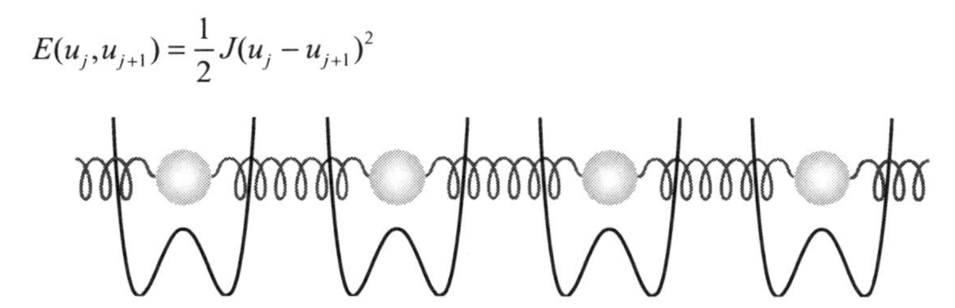

Figure 22. Atom and spring model with double wells, as described by Bruce and Cowley (1980).

This function is easily generalised to three dimensions. In a RUM system, the spring stiffness J is exactly analogous to the stiffness of the tetrahedra, and there is, in principle, an exact mapping of the RUM model onto this atom and spring model (Dove 1997a,b; Dove et al. 1999a). However, some care is needed in forming this mapping operation. The simple model has a very particular shape to the phonon dispersion curve, with a soft mode at $\mathbf{k} = 0$, and softening of other phonons only in the vicinity of $\mathbf{k} = 0$. On the other hand, for many RUM systems there is a softening of branches of phonons along lines or planes in reciprocal space, and these need to be taken into account, as we will discuss below.

Theoretical analysis of the simple model has shown that the transition temperature is determined by the parameters in the model as

$$RT_c = 1.32\frac{J\kappa_2}{\kappa_4}$$

(Bruce and Cowley 1980). The ratio κ_2 / κ_4 is simply equal to the displacement that corresponds to the minimum of the local potential energy well. The numerical prefactor arises from casting the model into its reciprocal space form, and integrating over the phonon surface that is determined by the interatomic springs (Sollich et al. 1994). If account is taken of changes to the phonon dispersion surface caused by the presence of lines or planes of RUMs, the effect is simply to lower the size of the numerical prefactor (Dove et al. 1999a). In some cases this is quite significant, and blind application of this model without taking account of the shape of the RUM surface will give meaningless results (Dove 1997a). Nevertheless, the changes to the phonon dispersion surface caused by the existence of the RUM does not alter the fact that the transition temperature is determined by the value of J, which, as we have noted above, is analogous to the stiffness of the tetrahedra. This gives a basis for understanding the origin of the size of the transition temperature, which has been explored in some detail elsewhere (Dove et al. 1999a).

Negative thermal expansion

The RUM model provides a natural explanation for the phenomenon of negative thermal expansion, the property in which materials *shrink* when they are heated. Recently the study of materials with this property has become quite active, particularly since the publication on detailed work on ZrW_2O_8 (Mary et al. 1996, Evans 1999), for which the (negative) thermal expansion tensor is isotropic and roughly constant over a very wide range of temperatures.

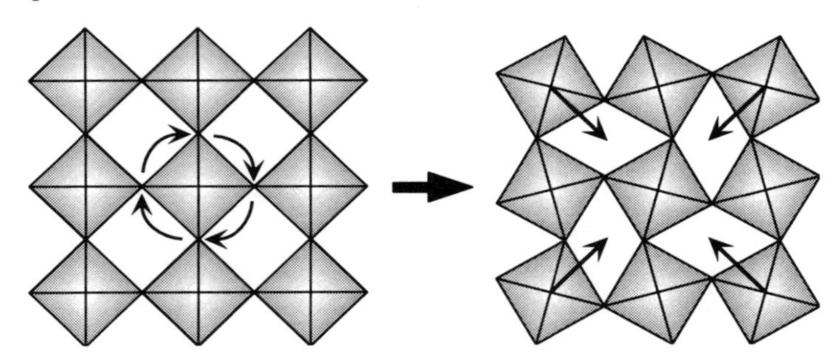

Figure 23. Rotating octahedra, showing the effect on the crystal volume and hence the origin of negative thermal expansion when the rotations are dynamic phonon modes rather than static distortions.

The idea is best illustrated with a two-dimensional representation of the network of linked octahedra in the cubic perovskite structure, as shown in Figure 23 (Welche et al. 1998, Heine et al. 1999). The rotation of any octahedron will cause its neighbors to rotate and to be dragged inwards. If this rotation is static, it simply describes what happens at a rotational phase transition in a perovskite structure. However, the octahedra will always be rotating in a dynamic sense in the high-symmetry phase, and when we account for all octahedra rotating due to the RUMs we can envisage a net reduction in the volume of the structure. This is exactly the same as for the case of cristobalite, but without the need for a driving force to bend the bonds. Thus we expect the mean square amplitude of rotation to simply scale as the temperature:

$$\langle \theta^2 \rangle \propto RT / \omega^2$$

The scaling with the inverse of the square of the phonon frequency follows from harmonic phonon theory, such that low frequency modes have the larger amplitudes. The rotations of the octahedra will give a net reduction in volume:

$$\Delta V \propto -\langle \theta^2 \rangle$$

By combining these two equations, we have the simple result:

$$\Delta V \propto -RT / \omega^2$$

This is negative thermal expansion, and the greatest contribution to this effect will arise from the phonons with the lower frequencies and which cause whole-body rotations of the polyhedra. These are the RUMs!

Even if the material does not have negative thermal expansion, the mechanism discussed here will always lead to the volume of the crystal being lower than would be calculated using a static model. For example, the volume of quartz at high temperatures is lower than would be calculated from the true Si–O bond lengths (Tucker et al. 2000a). The volumes of the leucite structures are clearly lower than would be given by true bond lengths, and the volumes of leucite structures containing different cations in the structure cavities are clearly affected by the way that these cations limit the amplitudes of RUM fluctuations (Hammonds et al. 1996). Another clear example is cristobalite (Swainson and Dove 1995b). The true thermal expansion of the Si–O bond can only be determined by measurements of the $T(r)$ function (Tucker et al. 2000e).

The RUM model of thermal expansion has been explored in some detail in general theoretical terms (Heine et al. 1999), and has been used to explain the occurrence of negative thermal expansion in both quartz (Welche et al. 1998) and ZrW_2O_8 (Pryde et al. 1996, 1998).

Localised deformations in zeolites

One of the most surprising results of our RUM studies is that some zeolite structures have one or more RUMs for each wave vector (Hammonds et al. 1996). Previously it was thought that the RUMs would be restricted to lines or planes of wave vectors because to have one RUM per wave vector would be a massive violation of the Maxwell condition. Nevertheless, we now have several examples of zeolite structures with several RUMs for each wave vector (Hammonds et al. 1997a,b; 1998a).

The existence of more than one RUM per wave vector allows for the formation of localised RUM distortions formed as linear combinations of RUMs for many different wave vectors. If a RUM of wave vector $\mathbf{k}$ gives a periodic distortion $\mathbf{U}(\mathbf{k})$, a localised distortion can be formed as

$$\mathbf{U}(\mathbf{r}) = \sum_{\mathbf{k},j} c_j(\mathbf{k}) \mathbf{U}(\mathbf{k},j) \exp(i\mathbf{k} \cdot \mathbf{r})$$

where j denotes different RUMs for a given wave vector. The values of the coefficients $c_j(\mathbf{k})$ are arbitrary. We have developed a method that allows the values of $c_j(\mathbf{k})$ to be chosen automatically in a way that gives the greatest degree of localisation of the distortion, centred on a pre-selected portion of the crystal structure (Hammonds et al. 1997a,b; 1998a). Although there is complete freedom in the way in which the RUMs can be combined to form localised distortions, the possible localised distortions are still tightly constrained by the form of the RUM eigenvectors. As a result, it is possible to show that some distortions can be localised to a much higher degree than others. This gives the whole method some power in being able to locate favoured adsorption sites within the zeolite structure. An example of a calculation of localised distortions is given in Figure 24.

RUMs in network glasses

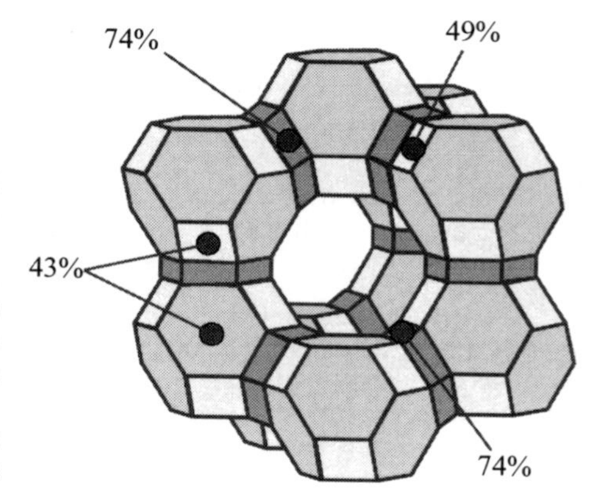

Figure 24. Representation of the zeolite faujasite, showing adsorption sites and the maximum degree of localisation of RUM distortions. The plot shows the linkages between tetrahedral sites.

We argued earlier that some of the Maxwell constraints can be removed by the action of symmetry, and this allows for the possibility of a structure having some RUM flexibility. This, however, is not the whole story. A network glass, such as silica, has no internal symmetry, and might therefore be thought incapable of supporting RUMs. Recent calculations of the RUM density of states of silica glass have shown that this is not the case. Instead, it appears that silica glass has the same RUM flexibility as β-cristobalite (Dove et al. 2000a, Trachenko et al. 1998, 2000). The RUM density of states plots for silica glass and β-cristobalite are shown in Figure 25. The important point, as noted earlier, is that the RUM density of states tends towards a constant value as $\omega \to 0$. We do not yet understand the origin of this RUM flexibility.

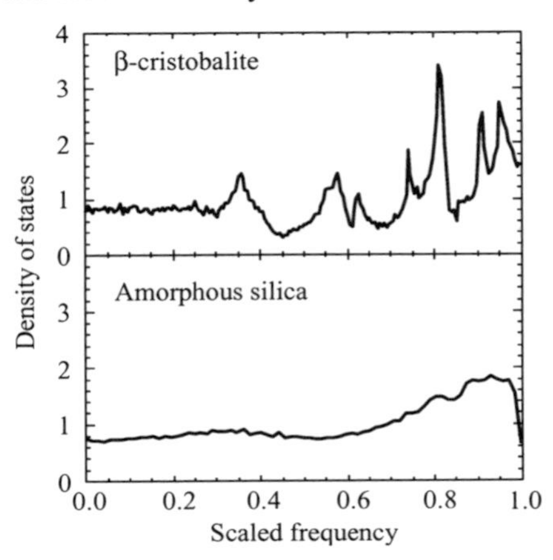

Figure 25. RUM density of states of β-cristobalite and silica glass.

It is interesting to compare inelastic neutron scattering measurements of $S(Q,E)$ for silica glass with those of the crystalline phases shown earlier in Figure 9. In Figure 26 we show the Q-E contour map for the low-energy excitations in silica glass (unpublished). The map looks remarkably similar to that for α-cristobalite, but there are now more low-energy excitations. These are the additional RUMs.

There is another aspect to the RUM flexibility that is particular to glasses, and that is the possibility for large-amplitude rearrangements of the structure (Trachenko et al. 1998, 2000). Here we envisage that a large group of tetrahedra can undergo a structural rearrangement without breaking its topology, moving from one minimum of the potential energy to another. This is possible since there is no symmetry acting as a constraint on the mean positions of the tetrahedra. We have observed such changes in molecular dynamics simulations of a set of samples of amorphous silica, and one is shown in Figure 27. These types of changes may correspond to those envisaged in the model of two-level tunnelling states that gives rise to the anomalous low-temperature thermal properties of silica glass.

CONCLUSIONS

The RUM model is now around 10 years old, and has developed in many ways that were not originally anticipated. The initial hope had been that the RUM model would help explain the origin of phase transitions in framework structures. In this it has been

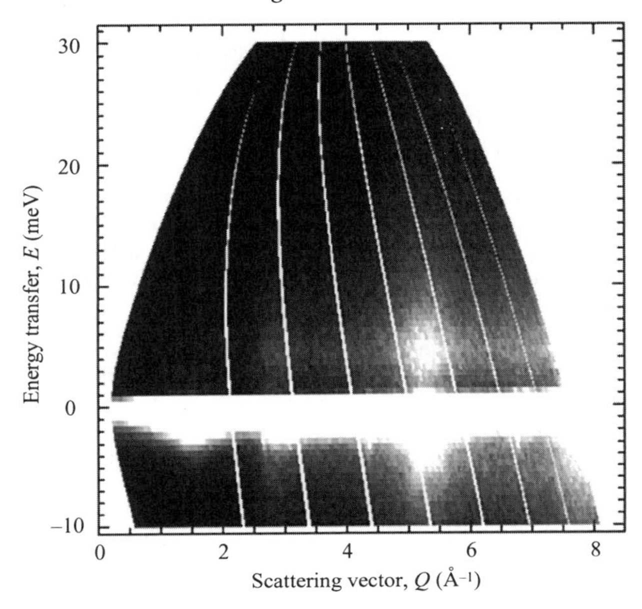

Figure 26. Inelastic neutron scattering Q-E map of silica glass, to be compared with those of cristobalite and tridymite in Figure 9.

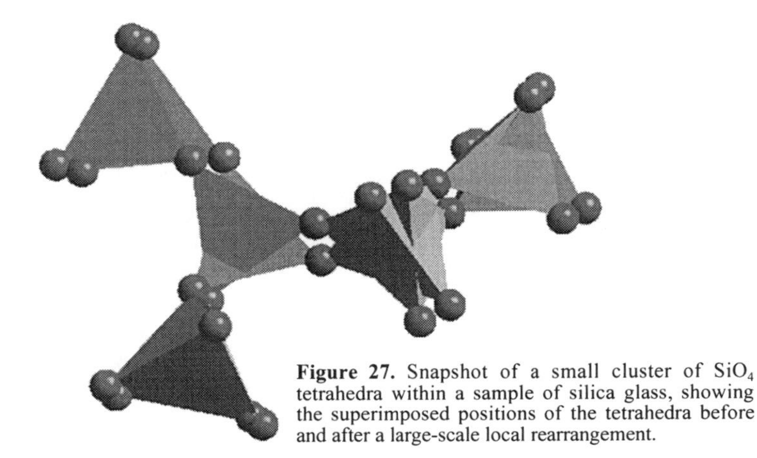

Figure 27. Snapshot of a small cluster of SiO_4 tetrahedra within a sample of silica glass, showing the superimposed positions of the tetrahedra before and after a large-scale local rearrangement.

very successful. It was quickly realised, through the mapping onto the ball and spring model, that the RUM model could also provide insights into the behavior of the phase transitions also. That the RUM model could also provide insights into thermal expansion and the behavior of zeolites were added bonuses!

Many of these features have been described in earlier reviews (Dove 1997b, Dove et al. 1998, 1999b), and we have sought to avoid going over the same ground as these. Subsequent to these reviews have been several developments in both theory and experiments. The recent theoretical advances include the recognition of the importance of curved surfaces of RUMs (Dove et al. 2000b), the demonstration that the simultaneous excitation of many RUMs does not damage the RUM picture (Gambhir et al. 1999), formal developments in the theory of negative thermal expansion (Welche et al. 1998,

Heine et al. 1999), a tighter understanding of the role of the RUMs in determining the transition temperature (Dove et al. 1999a), and the recognition that RUMs are important in network glasses (Trachenko et al. 1998). The new experimental work has been the measurements of the inelastic neutron scattering $S(Q,E)$ over a range of both Q and energies, and the use of new RMC methods to analyse total scattering $S(Q)$ data (Dove and Keen 1999, Keen and Dove 1999). The latter work has led to a way to visualise the action of the RUMs in a way that is based on experiment and not pure simulation, and has led to a deeper understanding of the nature of high-temperature phases. Moreover, the new experimental methods have led to the recognition that silicate glasses and crystals have many more similarities than previously thought, and the role of RUMs in facilitating these similarities has been recognised (Keen and Dove 1999, 2000).

The main theoretical challenge is to understand the origin of the curved surfaces, to understand why zeolites can have one or more RUMs per wave vector, and to understand why RUMs are so important in network glasses. Each of these challenges has come as a surprise. Another challenge is to enable the application of the RUM analysis to become a routine tool in the study of phase transitions. The programs are available for any worker to use (and can be downloaded from http://www.esc.cam.ac.uk/rums). We have shown that the RUM model can give many insights into the behavior of phase transitions, and many studies of phase transitions could be enhanced by some simple RUM calculations.

ACKNOWLEDGMENTS

This research has been supported by both EPSRC and NERC. We acknowledge the important contributions of Volker Heine, Kenton Hammonds, Andrew Giddy, Patrick Welche, Manoj Gambhir and Alix Pryde to many aspects of this work.

REFERENCES

Axe JD (1971) Neutron studies of displacive phase transitions. Trans Am Crystallogr Assoc 7:89–103

Berge B, Bachheimer JP, Dolino G, Vallade M, Zeyen CME. (1986) Inelastic neutron scattering study of quartz near the incommensurate phase transition. Ferroelectrics 66:73–84

Bethke J, Dolino G, Eckold G, Berge B, Vallade M, Zeyen CME, Hahn T, Arnold H, Moussa F (1987) Phonon dispersion and mode coupling in high-quartz near the incommensurate phase transition. Europhys Letters 3:207–212

Boysen H (1990) Neutron scattering and phase transitions in leucite. *In* Salje EKH, Phase transitions in ferroelastic and co-elastic crystals, Cambridge University Press, Cambridge, p 334–349

Boysen H, Dorner B, Frey F, Grimm H (1980) Dynamic structure determination for two interacting modes at the M-point in α- and β-quartz by inelastic neutron scattering. J Phys C: Solid State Phys 13:6127–6146

Bruce AD, Cowley RA (1980). Structural Phase Transitions. Taylor & Francis Ltd., London

Cai Y, Thorpe MF (1989) Floppy modes in network glasses. Phys Rev B 40:10535–10542

Carpenter MA, Salje EKH, Graeme-Barber A, Wruck B, Dove MT, Knight KS (1998) Calibration of excess thermodynamic properties and elastic constant variations due to the α–β phase transition in quartz. Am Mineral 83:2–22

Dolino G (1990) The α–inc–β transitions of quartz: A century of research on displacive phase transitions. Phase Transitions 21:59–72

Dolino G, Berge B, Vallade M, Moussa F (1989) Inelastic neutron scattering study of the origin of the incommensurate phase of quartz. Physica B 156:15–16

Dolino G, Berge B, Vallade M, Moussa F (1992) Origin of the incommensurate phase of quartz: I. Inelastic neutron scattering study of the high temperature β phase of quartz. J Physique, I. 2:1461–1480

Dove MT, Giddy AP, Heine V (1992) On the application of mean-field and Landau theory to displacive phase transitions. Ferroelectrics 136:33–49

Dove MT (1993) Introduction to Lattice Dynamics. Cambridge University Press, Cambridge

Dove MT, Giddy AP, Heine V (1993) Rigid unit mode model of displacive phase transitions in framework silicates. Trans Am Crystallogr Assoc 27:65–74

Dove MT, Heine V, Hammonds KD (1995) Rigid unit modes in framework silicates. Mineral Mag 59:629–639

Dove MT, Hammonds KD, Heine V, Withers RL, Kirkpatrick RJ (1996) Experimental evidence for the existence of rigid unit modes in the high-temperature phase of SiO_2 tridymite from electron diffraction. Phys Chem Minerals 23:55–61

Dove MT (1997a) Theory of displacive phase transitions in minerals. Am Mineral 82:213–244

Dove MT (1997b) Silicates and soft modes. *In* Thorpe MF, Mitkova MI (eds) Amorphous Insulators and Semiconductors. Kluwer, Dordrecht, p 349–383

Dove MT, Keen DA, Hannon AC, Swainson IP (1997) Direct measurement of the Si–O bond length and orientational disorder in β-cristobalite. Phys Chem Minerals 24:311–317

Dove MT, Heine V, Hammonds KD, Gambhir M and Pryde AKA (1998) Short-range disorder and long-range order: implications of the 'Rigid Unit Mode' model. *In* Thorpe MF, Billinge SJL (eds) Local Structure from Diffraction. Plenum, New York, p 253–272

Dove MT and Keen DA (1999) Atomic structure of disordered materials. *In* Catlow CRA, Wright K (eds) Microscopic Processes in Minerals. Kluwer, Dordrecht, p 371–387

Dove MT, Gambhir M, Heine V (1999a) Anatomy of a structural phase transition: Theoretical analysis of the displacive phase transition in quartz and other silicates. Phys Chem Minerals 26:344–353

Dove MT, Hammonds KD, Trachenko K (1999b) Floppy modes in crystalline and amorphous silicates. *In* MF Thorpe, Duxbury PM (eds) Rigidity Theory and Applications. Plenum, New York, p 217–238

Dove MT, Hammonds KD, Harris MJ, Heine V, Keen DA, Pryde AKA, Trachenko K, Warren MC (2000a) Amorphous silica from the Rigid Unit Mode approach. Mineral Mag 64:377–388

Dove MT, Pryde AKA, Keen DA (2000b) Phase transitions in tridymite studied using "Rigid Unit Mode" theory, Reverse Monte Carlo methods and molecular dynamics simulations. Mineral Mag 64:267–283

Dove MT, Craig MS, Keen DA, Marshall WG, Redfern SAT, Trachenko KO, Tucker MG (2000c) Crystal structure of the high-pressure monoclinic phase-II of cristobalite, SiO_2. Mineral Mag 64:569–576

Evans JSO (1999) Negative thermal expansion materials. J Chem Soc: Dalton Trans 1999:3317-3326

Gambhir M, Dove MT, Heine V (1999) Rigid Unit Modes and dynamic disorder: SiO_2 cristobalite and quartz. Phys Chem Minerals 26:484–495

Giddy AP, Dove MT, Pawley GS, Heine V (1993) The determination of rigid unit modes as potential soft modes for displacive phase transitions in framework crystal structures. Acta Crystallogr A49:697–703

Grimm H, Dorner B (1975) On the mechanism of the α–β phase transformation of quartz. Phys Chem Solids 36:407–413

He H, Thorpe MF (1985) Elastic properties of glasses. Phys Rev Letters 54:2107–2110

Hammonds KD, Dove MT, Giddy AP, Heine V (1994) CRUSH: A FORTRAN program for the analysis of the rigid unit mode spectrum of a framework structure. Am Mineral 79:1207–1209

Hammonds KD, Dove MT, Giddy AP, Heine V, Winkler B (1996) Rigid unit phonon modes and structural phase transitions in framework silicates. Am Mineral 81:1057–1079

Hammonds KD, H Deng, Heine V, Dove MT (1997a) How floppy modes give rise to adsorption sites in zeolites. Phys Rev Letters 78:3701–3704

Hammonds KD, Heine V, Dove MT (1997b) Insights into zeolite behavior from the rigid unit mode model. Phase Transitions 61:155–172

Hammonds KD, Heine V, Dove MT (1998a) Rigid Unit Modes and the quantitative determination of the flexibility possessed by zeolite frameworks. J Phys Chem B 102:1759–1767

Hammonds KD, Bosenick A, Dove MT, Heine V (1998b) Rigid Unit Modes in crystal structures with octahedrally-coordinated atoms. Am Mineral 83:476–479

Harris MJ, Dove MT, Parker JM (2000) Floppy modes and the Boson peak in crystalline and amorphous silicates: an inelastic neutron scattering study. Mineral Mag 64:435–440

Hatch DM, Ghose S (1991) The α–β phase transition in cristobalite, SiO_2: Symmetry analysis, domain structure, and the dynamic nature of the β-phase. Phys Chem Minerals 17:554–562

Hayward SA, Pryde AKA, de Dombal RF, Carpenter MA, Dove MT (2000) Rigid Unit Modes in disordered nepheline: a study of a displacive incommensurate phase transition. Phys Chem Minerals 27:285–290

Heaney PJ (1994) Structure and chemistry of the low-pressure silica polymorphs. Rev Mineral 29:1–40

Heine V, Welche PRL, Dove MT (1999) Geometric origin and theory of negative thermal expansion in framework structures. J Am Ceramic Soc 82:1793–1802

Hua GL, Welberry TR, Withers RL, Thompson JG (1988) An electron-diffraction and lattice-dynamical study of the diffuse scattering in β-cristobalite, SiO_2. Journal of Applied Crystallogr 21:458–465

Keen DA (1997) Refining disordered structural models using reverse Monte Carlo methods: Application to vitreous silica. Phase Trans 61:109–124

Keen DA (1998) Reverse Monte Carlo refinement of disordered silica phases. *In* Thorpe MF, Billinge SJL (eds) Local Structure from Diffraction. Plenum, New York, p 101–119

Keen DA, Dove MT (1999) Comparing the local structures of amorphous and crystalline polymorphs of silica. J Phys: Condensed Matter 11:9263–9273
Keen DA, Dove MT (2000) Total scattering studies of silica polymorphs: similarities in glass and disordered crystalline local structure. Mineral Mag 64:447–457
Mary TA, Evans JSO, Vogt T, Sleight AW (1996) Negative thermal expansion from 0.3 to 1050 Kelvin in ZrW_2O_8. Science 272:90–92
Maxwell JC (1864) On the calculation of the equilibrium and stiffness of frames. Phil Mag 27:294–299
McGreevy RL, Pusztai L (1988) Reverse Monte Carlo simulation: A new technique for the determination of disordered structures. Molec Simulations 1:359–367
McGreevy RL (1995) RMC—Progress, problems and prospects. Nuclear Instruments Methods A 354:1–16
Palmer DC, Finger LW (1994) Pressure-induced phase transition in cristobalite: an x-ray powder diffraction study to 4.4 GPa. Am Mineral 79:1–8
Pawley GS (1972) Analytic formulation of molecular lattice dynamics based on pair potentials. Physica Status Solidi 49b:475–488
Pryde AKA, Hammonds KD, Dove MT, Heine V, Gale JD, Warren MC (1996) Origin of the negative thermal expansion in ZrW_2O_8 and ZrV_2O_7. J Phys: Condensed Matter 8:10973–10982
Pryde AKA, Dove MT, Heine V (1998) Simulation studies of ZrW_2O_8 at high pressure. J Phys: Condensed Matter 10:8417–8428
Pryde AKA, Dove MT (1998) On the sequence of phase transitions in tridymite. Phys Chem Minerals 26:267–283
Schmahl WW, Swainson IP, Dove MT, Graeme-Barber A (1992) Landau free energy and order parameter behavior of the α–β phase transition in cristobalite. Z Kristallogr 201:125–145
Sollich P, Heine V, Dove MT (1994) The Ginzburg interval in soft mode phase transitions: Consequences of the Rigid Unit Mode picture. J Phys: Condensed Matter 6:3171–3196
Strauch D, Dorner B (1993) Lattice dynamics of α-quartz. 1. Experiment. J Phys: Condensed Matter 5:6149–6154
Stirling WG (1972) Neutron inelastic scattering study of the lattice dynamics of strontium titanate: harmonic models. J Phys C: Solid State Physics 5:2711–2730
Swainson IP, Dove MT (1993) Low-frequency floppy modes in β-cristobalite. Phys Rev Letters 71:193–196
Swainson IP, Dove MT (1995a) Molecular dynamics simulation of α- and β-cristobalite. J Phys: Condensed Matter 7:1771–1788
Swainson IP, Dove MT (1995b) On the thermal expansion of β-cristobalite. Phys Chem Minerals 22:61–65
Tautz FS, Heine V, Dove MT, Chen X (1991) Rigid unit modes in the molecular dynamics simulation of quartz and the incommensurate phase transition. Phys Chem Minerals 18:326–336
Tezuka Y, Shin S, Ishigame M (1991) Observation of the silent soft phonon in β-quartz by means of hyper-raman scattering. Phys Rev Lett 66:2356–2359
Thorpe MF, Djordjevic BR, Jacobs DJ (1997) The structure and mechanical properties of networks. *In* Thorpe MF, Mitkova MI (eds) Amorphous Insulators and Semiconductors. Kluwer, Dordrecht, The Netherlands, p 83–131
Trachenko K, Dove MT, Hammonds KD, Harris MJ, Heine V (1998) Low-energy dynamics and tetrahedral reorientations in silica glass. Phys Rev Letters 81:3431–3434
Trachenko KO, Dove MT, Harris MJ (2000) Two-level tunnelling states and floppy modes in silica glass. J Phys: Condensed Matter (submitted)
Tucker MG, Dove MT, Keen DA (2000) Direct measurement of the thermal expansion of the Si–O bond by neutron total scattering. J Phys: Condensed Matter 12:L425-L430
Tucker MG, Dove MT, Keen DA (2000) Simultaneous measurements of changes in long-range and short-range structural order at the displacive phase transition in quartz. Phys Rev Letters (submitted)
Tucker MG, Dove MT, Keen DA (2000) A detailed structural characterisation of quartz on heating through the α–β phase transition. Phys Rev B (submitted)
Tucker MG, Squires MD, Dove MT, Keen DA (2000) Reverse Monte Carlo study of cristobalite. J Phys: Condensed Matter (submitted)
Tucker MG, Dove MT, Keen DA (2000) Application of the Reverse Monte Carlo method to crystalline materials. J Applied Crystallogr (submitted)
Vallade M, Berge B, Dolino G (1992) Origin of the incommensurate phase of quartz: II. Interpretation of inelastic neutron scattering data. J Phys I. 2:1481–1495
Welberry TR, Hua GL, Withers RL (1989). An optical transform and Monte Carlo study of the disorder in β-cristobalite SiO_2. J Applied Crystallogr 22:87–95
Welche PRL, Heine V, Dove MT (1998) Negative thermal expansion in β-quartz. Phys Chem Minerals 26:63–77

Withers RL, Thompson JG, Welberry TR (1989) The structure and microstructure of α-cristobalite and its relationship to β-cristobalite. Phys Chem Minerals 16:517–523

Withers RL, Thompson JG, Xiao Y, Kirkpatrick RJ (1994) An electron diffraction study of the polymorphs of SiO_2-tridymite. Phys Chem Minerals 21:421–433

Withers RL, Tabira Y, Valgomar A, Arroyo M, Dove MT (2000) The inherent displacive flexibility of the hexacelsian tetrahedral framework and its relationship to polymorphism in Ba-hexacelsian. Phys Chem Minerals (submitted)

Wright AC (1993) Neutron and X-ray amorphography. *In* Simmons CJ, El-Bayoumi OH (eds) Experimental Techniques of Glass Science. Ceramic Trans, Am Ceram Soc, Westerville, p 205–314

Wright AC (1997) X-ray and neutron diffraction. *In* Thorpe MF, Mitkova MI (eds) Amorphous Insulators and Semiconductors. Kluwer, Dordrecht, The Netherlands, p 83–131

2

Strain and Elasticity at Structural Phase Transitions in Minerals

Michael A. Carpenter

Department of Earth Sciences
University of Cambridge
Downing Street
Cambridge CB2 3EQ U.K.

INTRODUCTION

Almost any change in the structure of a crystal, due to small atomic displacements, atomic ordering, magnetisation, etc., is usually accompanied by changes in lattice parameters. If suitable reference states are defined, such lattice parameter variations can be described quantitatively as a combination of linear and shear strains. What was originally a geometrical description then becomes a thermodynamic description if the relationships of the strains to the elastic constants are also defined. For a phase transition in which the parent and product phases are related by symmetry, only very specific combinations of strain will be consistent with the symmetry change, and the magnitudes of each strain will depend directly on the extent of transformation. In other words, such strains provide information on the evolution of the order parameter and the thermodynamic character of a phase transition. Spontaneous strains due to phase transitions in silicate minerals can be as large as a few percent, and, with modern high resolution diffractometry using neutron or X-ray sources, variations as small as ~0.1‰ can be detected routinely. This level of resolution is illustrated by the lattice parameter and strain variations through a cubic → tetragonal → orthorhombic sequence for the perovskite $NaTaO_3$, for example (Fig. 1).

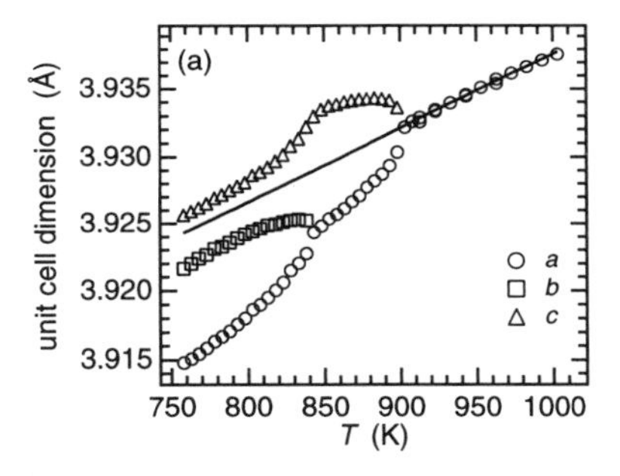

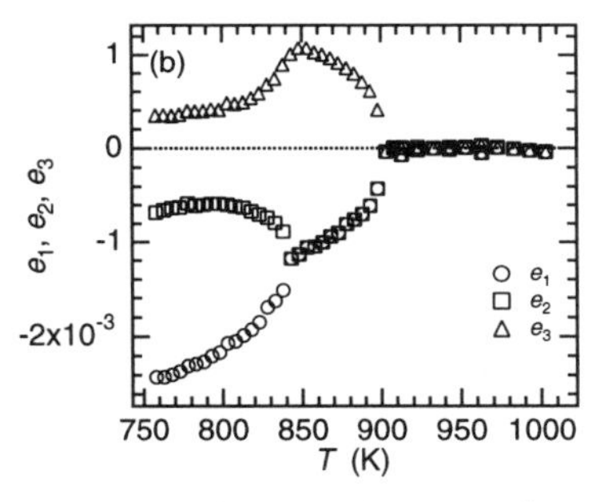

Figure 1. (a) Lattice parameters (expressed in terms of reduced, pseudocubic unit cell) as a function of temperature through the cubic → tetragonal → orthorhombic sequence of $NaTaO_3$ perovskite (powder neutron diffraction data of Darlington and Knight 1999). A straight line through the data for the cubic structure gives the reference parameter a_0. (b) Linear strains e_1, e_2, e_3, parallel to reference axes X, Y and Z, respectively, derived from the data in (a).

Conversion from a purely geometrical description, in terms of lattice parameters, to a thermodynamic description, in terms of strain, leads through to the elastic properties. Whereas there are up to six independent strains, e_i (i = 1-6), for a crystal, there are up to 21 independent elastic constants, C_{ik} (i, k = 1-6). Add to this the fact that strain states represent minima in a free energy surface, given by $\partial G/\partial e_i = 0$, while elastic constants

1529-6466/00/0039-0002$05.00

represent the shape of the free energy surface around the minima, given by $\partial^2 G/\partial e_i \partial e_k = 0$, and the expectation is immediately that elastic properties should be particularly sensitive to transition mechanism. A schematic representation of different elastic anomalies associated with a second order transition from cubic to tetragonal symmetry is shown in Figure 2 to illustrate this sensitivity, for example. Note that elastic anomalies tend to be very much larger than strain anomalies, that they can develop in both the high symmetry and low symmetry phases, and that they frequently extend over a substantial pressure and temperature range away from a transition point itself. Any other property of a material which depends on elasticity, such as the velocity of acoustic waves through it, will also be highly sensitive to the existence of structural phase transitions. Matching observed and predicted elastic anomalies should provide a stringent test for any model of a phase transition in a particular material.

Table 1. Phase transitions in minerals for which elastic constant variations should conform to solutions of a Landau free-energy expansion (after Carpenter and Salje 1998).

Selected phase transitions in minerals	
Proper ferroelastic behaviour	Co-elastic behaviour
albite: $C2/m \rightleftharpoons C\bar{1}$ ($Q_{od} = 0$)	quartz: $P6_422$ or $P6_222 \rightleftharpoons P3_12$ or $P3_22$
Sr-anorthite: $I2/c \rightleftharpoons I\bar{1}$	leucite: $I4_1/acd \rightleftharpoons I4_1/a$
leucite: $Ia3d \rightleftharpoons I4_1/acd$	pigeonite: $C2/c \rightleftharpoons P2/c$
	anorthite: $I\bar{1} \rightleftharpoons P\bar{1}$
Pseudo-proper ferroelastic behaviour	calcite: $R\bar{3}m \rightleftharpoons R\bar{3}c$
tridymite: $P6_322 \rightleftharpoons C222_1$	tridymite: $P6_3/mmc \rightleftharpoons P6_322$
vesuvianite: $P4/nnc \rightleftharpoons P2/n$ (?)	kaliophilite: $P6_322 \rightleftharpoons P6_3$
stishovite: $P4_2/mnm \rightleftharpoons Pnnm$	kalsilite: $P6_3mc \rightleftharpoons P6_3$
	$P6_3mc \rightleftharpoons P6_3mc$ (superlattices)
Improper ferroelastic behaviour	$P6_3 \rightleftharpoons P6_3$ (superlattices)
$(Mg,Fe)SiO_3$ perovskite: cubic $\rightleftharpoons$ tetragonal (?)	cummingtonite: $C2/m \rightleftharpoons P2_1/m$
tetragonal $\rightleftharpoons$ orthorhombic (?)	lawsonite: $Cmcm \rightleftharpoons Pmcn \rightleftharpoons P2_1cn$
neighborite: $Pm\bar{3}m \rightleftharpoons Pbnm$	titanite: $A2/a \rightleftharpoons P2_1/a$
$CaTiO_3$ perovskite: $Pm\bar{3}m \rightleftharpoons I4/mcm \rightleftharpoons Pbnm$	
cristobalite: $Fd\bar{3}m \rightleftharpoons P4_32_12$ or $P4_12_12$	
calcite: $R\bar{3}c \rightleftharpoons P2_1/c$	

Table 1 (after Carpenter and Salje 1998) contains a list of some of the structural phase transitions which have been observed in silicate minerals, as classified according to their elastic behaviour. "True proper" ferroelastic transitions, in which the structural instability arises purely as a consequence of some acoustic mode tending to zero frequency and the order parameter is itself a spontaneous strain, are relatively rare. The classic example among other materials is the pressure-dependent, tetragonal ⇔ orthorhombic transition in TeO_2 (Peercy and Fritz 1974, Peercy et al. 1975, Worlton and Beyerlein 1975, McWhan et al. 1975, Skelton et al. 1976, and see Carpenter and Salje 1998 for further references). A definitive test for this transition mechanism is that an individual elastic constant, or some specific combination of elastic constants (C_{11}–C_{12} in TeO_2) tends to zero linearly when the transition point is approached from both the high symmetry side and the low symmetry side. The transformation behaviour can be described quite simply using a free energy expansion such as

$$G = \frac{1}{2}a(P - P_c)(e_1 - e_2)^2 + \frac{1}{4}b(e_1 - e_2)^4. \tag{1}$$

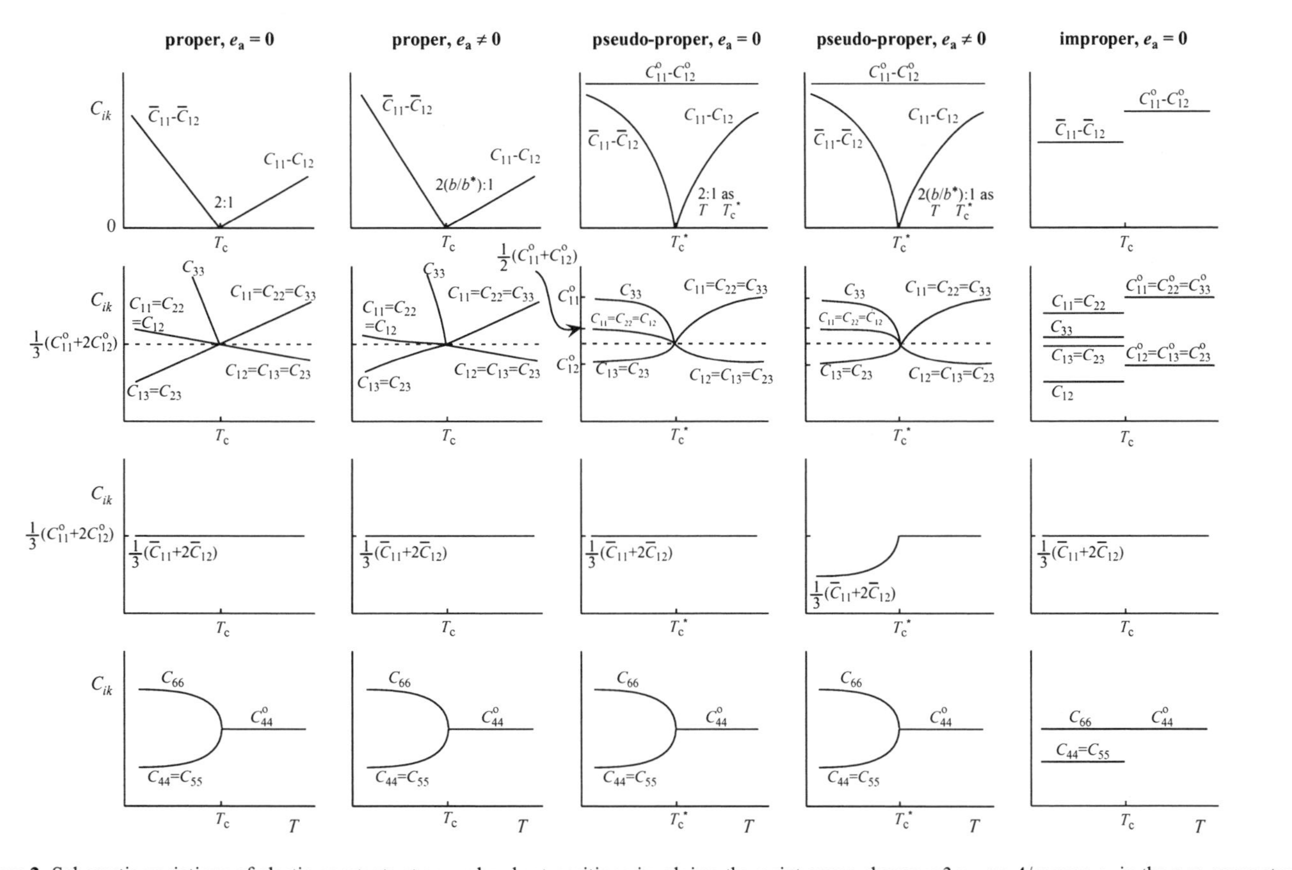

Figure 2. Schematic variations of elastic constants at second-order transitions involving the point-group change $m3m \Leftrightarrow 4/mmm$; e_a is the non-symmetry-breaking strain. For the proper and pseudo-proper cases, it has been assumed that the third-order term is negligibly small; in the improper case ($Pm3m \Leftrightarrow I4/mcm$), this term is strictly zero by symmetry (from Carpenter and Salje 1998). Note:

$$(\bar{C}_{11} - \bar{C}_{12}) = \frac{1}{3}(C_{11} + C_{12} + 2C_{33} - 4C_{13}),\ \frac{1}{3}(\bar{C}_{11} + 2\bar{C}_{12}) = \frac{1}{9}(2C_{11} + C_{33} + 2C_{12} + 4C_{13}).$$

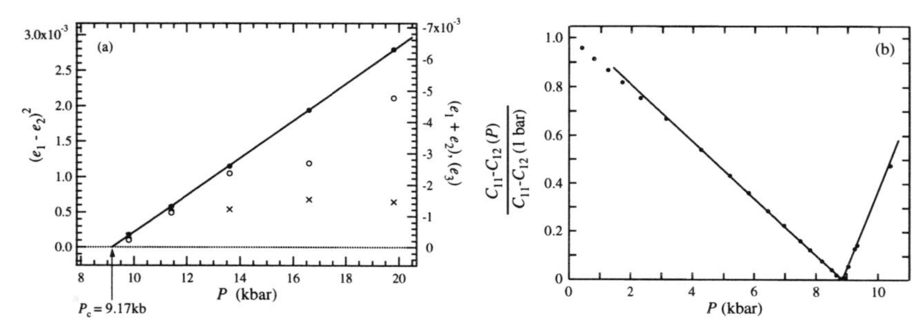

Figure 3. Spontaneous strains and elastic properties at the 422 ⇔ 222 transition in TeO_2. (a) Spontaneous strain data extracted from the lattice-parameter data of Worlton and Beyerlein (1975). The linear pressure dependence of $(e_1 - e_2)^2$ (filled circles) is consistent with second-order character for the transition. Other data are for non-symmetry-breaking strains: $(e_1 + e_2)$ (open circles), e_3 (crosses). (b) Variation of the symmetry-adapted elastic constant $(C_{11} - C_{12})$ at room temperature (after Peercy et al. 1975). The ratio of slopes above and below P_c is ~3:1 and deviates from 2:1 due to the contribution of the non-symmetry-breaking strains. (After Carpenter and Salje 1998).

At equilibrium, the crystal must be stress free ($\partial G/\partial(e_1-e_2) = 0$), giving

$$(e_1 - e_2)^2 = \frac{a}{b}(P_c - P) \tag{2}$$

and the linear strain relationship shown in Figure 3a. By definition, the elastic constants are the second derivative of free energy with respect to strain, giving

$$\begin{aligned}\left(C_{11} - C_{12}\right) &= \frac{\partial^2 G}{\partial(e_1 - e_2)^2} \\ &= a(P - P_c) \qquad \text{(at } P < P_c\text{)} \\ &= 2a(P_c - P) \qquad \text{(at } P > P_c\text{)}\end{aligned} \tag{3}$$

and the linear elastic constant variations shown in Figure 3b. Critical elastic constants and soft acoustic modes for all possible symmetry changes according to this mechanism are listed elsewhere (Cowley 1976, Carpenter and Salje 1998). Amongst minerals, transitions in albite, Sr-anorthite and leucite perhaps conform to this behaviour, but the necessary elastic measurements have not yet been made to confirm this.

By far and away the majority of phase transitions in silicates arise by some other mechanism, such as the softening of an optic mode. In these cases, it can be convenient and instructive to think of the total excess free energy due to the transition as being composed of three parts:

$$\begin{aligned}G = &\frac{1}{2}a(T - T_c)Q^2 + \frac{1}{4}bQ^4 + \ldots && (G_Q) \\ &+\lambda_1 e_i Q + \lambda_2 e_i Q^2 + \ldots && (G_{\text{coupling}}) \\ &+\frac{1}{2}\sum_{i,k} C_{ik}^{o} e_i e_k && (G_{\text{elastic}}).\end{aligned} \tag{4}$$

One contribution, G_Q, arises from changes in the order parameter, Q, and is shown here as a typical Landau potential. Spontaneous strains, e_i, arise because the structure relaxes in response to the change in Q. A coupling energy, G_{coupling}, describing this interaction is given by terms in e_i and Q with coefficients, λ_1, λ_2, ... Finally, if a change in strain occurs, there will also be an elastic energy, G_{elastic}, derived from Hooke's law. C_{ik}^{o} represent the bare elastic constants of the crystal and have values which exclude the effects of the phase transition. "Pseudo-proper" ferroelastic transitions are those in which one strain has the same symmetry as the order parameter, while "improper" ferroelastic and "co-elastic" transitions are those in which all the strains have symmetry which is different from that of the order parameter.

Direct experiments to measure the elastic constants of a crystal involve the application of a stress and measurement of the resulting strain. In general, the observed elastic constants will simply be those of the bare elastic constants. If, however, the experiment is performed at a pressure and temperature which lies in the vicinity of a transition point, there is the additional possibility that a change in strain, e_i, will induce a change in Q via the strain/order parameter coupling. In other words, a small additional energy change will occur and the crystal will be softer than expected. This softening can be described quantitatively using the well known expression (after Slonczevski and Thomas 1970)

$$C_{ik} = C_{ik}^{\text{o}} - \sum_{m,n} \frac{\partial^2 G}{\partial e_i \partial Q_m} \cdot \left(\frac{\partial^2 G}{\partial Q_m \partial Q_n} \right)^{-1} \cdot \frac{\partial^2 G}{\partial e_k \partial Q_n} . \tag{5}$$

Here Q_m and Q_n can be separate order parameters or order parameter components. In combination, Equations (4) and (5) provide a means of predicting the elastic constant variations associated with any phase transition in which the relaxation of Q in response to an applied stress is rapid relative to the time scale of the experimental measurement. A classic example of the success that this approach can have for describing the elastic behaviour of real materials is provided by the work of Errandonea (1980) for the orthorhombic ⇔ monoclinic transition in LaP_5O_{14} (Fig. 4).

A final general consideration is that Landau theory is expected to give an accurate representation of changes in the physical and thermodynamic properties over wide *PT* intervals when a transition is accompanied by significant spontaneous strains. This is because the relatively long ranging influence of strain fields acts to suppress order parameter fluctuations.

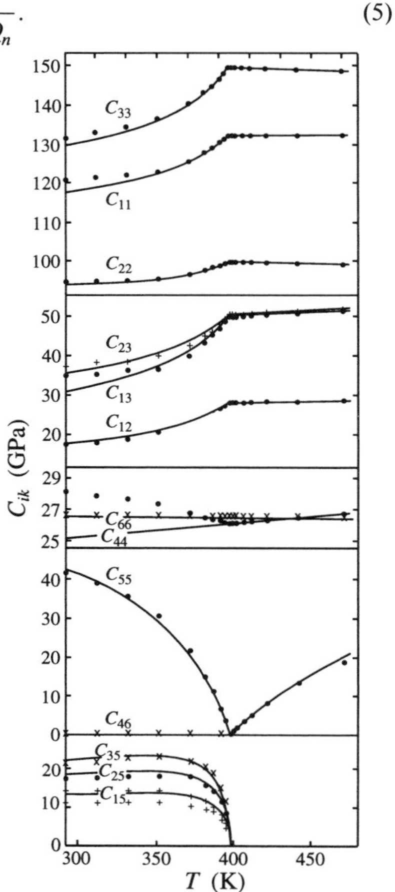

Figure 4. Variation with temperature of the complete set of elastic constants for LaP_5O_{14} at the *mmm* ⇔ *2/m* transition (from Carpenter and Salje 1998, after Errandonea 1980). Solid curves are solutions derived from a Landau free energy expansion for a pseudo-proper ferroelastic transition.

Equations of the form of Equation (4) form the basis of the analysis of strain and elasticity reviewed in this chapter. The issues to be addressed are (a) the geometry of strain, leading to standard equations for strain components in terms of lattice parameters, (b) the relationship between strain and the driving order parameter, and (c) the elastic anomalies which can be predicted on the basis of the resulting free energy functions. The overall approach is presented as a series of examples. For more details of Landau theory and an introduction to the wider literature, readers are referred to reviews by Bruce and Cowley (1981), Wadhawan (1982), Tolédano et al. (1983), Bulou et al. (1992), Salje (1992a,b; 1993), Redfern (1995), Carpenter et al. (1998a), Carpenter and Salje (1998).

LATTICE GEOMETRY AND REFERENCE STATES

A standard set of reference axes and equations to describe spontaneous strains is now well established (Schlenker et al. 1978, Redfern and Salje 1987, Carpenter et al. 1998a). The orthogonal reference axes, X, Y and Z, are selected so that Y is parallel to the crystallographic y-axis, Z is parallel to the normal to the (001) plane (i.e. parallel to ***c****) and X is perpendicular to both. The +X direction is chosen to conform to a right-handed coordinate system. Strain is a second rank tensor; three linear components, e_{11}, e_{22} and e_{33} are tensile strain parallel to X, Y and Z respectively and e_{13}, e_{23}, e_{12} are shear strains in the XZ, YZ and XY planes, respectively. The general equations of Schlenker et al. (1978) define the strains in terms of the lattice parameters of a crystal (a, b, c, α, β^*, γ, where β^* is the reciprocal lattice angle) with respect to the reference state for the crystal (a_o, b_o, c_o, α_o, β_o^*, γ_o):

$$e_1 = e_{11} = \frac{a\sin\gamma}{a_o\sin\gamma_o} - 1 \qquad (6) \qquad e_2 = e_{22} = \frac{b}{b_o} - 1 \qquad (7)$$

$$e_3 = e_{33} = \frac{c\sin\alpha\sin\beta^*}{c_o\sin\alpha_o\sin\beta_o^*} - 1 \qquad (8)$$

$$\frac{1}{2}e_4 = e_{23} = \frac{1}{2}\left[\frac{c\cos\alpha}{c_o\sin\alpha_o\sin\beta_o^*} - \frac{b\cos\alpha_o}{b_o\sin\alpha_o\sin\beta_o^*} + \frac{\cos\beta_o^*}{\sin\beta_o^*\sin\gamma_o}\left(\frac{a\cos\gamma}{a_o} - \frac{b\cos\gamma_o}{b_o}\right)\right] \qquad (9)$$

$$\frac{1}{2}e_5 = e_{13} = \frac{1}{2}\left(\frac{a\sin\gamma\cos\beta_o^*}{a_o\sin\gamma_o\sin\beta_o^*} - \frac{c\sin\alpha\cos\beta^*}{c_o\sin\alpha_o\sin\beta_o^*}\right) \qquad (10)$$

$$\frac{1}{2}e_6 = e_{12} = \frac{1}{2}\left(\frac{a\cos\gamma}{a_o\sin\gamma_o} - \frac{b\cos\gamma_o}{b_o\cos\gamma_o}\right). \qquad (11)$$

For a phase transition, a, b, c, ..., refer to the low symmetry form of a crystal at a given pressure, temperature, P_{H2O}, f_{O2}, ..., while a_o, b_o, c_o, ..., refer to the high symmetry form of the crystal under identical conditions. Voigt notation (e_i = 1-6) is generally more convenient than the full tensor notation (e_{ik}, i,k = 1-3). The strain equations are greatly simplified when any of the lattice angles are 90°. For example, the spontaneous strains accompanying a cubic ⇔ tetragonal phase transition are

$$e_1 = e_2 = \frac{a}{a_o} - 1 \qquad (12) \qquad e_3 = \frac{c}{c_o} - 1. \qquad (13)$$

As discussed in detail in Carpenter et al. (1998a), equations for other changes in crystal system are trivial to derive from Equations (6)-(11).

Several strategies are available for determining the reference parameters. A common assumption is that the volume change accompanying the transition is negligibly small. In the case of a cubic ⇔ tetragonal phase transition, for example, the reference parameter can be expressed as $a_o = (a^2c)^{1/3}$. For most phase transitions, however, there is a significant change in volume and, for some, a change in volume is the dominant strain effect. It is usually advisable to obtain the reference parameters by extrapolation from the stability field of the high symmetry phase therefore. For this purpose, high quality data must be collected over a pressure and temperature interval which extends well into the high symmetry field. If temperature is the externally applied variable, experience suggests that linear extrapolations are usually adequate (e.g. Fig. 1a). If pressure is the externally applied variable, it is necessary to use an equation of state which includes non-linear contributions (see Chapter 4 by R.J.Angel in this volume).

SYMMETRY-ADAPTED STRAIN, SYMMETRY-BREAKING STRAIN, NON-SYMMETRY-BREAKING STRAIN AND SOME TENSOR FORMALITIES

Spontaneous strain is a symmetrical second rank tensor property and must conform to Neumann's principle in relation to symmetry. A general spontaneous strain with all six of the independent strain components having non-zero values can be referred to an alternative set of axes by diagonalisation to give

$$\begin{bmatrix} e_1 & e_6 & e_5 \\ \cdot & e_2 & e_4 \\ \cdot & \cdot & e_3 \end{bmatrix} \rightarrow \begin{bmatrix} \varepsilon_1 & 0 & 0 \\ 0 & \varepsilon_2 & 0 \\ 0 & 0 & \varepsilon_3 \end{bmatrix},$$

where the eigenvalues ε_1, ε_2 and ε_3 are tensile strains along the principal axes of the strain ellipsoid. Orientation relationships between the X-, Y- and Z-axes and the new reference axes are given by the eigenvectors. The sum of the diagonal components is now equivalent, for small strains, to the volume strain, V_s, which is defined as

$$V_s = \frac{V - V_o}{V} \approx \varepsilon_1 + \varepsilon_2 + \varepsilon_3. \qquad (14)$$

This is different from the total strain because some of the terms ε_i may be positive and some negative. The total strain can be defined as a scalar quantity, ε_{ss}, and, while several equally valid definitions for this are given in the literature, the two most commonly used in the earth sciences literature (e.g. Redfern and Salje 1987, Salje 1993) are

$$\varepsilon_{ss} = \sqrt{\sum_i e_i^2} \quad (i = 1\text{-}6), \qquad (15) \qquad\qquad \varepsilon_{ss} = \sqrt{\sum_i \varepsilon_i^2} \quad (i = 1\text{-}3). \qquad (16)$$

The symmetry properties of spontaneous strains are most conveniently understood by referring to the irreducible representations and basis functions for the point group of the high symmetry phase of a crystal of interest. These are given in Table 2 for the point group $4/mmm$ as an example. Basis functions $(x^2 + y^2)$ and z^2 are associated with the identity representation and are equivalent to $(e_1 + e_2)$ and e_3 respectively. This is the same as saying that both strains are consistent with $4/mmm$ symmetry ($e_1 = e_2$). The shear strain $(e_1 - e_2)$ is equivalent to the basis function $(x^2 - y^2)$ which is associated with the B_{1g} representation, the shear strain e_6 is equivalent to xy (B_{2g}) and shear strains e_4 and e_5 to xz, yz respectively (E_g). The combinations $(e_1 + e_2)$ and $(e_1 - e_2)$ are referred to as symmetry-adapted strains because they have the form of specific basis functions of the irreducible representations.

Table 2. Irreducible representations and basis functions for point group *4/mmm*.

	E	$2C_4$	C_2	$2C_2'$	$2C_2''$	i	$2S_4$	σ_h	$2\sigma_v$	$2\sigma_d$	Basis functions
A_{1g}	1	1	1	1	1	1	1	1	1	1	$x^2 + y^2$, z^2
A_{2g}	1	1	1	−1	−1	1	1	1	−1	−1	R_z
B_{1g}	1	−1	1	1	−1	1	−1	1	1	−1	$x^2 - y^2$
B_{2g}	1	−1	1	−1	1	1	−1	1	−1	1	xy
E_g	2	0	−2	0	0	2	0	−2	0	0	R_x, R_y, xz, yz
A_{1u}	1	1	1	1	1	−1	−1	−1	−1	−1	
A_{2u}	1	1	1	−1	−1	−1	−1	−1	1	1	z
B_{1u}	1	−1	1	1	−1	−1	1	−1	−1	1	
B_{2u}	1	−1	1	−1	1	−1	1	−1	1	−1	
E_u	2	0	−2	0	0	−2	0	2	0	0	x, y

They can also be referred to as non-symmetry-breaking and symmetry-breaking strains, respectively; if the strain $(e_1 + e_2)$ takes on some value other than zero, the symmetry of the crystal is not affected, but if the strain $(e_1 - e_2)$ becomes non-zero in value, tetragonal symmetry is broken and the crystal becomes orthorhombic. In general, the total spontaneous strain due to a phase transition, $[e_{tot}]$ may be expressed as the sum of two tensors

$$[e_{tot}] = [e_{sb}] + [e_{nsb}] \tag{17}$$

where $[e_{sb}]$ is the symmetry-breaking strain and $[e_{nsb}]$ is the non-symmetry-breaking strain. This is not merely a matter of semantics because the two types of strains have different symmetry properties. All volume strains are associated with the identity representation and are non-symmetry-breaking. All symmetry-breaking strains are pure shear strains.

Only high symmetry point groups have second order basis functions consisting of more than one quadratic term ($x^2 - y^2$, $x^2 + y^2$, etc.). Symmetry-adapted strains are therefore restricted to these. The most commonly used combinations refer to a cubic parent structure for which there are three symmetry-adapted strains

$$e_a = e_1 + e_2 + e_3 \tag{18}$$

$$e_t = \frac{1}{\sqrt{3}}(2e_3 - e_1 - e_2) \tag{19}$$

$$e_o = e_1 - e_2. \tag{20}$$

Here e_t is a tetragonal shear, e_o is an orthorhombic shear and, for small strains, e_a is the volume strain. Symmetry-adapted strains for the sequence of phase transitions $Pm\bar{3}m \Leftrightarrow P4/mbm \Leftrightarrow Cmcm$ are shown for $NaTaO_3$ in Figure 5, though, for reasons which are given later, a different orientation relationship between crystallographic x-, y- and z-axes with respect to reference X-, Y- and Z-axes was chosen for this system.

COUPLING BETWEEN STRAIN AND THE ORDER PARAMETER

Both the order parameter and the spontaneous strain for a phase transition have symmetry properties. The relationship between them is therefore also constrained by symmetry. Only in the case of proper ferroelastic transitions is the symmetry-breaking strain itself the order parameter. For most transitions, the order parameter relates to some

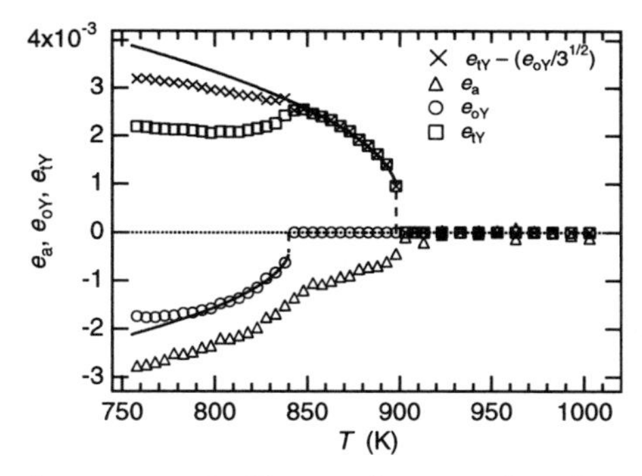

Figure 5. Variation of the symmetry-adapted spontaneous strains, derived from the data in Figure 1, for the cubic → tetragonal → orthorhombic sequence of $NaTaO_3$ perovskite. Solid lines are standard Landau solutions for weakly first order transitions (cubic ⇔ tetragonal, tetragonal ⇔ orthorhombic).

other structural effect, such as a soft optic mode, which provides the driving mechanism. Lattice distortions then occur by coupling of the spontaneous strain to the order parameter. Under these conditions, the general form of Equation (4) may be given as

$$G(Q,e) = L(Q) + \sum_{i,m,n} \lambda_{i,m,n} e_i^m Q^n + \frac{1}{2}\sum_{i,k} C_{ik}^{o} e_i e_k \tag{21}$$

where λ_i are coupling coefficients, m and n are positive integers and the free energy contribution from the order parameter alone is given by the normal Landau expansion

$$L(Q) = \frac{1}{2}a(T - T_c)Q^2 + \frac{1}{3}bQ^3 + \frac{1}{4}cQ^4 + \ \ldots \tag{22}$$

The coupling terms and elastic energy terms must comply with standard symmetry rules, which require that each term in the free energy expansion transforms (in the group theoretical sense) as the identity representation, $\Gamma_{identity}$. The order parameter, by definition, transforms as the active representation, R_{active}. The rules for coupling terms are then as follows:

(a) If e_i transforms as R_{active}, all terms in e_iQ^n with $n \geq 1$ are allowed when Q^{n+1} is present in $L(Q)$.

(b) If e_i transforms as $\Gamma_{identity}$, all terms in e_iQ^n with $n \geq 2$ are allowed when Q^n is present in $L(Q)$.

(c) If e_i does not transform as either R_{active} or as $\Gamma_{identity}$, no general rule applies and the full multiplication of representations must be completed to test for the presence of $\Gamma_{identity}$.

Fortunately, there is a widely available computer program, ISOTROPY (Stokes and Hatch, Brigham Young University) which can be used to perform all the group theoretical manipulations and generate the correct coupling terms for any change in symmetry. The tables of Stokes and Hatch (1988) are also invaluable in this respect. In practice it is also found that, for many materials, only the lowest order coupling term is needed to describe observed relationships between each strain component and the driving order parameter.

Three examples can be used to illustrate most of the features of coupling which have so far been observed. Stishovite, SiO_2, undergoes a structural phase transition with increasing pressure which involves the symmetry change $P4_2/mnm$ ⇔ $Pnnm$ ($4/mmm$ ⇔ mmm) (Nagel and O'Keeffe 1971, Hemley et al. 1985, Cohen 1987, 1992, 1994; Hemley 1987, Tsuchida and Yagi 1989, Yamada et al. 1992, Matsui and Tsuneyaki 1992, Lacks and Gordon 1993, Lee and Gonze 1995, Kingma et al. 1995, 1996; Karki et al. 1997a,b; Teter et al. 1998, Andrault et al. 1998, Carpenter et al. 2000a, Hemley et al. 2000). The driving order parameter is a zone centre soft optic mode with B_{1g} symmetry and, associated with the B_{1g} representation of point group 4/mmm, is the basis function $(x^2 - y^2)$ (Table 2). The symmetry-breaking strain $(e_1 - e_2)$ therefore has the same symmetry as the order parameter, giving the lowest order coupling term as $\lambda(e_1 - e_2)Q$. The non-symmetry-breaking strain $(e_1 + e_2)$ is associated with the identity representation (Table 2) and the lowest order coupling term permitted by symmetry is $\lambda(e_1 + e_2)Q^2$. Strains e_4 and e_5 are associated with the E_g representation and the lowest order coupling term is $\lambda(e_4{}^2 - e_5{}^2)Q$, while e_6 has B_{2g} symmetry and the lowest order coupling term is $\lambda e_6{}^2Q^2$. The full Landau expansion is then (from Carpenter et al. 2000a).

$$G = \frac{1}{2}a(P - P_c)Q^2 + \frac{1}{4}bQ^4 + \lambda_1(e_1 + e_2)Q^2 + \lambda_2(e_1 - e_2)Q + \lambda_3 e_3 Q^2$$
$$+\lambda_4\left(e_4^2 - e_5^2\right)Q + \lambda_6 e_6^2 Q^2 + \frac{1}{4}\left(C_{11}^{o} + C_{12}^{o}\right)(e_1 + e_2)^2 + \frac{1}{4}\left(C_{11}^{o} - C_{12}^{o}\right)(e_1 - e_2)^2$$
$$+C_{13}^{o}(e_1 + e_2)e_3 + \frac{1}{2}C_{33}^{o}e_3^2 + \frac{1}{2}C_{44}^{o}\left(e_4^2 + e_5^2\right) + \frac{1}{2}C_{66}^{o}e_6^2. \tag{23}$$

In the orthorhombic structure, the strains e_4, e_5 and e_6 are all zero, while the other strains must conform to the equilibrium condition, $\partial G/\partial e = 0$. This gives

$$(e_1 - e_2) = -\frac{\lambda_2}{\frac{1}{2}\left(C_{11}^{o} - C_{12}^{o}\right)}Q \tag{24}$$

$$(e_1 + e_2) = -\frac{\lambda_1}{\frac{1}{2}\left(C_{11}^{o} + C_{12}^{o}\right)}Q^2 \tag{25}$$

$$e_3 = -\frac{\lambda_3}{C_{33}^{o}}Q^2. \tag{26}$$

Thus, typically, $e_{sb} \propto Q$ applies when the symmetry-breaking strain has the same symmetry as the order parameter and $e \propto Q^2$ applies in all other cases. The volume strain should also scale as $V_s \propto Q^2$, therefore. Experimental data for the lattice parameters of stishovite as a function of pressure are shown in Figure 6. The symmetry-breaking strain is given by

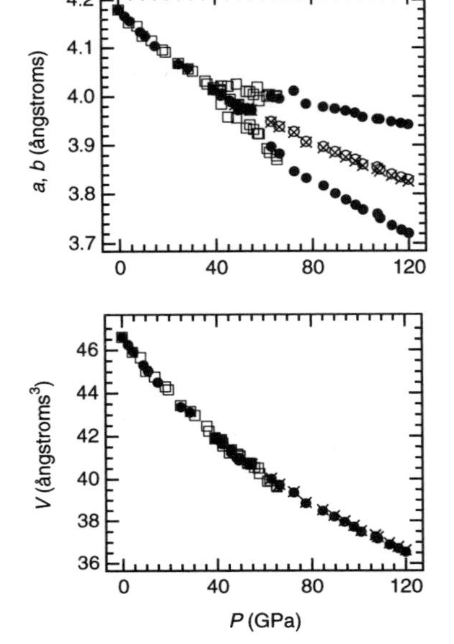

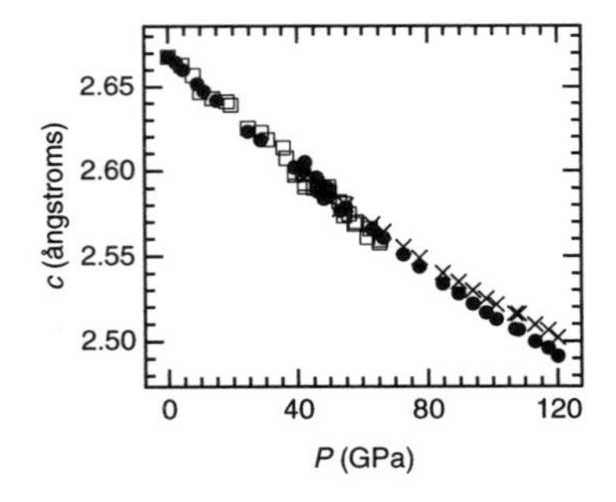

Figure 6. Lattice parameter variations through the tetragonal ⇔ orthorhombic transition in stishovite (data of Ross et al. 1990; Hemley et al. 1994; Mao et al. 1994; Andrault et al. 1998). Filled circles are the data of Ross et al., Hemley et al. and Andrault et al.; open squares are the data of Mao et al.; open circles represent the reference parameter a_o given by $(ab)^{1/2}$; crosses are reference parameters calculated using the experimental lattice parameters for orthorhombic stishovite and coupling parameters λ_1 = −8 GPa and λ_3 =17 GPa (after Carpenter et al. 2000a).

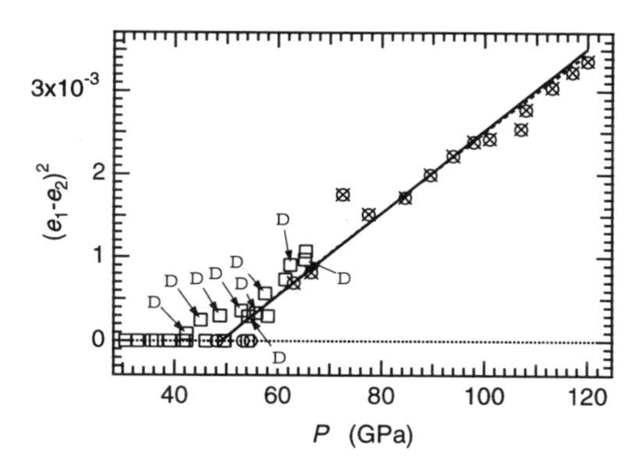

Figure 7. A linear variation of the square of the symmetry breaking strain with pressure implies second order character for the tetragonal ⇔ orthorhombic transition in stishovite. Circles represent strains calculated using $a_o = (ab)^{1/2}$ and the data of Andrault et al. (1998); crosses represent strains calculated using the alternative variation of a_o shown by crosses in Figure 6. Open squares are from single crystal X-ray diffraction data of Mao et al. (1994); D designates measurements of the orthorhombic phase made during decompression.

$$(e_1 - e_2) = \frac{a - a_o}{a_o} - \frac{b - a_o}{a_o} = \frac{a - b}{a_o} \tag{27}$$

and is not especially sensitive to the choice of reference parameter. As an approximation, a_o can be taken as $(ab)^{1/2}$, where a and b are lattice parameters of the orthorhombic structure. The square of the symmetry-breaking strain is found to vary linearly with pressure (Fig. 7), implying $Q^2 \propto (P - P_c)$ and, therefore, second order character for the transition. The experimental data are not sufficiently precise to determine with confidence the magnitude of any volume strain.

The Landau free energy expansion to describe the β ⇔ α ($P6_422$ ⇔ $P3_121$) transition in quartz has been constructed in the same manner to include lowest order coupling terms for all the possible strain components (Grimm and Dorner 1975, Bachheimer and Dolino 1975, Banda et al. 1975, Dolino and Bachheimer 1982). It is reproduced here from Carpenter et al. (1998b):

$$\begin{aligned} G = {} & \frac{1}{2}a(T - T_c)Q^2 + \frac{1}{4}bQ^4 + \frac{1}{6}cQ^6 + \frac{1}{8}dQ^8 + \lambda_1(e_1 + e_2)Q^2 + \lambda_3 e_3 Q^2 \\ & + \lambda_4\left(e_4^2 + e_5^2\right)Q^2 + \lambda_5(e_1e_4 - e_2e_4 + e_5e_6)Q + \lambda_6\left[e_6^2 + (e_1 - e_2)^2\right]Q^2 \\ & + \lambda_7(e_1 + e_2)Q^4 + \lambda_8 e_3 Q^4 + \lambda_9(e_1e_4 - e_2e_4 + e_5e_6)Q^3 \\ & + \frac{1}{4}\left(C_{11}^o + C_{12}^o\right)(e_1 + e_2)^2 + \frac{1}{4}\left(C_{11}^o - C_{12}^o\right)(e_1 - e_2)^2 + C_{13}^o(e_1 + e_2)e_3 \\ & + \frac{1}{2}C_{33}^o e_3^2 + \frac{1}{2}C_{44}^o\left(e_4^2 + e_5^2\right) + \frac{1}{2}C_{66}^o e_6^2. \end{aligned} \tag{28}$$

In this case, some higher order coupling terms have been included for $(e_1 + e_2)$ and e_3. Lattice parameter variations through the transition (from Carpenter et al. 1998b) are shown in Figure 8. Linear extrapolation of lattice parameters for β-quartz gives the variations of a_o and c_o for calculation of the linear strains, $e_1 = e_2 = (a - a_o)/a_o$, $e_3 = (c - c_o)/c_o$. The linear strains exceed 1 % at low temperatures and the volume strain reaches ~5% (Fig. 8d). The evolution of the order parameter in α-quartz can be determined independently from measurements of optical properties (Bachheimer and Dolino 1975), and reveals that the expected relationship $(e_1 + e_3) \propto e_3 \propto Q^2$ is not observed (Fig. 8e). Instead, the variation can be described by inclusion of the higher order coupling terms in Equation (28) which gives

$$\left(e_1 + e_2\right) = \left[\frac{2\lambda_3 C_{13}^{o} - 2\lambda_1 C_{33}^{o}}{\left(C_{11}^{o} + C_{12}^{o}\right)C_{33}^{o} - 2C_{13}^{o\,2}}\right]Q^2 + \left[\frac{2\lambda_8 C_{13}^{o} - 2\lambda_7 C_{33}^{o}}{\left(C_{11}^{o} + C_{12}^{o}\right)C_{33}^{o} - 2C_{13}^{o\,2}}\right]Q^4 \tag{29}$$

$$e_3 = \left[\frac{2\lambda_1 C_{13}^{o} - \lambda_3\left(C_{11}^{o} + C_{12}^{o}\right)}{\left(C_{11}^{o} + C_{12}^{o}\right)C_{33}^{o} - 2C_{13}^{o\,2}}\right]Q^2 + \left[\frac{2\lambda_7 C_{13}^{o} - \lambda_8\left(C_{11}^{o} + C_{12}^{o}\right)}{\left(C_{11}^{o} + C_{12}^{o}\right)C_{33}^{o} - 2C_{13}^{o\,2}}\right]Q^4. \tag{30}$$

Fits to the data on this basis are shown in Figure 8e. Values for the coupling coefficients can be extracted from experimentally determined relationships of this type if the bare elastic constants are known.

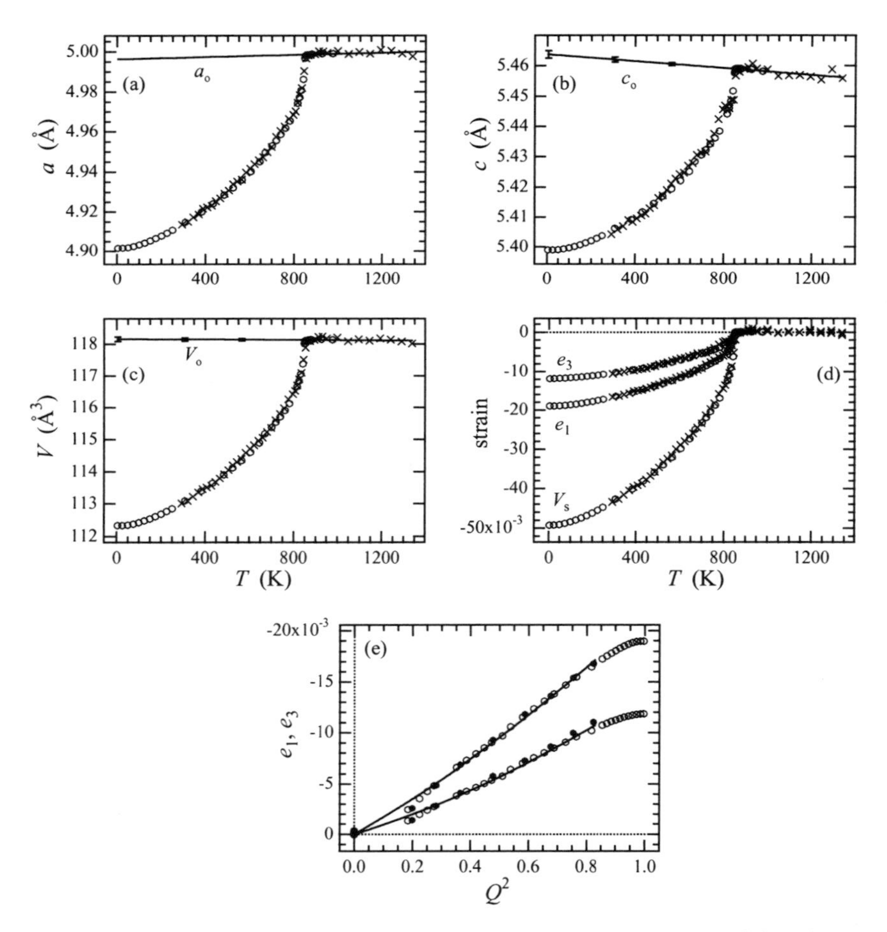

Figure 8. (a, b, c) Lattice parameters of quartz showing linear extrapolations of the reference parameters a_o, c_o, V_o. (d) Spontaneous strains derived from the lattice parameters. (e) Variations of e_1 and e_3 as a function of Q^2 derived from optical data of Bachheimer and Dolino (1975). The solid lines are fits to the data, excluding data from below room temperature which are influenced by order parameter saturation (after Carpenter et al. 1998b).

The $\beta \Leftrightarrow \alpha$ quartz transition is just first order in character, and a solution of the form

$$Q^2 = \frac{2}{3} Q_o^2 \left\{ 1 + \left[1 - \frac{3}{4}\left(\frac{T - T_c}{T_{tr} - T_c} \right) \right]^{1/2} \right\} \tag{31}$$

describes the equilibrium evolution of the order parameter. The magnitude of the discontinuity at the transition point is given by $Q_o = 0.38$ and the temperature difference ($T_{tr} - T_c$) is only 7K, which means that the transition is close to being tricritical in character. The strain evolution for this first order solution is shown in Figure 9. It should be noted that the stability field of the incommensurate phase is not resolved in these data.

For high symmetry systems, strain coupling needs to take account of the separate components of the order parameter. For example, cubic $\Leftrightarrow$ tetragonal, cubic $\Leftrightarrow$ orthorhombic and tetragonal $\Leftrightarrow$ orthorhombic transitions in perovskites are associated with the M_3 and R_{25} points of the reciprocal lattice of space group $Pm\bar{3}m$ and there are two separate order parameters, each with three components. The full Landau expansion is (from Carpenter et al. 2000b).

$$\begin{aligned}
G = {} & \frac{1}{2}a_1(T - T_{c1})(q_1^2 + q_2^2 + q_3^2) + \frac{1}{2}a_2(T - T_{c2})(q_4^2 + q_5^2 + q_6^2) \\
& + \frac{1}{4}b_1(q_1^2 + q_2^2 + q_3^2)^2 + \frac{1}{4}b_1'(q_1^4 + q_2^4 + q_3^4) + \frac{1}{4}b_2(q_4^2 + q_5^2 + q_6^2)^2 \\
& + \frac{1}{4}b_2'(q_4^4 + q_5^4 + q_6^4) + \frac{1}{6}c_1(q_1^2 + q_2^2 + q_3^2)^3 + \frac{1}{6}c_1'(q_1 q_2 q_3)^2 \\
& + \frac{1}{6}c_1''(q_1^2 + q_2^2 + q_3^2)(q_1^4 + q_2^4 + q_3^4) + \frac{1}{6}c_2(q_4^2 + q_5^2 + q_6^2)^3 + \frac{1}{6}c_2'(q_4 q_5 q_6)^2 \\
& + \frac{1}{6}c_2''(q_4^2 + q_5^2 + q_6^2)(q_4^4 + q_5^4 + q_6^4) + \lambda_q(q_1^2 + q_2^2 + q_3^2)(q_4^2 + q_5^2 + q_6^2) \\
& + \lambda_q'(q_1^2 q_4^2 + q_2^2 q_5^2 + q_3^2 q_6^2) + \lambda_1 e_a(q_1^2 + q_2^2 + q_3^2) + \lambda_2 e_a(q_4^2 + q_5^2 + q_6^2) \\
& + \lambda_3\left[\sqrt{3}e_o(q_2^2 - q_3^2) + e_t(2q_1^2 - q_2^2 - q_3^2)\right] \\
& + \lambda_4\left[\sqrt{3}e_o(q_5^2 - q_6^2) + e_t(2q_4^2 - q_5^2 - q_6^2)\right] + \lambda_5(e_4 q_4 q_6 + e_5 q_4 q_5 + e_6 q_5 q_6) \\
& + \lambda_6(q_1^2 + q_2^2 + q_3^2)(e_4^2 + e_5^2 + e_6^2) + \lambda_7(q_1^2 e_6^2 + q_2^2 e_4^2 + q_3^2 e_5^2) \\
& + \frac{1}{4}(C_{11}^o - C_{12}^o)(e_o^2 + e_t^2) + \frac{1}{6}(C_{11}^o + 2C_{12}^o)e_a^2 + \frac{1}{2}C_{44}^o(e_4^2 + e_5^2 + e_6^2).
\end{aligned} \tag{32}$$

Different product phases with space groups which are subgroups of $Pm\bar{3}m$ have different combinations of order parameter components with non-zero values, as listed in Table 3. From a geometrical perspective, these order parameter components can be understood in terms of the octahedral tilt description of perovskite phase transitions due to Glazer (1972, 1975) (and see Woodward 1997), such that each component corresponds to a tilt of the octahedra about a symmetry axis (Howard and Stokes 1998). Unfortunately, three different reference systems have been used in the past—the reference axes for the octahedral tilt descriptions, reference axes for the group theory program ISOTROPY and the reference axes of Schlenker et al. (1978) for defining spontaneous strains. In order to relate order parameter components to strains, as set out in Equation (32), which is derived from ISOTROPY, it is necessary to redefine the strain axes. Linear strains e_1, e_2 and e_3 remain

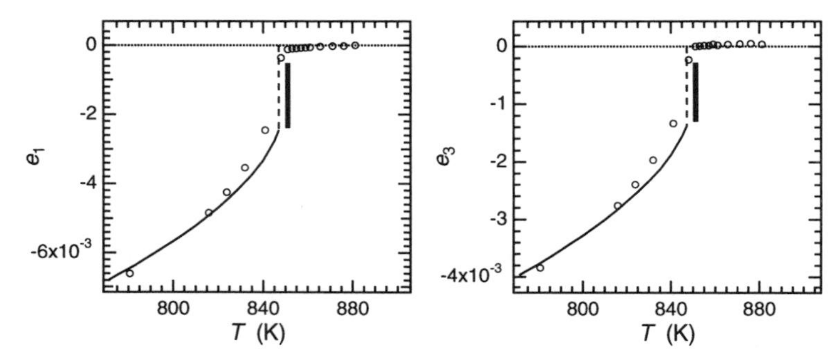

Figure 9. Strain variations in quartz close to the β ⇔ α phase transition. Solid lines are the Landau solution for a first order phase transition with a small discontinuity in the order parameter at the transition temperature. Solid bars represent the discontinuities expected on the basis of linear expansion data of Bachheimer (1980, 1986) and Mogeon (1988). From Carpenter et al. (1998b).

parallel to X, Y and Z, respectively, but the crystallographic axes of crystals with space groups $Pm\bar{3}m$, $P4/mbm$, $I4/mcm$, $Pnma$ and $Cmcm$ are set to the orientations shown in Figure 10. In order to follow the single non-zero order parameter component of $P4/mbm$ (q_1) through to the equivalent component of $Cmcm$ (q_3), it is necessary to choose a twin component of $P4/mbm$ which has **c** parallel to **Y**. The non-zero component of $P4/mbm$ is then q_3 and the strains are given by

$$e_1 = e_3 = \frac{\frac{a}{\sqrt{2}} - a_o}{a_o} \quad (33) \qquad e_2 = \frac{c - a_o}{a_o}. \quad (34)$$

The symmetry-adapted strains e_t and e_o are replaced by e_{tY} and e_{oY}, where

$$e_{tY} = \frac{1}{\sqrt{3}}(2e_2 - e_1 - e_3) \quad (35) \qquad e_{oY} = e_1 - e_3 \quad (36)$$

and, under equilibrium conditions,

$$e_a = -\frac{\lambda_1 q_3^2}{\frac{1}{3}\left(C_{11}^o + 2C_{12}^o\right)} \quad (37) \qquad e_{oY} = e_4 = e_5 = e_6 = 0 \quad (38)$$

$$e_{tY} = -\frac{2\lambda_3 q_3^2}{\frac{1}{2}\left(C_{11}^o - C_{12}^o\right)} \quad .(39)$$

For $Cmcm$ structures, the strains are defined as

$$e_1 = \frac{\frac{a}{2} - a_o}{a_o} \quad (40) \qquad e_2 = \frac{\frac{c}{2} - a_o}{a_o} \quad (41)$$

$$e_3 = \frac{\frac{b}{2} - a_o}{a_o}, \quad (42)$$

Table 3. Order parameter components for the subgroups of $Pm\bar{3}m$ associated with special points M_3^+ and R_4^+ (after Howard and Stokes 1998). The system of reference axes for these components is that used in Stokes and Hatch (1988) and the group theory program ISOTROPY.

Space Group	Order parameter components	Relationships between order parameter components
$Pm\bar{3}m$	000 000	
$P4/mbm$	$q_1$00 000	
$I4/mmm$	$q_1 0 q_3$ 000	$q_1 = q_3$
$Im\bar{3}$	$q_1 q_2 q_3$ 000	$q_1 = q_2 = q_3$
$Immm$	$q_1 q_2 q_3$ 000	$q_1 \neq q_2 \neq q_3$
$I4/mcm$	000 $q_4$00	
$Imma$	000 $q_4 0 q_6$	$q_4 = q_6$
$R\bar{3}c$	000 $q_4 q_5 q_6$	$q_4 = q_5 = q_6$
$C2/m$	000 $q_4 0 q_6$	$q_4 \neq q_6$
$C2/c$	000 $q_4 q_5 q_6$	$q_4 = q_6 \neq q_5$
$P\bar{1}$	000 $q_4 q_5 q_6$	$q_4 \neq q_5 \neq q_6$
$Cmcm$	00q_3 $q_4$00	$q_3 \neq q_4$
$Pnma$	0$q_2$0 $q_4 0 q_6$	$q_2 \neq q_4 = q_6$
$P2_1/m$	0$q_2$0 $q_4 0 q_6$	$q_2 \neq q_4 \neq q_6$
$P4_2/nmc$	0$q_2 q_3$ $q_4$00	$q_2 = q_3 \neq q_4$

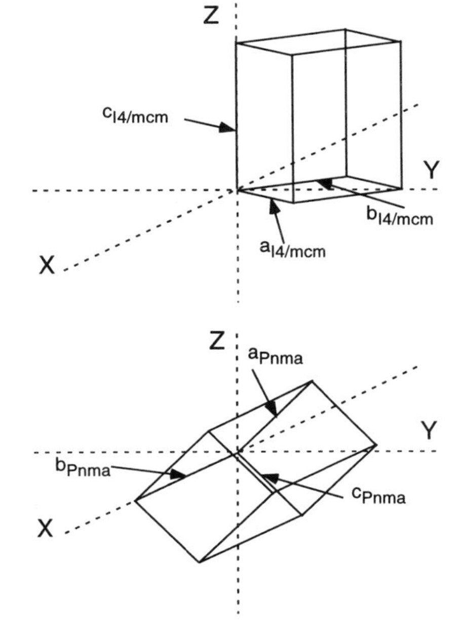

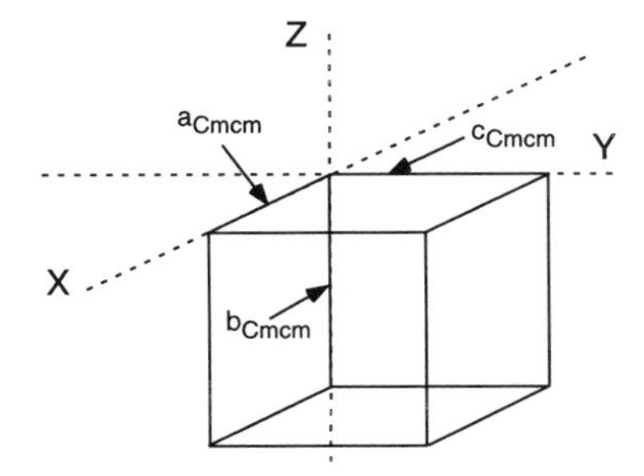

Figure 10. Relationships between unit cell orientations for *I4/mcm*, *Pnma* and *Cmcm* structures and the reference axes, X, Y, Z. Cell edges of the $Pm\bar{3}m$ unit cell are parallel to the reference axes. The *P4/mbm* structure has the same unit cell as the *I4/mcm* structure, except that the c repeat of the former is half that of the latter (from Carpenter et al. 2000b).

and, under equilibrium conditions,

$$e_a = -\frac{\left(\lambda_1 q_3^2 + \lambda_2 q_4^2\right)}{\frac{1}{3}\left(C_{11}^o + 2C_{12}^o\right)} \quad (43) \qquad e_{oY} = +\frac{\sqrt{3}\lambda_4 q_4^2}{\frac{1}{2}\left(C_{11}^o - C_{12}^o\right)} \quad (44)$$

$$e_{tY} = -\frac{\left(2\lambda_3 q_3^2 - \lambda_4 q_4^2\right)}{\frac{1}{2}\left(C_{11}^o - C_{12}^o\right)}. \quad (45)$$

The spontaneous strains shown in Figure 5 were derived from the lattice parameter data for $NaTaO_3$ in Figure 1 according to these definitions. With the additional information relating each order parameter component to the different strain components, it is now possible to use the strain evolution as a proxy for the order parameter evolution (q_3 corresponds to the development of the first octahedral tilt system and q_4 to the second tilt system, which is orthogonal to the first). Rearranging Equations (44) and (45) gives

$$\left(e_{tY} - \frac{e_{oY}}{\sqrt{3}}\right) = -\frac{2\lambda_3 q_3^2}{\frac{1}{2}\left(C_{11}^o - C_{12}^o\right)} \quad (46)$$

so that the variations of q_3^2 and q_4^2 can be separated. The strain combination given in Equation (46) has also been plotted on Figure 5. Assuming that there is no higher order coupling between the shear strain and the order parameter, the $Pm\bar{3}m \Leftrightarrow P4/mbm$ transition is marked by an abrupt increase in q_3^2 with a temperature evolution which is consistent with first order character (Eqn. 31) and $(T_{tr} - T_c) = 5$ K. The fit shown for the $P4/mbm \Leftrightarrow Cmcm$ transition is also for a first order transition, but with $(T_{tr} - T_c) = 2$ K, which is almost indistinguishable from tricritical character ($q_4^4 \propto (T - T_c)$). At this transition, q_3^2 continues to increase but with a slightly different trend, due probably to biquadratic coupling between q_3 and q_4. It is interesting to note that changes in the volume strain, e_a, anticipate the tetragonal $\Leftrightarrow$ orthorhombic transition by about ~10 K in the tetragonal field. The variation of q_3^2 as a function of q_4^2 is represented by the strain variations in Figure 11a and is typical of a system with two order parameters which have weak biquadratic coupling (Salje and Devarajan 1986). The variation of e_{oY} and $(e_{tY} - e_{oY}/\sqrt{3})$ with e_a is shown in Figure 11b, showing internal consistency for most of the temperature interval down to ~800 K. Below ~800 K, the volume strain deviates from the expected linear relationship with the shear strains suggesting additional contributions, probably from higher order coupling of the strain with either (or both) of the order parameter components. Such higher order coupling would also cause a change in the equilibrium evolution of q_4 below ~800 K, as seen in Figure 5.

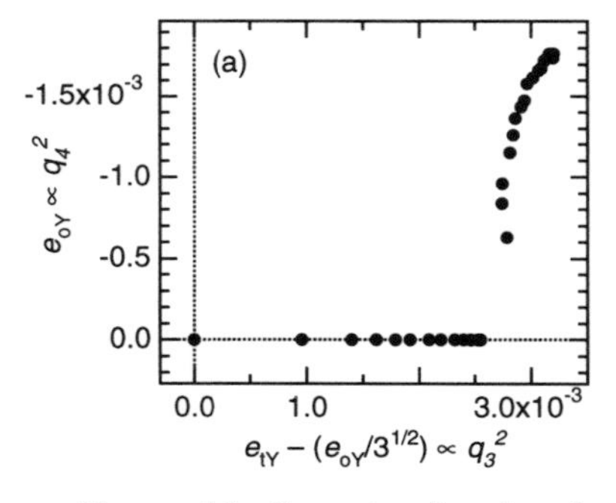

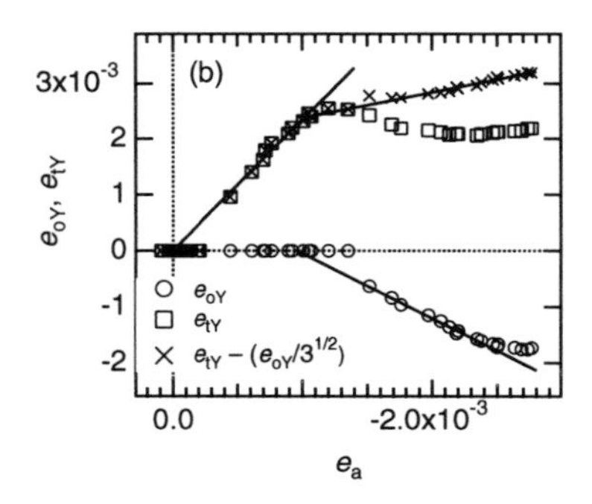

Figure 11. Symmetry-adapted strain variations in the cubic → tetragonal → orthorhombic sequence of $NaTaO_3$ perovskite.

THERMODYNAMIC CONSEQUENCES OF STRAIN/ORDER PARAMETER COUPLING

The most general consequence of strain/order parameter coupling is that any local interactions which generate a strain can influence the state of neighbouring local regions over the full length scale of the strain field. Interaction lengths of this order are sufficient to promote mean-field behaviour except in a temperature interval of, perhaps, less than 1 K close to the transition point for a second order transition. Landau theory is expected to provide rather accurate descriptions of phase transitions when they are accompanied by significant strain in most minerals, therefore. The second well established consequence of such coupling is the renormalisation of the coefficients in a Landau expansion, leading to changes in transition temperature (or pressure) and thermodynamic character for the transition. These effects can again be illustrated using the examples of stishovite, quartz and perovskite.

Bilinear coupling in stishovite, as described by the term $\lambda_2(e_1 - e_2)Q$ in Equation (23), leads to a change in the transition pressure from P_c to P_c^* where

$$P_c^* = P_c + \frac{\lambda_2^2}{a\frac{1}{2}\left(C_{11}^o - C_{12}^o\right)}. \tag{47}$$

Linear-quadratic coupling, as described by the terms $\lambda_1(e_1 + e_2)Q^2$ and $\lambda_3 e_3 Q^2$, leads to a renormalisation of the fourth order coefficient from b to b^* where

$$b^* = b - 2\left[\frac{\lambda_3^2\left(C_{11}^o + C_{12}^o\right) + 2\lambda_1^2 C_{33}^o - 4\lambda_1\lambda_3 C_{13}^o}{\left(C_{11}^o + C_{12}^o\right)C_{33}^o - 2C_{13}^{o\,2}}\right]. \tag{48}$$

Equation (23) would then be reduced to its simplest form as

$$G = \frac{1}{2}a\left(P - P_c^*\right)Q^2 + \frac{1}{4}b^* Q^4. \tag{49}$$

Some insight into the mechanism by which P_c is changed is provided by the frequency of the soft mode responsible for the phase transition. In the stability field of the high symmetry (tetragonal) phase, the inverse order parameter susceptibility, χ^{-1}, of the order parameter varies as

$$\chi^{-1} = \frac{\partial^2 G}{\partial Q^2} = a\left(P - P_c\right) \tag{50}$$

and, since the square of the frequency of the soft mode, ω^2, is expected to scale with χ^{-1} (e.g. Bruce and Cowley 1981, Dove 1993), this means that ω^2 should go linearly to zero at $P = P_c$. As seen from the experimental data of Kingma et al. (1995) in Figure 12, ω^2 would go to zero at a transition pressure above 100 GPa if there was no coupling with strain. Coupling with the shear strain stabilises the orthorhombic structure and displaces the transition point to ~50 GPa (Fig. 12). Once the transition has occurred, the soft mode recovers in the orthorhombic structure.

The energy contributions from the order parameter, G_Q, strain/order parameter coupling, $G_{coupling}$, and the Hooke's law elastic energy, $G_{elastic}$, can be compared if, as is the case for stishovite, numerical values have been measured or estimated for all the coupling terms and bare elastic constants. Separation of the energies in this way is slightly artificial in the context of the overall transition mechanism, but is also quite instructive.

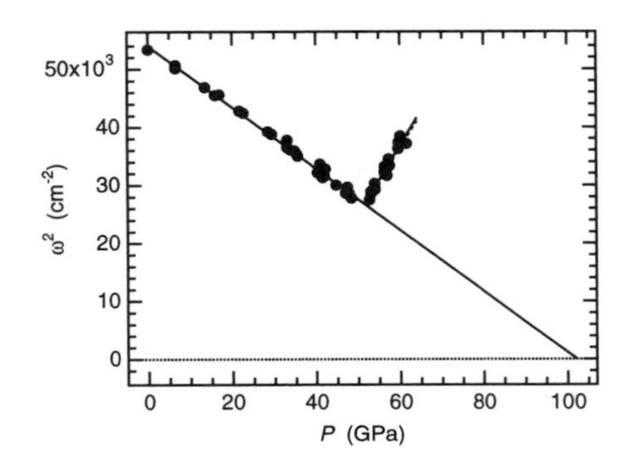

Figure 12. Variation of frequency squared for the soft optic mode through the tetragonal ⇔ orthorhombic transition in stishovite (data from Kingma et al. 1995). Straight lines through the data intersect at $P_c^* = 51.6$ GPa; the straight line for tetragonal stishovite extrapolates to zero at $P_c = 102.3$ GPa. A broken line for the orthorhombic phase includes the pressure dependence of b^*, but is barely distinguishable from the straight line fit to the data (solid line). After Carpenter et al. (2000a).

Taking a pressure of 100 GPa, for example, the total excess free energy of *Pnnm* stishovite with respect to $P4_2/mnm$ stishovite at this pressure is –1.57 kJ.mole^{-1} based on the calibration of Equation (23). Of this, G_Q accounts for +1.92, $G_{coupling}$ for –6.99 and $G_{elastic}$ for +3.50 kJ.mole^{-1}. It is clear that, while the symmetry-breaking mechanism is due to the soft optic mode, the main driving energy for the transition actually arises from the coupling of the soft mode with lattice strains.

In the case of quartz, the only strains are non-symmetry-breaking and would normally be expected to lead to renormalisation only of the fourth order Landau coefficient. Higher order coupling leads to renormalisation of higher order terms as well, however. A renormalised version of the Landau expansion for the β ⇔ α quartz transition (Eqn. 28) would be

$$G = \frac{1}{2}a(T - T_c)Q^2 + \frac{1}{4}b^*Q^4 + \frac{1}{6}c^*Q^6 + \frac{1}{8}d^*Q^8 \tag{51}$$

where

$$b^* = b - 2\left[\frac{\lambda_3^2\left(C_{11}^o + C_{12}^o\right) + 2\lambda_1^2 C_{33}^o - 4\lambda_1\lambda_3 C_{13}^o}{\left(C_{11}^o + C_{12}^o\right)C_{33}^o - 2C_{13}^{o\,2}}\right] \tag{52}$$

$$c^* = c - 6\left[\frac{\lambda_3\lambda_8\left(C_{11}^o + C_{12}^o\right) + 2\lambda_1\lambda_7 C_{33}^o - 2\lambda_1\lambda_8 C_{13}^o - 2\lambda_3\lambda_7 C_{13}^o}{\left(C_{11}^o + C_{12}^o\right)C_{33}^o - 2C_{13}^{o\,2}}\right] \tag{53}$$

$$d^* = d - 4\left[\frac{2\lambda_7^2 C_{33}^o - 4\lambda_7\lambda_8 C_{13}^o + \lambda_8^2\left(C_{11}^o + C_{12}^o\right)}{\left(C_{11}^o + C_{12}^o\right)C_{33}^o - 2C_{13}^{o\,2}}\right]. \tag{54}$$

Data of Tezuka et al. (1991) for the soft mode in β-quartz extrapolate as $\omega^2 \propto T$ down to zero at T_c = ~840 K. The transition occurs at T_{tr} = ~847 K, however, where, from the solution for a first order transition (with $d^* = 0$)

$$\left(T_{tr} - T_c\right) = \frac{3\left(b^*\right)^2}{16ac^*}. \tag{55}$$

The transition is only first order in character because of the strain/order parameter coupling, which causes b^* to be negative (–1931 J.mole^{-1}) while b is positive (+4875 J.mole^{-1}), however. The phase transition would actually be second order in character in a crystal of quartz which was clamped in such a way as to prevent any strain from developing. Splitting the energies into three parts gives, at 424 K for example, $G_{excess} = -1085$, $G_Q = +358$, $G_{coupling} = -2885$ and $G_{elastic} = +1442$ J.mole^{-1}. The implication is again that, while the soft mode provides the symmetry-breaking mechanism, much of the energy advantage of lowering the symmetry comes from the coupling of the soft mode with lattice strains.

A change in thermodynamic character of a phase transition may occur across a solid solution if the strength of coupling between the order parameter and strains which couple as λeQ^2 occurs in response to changing composition. This can be illustrated using data collected at room temperature for the $CaTiO_3$-$SrTiO_3$ solid solution. Lattice parameters and strains derived from them are shown in Figure 13 (from Carpenter et al. 2000b). There are significant shear strains in both the *Pnma* and *I4/mcm* stability fields, but the volume strain is significant only at the Ca-rich end of the solid solution. Renormalisation of the fourth order coefficient due to the term in $\lambda_2 e_a q_4^2$ from Equation (32) will therefore be reduced with increasing Sr-content across the solid solution. The $Pm\bar{3}m \Leftrightarrow I4/mcm$ transition in $CaTiO_3$ is close to being tricritical in character ($b^* \approx 0$; Fig. 14a) but is described by a 246 Landau potential (Eqn. 51, $d^* = 0$) with positive b^* in $SrTiO_3$ (Salje et al. 1998, Hayward and Salje 1999). As a function of composition, the variation of e_{tZ}^2 (where e_{tZ} is equivalent to e_t in Eqn. 19) at Sr-rich compositions is also consistent with a 246 potential (Fig. 14b). Reduced coupling of the driving order parameter with e_a as Sr is substituted for Ca would certainly contribute to this change in character of the cubic $\Leftrightarrow$ tetragonal transition. It is interesting to note that replacing Ca by Sr in the structure causes a volume increase, while the displacive phase transition causes a volume decrease. In other words, the chemical effect is in direct opposition to the direction of the spontaneous strain, with the result that the latter is suppressed.

It is also possible that the coupling coefficients can be temperature-dependent, in which case the apparent thermodynamic character of a phase transition might appear to vary with temperature. This effect would be detectable as an unusual evolution pattern for the order parameter, as shown, perhaps, by $NaMgF_3$ perovskite. In this perovskite, the orthorhombic structure appears to develop directly from the cubic structure according to a transition $Pm\bar{3}m \Leftrightarrow Pnma$ (Zhao et al. 1993a,b, 1994; Topor et al. 1997). The orthorhombic structure has $q_2 \neq q_4 = q_6 \neq 0$ and $q_1 = q_3 = q_5 = 0$. Equation (32) yields

$$e_a = -\frac{\left(\lambda_1 q_2^2 + 2\lambda_2 q_4^2\right)}{\frac{1}{3}\left(C_{11}^o + 2C_{12}^o\right)} \quad (56) \qquad e_{tX} = -\frac{2\left(\lambda_3 q_2^2 - \lambda_4 q_4^2\right)}{\frac{1}{2}\left(C_{11}^o - C_{12}^o\right)} \quad (57)$$

$$e_4 = -\frac{\lambda_5}{C_{44}^o} q_4^2, \quad (58)$$

where e_{tX} is a tetragonal shear strain defined as

$$e_{tX} = \frac{1}{\sqrt{3}}(2e_1 - e_2 - e_3), \quad (59)$$

and e_4 is an orthorhombic shear which is given by

$$|e_4| = \left| \frac{\frac{a}{\sqrt{2}} - a_o}{a_o} - \frac{\frac{c}{\sqrt{2}} - a_o}{a_o} \right|. \quad (60)$$

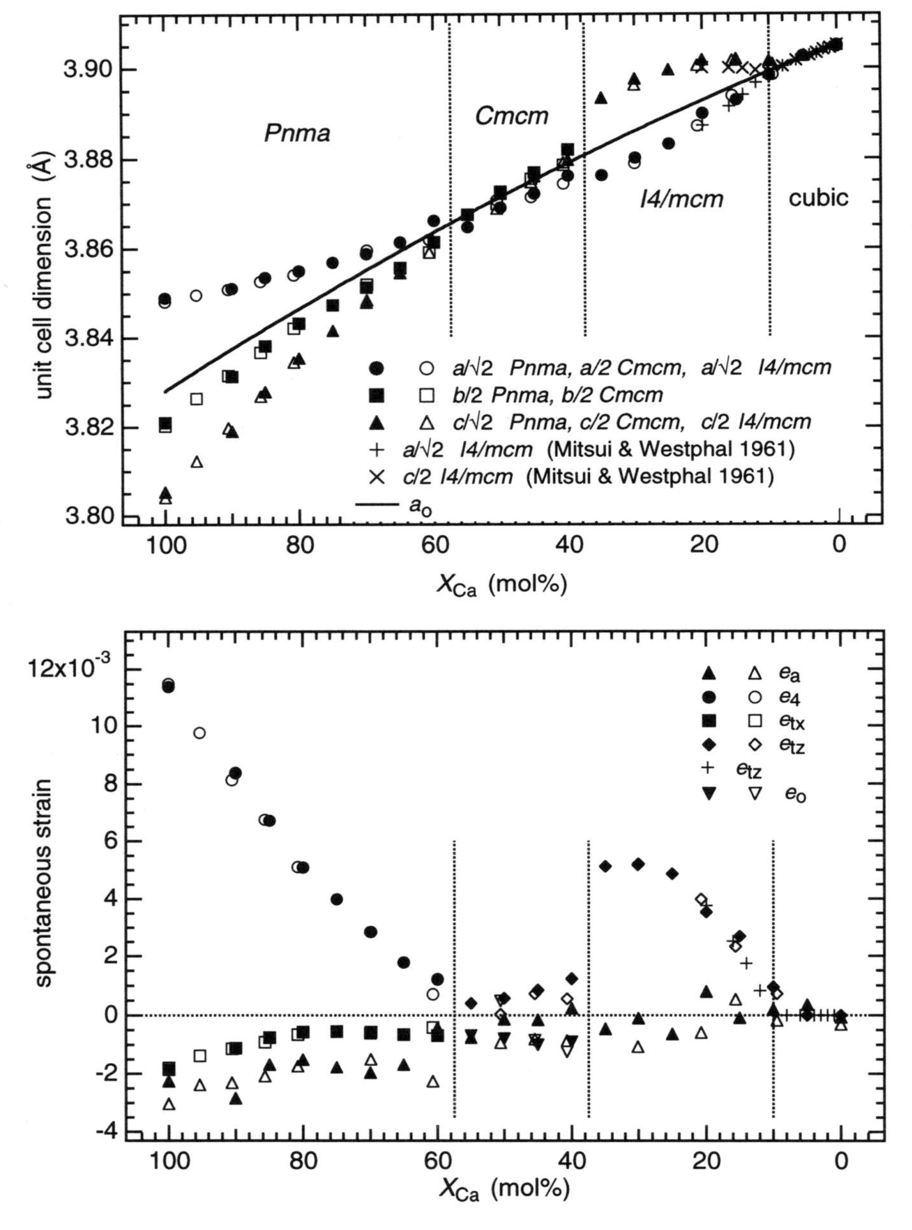

Figure 13. Lattice parameters of the $CaTiO_3$-$SrTiO_3$ solid solution at room temperature (top), expressed in terms of the reduced pseudocubic unit cell. Open symbols: data of Qin et al. (2000); filled symbols: data of Ball et al. (1998). In the cubic field, data of Ball et al. are shown by filled circles and data of Mitsui and Westphal (1961) are shown by stars. The curve for a_o includes a symmetric excess volume of mixing. Spontaneous strains (bottom) calculated from lattice parameter data of Mitsui and Westphal (1961), Ball et al. (1998) and Qin et al. (2000) (crosses, filled symbols, open symbols, respectively). From Carpenter et al. (2000b).

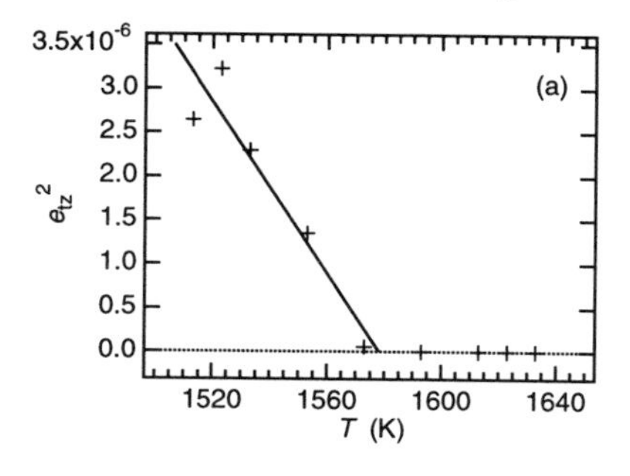

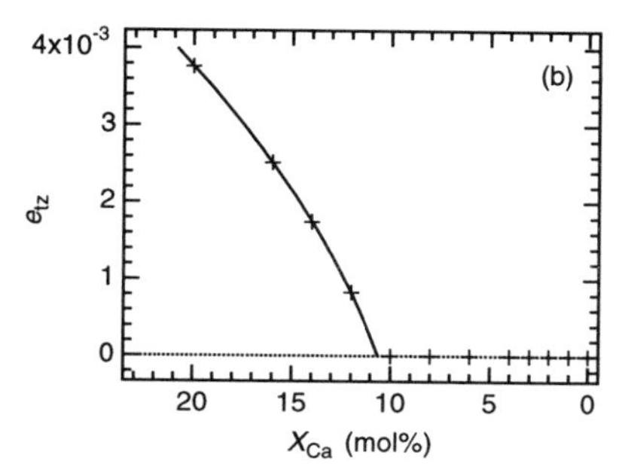

Figure 14. (a) The square of the tetragonal shear strain for the *I4/mcm* structure of $CaTiO_3$ is an approximately linear function of temperature, consistent with tricritical character for the $Pm\bar{3}m \Leftrightarrow I4/mcm$ transition. (b) Data of Mitsui and Westphal (1961) for the $Pm\bar{3}m \Leftrightarrow I4/mcm$ transition as a function of composition at 296 K in the $CaTiO_3$-$SrTiO_3$ solid solution. The curve is a fit to the data using a standard 246 Landau potential. From Carpenter et al. (2000b).

e_4 is positive for $a > c$ and negative for $a < c$. Variations of these strains, from the lattice parameter data of Zhao et al. (1993a), are shown as a function of temperature in Figure 15. The orthorhombic shear strain, e_4, evolves as $e_4^2 \propto (T - T_c)$ in an interval of ~250 K below the transition point of ~780 K (Zhao et al. 1993b, Topor et al. 1997, Carpenter et al. 2000b). Below this temperature, e_4 becomes a more linear function of temperature, suggesting a change from tricritical ($q_4^4 \propto (T_c - T)$) to second order ($q_4^2 \propto (T_c - T)$) character. The volume strain increases in magnitude normally ($e_a \propto e_4$) over the tricritical range but then gets smaller. The relevant coupling coefficients are λ_1 and λ_2, which, if they become smaller with falling temperature, could contribute to the change from $b^* \approx 0$ close to the transition to $b^* > 0$ at lower temperatures. From Equation (57) it is easy to understand why the tetragonal shear strain, e_{tX}, is small if the octahedral tilts corresponding to q_2 and q_4 have about the same magnitude.

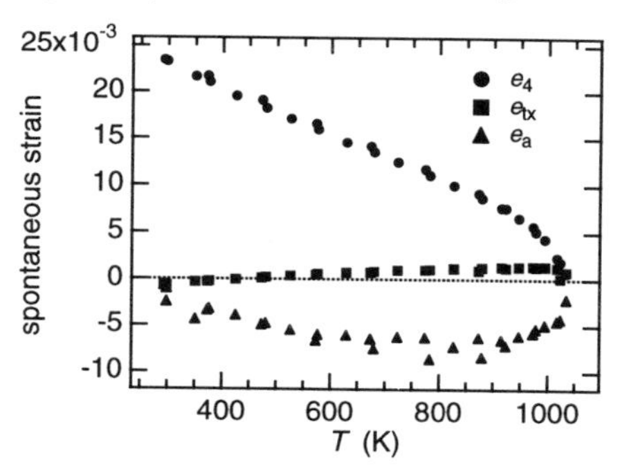

Figure 15. Variations of symmetry-adapted strains associated with the $Pm\bar{3}m \Leftrightarrow Pnma$ transition in $NaMgF_3$, as calculated using lattice parameter data from Zhao et al. (1993a).

ELASTIC CONSTANT VARIATIONS

As with spontaneous strains, the elastic constants of a crystal have symmetry properties. Symmetry-adapted combinations of the elastic constants are obtained by diagonalising the elastic constant matrix for a given crystal class, and the eigenvalues are then associated with different irreducible representations of the point group. Each is then also associated with a particular symmetry-adapted strain. Manipulations of this type only need to be done once for all possible changes in crystal class and the results are available in tabulated form in the literature (e.g. Table 6 of Carpenter and Salje 1998). In practice, the most important process is the derivation of the Landau free energy expansion for a phase transition. Once this has been done, it is possible to predict the variation of all the individual elastic constants. If all the coefficients and bare elastic constants are known, the predictions are quantitative. When the time scale for adjustments of the order parameter of a crystal in response to an applied stress is short relative to the time scale of the application of that stress, it is only necessary to apply Equation (5) to the expansion. It is generally assumed that the bare elastic constants, C^{o}_{ik}, which exclude the effects of the phase transition, vary only weakly with P and T. An explicit pressure and temperature dependence for these can be included if required, however. The examples of stishovite, quartz and perovskite are again used to illustrate some of the patterns of behaviour in materials where spontaneous strains arise by coupling with the order parameter due to a soft optic mode.

Expressions for the elastic constant variations due to the tetragonal ⇔ orthorhombic transition in stishovite derived by applying Equation (5) to Equation (23) are listed in Table 4. The most important factors in determining the form of evolution of the elastic constants are, self-evidently, the strength of coupling between the order parameter and strain (λ_1, λ_2, ...), the order parameter susceptibility, χ^{-1}, given by

$$\chi^{-1} = a\left(P - P_c\right) \qquad \left(\text{at } P < P_c^*\right) \tag{61}$$

$$\chi^{-1} = 2a\frac{b}{b^*}\left(P_c^* - P\right) + a\left(P_c^* - P_c\right) \qquad \left(\text{at } P > P_c^*\right) \tag{62}$$

and the equilibrium evolution of Q, which is given by

$$Q^2 = \frac{a}{b^*}\left(P_c^* - P\right). \tag{63}$$

The bilinear coupling term $\lambda(e_1 - e_2)Q$ causes the related symmetry-adapted elastic constant $(C_{11} - C_{12})$ to tend to zero as $P \rightarrow P_c^*$ with a pronounced curvature in the stability fields of both the low and high symmetry forms. The ratio of slopes as $P \rightarrow P_c^*$ from $P > P_c^*$ and $P < P_c^*$ is given by $2(b/b^*)$:1 for a second order transition. This ratio varies according to the equilibrium evolution of Q and is $4(b/b^*)$:1 in the case of tricritical character, for example. Each of C_{11} and C_{12} depends on a term $\lambda_2^2\chi$ and therefore varies in the high symmetry phase ahead of the phase transition. On the other hand, C_{13} depends on terms which all include Q and therefore does not deviate from C^{o}_{13} in the high symmetry phase. C_{44} and C_{55} vary linearly with Q in the stability field of the low symmetry phase while C_{66} varies with Q^2. These variations are all shown in Figure 16, from Carpenter et al. (2000a).

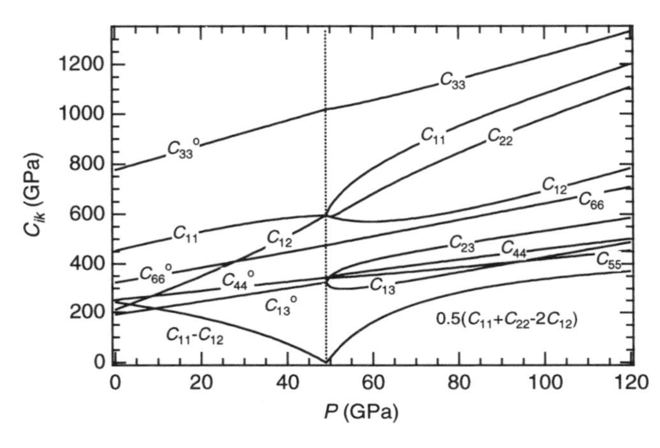

Figure 16. Variations of the elastic constants through the tetragonal ⇔ orthorhombic transition in stishovite, given by the expressions in Table 4 (after Carpenter et al. 2000a).

Table 4. Expressions for the elastic constants of stishovite (from Carpenter et al. 2000a).

tetragonal structure ($P4_2/mnm$)	orthorhombic structure ($Pnnm$)
$C_{11} = C_{22} = C_{11}^{o} - \lambda_2^2\chi$	$C_{11} = C_{11}^{o} - (4\lambda_1^2 Q^2 + \lambda_2^2 + 4\lambda_1\lambda_2 Q)\chi$
	$C_{22} = C_{11}^{o} - (4\lambda_1^2 Q^2 + \lambda_2^2 - 4\lambda_1\lambda_2 Q)\chi$
$C_{33} = C_{33}^{o}$	$C_{33} = C_{33}^{o} - 4\lambda_3^2 Q^2\chi$
$C_{12} = C_{12}^{o} + \lambda_2^2\chi$	$C_{12} = C_{12}^{o} - (4\lambda_1^2 Q^2 - \lambda_2^2)\chi$
$C_{13} = C_{23} = C_{13}^{o}$	$C_{13} = C_{13}^{o} - (4\lambda_1\lambda_3 Q^2 + 2\lambda_2\lambda_3 Q)\chi$
	$C_{23} = C_{13}^{o} - (4\lambda_1\lambda_3 Q^2 - 2\lambda_2\lambda_3 Q)\chi$
$C_{11} - C_{12} = (C_{11}^{o} - C_{12}^{o}) - 2\lambda_2^2\chi$	$C_{11} - C_{12} = (C_{11}^{o} - C_{12}^{o}) - (2\lambda_2^2 + 4\lambda_1\lambda_2 Q)\chi$
	$\overline{C}_{11} - \overline{C}_{12} = \frac{1}{2}(C_{11} + C_{22} - 2C_{12})$
	$= (C_{11}^{o} - C_{12}^{o}) - 2\lambda_2^2\chi$
$C_{11} + C_{12} = C_{11}^{o} + C_{12}^{o}$	$C_{11} + C_{12} = (C_{11}^{o} + C_{12}^{o}) - (8\lambda_1^2 Q^2 + 4\lambda_1\lambda_2 Q)\chi$
	$\overline{C}_{11} + \overline{C}_{12} = \frac{1}{2}(C_{11} + C_{22} + 2C_{12})$
	$= (C_{11}^{o} + C_{12}^{o}) - 8\lambda_1^2 Q^2\chi$
$C_{44} = C_{55} = C_{44}^{o}$	$C_{44} = C_{44}^{o} + 2\lambda_4 Q$
	$C_{55} = C_{44}^{o} - 2\lambda_4 Q$
$C_{66} = C_{66}^{o}$	$C_{66} = C_{66}^{o} + 2\lambda_6 Q^2$

Values for most of the coefficients in Equation (23) were extracted from experimental data. Values were assigned to λ_4 and λ_6 arbitrarily in the absence of the any relevant experimental observations. The bare elastic constants were given a linear pressure dependence based on the variations calculated by Karki et al. (1997a). When experimental data become available, a comparison between observed and predicted elastic constant variations will provide a stringent test for the model of this phase transition as represented by Equation (23).

Bulk properties of an aggregate of stishovite crystals are also predicted to be substantially modified by the phase transition. These are obtained from the variations of the individual elastic constants using the average of Reuss and Voigt limits (Hill 1952, Watt 1979). The bulk modulus, K, is not sensitive to the transition but the shear modulus, G, is expected to show a large anomaly over a wide pressure interval (Fig. 17a). Consequently, the velocities of P and S waves should also show a large anomaly (Fig. 17b), with obvious implications for the contribution of stishovite to the properties of the earth's mantle if free silica is present (Carpenter et al. 2000a, Hemley et al. 2000).

The form of the elastic constant variations of quartz can be predicted from Equation (28) in the same way, and the resulting expressions are listed in Table 5 (from Carpenter et al. 1998b). An important difference between stishovite and quartz is that, in the latter, there is no bilinear coupling of the form λeQ. As a consequence, all the deviations of the elastic constants from their bare values depend explicitly on Q. This means that there should be no deviations associated with the phase transition as the transition point is approached from the

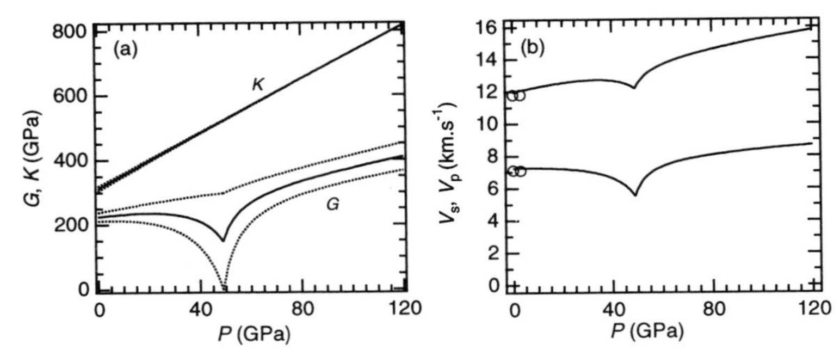

Figure 17. Variations of bulk properties derived from variations of the individual elastic constants of stishovite. (a) Bulk modulus (K) and shear modulus (G). Solid lines are the average of Reuss and Voigt limits; the latter are shown as dotted lines (Voigt limit > Reuss limit). (b) Velocities of P (top) and S (lower) waves through a polycrystalline aggregate of stishovite. Circles indicate experimental values obtained by Li et al. (1996) at room pressure and 3GPa. After Carpenter et al. (2000a).

high temperature side. Some softening is observed in β-quartz (Fig. 18) but this arises by a different mechanism, which involves coupling between different vibrational modes along branches of the soft mode (Pytte 1970, 1971; Axe and Shirane 1970, Höchli 1972, Rehwald 1973, Cummins 1979, Lüthi and Rehwald 1981, Yao et al. 1981, Fossum 1985) and has been described in detail by Carpenter and Salje (1998). The temperature dependence of this fluctuation-induced softening is generally described by

$$C_{ik} - C_{ik}^{o} = \Delta C_{ik} = A_{ik}|T - T_c|^{K_{ik}}. \tag{64}$$

A_{ik} and K_{ik} are properties of the material of interest and have to be measured. Their subscripts are retained as labels to match with the corresponding C_{ik} terms. The values of K_{ik} are sensitive both to the degree of anisotropy of dispersion curves about the reciprocal lattice vector of the soft mode and to the extent of softening along each branch. The effects

Table 5. Expressions for the elastic constant variations in quartz due to the β ⇌ α transition (from Carpenter et al. 1998b).

β-quartz (622)	α-quartz (32)
$C_{11} = C_{22} = C_{11}^{o}$	$C_{11} = C_{22} = C_{11}^{o} + 2\lambda_6 Q^2 - [2\lambda_1 Q + 4\lambda_7 Q^3]^2 \chi$
$C_{33} = C_{33}^{o}$	$C_{33} = C_{33}^{o} - [2\lambda_3 Q + 4\lambda_8 Q^3]^2 \chi$
$C_{12} = C_{12}^{o}$	$C_{12} = C_{12}^{o} - 2\lambda_6 Q^2 - [2\lambda_1 Q + 4\lambda_7 Q^3]^2 \chi$
$C_{13} = C_{23} = C_{13}^{o}$	$C_{13} = C_{23} = C_{13}^{o} - [2\lambda_1 Q + 4\lambda_7 Q^3] . [2\lambda_3 Q + 4\lambda_8 Q^3] \chi$
$C_{11} - C_{12} = C_{11}^{o} - C_{12}^{o}$	$C_{11} - C_{12} = (C_{11}^{o} - C_{12}^{o}) + 4\lambda_6 Q^2$
$C_{11} + C_{12} = C_{11}^{o} + C_{12}^{o}$	$C_{11} + C_{12} = (C_{11}^{o} + C_{12}^{o}) - 2[2\lambda_1 Q + 4\lambda_7 Q^3]^2 \chi$
$C_{14} = -C_{24} = C_{56} = 0$	$C_{14} = -C_{24} = C_{56} = \lambda_5 Q + \lambda_9 Q^3$
$C_{44} = C_{55} = C_{44}^{o}$	$C_{44} = C_{55} = C_{44}^{o} + 2\lambda_4 Q^2$
$C_{66} = C_{66}^{o} = \frac{1}{2}(C_{11}^{o} - C_{12}^{o})$	$C_{66} = C_{66}^{o} + 2\lambda_6 Q^2 = \frac{1}{2}(C_{11} - C_{12})$

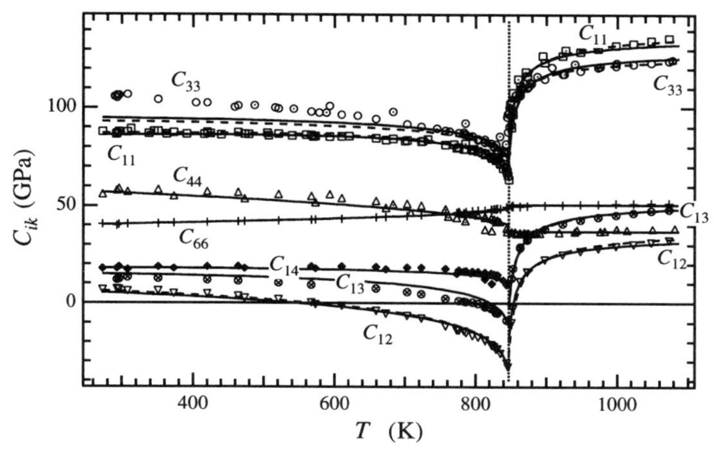

Figure 18. Comparison between calculated elastic-constant variations of quartz and experimental data from the literature. For C_{11}, C_{33} and C_{44} a distinction has been made between data from ultrasonic experiments (open symbols) and data from Brillouin scattering (open symbols containing a dot). Two sets of calculated variations are shown for C_{11}, C_{12}, C_{13} and C_{33}, depending on how the bare elastic constants, C^{o}_{ik}, were determined. In the case of C_{13} the two curves are almost superimposed. Fits to the data for C_{14}, C_{44} and C_{66} in α-quartz, using equations listed in Table 5, are shown as solid lines; C^{o}_{44} and C^{o}_{66} were assumed to be constant (after Carpenter et al. 1998b).

are restricted to elastic constants which transform as the identity representation and, for an elastically uniaxial material such as quartz, the ΔC_{ik} variations are related by $\Delta C_{11} = \Delta C_{12}$, $(\Delta C_{13})^2 = \Delta C_{11}\Delta C_{33}$ (Axe and Shirane 1970, Höchli 1972, Yamamoto 1974, Carpenter et al. 1998b). Values of K for quartz are ~-0.6, which is close to the value of -1/2 expected for a system that behaves as a more or less isotropic material with respect to the phase transition.

There is a marked curvature to all the elastic constants below the transition point, which reflects the variation of the order parameter and the order parameter susceptibility. No individual elastic constant or symmetry-adapted combination of elastic constants is expected to tend to zero at the transition point. In this case the agreement between observed and calculated values shown in Figure 18 implies that the model represented by Equation (28) provides a good description of the phase transition. Agreement is not as close for C_{33} as it is for the other elastic constants, however, suggesting that the causes of strain parallel to [001] of α-quartz have not yet been fully explained.

Of some geological interest and significance are possible phase transitions in $(Mg,Fe)SiO_3$ perovskite in the earth's mantle (e.g. Hemley and Cohen 1992, Wentzcovitch et al. 1995, Warren and Ackland 1996, Carpenter and Salje 1998, and references therein). Many of the properties of this perovskite might be only slightly affected by a cubic ⇔ orthorhombic or tetragonal ⇔ orthorhombic transition but the elastic properties would be expected to show large variations. The form of these can now be predicted in general terms. Consider a $Pm\bar{3}m \Leftrightarrow I4/mcm$ transition, for example, which would be described by a Landau expansion derived from Equation (32) such as

$$G = \frac{1}{2}a(T - T_c)(q_4^2 + q_5^2 + q_6^2) + \frac{1}{4}b(q_4^2 + q_5^2 + q_6^2)^2 + \frac{1}{4}b'(q_4^4 + q_5^4 + q_6^4)$$
$$+ \frac{1}{6}c(q_4^2 + q_5^2 + q_6^2)^3 + \frac{1}{6}c'(q_4q_5q_6)^2 + \frac{1}{6}c''(q_4^2 + q_5^2 + q_6^2)(q_4^4 + q_5^4 + q_6^4)$$
$$+ \lambda_2 e_a(q_4^2 + q_5^2 + q_6^2) + \lambda_3\left[\sqrt{3}e_o(q_5^2 - q_6^2) + e_t(2q_4^2 - q_5^2 - q_6^2)\right]$$
$$+ \lambda_5(e_4q_4q_6 + e_5q_4q_5 + e_6q_5q_6) + \frac{1}{4}(C_{11}^o - C_{12}^o)(e_o^2 + e_t^2) + \frac{1}{6}(C_{11}^o + 2C_{12}^o)e_a^2$$
$$+ \frac{1}{2}C_{44}^o(e_4^2 + e_5^2 + e_6^2). \tag{65}$$

In order to reduce algebraic complexity, it is assumed that the volume strain is small ($\lambda_2 \approx 0$). Many transitions between cubic, tetragonal and orthorhombic perovskite structures are close to tricritical in character, in which case the renormalised fourth order coefficient b^* can be taken as zero, i.e.

$$b^* = b + b' - \frac{16\lambda_3^2}{(C_{11}^o - C_{12}^o)} = 0, \tag{66}$$

for the equilibrium conditions $q_5 = q_6 = e_o = 0$, and $q_4 \neq 0$ at $T < T_c$. The susceptibility with respect to q_4 is

$$\left(\frac{\partial^2 G}{\partial q_4^2}\right)^{-1} = \left[2(b + b') + 4(c + c'')q_4^4\right]^{-1}. \tag{67}$$

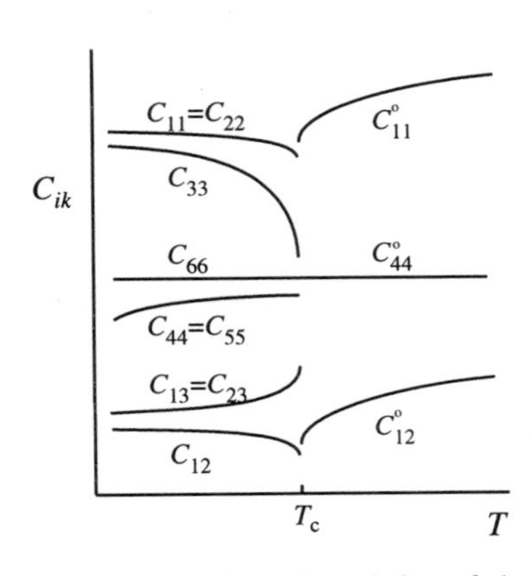

Figure 19. Schematic variation of the elastic constants at a tricritical transition involving the symmetry change $Pm\bar{3}m \Leftrightarrow I4/mcm$, based on the expressions given in Table 6.

Expressions for all elastic constants derived from Equation (65) are given in Table 6. As with quartz, there is likely to be an additional softening above T_c due to thermal fluctuations in the cubic structure. These can be described by Equation (64) and have been observed in $KMnF_3$ by Aleksandrov et al. (1966), Reshchikova et al. (1970), Melcher and Plovnik (1971), Cao and Barsch (1988). Experimental difficulties, particularly associated with the effects of transformation twinning, have so far limited the amount of experimental data available for comparison, but predicted variations are likely to be qualitatively similar to the illustrations in Figure 19. In addition to the elastic softening over a wide temperature and pressure interval about the transition point, any such phase transition would also be expected to cause strong attenuation of acoustic waves (e.g. Schwabl 1985, Schwabl and Täuber 1996, Carpenter et al. 2000a, and references therein).

Each of these three examples of elastic anomalies associated with a phase transition is for a pure phase as a function of pressure or temperature. As seen for the $CaTiO_3$-$SrTiO_3$ system (Fig. 13), however, changes in bulk composition across a solid solution also give rise to phase transitions. For these, the elastic anomalies accompanying changes in

Table 6. Predicted variations for elastic constants of a material subject to a phase transition involving the symmetry change $Pm\bar{3}m \rightleftharpoons I4/mcm$ when the transition is tricritical in character ($b^* = 0$) and the volume strain is small ($\lambda_2 = 0$).

$Pm\bar{3}m$ structure	$I4/mcm$ structure, $e_a = 0$
$q_4 = q_5 = q_6 = 0$	$q_5 = q_6 = 0,\quad q_4^4 = \left[\dfrac{a(T_c - T)}{(c + c'')}\right]$
	$A = 3\left[(b + b') + 2(c + c'')q_4^2\right]$
$C_{11} = C_{22} = C_{33} = C_{11}^o$	$C_{11} = C_{22} = C_{11}^o - \left(\dfrac{8\lambda_3^2}{A}\right)$
	$C_{33} = C_{11}^o - \left(\dfrac{32\lambda_3^2}{A}\right)$
$C_{12} = C_{13} = C_{23} = C_{12}^o$	$C_{12} = C_{12}^o - \left(\dfrac{8\lambda_3^2}{A}\right)$
	$C_{13} = C_{23} = C_{12}^o + \left(\dfrac{16\lambda_3^2}{A}\right)$
$C_{11} - C_{12} = C_{11}^o - C_{12}^o$	$C_{11} - C_{12} = C_{11}^o - C_{12}^o$
	$\overline{C}_{11} - \overline{C}_{12} = \dfrac{1}{3}(C_{11} + C_{12} + 2C_{33} - 4C_{13})$
	$= (C_{11}^o - C_{12}^o) - \left(\dfrac{48\lambda_3^2}{A}\right)$
$C_{11} + 2C_{12} = C_{11}^o + 2C_{12}^o$	$\overline{C}_{11} + 2\overline{C}_{12} = \dfrac{1}{3}(C_{33} + 2C_{11} + 2C_{12} + 4C_{13})$
	$= C_{11}^o + 2C_{12}^o$
$C_{44} = C_{55} = C_{66} = C_{44}^o$	$C_{44} = C_{55} = C_{44}^o - \left[\dfrac{6\lambda_5^2}{3(3b + b') - 4c''q_4^2}\right]$
	$C_{66} = C_{44}^o$

composition will be essentially the same as when pressure or temperature is the applied variable. It must be expected that the elastic properties of mineral solid solutions will display large variations if there is any possibility of a structural phase transition occurring within them.

ACKNOWLEDGMENTS

I am grateful to Dr. C.N.W.Darlington and Dr. S.A.T.Redfern for kindly providing lattice parameter data for $NaTaO_3$ and $CaTiO_3$, respectively. Much of the work described here was carried out within the TMR Network on Mineral Transformations funded by the European Union (contract number ERB-FMRX-CT97-0108).

REFERENCES

Aleksandrov KS, Reshchikova LM, Beznosikov BV (1966) Behaviour of the elastic constants of $KMnF_3$ single crystals near the transition of puckering. Phys Stat Sol 18:K17-K20

Axe JD, Shirane G (1970) Study of the α-β quartz phase transformation by inelastic neutron scattering. Phys Rev B 1:342-348

Andrault D, Fiquet G, Guyot F, Hanfland M (1998) Pressure-induced Landau-type transition in stishovite. Science 282:720-724

Bachheimer JP (1980) An anomaly in the β phase near the α-β transition of quartz. J Physique Lett 41:L559-L561

Bachheimer JP (1986) Optical rotatory power and depolarisation in the α-, incommensurate and β-phases of quartz (20 to 600°C). J Phys C 19:5509-5517

Bachheimer JP, Dolino G (1975) Measurement of the order parameter of α-quartz by second-harmonic generation of light. Phys Rev B 11:3195-3205

Banda EJKB, Craven RA, Parks RD (1975) α-β transition in quartz: classical versus critical behavior. Solid St Comm 17:11-14

Ball CJ, Begg BD, Cookson DJ, Thorogood GJ, Vance ER (1998) Structures in the system $CaTiO_3/SrTiO_3$. J Sol St Chem 139:238-247

Bruce AD, Cowley RA (1981) Structural Phase Transitions. Taylor and Francis, London, 326 p

Bulou A, Rousseau M, Nouet J (1992) Ferroelastic phase transitions and related phenomena. Key Eng Mat 68:133-186

Cao W, Barsch GR (1988) Elastic constants of $KMnF_3$ as functions of temperature and pressure. Phys Rev B 38:7947-7958

Carpenter MA, Salje EKH (1998) Elastic anomalies in minerals due to structural phase transitions. Eur J Mineral 10:693-812

Carpenter MA, Salje EKH, Graeme-Barber A (1998a) Spontaneous strain as a determinant of thermodynamic properties for phase transitions in minerals. Eur J Mineral 10:621-691

Carpenter MA, Salje EKH, Graeme-Barber A, Wruck B, Dove MT, Knight KS (1998b) Calibration of excess thermodynamic properties and elastic constant variations due to the α ↔ β phase transition in quartz. Am Mineral 83:2-22

Carpenter MA, Hemley RJ, Mao HK (2000a) High-pressure elasticity of stishovite and the $P4_2/mnm$ ↔ *Pnnm* phase transition. J Geophys Res 105:10807-10816

Carpenter MA, Becerro AI, Seifert F (2000b) Strain analysis of phase transitions in $(Ca,Sr)TiO_3$ perovskites. Am Mineral (submitted)

Cohen RE (1987) Calculation of elasticity and high pressure instabilities in corundum and stishovite with the potential induced breathing model. Geophys Res Lett 14:37-40

Cohen RE (1992) First-principles predictions of elasticity and phase transitions in high pressure SiO_2 and geophysical implications. *In* Y Syono, MH Manghnani (eds) High-Pressure Research: Application to Earth and Planetary Sciences. Geophys Monogr Ser 67:425-431 Am Geophys Union, Washington, DC.

Cohen RE (1994) First-principles theory of crystalline SiO_2. *In* PJ Heaney, CT Prewitt, GV Gibbs (eds) Silica - Physical Behavior, Geochemistry, and Materials Applications. Rev Mineral 29:369-402

Cowley RA (1976) Acoustic phonon instabilities and structural phase transitions. Phys Rev B 13:4877-4885

Cummins HZ (1979) Brillouin scattering spectroscopy of ferroelectric and ferroelastic phase transitions. Phil Trans Roy Soc Lond A 293:393-405

Darlington CNW, Knight KS (1999) High-temperature phases of $NaNbO_3$ and $NaTaO_3$. Acta Crystallogr B55:24-30

Dolino G, Bachheimer JP (1982) Effect of the α-β transition on mechanical properties of quartz. Ferroelectrics 43:77-86

Dove MT (1993) Introduction to lattice dynamics p 258 Cambridge University Press, Cambridge

Errandonea G (1980) Elastic and mechanical studies of the transition in LaP_5O_{14}: a continuous ferroelastic transition with a classical Landau-type behavior. Phys Rev B 21:5221-5236

Fossum JO (1985) A phenomenological analysis of ultrasound near phase transitions. J Phys C 18:5531-5548

Glazer AM (1972) The classification of tilted octahedra in perovskites. Acta Crystallogr B28:3384-3392

Glazer AM (1975) Simple ways of determining perovskite structures. Acta Crystallogr A31:756-762

Grimm H, Dorner B (1975) On the mechanism of the α-β phase transformation of quartz. J Phys Chem Solids 36:407-413

Hayward SA, Salje EKH (1999) Cubic-tetragonal phase transition in $SrTiO_3$ revisited: Landau theory and transition mechanism. Phase Trans 68:501-522

Hemley RJ (1987) Pressure dependence of Raman spectra of SiO_2 polymorphs: α-quartz, coesite, and stishovite. *In* MH Manghnani, Y Syono (eds) High-Pressure Research in Mineral Physics. Geophys Monogr Ser 39:347-359 Am Geophys Union, Washington, DC

Hemley RJ, Cohen RE (1992) Silicate perovskite. Ann Rev Earth Planet Sci 20:553-600

Hemley RJ, Jackson MD, Gordon RG (1985) Lattice dynamics and equations of state of high-pressure mineral phases studied with electron gas theory. Eos Trans Am Geophys Union 66:357

Hemley RJ, Prewitt CT, Kingma KJ (1994) High-pressure behavior of silica *In* PJ Heaney, CT Prewitt, GV Gibbs (eds) Silica—Physical Behavior, Geochemistry, and Materials Applications. Rev Mineral 29:41-81

Hemley RJ, Shu J, Carpenter MA, Hu J, Mao H-K, Kingma KJ (2000) Strain-order parameter coupling in the pressure-induced ferroelastic transition in dense SiO_2. Phys Rev Lett (accepted)

Hill R (1952) The elastic behaviour of a crystalline aggregate. Proc Phys Soc Lond A 65:349-354

Höchli UT (1972) Elastic constants and soft optical modes in gadolinium molybdate. Phys Rev B 6:1814-1823

Howard CJ, Stokes HT (1998) Group-theoretical analysis of octahedral tilting in perovskites. Acta Crystallogr B54:782-789

Karki BB, Stixrude L, Crain J (1997a) *Ab initio* elasticity of three high-pressure polymorphs of silica. Geophys Res Lett 24:3269-3272

Karki BB, Warren MC, Stixrude L, Ackland GJ, Crain J (1997b) *Ab initio* studies of high-pressure structural transformations in silica. Phys Rev B 55:3465-3471

Kingma KJ, Cohen RE, Hemley RJ, Mao HK (1995) Transformation of stishovite to a denser phase at lower-mantle pressures. Nature 374:243-245

Kingma KJ, Mao HK, Hemley RJ (1996) Synchrotron x-ray diffraction of SiO_2 to multimegabar pressures. High Pressure Res 14:363-374

Lacks DJ, Gordon RG (1993) Calculations of pressure-induced phase transitions in silica. J Geophys Res 98:22147-22155

Lee C, Gonze X (1995) The pressure-induced ferroelastic phase transition of SiO_2 stishovite. J Phys Condens Matter 7:3693-3698

Li B, Rigden SM, R.C. Liebermann RC (1996) Elasticity of stishovite at high pressure. Phys Earth Planet Inter 96:113-127

Lüthi B, Rehwald W (1981) Ultrasonic studies near structural phase transitions. *In* KA Müller, H Thomas (eds) Structural Phase Transitions I. Topics in Current Physics 23:131-184. Springer Verlag, Berlin, Heidelberg, New York

Mao HK, Shu J, Hu J, Hemley RJ (1994) Single-crystal X-ray diffraction of stishovite to 65 GPa. Eos Trans Am Geophys Union 75:662

Matsui Y, Tsuneyuki S (1992) Molecular dynamics study of rutile-$CaCl_2$-type phase transitions of SiO_2. *In* High-Pressure Research: Application to Earth and Planetary Sciences. Y Syono, MH Manghnani (eds). p 433-439 Am Geophys Union, Washington, DC

McWhan DB, Birgeneau RJ, Bonner WA, Taub H, Axe JD (1975) Neutron scattering study at high pressure of the structural phase transition in paratellurite. J Phys C 8:L81-L85

Melcher RL, Plovnik RH (1971) The anomalous elastic behavior of $KMnF_3$ near a structural phase transition. *In* MA Nusimovici (ed) Phonons. Flammarion, Paris, p 348-352

Mitsui, T, Westphal, WB (1961) Dielectric and X-ray studies of $Ca_xBa_{1-x}TiO_3$ and $Ca_xSr_{1-x}TiO_3$. Phys Rev 124:1354-1359

Mogeon F (1988) Propriétés statiques et comportement irreversible de la phase incommensurable du quartz. Doctoral thesis, Université Joseph Fourier, Grenoble I

Nagel L, O'Keeffe M (1971) Pressure and stress induced polymorphism of compounds with rutile structure. Mater Res Bull 6:1317-1320

Peercy PS, Fritz IJ (1974) Pressure-induced phase transition in paratellurite (TeO_2). Phys Rev Lett 32:466-469

Peercy PS, Fritz IJ, Samara GA (1975) Temperature and pressure dependences of the properties and phase transition in paratellurite (TeO_2): ultrasonic, dielectric and Raman and Brillouin scattering results. J Phys Chem Solids 36:1105-1122

Pytte E (1970) Soft-mode damping and ultrasonic attenuation at a structural phase transition. Phys Rev B 1:924-930

Pytte E (1971) Acoustic anomalies at structural phase transitions. *In* EJ Samuelsen, E Anderson J Feder (eds) Structural phase transitions and soft modes. NATO ASI, Scandinavian Univ Books, Oslo, p 151-169

Qin S, Becerro AI, Seifert F, Gottsmann J, Jiang J (2000) Phase transitions in $Ca_{1-x}Sr_xTiO_3$ perovskites: effects of composition and temperature. J Mat Chem 10:1-8

Redfern SAT (1995) Relationship between order-disorder and elastic phase transitions in framework minerals. Phase Trans 55:139-154

Redfern SAT, Salje E (1987) Thermodynamics of plagioclase II: temperature evolution of the spontaneous strain at the $I\bar{1} \leftrightarrow P\bar{1}$ phase transition in anorthite. Phys Chem Minerals 14:189-195

Rehwald W (1973) The study of structural phase transitions by means of ultrasonic experiments. Adv Phys 22:721-755

Reshchikova LM, Zinenko VI, Aleksandrov KS (1970) Phase transition in $KMnF_3$. Sov Phys Solid St 11:2893-2897
Ross NL, Shu JF, Hazen RM, Gasparik T (1990) High-pressure crystal chemistry of stishovite. Am Mineral 75:739-747
Salje EKH (1992a) Application of Landau theory for the analysis of phase transitions in minerals. Phys Rep 215:49-99
Salje EKH (1992b) Phase transitions in minerals: from equilibrium properties towards kinetic behaviour. Ber Bunsenge Phys Chem 96:1518-1541
Salje EKH (1993) Phase Transitions in Ferroelastic and Co-elastic Crystals (student edition). Cambridge University Press, Cambridge, 229 p
Salje EKH, Devarajan V (1986) Phase transitions in systems with strain-induced coupling between two order parameters. Phase Trans 6:235-248
Salje E.H, Gallardo MC, Jiménez J, Romero FJ, del Cerro J (1998) The cubic-tetragonal phase transition in strontium titanate: excess specific heat measurments and evidence for a near-tricritical, mean field type transition mechanism. J Phys Cond Matter 10:5535-5543
Schlenker JL, Gibbs GV, Boisen Jr MB (1978) Strain-tensor components expressed in terms of lattice parameters. Acta Crystallogr A34:52-54
Schwabl F (1985) Propagation of sound at continuous structural phase transitions. J Stat Phys 39:719-737
Schwabl F, Täuber UC (1996) Continuous elastic phase transitions in pure and disordered crystals. Phil Trans Roy Soc Lond A 354:2847-2873
Skelton EF, Feldman JL, Liu CY, Spain IL (1976) Study of the pressure-induced phase transition in paratellurite (TeO_2). Phys Rev B 13:2605-2613
Slonczewski JC, Thomas H (1970) Interaction of elastic strain with the structural transition of strontium titanate. Phys Rev B 1:3599-3608
Stokes HT, Hatch DM (1988) Isotropy Subgroups of the 230 Crystallographic Space Groups. World Scientific, Singapore
Teter DM, Hemley RJ, Kresse G, Hafner J (1998) High pressure polymorphism in silica. Phys Rev Lett 80:2145-2148
Tezuka Y, Shin S, Ishigame M (1991) Observation of the silent soft phonon in β-quartz by means of hyper-Raman scattering. Phys Rev Lett 18:2356-2359
Tolédano P, Fejer MM, Auld BA.(1983) Nonlinear elasticity in proper ferroelastics. Phys Rev B 27:5717-5746
Topor L, Navrotsky A, Zhao Y, Weidner DJ (1997) Thermochemistry of fluoride perovskites: heat capacity, enthalpy of formation, and phase transition of $NaMgF_3$. J Solid St Chem 132:131-138
Tsuchida Y, and T Yagi (1989) A new, post-stishovite high-pressure polymorph of silica. Nature 340:217-220
Wadhawan VK (1982) Ferroelasticity and related properties of crystals. Phase Trans 3:3-103
Woodward PM (1997) Octahedral tilting in perovskites. I. Geometrical considerations. Acta Crystallogr B53:32-43
Warren MC, Ackland GJ (1996) *Ab initio* studies of structural instabilities in magnesium silicate perovskite. Phys Chem Minerals 23:107-118
Watt JP (1979) Hashin-Shtrikman bounds on the effective elastic moduli of polycrystals with orthorhombic symmetry. J Appl Phys 50:6290-6295
Wentzcovitch RM, Ross NL, Price GD (1995) *Ab initio* study of $MgSiO_3$ and $CaSiO_3$ perovskites at lower mantle pressures. Phys Earth Planet Int 90:101-112
Worlton TG, Beyerlein RA (1975) Structure and order parameters in the pressure-induced continuous transition in TeO_2. Phys Rev B 5:1899-1907
Yamada Y, Tsuneyuki S, Matsui Y (1992) Pressure-induced phase transitions in rutile-type crystals. *In* Y Syono, MH Manghnani (eds) High Pressure Research: Application to Earth and Planetary Sciences. Geophys Monogr Ser 67:441-446, Am Geophys Union, Washington DC
Yamamoto A (1974) Lattice-dynamical theory of structural phase transition in quartz. J Phys Soc Japan 37:797-808
Yao W, Cummins HZ, Bruce RH (1981) Acoustic anomalies in terbium molybdate near the improper ferroelastic-ferroelectric phase transition. Phys Rev B 24:424-444
Zhao Y, Weidner DJ, Parise JB, Cox DE (1993a) Thermal expansion and structural distortion of perovskite—data for $NaMgF_3$ perovskite. Part I. Phys Earth Planet Int 76:1-16
Zhao Y, Weidner DJ, Parise JB, Cox DE (1993b) Critical phenomena and phase transition of perovskite—data for $NaMgF_3$ perovskite. Part II. Phys Earth Planet Int 76:17-34
Zhao Y, Weidner DJ, Ko J, Leinenweber K, Liu X, Li B, Meng Y, Pacalo REG, Vaughan MT, Wang Y, Yeganeh-Haeri A (1994) Perovskite at high P-T conditions: an *in situ* synchrotron X ray diffraction study of $NaMgF_3$ perovskite. J Geophys Res 99:2871-2885

3

Mesoscopic Twin Patterns in Ferroelastic and Co-Elastic Minerals

Ekhard K. H. Salje

Department of Earth Sciences
University of Cambridge
Downing Street,
Cambridge CB2 3EQ U.K.

INTRODUCTION

Minerals are often riddled with microstructures when observed under the electron microscope. Although the observation and classification of microstructures such as twin boundaries, anti-phase boundaries, exsolution lamellae etc., has been a longstanding activity of mineralogists and crystallographers it has only been very recently that we started to understand the enormous importance of microstructures for the physical and chemical behaviour of minerals.

Let us take a twin boundary as an example. Illustrations of twinned minerals have decorated books on mineralogy for more than a century. A first major step forward was made when Burger (1945) noticed that twinning can have different origins, namely that it may be a growth phenomenon, created mechanically or be the outcome of a structural phase transition. Much research was then focussed on the latter type of twins, which Burger called 'transformation twins.' Transformation twin patterns can be reproduced rather well in a larger number of technically important materials, such as martensitic steels, and it was understood very quickly that they dominate much of the mechanical behaviour. Ferroelectric and ferromagnetic domain patterns are equally important for electronic memory devices while ferroelastic, hierarchical domain structures have fascinated mathematicians and physicists because of the apparent simplicity and universality of their elastic interactions.

A novel aspect relates to the chemical properties of such 'mesoscopic' structures. The term 'mesoscopic' stands for 'in-between scales,' as bracketed by atomistic and macroscopic length scales. The two key observations are the following. Firstly, it was shown experimentally that twin boundaries can act as fast diffusion paths. Aird and Salje (1998, 2000) succeeded in showing enhanced oxygen and sodium diffusion along twin walls in the perovskite like structure $Na_xWO_{3-\delta}$. They also showed that the electronic structure of the twin boundary is different from that of the bulk, results echoed by structural investigations into the twin structures of cordierite by Blackburn and Salje (1999a,b,c). Enhanced concentration of potassium and its transport along twin boundaries in alkali feldspar was found by Camara et al. (2000) and Hayward and Salje (2000).

The second key observation is that the intersections between a twin boundary and a crystal surface represent chemically activated sites (and mechanically soft areas) (Novak and Salje 1998a, 1998b). It appears safe to assume that similarly activated sites exist also at the intersection of APBs and dislocations with the surface (e.g. Lee et al. 1998, Hochella and Banfield 1995). Besides the obvious consequences for the leaching behaviour of minerals, these key observations lead to the hypothesis of 'confined chemical reactions' inside mesoscopic patterns. The idea is as follows: as the surface energy is changed near mesoscopic interfaces, dopant atoms and molecules can be anchored near such interfaces. Some particles will diffuse into the mineral and react with

1529-6466/00/0039-0003$05.00

other dopants to form new chemical compounds. These compounds may well be very different from those formed outside the mineral because the structural confinement leads to modified chemical potentials, reduced chiral degrees of freedom, etc. The new compounds are likely to possess novel properties (e.g. superconducting wall structures in an insulating matrix as observed by Aird and Salje 1998) and may also exist in large quantities in the natural environment. Their detection in any sample is difficult, however, because the cores of mesoscopic structures contain little volume (e.g. some 10 ppm), although the extraction of the reaction products is not dissimilar to the extraction of man-made catalytic reactants and may hence be deemed possible.

Besides the outstanding chemical characteristics of certain mesoscopic structures, they also possess a number of surprising physical characteristics. Typical examples are the initiation of premelting near dislocations, twin boundaries or grain boundaries (e.g. Raterron et al. 1999, Jamnik and Maier 1997) and the movement of twin boundaries under external stress which leads to non-linear strain-stress relationships. It is the purpose of this review to focus on some of the characteristic features of mesocopic structures and to illustrate the generic results for the case of ferroeleastic twin patterns (Salje 1993).

FERROELASTIC TWIN WALLS

When shear stresses are applied to crystals one often observes a highly non-linear elastic response. Hooke's law is not valid in these cases and the strain-stress relationships show hysterersis behaviour (Fig. 1). Such hysteresis is the hallmark of the "ferroelastic effect" which is defined as the mechanical switching between at least two orientation states of the crystal. The hysteresis in Figure 1 was measured in $Pb_3(PO_4)_2$, (Salje and Hoppmann 1976, Salje 1993). The orientational states in $Pb_3(PO_4)_2$ relate to the displacement of Pb inside its oxygen coordination (Fig. 1). The lead atom establishes chemical bonds with two oxygen positions with shorter bond distances compared with the distances between lead and the remaining 4 oxygen atoms. This leads to a structural deformation which shears the structural network and lowers the symmetry of the crystal structure from trigonal to monoclinic. The different orientations of short bond distances correspond to the different orientational strain states and the mechanical switching occurs between these states (Salje and Devarajan 1981, Bismayer et al. 1982, Salje and Wruck 1983, Salje et al.1983, Bleser et al. 1994, Bismayer et al. 1994).

Although the atomic mechanism captures the essential reason why a hysteresis can be observed, it has little to do with the actual switching mechanism. The energy stored in one hysteresis loop is much smaller than the energy required to switch all Pb-atoms from one configuration to another, for example. The greatly reduced hysteresis energy indicates that not all atoms take part in the switching process simultaneously. Instead the switch occurs sequentially with domain walls propagating through the crystal. Only atoms inside the domain wall change their configuration, while all atoms in front are in the old configuration and all atoms behind the wall are in the new configuration. Once a domain wall has swept through a crystal, the switch has been completed with all configurations changed. The movement of the domain wall essentially determines the dynamical properties of the switching process (Salje 1993).

The size of the ferroelastic hysteresis depends sensitively on thermodynamic parameters such a temperature, T, pressure, P, or chemical composition, N. Most ferroelastic materials show phase transitions between a ferroelastic phase and a paraelastic phase. As it is rather a major experimental undertaking to measure ferroelastic hysteresis with any acceptable degree of accuracy, it has become customary to call a material ferroelastic if a phase transition occurs (or may be thought to occur) which may

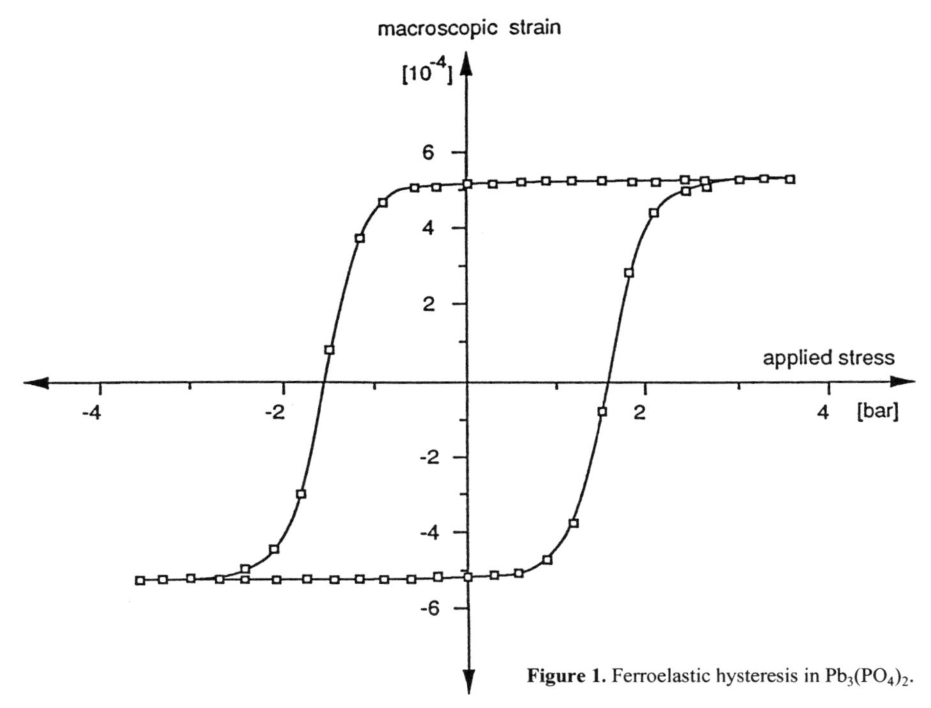

Figure 1. Ferroelastic hysteresis in $Pb_3(PO_4)_2$.

conceivably generate ferroelasticity. The essential parameter is then the 'spontaneous strain,' i.e. the deformation of the crystal generated by the phase transition, which has to have at least two orientations between which switching may occur. If a spontaneous strain is generated but no switching is possible (often for symmetry reasons) then the distorted phase is called co-elastic rather than ferroelastic (Salje 1993, Carpenter et al. 1998).

Twin walls are easily recognisable under the optical microscope and in the transmission electron microscope (TEM). None of these techniques has succeeded in characterising the internal structure of a twin wall, however. This is partly due to experimental limitations of TEM imaging and partly due to the fact that any fine structure may be significantly changed when samples are prepared as atomically thin slabs or wedges for TEM studies. A more subtle mode of investigation uses X-ray diffraction. The progress in both computing of large data sets and X-ray detector technology has led to extremely powerful systems which can measure very weak diffuse scattering from twin walls next to strong Bragg scattering from the ferroelastic domains (Salje 1995, Locherer et al. 1996, 1998; Chrosch and Salje 1997).

An ideal situation for the observation of twin structures is an array of parallel, non-periodically spaced walls. The diffraction signal has then a 'dog-bone structure' as shown in Figure 2. The equi-intensity surface shows diffraction signals at an intensity of 10^{-4} compared with the intensity of the two Bragg peaks which are located inside the two thicker ends of the dog bone (Locherer et al. 1998). The intensity in the handle of the dog bone is then directly related to the thickness and internal structure of the twin wall. In order to calibrate the wall thickness the following theoretical framework is chosen. The specific energy of a wall follows in lowest order theory from a Landau potential

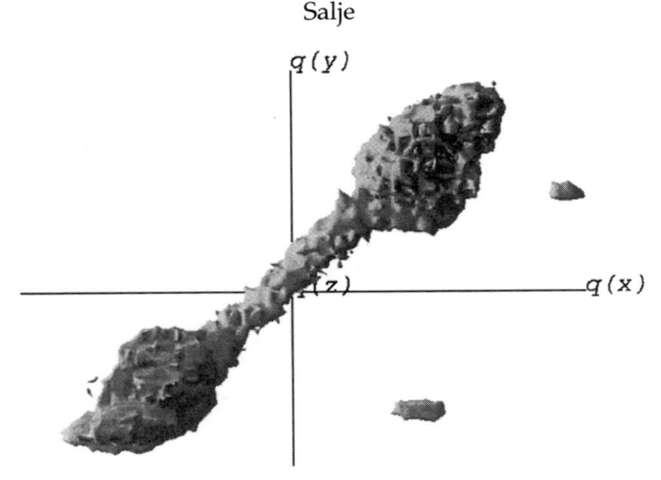

Figure 2. Isointensity surface plot of the region of the 400/040 Bragg peak at WO_3 at room temperature. The Bragg peak is "inside" the isosurface; the dog bone structure stems from the scattering from twin walls.

$$G = \tfrac{1}{2} A\theta_s \left[coth \frac{\theta_s}{T} - coth \frac{\theta_s}{T_c} \right] Q^2 + \tfrac{1}{4} BQ^4 + \ldots \tfrac{1}{2} g(\nabla Q)^2 \tag{1}$$

where θ_s is the quantum mechanical saturation temperature, Q is the order parameter, g is the Ginzburg (or dispersion) parameter and A, B are parameters related to the local effective potential.

For this excess Gibbs free energy the wall profile follows from energy minimization as:

$$Q = Q_o \tanh \frac{X}{W}$$

$$W^2 = 2g / \left[A\theta_s (\coth \frac{\theta_s}{T} - \cos \frac{\theta_s}{T_c} \right] \tag{2}$$

where Q_o is the value of the order parameter in the bulk and $2W$ is the wall thickness. The dog bone pattern allows the determination of Q_o which is related to the length of the dogbone, and W from the intensity distribution in the centre of the dogbone. Typical wall thicknesses in units of the crystallographic repetition length normal to the twin plane vary between 2 and 10 (Table 1).

The wall thickness is predicted in Equation (2) to increase in a second order phase transitions when the temperature approaches T_c. This effect has been studied in detail in the case of $LaAlO_3$ (Chrosch and Salje 1999). This material crystallises in the perovskite structure and undergoes a second order phase transition at $T_c \approx 850$ K from a cubic to a rhombohedral phase. The transformation is related to rotations of alternative AlO_6 octahedra around one of the cubic threefold axes, through an angle which directly relates to the order parameter of the transition. The twin angle is very small ($\psi = 0.19°$ at room temperature), i.e. the dogbone is rather short. In this case the diffuse scattering at the centre of the dogbone is very difficult to detect. Careful experiments (Chrosch and Salje 1999) showed a temperature evolution of the diffuse intensity and wall thickness as shown in Figure 3. The solid line in this graph shows the predicted temperature evolution, the steep increase of W at temperatures close to the transition point is clearly visible.

Table 1. Twin walls widths in units of the crystallographic repetition length normal to the twin plane.

Material	*Method*	*2W/A*	*References*
$LaAlO_3$	X-ray diffraction	10	Chrosch and Salje (1999)
WO_3	X-ray diffraction	12	Locherer et al. (1996)
$Gd_2(MoO_4)_3$	Electron microscopy	<10	Yamamoto et al. (1997)
$Pb_3(PO_4)_2$	Electron microscopy	≤8	Roucau et al. (1979)
	X-ray diffraction	<12.7	Wruck et al. (1994)
Sm_2O_3	Electron microscopy	2-3	Boulesteix et al. (1983)
KH_2PO_4	X-ray diffraction	4.2	Andrews and Cowley (1986)
$BaTiO_3$	Electron microscopy	4-10	Stemmer et al. (1995)
$YBa_2Cu_3O_{7-\delta}$	X-ray diffraction	2	Chrosch and Salje (1994)
$(Na,K)Al1Si_3O_8$	X-ray diffraction	2.75	Hayward et al. (1996)

W is the parameter in Equation (2), thus the width of the twin wall is approximately $2W$. The repetition unit is calculated from the smallest unit cell without cell doubling in $LaAlO_3$ and WO_3. Taking the cell doubling into account the ratio $2W/A$ is half the value given in the table.

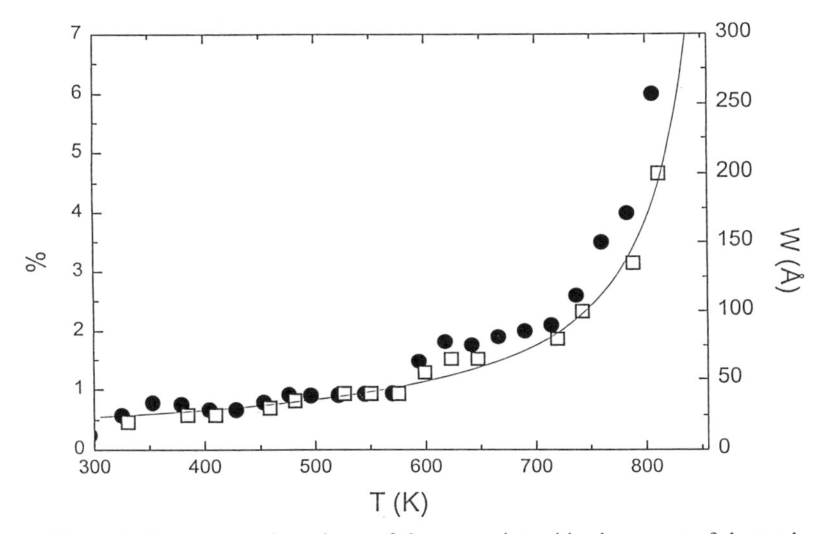

Figure 3. Temperature dependence of the excess intensities in percent of the total scattering signal (circles) and of the derived wall widths W (squares). The solid line represents a least-square fit to $W(T) \propto 1/|T-T_c|$ and yields $T_c = 875(5)$ K.

We now turn to the discussion of the wall energies in perovskite structures and other materials. It was observed in this study that the wall patterns change spontaneously at temperatures below the transition point. A similar effect was reported for $SrTiO_3$ where the domain wall density changed between 40 K and $T_c \approx 105$ K but not at lower temperatures (Cao and Barsch 1990). In other materials such as $Pb_3(PO_4)_2$ or feldspar (Yamamoto et al. 1997, Chrosch and Salje 1994, Hayward et al.1996) no such changes were observed on a laboratory timescale. We now test the hypothesis that the tendency of perovskite type structures to form domain walls spontaneously relates to their low energy. The wall energy for $T << T_c$ is given for a 2-4 Landau potential as (Salje 1993)

$$E_{wall} = W (A\ T_c)^2 / B \tag{3}$$

and for tricritical 2-6 potential as (Hayward et al. 1996)

$$E_{wall} = 2\ W (A\ T_c)^{2/3} / C^{1/2}. \tag{4}$$

Extended calculations using explicit strain coupling with the order parameter of $SrTiO_3$ were undertaken by Cao and Barsch (1990). As no direct experimental determinations of W exist for $SrTiO_3$, these authors estimated W = 1 nm from the dispersion of the acoustic phonon. This value is somewhat smaller than the values of $LaAlO_3$ (W = 2 nm) and WO_3 (W = 3.2 nm) (Locherer et al. 1998), although the uncertainties of the phonon dispersion and the rocking experiments may well account for such differences. An even smaller wall width of W = 0.5nm was estimated by Stemmer et al. (1995) from filtered TEM images of $PbTiO_3$. This method may underestimate the extent of the strain fields which were considered in $SrTiO_3$ and $LaAlO_3$. Furthermore, the order parameters in $SrTiO_3$ and $LaAlO_3$ relate to the rotation and flattening of oxygen octahedra (Rabe and Waghmare 1996, Hayward and Salje 1998) while Ti off-centre shifts become important in $PbTiO_3$. It is plausible from geometrical and energetic considerations that the off-centering decays over shorter distances than the rotation of the octahedra tilt axes.

The wall energy at $T << T_c$ in $PbTiO_3$ was estimated to be 5×10^{-2} J.m^{-2} while the wall energy in $SrTiO_3$ was predicted theoretically (Cao and Barsch 1990) as 4×10^{-4} J.m^{-2}. This latter value is smaller than those for other ferroelastic materials such as $In_{0.79}Tl_{0.21}$ ($E_{wall} = 1.1 \times 10^{-3}$ J.m^{-2} at 206 K) (Barsch and Krumhansl, 1988) or $BaTiO_3$ ($E_{wall} \approx 4 \times 10^{-3}$ J.m^{-2} for 90° walls and $E_{wall} \approx 10 \times 10^{-3}$ J.m^{-2} for 180° walls) (Bulgaevskii 1964). Using the thermodynamic data for the phase transitions at 100K in $SrTiO_3$ (Salje et al. 1998), with a tricritical potential and A = 0.504 J.mol^{-1}K^{-1}, C = 50.65 J K^{-1}, T_c = 100.604 K and W = 1 nm, a wall energy of $E_{wall} = 1.3 \times 10^{-3}$ J.m^{-2} is found. This value is much bigger than the estimate by Cao and Barsch (1990) and closer to the observations in other perovskite type structures. In comparison, the wall energy in alkali feldspar using the data published by Hayward et al. (1996) is $E_{wall} = 10.3 \times 10^{-3}$ J.m^{-2} at room temperature, i.e. this wall energy is similar to that of 180° twin boundaries in $BaTiO_3$ and greater than the wall energy in $SrTiO_3$.

Using these energies as reference points we may now estimate the wall energy in $LaAlO_3$ at low temperatures. The Landau potential follows from our observation that $Q^2 \propto 1/|T-T_c|$ ($\theta_s << T$) with

$$G = 1/2\ A\ (T-T_c)Q^2 + 1/4\ BQ^4 + 1/2\ g\ (\nabla Q)^2 \tag{5}$$

The step of the specific heat at $T = T_c$ is $A^2T_c/(2B)$ = 0.0105 J.g^{-1}.K^{-1}, which leads to A = 0.021 J.g^{-1}.K^{-1} and B = 17.16 J.g^{-1}. With W = 2nm one finds the wall energy to be 28×10^{-3} J.m^{-2}. This value is comparable with the wall energies of $BaTiO_3$ and feldspars but significantly greater than the one of $SrTiO_3$. The origin of this difference is the surprisingly large Landau step of ΔC_p in $LaAlO_3$ (1.05×10^{-2} J.g^{-1}.K^{-1}) (Bueble et al. 1998) as compared with the value of 3.5×10^{-3} J.g^{-1}.K^{-1} in $SrTiO_3$ (Salje et al. 1998). This means that the excess Gibbs free energy is much greater in $LaAlO_3$ than in $SrTiO_3$ and, hence, the wall energies differ significantly even between materials of the same structure type. In summary, we find that the order of magnitude for twin wall energies is typically several millijoule per square meter.

BENDING OF TWIN WALLS AND FORMATION OF NEEDLE DOMAINS

The bending of twin walls is the essential ingredient for the formation of needle twins. Three energy contributions for the bending of twin walls were identified by Salje and Ishibashi (1996). The first contribution is the elastic "anisotropy energy," which is the energy required for the rotation of a twin wall. The rotation axis lies inside the wall. Such rotation would lead to the formation of dislocations if the released energy becomes comparable with the dislocation energy. In displacive phase transitions, such dislocations have not been observed experimentally so the energy must be dissipated without topological defects. In this case, the energy density is, for small rotation angles $\alpha = 2\mathrm{d}y/\mathrm{d}x$:

$$E_{\mathrm{anisotropy}} = U\left(\frac{dy}{dx}\right)^2 \tag{6}$$

where the y-axis is again perpendicular to the unperturbed wall, the x-axis is parallel to the unperturbed wall segment (i.e., without the rotation), and U is a constant derived by Salje and Ishibashi (1996).

The second energy contribution stems from the fact that a wall with a finite thickness will resist bending because bending implies compression of the wall on one side and extension on the other. The energy density of this "bending" is

$$E_{\mathrm{bending}} = S\left(\frac{d^2y}{dx^2}\right)^2 \tag{7}$$

where S is a constant defined by Salje and Ishibashi (1996).

Finally, lateral movement of the wall is resisted by pinning as described by the "Peierls energy:"

$$E_{\mathrm{pinning}} = Py^2 \tag{8}$$

which holds for small values of y. The parameter P is a measure of the Peierls energy. More complex Peierls energies were discussed by Salje and Ishibashi (1996).

The shape of a needle twins, i.e. the trajectory of the wall position $y(x)$, is determined by the minimum of the total energy

$$\delta E = \delta \int \left(E_{anisotropy} + E_{bending} + E_{pinning}\right) dx = 0. \tag{9}$$

The following solutions were found by Salje and Ishibashi (1996) for the coordination system shown in Figure 4. For case 1, bending dominated needles without lattice relaxation, the wall trajectory is close to a parabolic shape

$$y = \frac{y_{\max}}{2\lambda^3}(\lambda - x)^2(2\lambda + x) \tag{10}$$

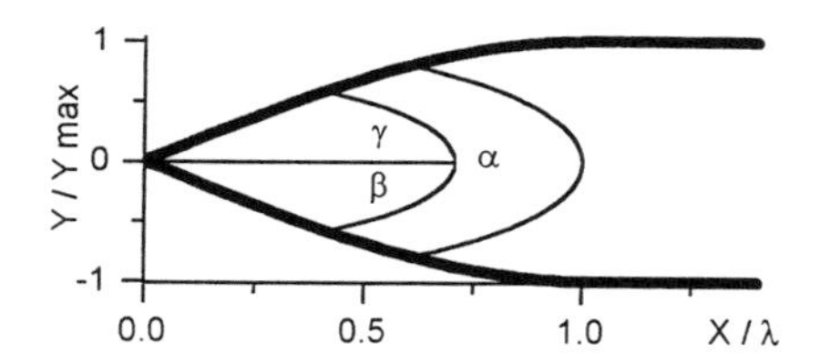

Figure 4. Wall profile as calculated for needles with large curvature energy and no lattice relaxation (see Salje and Ishibashi 1996). The coordinate systems used in the text is shown. For single needles $\gamma = \beta = \alpha/2$, for forked needles we observed $\beta \le \gamma$ where β is the interior angle of a forked pair.

where y_{max} is the position of the needle tip and λ is the distance (along the x-axis) between the needle tip and the shaft of the needle. This trajectory contains no adjustable parameters and has no characteristic length scale. Needles of this type are, thus, universal, i.e., their shape does not depend on temperature, pressure, or the actual mineral in which they occur. For case 2, anisotropy dominated needles without lattice relaxation, the trajectory is linear with

$$y = y_{max}\,(1 - x/\lambda). \tag{11}$$

This trajectory is also universal. For case 3, anisotropy dominated needles with elastic lattice relaxation or superposition of anisotropy energy and bending energy, we expect an exponential trajectory

$$y = y_{max}\,\exp(-x/\lambda). \tag{12}$$

The value of λ depends on the lattice relaxation $\lambda^2 = U/P$, which is no longer a simple geometrical parameter. In particular, the ratio U/P may depend on temperature and pressure so that small Peierls energies will favour long, pointed needles whereas large Peierls energies lead to short needles. In the case of superposition of anisotropy and bending energies the length scale is set by $\lambda^2 = S/U$, which is also a non-universal parameter.

A large variety of other functional forms of the wall trajectories were discussed by Salje and Ishibashi (1996). Only the three cases above were so far used to discuss the experimental observations in this paper.

Comparison with experimental observations

Needle twins with linear trajectories close to the needle tip were found by Salje et al. (1998) in several ferroelastic materials such as $PbZrO_3$, WO_3, $BiVO_4$, $GdBa_2Cu_3O_7$, $[N(CH_3)_4]_2{\cdot}ZnBr_4$, and the alloy CrAl. Some experimental data are listed in Table 2, and typical examples are shown in Figure 5. In all of these materials, the change between the shaft of the needle and the tip is abrupt. The straight trajectory of the needle tip indicates that the anisotropy energy of these crystals is much larger than the bending energy and that Peierls energies are unimportant (apart from the pinning of the shaft). The three needles measured from the same specimen of CrAl each have the same needle tip angle; two of these terminate against the same 90° twin wall. Although there is a general relationship between the angle at the needle tip and the needle width, for a particular sample, such as CrAl, it appears that the tip angle is uniform and the variation in needle

Table 2. Parameters of wall trajectories for selected materials.

Material	*Reference*	*Length (λ)*	*Tip-angle*	y_{max}
WO_3	Microanalysis Lab, Cambridge	680 μm	1.6	9 μm
$[N(CH_3)_4]_2{\cdot}ZnBr_4$	Sawada (pers. comm.)	no scale	4	12
CrAl	Van Tendeloo (pers. comm.)	300 nm	6.8	18 nm
CrAl	Van Tendeloo (pers. comm.)	367 nm	6.8	22 nm
CrAl	Van Tendeloo (pers. comm.)	375 nm	6.8	22 nm
$BiVO_4$	Van Tendeloo (pers. comm.)	320 nm	8.9	25 nm
$GdBa_2Cu_3O_7$	Shmytko et al. 1989	no scale	10.3	
$PbZrO_3$	Dobrikov and Presnyakova 1980	300 nm	12.5	33 nm

Note: Fitting parameters are y_{max} and tip angle; λ calculated from these.

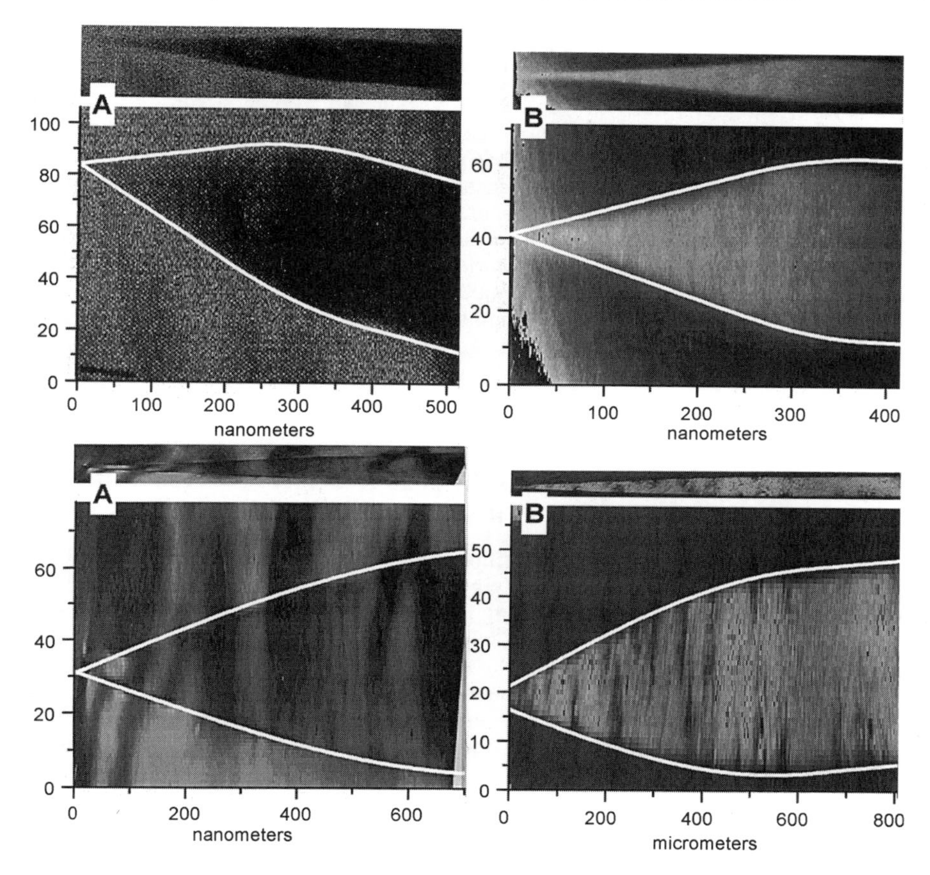

Figure 5. (A,B–upper) Images of linear needle tips. The lines drawn are straight line fits along the shaft and tip, the two lines being joined with a parabola. The upper images show the true aspect ratio, the lower images show the *y*-axis expanded to demonstrate the linear nature of the needle tip. The apparent asymmetry in the latter is due to the non-linear expansion. left: $PbZrO_3$, right: $BiVO_4$. Both axes have the same units. (A,B–lower) Needle twins with curved trajectories. The upper images show the true aspect ratio, the lower images show the *y* axis expanded to demonstrate the curvature of the needle tip. In both cases the tip of the needle is truncated against another feature so the needle origin must be estimated from the position of the centre line, to determine λ. (A) GeTe, (B) $K_2Ba\ (NO_2)_4$. Both axes have the same units.

width is accommodated more by a corresponding increase in λ. This is in contrast to the parabolic and exponential cases in which the angle at the tip varies systematically with the width of the twin. Bending dominated wall energies are predicted to lead to trajectories of a modified parabolic shape. Such needles have been observed in the alloy GeTe and in $K_2Ba(NO_2)_4$.

Needle twin walls with exponential trajectories were observed in many ferroelastic materials such as $Pb_3(PO_4)_2$ and KSCN and for twin walls in $BaTiO_3$. Typical length scales, λ, are 85 μm (KSCN), 55 nm to 3μm [$Pb_3(PO_4)_2$] and 270 nm ($BaTiO_3$). Furthermore, although the parabolic and linear cases show a correlation between the tip angle and needle width, this is less apparent for the exponential needles. The tip angle for angle and needle width, this is less apparent for the exponential needles. The tip angle for these materials is considerably greater than for parabolic materials of similar width.

In several materials, needles were observed that had formed a 'tuning fork' pair rather than a single needle. Examples of this are $GdBa_2Cu_3O_7$, $YBa_2Cu_3O_7$, $Pb_3(PO_4)_2$ and feldspar (Smith et al. 1987), and $La_{2-x}Sr_xCuO_4$ (Chen et al. 1991). In all of these cases the two halves of the split needle were the same but each half was asymmetric, the needle tips being displaced towards the centre line of the pair. $GdBa_2Cu_3O_7$ has linear needle tip trajectories (as was found for the simple needles of this material) and the inner and outer tip angles were the same. The other materials measured by Salje et al. (1998b) had exponential needle tips and the inner tip angle was less than the outer. Single needles of comparable width were found from the same samples as each of the four split needles examined in detail so a direct comparison can be made between these. The split needle twins were always the widest in the sample, and there appears to be a limiting tip angle for a particular material above which splitting occurs to produce a joint pair of needles with a much reduced tip angle.

NUCLEATION OF TWIN BOUNDARIES FOR RAPID TEMPERATURE QUENCH: COMPUTER SIMULATION STUDIES

When minerals undergo phase transitions involving atomic ordering on different crystallographic sites the ordering is often slow enough so that the early stages of twinning are conserved even over geological times. A typical example is the Al,Si ordering in alkali feldspars which still contains large strain energies so that the resulting mesoscopic twin patterns still show all the essentials of ferroelastic domain structures (Salje, 1985a, Salje et al. 1985b). The time evolution of the formation of pericline twins in alkali feldspar was studied in detail by computer simulation (Tsatskis and Salje 1996). Some details are now discussed because they illustrate generic simulation techniques which have subsequently been used for more complex structures such as the simulation of twin structures in cordierite (Blackburn and Salje 1999a,b,c).

The underlying physical picture is an order-disorder phase transition that generates spontaneous strain. The ordering of atoms of types A and B (here Al and Si) occupy certain sets of positions in a perfect crystalline structure or host matrix. The phase transition is, as usual, the result of interaction between the ordering atoms. It is assumed that they interact only elastically (through the host matrix), i.e. indirectly, and that direct chemical (short-range) and interactions can be neglected. The origin of the elastic interatomic interaction is the electro-static distortion of the host matrix by the ordering atoms. In the absence of the ordering atoms, the host matrix is an ideal crystal, and all its atoms are in mechanical equilibrium. When the ordering atoms are inserted into the host matrix, each ordering atom produces external stress with respect to the host matrix, shifting its atoms away from the initial equilibrium positions. As a result, internal forces arise in the host matrix that tend to return the host atoms to the initial positions. A new state of equilibrium corresponding to a given distribution of ordering atoms is then reached, in which the sum of external and internal forces acting on each atom of the host matrix vanishes. In this new equilibrium state the host atoms are displaced from the initial positions, and the host matrix is distorted. Different configurations of ordering atoms correspond to different sets of displacements. In the simplest possible approximation the resulting displacement of a host atom is a superposition of displacements caused by individual ordering atoms. One atom "feels" the distortion of the host matrix created by another atom, and this results in the effective long-range elastic interaction between these two atoms.

More formally, if a system consists of two subsystems that are not independent, i.e. the subsystems interact with each other, then the Hamiltonian of this system, generally speaking, should be a sum of three contributions. The first two terms are the Hamiltonians of the isolated subsystems, and the third term represents the interactions between these subsystems. In our case the crystal consists of the ordering atoms and the host matrix. Therefore, the Hamiltonian under consideration has the form

$$H = H_{\text{host}} + H_{\text{ord}} + H_{\text{int}} \tag{13}$$

where H_{host} and H_{ord} are Hamiltonians of the host matrix and the ordering atoms, respectively, and H_{int} is the interaction Hamiltonian. We now specify the form of all three Hamiltonians. The Hamiltonian of the host matrix, which is its potential energy, is a function of the static displacements u of host atoms and describes the energy increase when the host matrix is pulled out of the mechanical equilibrium. In the case of sufficiently small displacements it is possible to expand the host-matrix energy in powers of the displacements and to retain only the first nonzero (quadratic) term. This is the harmonic approximation usually used in the theory of lattice dynamics (e.g. Ashcroft and Mermin 1976). The zero-order term, which does not depend on displacements, is ignored. Further, because it is supposed that the ordering atoms do not interact directly, their energy is configurationally independent, and the second contribution to the Hamiltonian (Eqn. 13), H_{ord}, is zero. Finally, the interaction Hamiltonian describes the effect of forces f with which an ordering atoms acts on neighbouring host atoms. We assume that these forces have constant values, regardless of the positions of the host atoms, and this means that the interaction term is the linear function of displacements, because of the relation

$$f_n^i = -\frac{\partial H_{int}}{\partial u_n^i} \tag{14}$$

where f_n^i is the ith Cartesian component of a force acting on the atom at site n of the host matrix, and u_n^i is the corresponding displacement. Obviously, an ordering atom of each kind has its own set of forces, and the resulting force on the host matrix is therefore a function of the configuration of the ordering atoms. A particular configuration of the ordering atoms is fully described by the set of occupation numbers $p_l^\alpha = \alpha = A, B$,

$$p_l^\alpha = \begin{Bmatrix} 1, \textit{atom of type } \alpha \textit{ at site } l \\ 0, \textit{otherwise} \end{Bmatrix} \tag{15}$$

and l is the position of an ordering atom. The force f_n^i (Eqn. 14) depends on the occupation numbers for the ordering atoms surrounding the atom at site n of the host matrix and can be conveniently written in their terms as

$$f_n^i = \sum_l \left(F_{nl}^{iA} p_l^A + F_{nl}^{iB} p_l^B \right) = \sum_l F_{nl}^{i\alpha} p_l^\alpha . \tag{16}$$

In this equation F_{nl}^i is the i-th Cartesian component of the so-called Kanzaki force (Khachaturyan 1983) with which the ordering atom of type α at site l acts on the host atom at site n. Taking into account Equations (14) and (15), we finally get the full Hamiltonian of the system in the following form:

$$\begin{aligned} H &= \tfrac{1}{2} uAu - uFp \\ &= \tfrac{1}{2} \sum_{nm} \sum_{ij} u_n^i A_{nm}^{ij} u_m^j - \sum_{nl} \sum_i \sum_n u_n^i F_{al}^{i\alpha} p_l^\alpha \end{aligned} \tag{17}$$

where A is the Born-von Kármán tensor of the host matrix. It is seen that in the Hamiltonian (Eqn. 17) the variables corresponding to the two subsystems (displacements of the host atoms and occupation numbers of the ordering atoms) are coupled bilinearly because of the interaction term H_{int}. The result (Eqn. 17) can also be obtained by representing the internal energy of the host matrix as a series in powers of the small displacements of the host atoms and disregarding third-order and higher terms (Krivoglaz 1969, Khachaturyan 1983). Similar Hamiltonians have been used to study transient tweed and twin patterns that arise in the process of ordering on simple lattices caused by elastic interactions (Marais et al. 1991, Salje and Parlinski 1991, Salje 1992, Parlinski et al. 1993a,b; Bratkovsky et al. 1994a,b,c). Unlike in the case of the harmonic approximation for the host matrix, here the first-order contribution is nonzero because of applied external forces. The actual positions of the ordering atoms are not specified in the model; the Hamiltonian (Eqn. 17) contains only the displacements of the host atoms. In fact, this Hamiltonian describes the host matrix subjected to external forces, and these forces depend on a particular configuration of the ordering atoms, which is described in terms of the occupation numbers.

Let us turn now to the quantitative description of the effective long-range interaction between the ordering atoms starting from the Hamiltonian (Eqn. 17). To find static displacements corresponding to a given configuration of the ordering atoms, it is necessary to minimize the energy of the system (i.e. the Hamiltonian, Eqn. 17) with respect to displacements u_n for a given set of the occupation numbers representing this configuration. The static displacements u^{st} are, therefore, solutions of the coupled equations

$$\begin{aligned}\frac{\partial H}{\partial u_n^i} &= \left(Au - Fp\right)_n^i \\ &= \sum_m \sum_j A_{nm}^{ij} u_m^j - \sum_l \sum_\alpha F_{nl}^{i\alpha} p_l^\alpha = 0\end{aligned} \tag{18}$$

and can be written as

$$\left(u^{st}\right)_n^i = \left(A^{-1}Fp\right)_n^i = \sum_{ml} \sum_j \sum_\alpha \left(A^{-1}\right)_{nm}^{ij} F_{ml}^{j\alpha?} p_l^\alpha \tag{19}$$

Decomposing an instantaneous displacement of the host atom into two parts,

$$u_n = u_n^{st} + \Delta u_n, \tag{20}$$

substituting this sum into the initial Hamiltonian (Eqn. 17), and using Equation (19), we arrive at the following expression:

$$\begin{aligned}H &= \tfrac{1}{2} Vp + \tfrac{1}{2} \Delta u A \Delta u \\ &= \tfrac{1}{2} \sum_{kl} \sum_{\alpha\beta} p_l^\alpha V_{lk}^{\alpha\beta} p_k^\beta + \tfrac{1}{2} \sum_{nm} \sum_{ij} \Delta u_n^i A_{nm}^{ij} \Delta u_m^j .\end{aligned} \tag{21}$$

The first term is the standard Hamiltonian used in the phenomenological theory of ordering (de Fontaine 1979, Ducastelle 1991), which contains only variables corresponding to the ordering atoms (i.e. occupation numbers), and the second term describes harmonic vibrations of the host atoms around new (displaced due to static external forces) equilibrium positions. The Born-von Kármán tensor A and, therefore, the phonon frequencies are the same as in the case of the undistorted host marix; in the harmonic approximation static deformations do not affect lattice vibrations. In this Hamiltonian the degrees of freedom corresponding to the two sub-systems are completely separated, and at finite temperatures the thermal vibrations of the host atoms are independent of the configuration of the ordering atoms. The effective interaction V between the ordering atoms has the form

$$V_{lk}^{\alpha\beta} = -\left(F^{\mathrm{T}} A^{-1} F\right)_{lk}^{\alpha\beta} = -\sum_{nm}\sum_{ij}\left(F^{\mathrm{T}}\right)_{ln}^{\alpha i}\left(A^{-1}\right)_{nm}^{ij} F_{mk}^{j\beta}. \tag{22}$$

Using spin variables s_i,

$$p_i^{\mathrm{A}} = \tfrac{1}{2}\left(1+s_i\right); \quad p_i^{\mathrm{B}} = \tfrac{1}{2}\left(1-s_i\right) \tag{23}$$

it is easy to show that the effective Hamiltonian for ordering in the giant canonical ensemble,

$$\tilde{H} = \tfrac{1}{2} pVp - \sum_{\alpha} \mu^{\alpha} N^{\alpha} \tag{24}$$

where μ^{α} and N^{α} are the chemical potential and total number of atoms of type α, respectively, is formally equivalent to that of the Ising model (de Fontaine 1979, Ducastelle 1991),

$$\tilde{H} = -\tfrac{1}{2}\sum_{lk} J_{lk} s_l s_k - \sum_{l} h_l s_l \tag{25}$$

where the effective exchange integral J_{lk} and the magnetic field h_l are given by

$$J_{lk} = \tfrac{1}{4}\left(2V_{lk}^{AB} - V_{lk}^{AA} - V_{lk}^{BB}\right) \tag{26}$$

$$h_l = \tfrac{1}{2}\left(\mu^{A} - \mu^{B}\right) - \tfrac{1}{4}\sum_{k}\left(V_{lk}^{\mathrm{AA}} - V_{lk}^{\mathrm{BB}}\right). \tag{27}$$

Inserting Expression (22) for the effective interaction between the ordering atoms in Equation (26) yields

$$\begin{aligned} J_{lk} &= \tfrac{1}{4}\left[\left(F^{A} - F^{B}\right)^{T} A^{-1}\left(F^{A} - F^{B}\right)\right]_{ik} \\ &= \tfrac{1}{4}\sum_{nm}\sum_{ij}\left(F^{\mathrm{A}} - F^{\mathrm{B}}\right)_{ln}^{i}\left(A^{-1}\right)_{nm}^{ij}\left(F^{A} - F^{B}\right)_{mk}^{j}. \end{aligned} \tag{28}$$

This equation shows that it is the difference $F^{\mathrm{A}} - F^{\mathrm{B}}$ between the Kanzaki forces for the two types of ordering atoms which matters, and not the values of F^{A} and F^{B}. It is important to note that the site-diagonal matrix elements $V_{ll}^{\alpha p}$ and J_{ll} of the effective interatomic interaction V and exchange integral J have nonzero values; in other words, there exists the effect of a self-intraction of the ordering atoms. The reason for the self-interaction is easy to understand. A single ordering atom placed into the empty host matrix distorts the latter and thereby changes the energy of the system. This energy change corresponds precisely to the diagonal matrix element of the effective interaction V. This self-interaction is important for the calculation of the energy difference corresponding to the interchange of a pair of atoms. It can be shown that in reciprocal space the effective spin interaction has the singularity at the point $\mathbf{k} = 0$ (e.g. de Fontaine 1979), similar to the singularity of the volocity of sound (Folk et al. 1976, Cowley 1976). This singularity is characteristic of the elastic interactions. The $\mathbf{k} \rightarrow 0$ limit of the Fourier transform of the effective interaction depends on the direction along which the point $\mathbf{k} = 0$ is approached.

In the case of alkali feldspars the unit cell is quite complicated. It contains four formula units, and 53 atoms in the asymmetric unit. As a result, the simulated sample of Tsatskis and Salje (1996) had the form of a very thin slab (or film); the computational unit cell defined for the whole slab contained slightly more than four formula units. In the simulation the slab had {101} orientation, which allowed the observation of only the

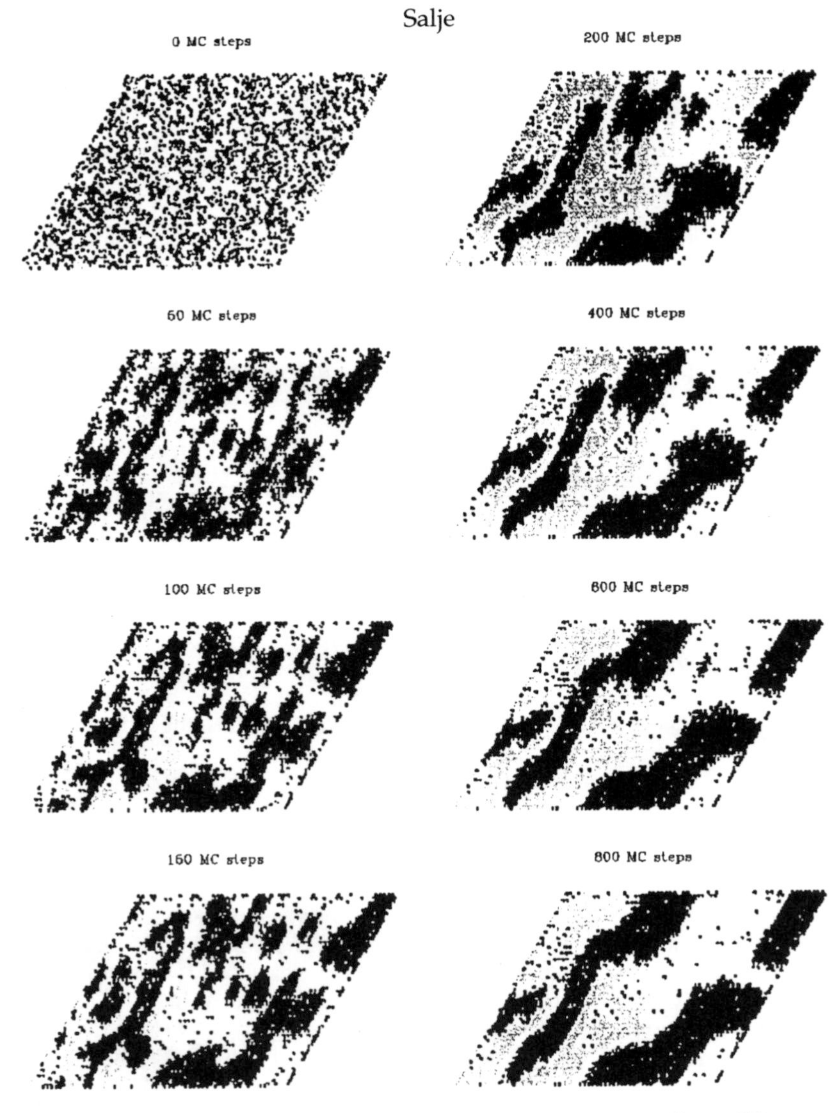

Figure 6. Sequences of snapshots of the simulated twin microstructure corresponding to different annealing times in alkali feldspars, t (indicated in Monte Carlo steps per ordering atom). Only Al atoms distributed over $T1$ positions of a single crankshaft are shown. Different symbols (heavy and light dots) are used to distinguish between $T1$o and $T1$m sites. The first snapshot corresponds to the initial, totally disordered Al-Si distribution. The annealing temperature is $0.42T_c$.

pericline twins; the simulated slab contains two crankshafts in the crystallographic y direction.

Snapshots of the twin microstructure in the simulated sample annealed below the transition are shown in Figure 6. Only Al atoms belonging to a single crankshaft and located at T1 sites are shown in Figure 6, and different symbols are used to represent Al atoms at T1o and T1m sites. The Al atoms at T2 positions and all Si and host atoms are not shown in order to clearly distinguish the two variants of the ordered phase. At early stages of the ordering kinetics, very fast local order and formation of the pericline twin domains were found. At the beginning, ordering and coarsening occurred simultaneously, and no clear distinction between these two processes was possible. After a short time, however, the local ordering is almost complete, and the pattern consists of a fine mixture of well-defined regions in which the order parameter Q_{od} acquires positive or negative values. The patches of the ordered phase continue to coarsen at later stages, and the preferential orientation of the domain walls gradually appears. At even later stages, the system arranges itself into a pattern of relatively wide stripes of pericline twins aligned mainly along the direction described by the pericline twin law at a macroscopic level. The evolution of the sample at this stage is very slow, making it problematic to monitor further development of the stripe pattern. However even in this regime the domain walls experience significant deviations from the soft direction at a local scale. As the temperature increases and approaches the transition point, the transitional areas between pericline twins becomes more and more diffuse. This is in complete agreement with the predictions of the continuum Landau-Ginzburg theory, according to which the width of a domain wall increases with temperature and finally diverges at the instability point (e.g. Salje 1993). The pericline wall has only one symmetry constraint, namely, it must contain the crystallographic y-axis. Its actual orientation depends on the ratio e_6/e_4 of the components of the spontaneous strain, i.e. the wall can rotate around the y-axis at the ratio e_6/e_4 changes. When the crystal is quenched through T_c, the Al and Si atoms order locally and build up local strain. The twin walls accommodate this local strain, so that their orientation corresponds to the lattice deformation on a length scale of a few unit cells. This lattice deformation deviates substantially from that of the uniformly ordered sample. Local segments of walls at the early stages are not well aligned, therefore. With increasing degree of order, the spontaneous strain becomes more uniform, and a global alignment of walls is observed.

The second main result of the simulation is that pericline walls are not smooth on an atomistic scale. In the simulation Tsatskis and Salje (1996) found that the wall thickness at temperatures is always ~ 0.1-1.5 nm. This result seems to support the idea that there is a minimum thickness for a pericline wall when the temperature approaches absolute zero. The origin of this minimum thickness appears to be geometrical in nature, namely, that an atomistically perfect pericline wall cannot be constructed along an arbitrary direction containing the crystallographic y-axis. The orientation of the wall is determined by the macroscopic compatibility condition, which contains no direct information about the underlying crystal structure. Only under exceptional circumstances do such walls coincide with crystallographic planes that are apt to form structural twin planes. Generally, the orientation of the twin walls does not correlate with the crystal structure, leading to faceting of the wall. This faceting is then the reason for the effective finite thickness of the twin wall (see also Blackburn and Salje 1999a,b,c).

INTERSECTION OF A DOMAIN WALL WITH THE MINERAL SURFACE

The termination of mesoscopic structures at mineral surfaces gives rise to modifications of surface relaxation and, thus, variations of the surface potential. In many cases, such as dislocations, ferroelectric domain walls, etc., simple etching techniques are traditionally used to visualise the resulting surface patterns. In the case of twin walls, Novak and Salje (1998a,b) have analysed the characteristic surface structures by minimisation of a generic point lattice with Lennard-Jones type interactions between the lattice points. Any interaction which is longer ranging than nearest neighbour distances lead to surface relaxations of the type

$$Q = Q_o\,(1 - e^{-y/\lambda}) \qquad (29)$$

where Q is here the relaxational coordinate which approaches the equilibrium value Q_o in the bulk. If we now consider Q_o to be the bulk order parameter of a ferroelastic transition we can consider the spatial variation of $Q(r)$ near the surface. In this particular study, Novak and Salje (1998a, 1998b) considered a twin wall that was oriented perpendicular to the surface. It was found that the characteristic relaxation length λ is at a minimum close to the twin domain wall and increases with distance away from it. This can be seen from Figure 7, where the measure of the surface relaxation is any contour of constant Q. At an infinite distance λ would reach its maximum value, λ_{max}, which is the surface relaxation depth of the lattice if no twin domain walls are present. Consequently, in materials with microstructure form by an array of periodic twin domain walls (Salje and Parlinski 1991), the depth of the surface relaxation λ_{array} is suppressed as a function of the domain wall density. The magnitude of the order parameter at the surface, Q_s, exhibits the opposite behaviour.

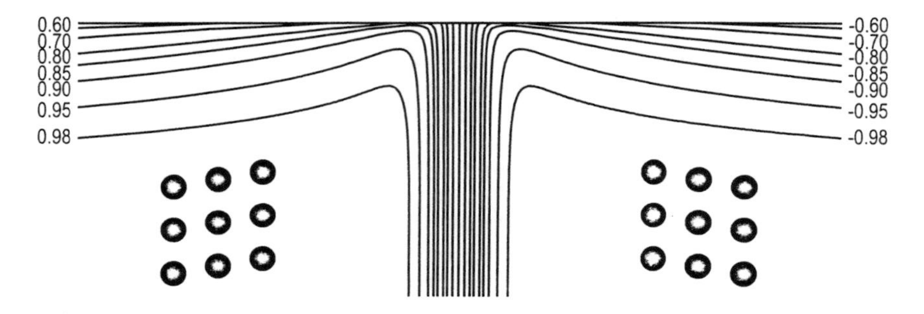

Figure 7. Distribution of the order parameter Q at the surface of the lattice (first 50 layers): Lines represent contact Q/Q_o. There are three lines in the middle of the twin domain wall that are not labelled: they represent the Q/Q_o values of 0.40, 0.00, and –0.40 respectively. Notice the steepness of the gradient of Q/Q_o through the twin domain wall. The two structures represent sheared twin atomic configurations in the bulk (far from the domain wall and surfaces).

The relation between W and W_s, the domain wall widths in the bulk and at the surface, can be seen in Figure 8. The effect of the surface relaxation is clearly visible as the order parameter at the surface Q_s never reaches the bulk value Q_o. The distribution of the square of the order parameter Q_s^2 at the surface shows the structure that some of the related experimental works have been reported (Tsunekawa et al. 1995, Tung Hsu and Cowley 1994), namely a groove centred at the twin domain wall with two ridges, one on each side.

In addition, the square of the surface order parameter is proportional to the chemical reactivity profile of the twin domain wall interface at the surface (Locherer et al. 1996, Houchmanzadeh et al.(1992). Intuitively, one would expect the chemical reactivity of the surface to be largest at the centre of the twin domain wall, falling off as the distance from the centre of the wall increases. Contrary to the expected behaviour, a more complex behaviour is found. The reactivity, a monotonic function of Q_s^2, is expected to fall off as the distance from the centre of the wall increases, but only after if has reached a maximum of a distance of $\sim 3W$ from the centre of the domain wall. If such a structure is expected to show particle adsorption (e.g. in the MBE growth of thin films on twinned substrates) we expect the sticking coefficient to vary spatially. In one scenario, adsorption may be enhanced on either side of the wall while being reduced at the centre. The real space topography of the surface is determined by both sources of relaxation—twin domain wall and the surface. These are distinct and, when considered separately, the wall

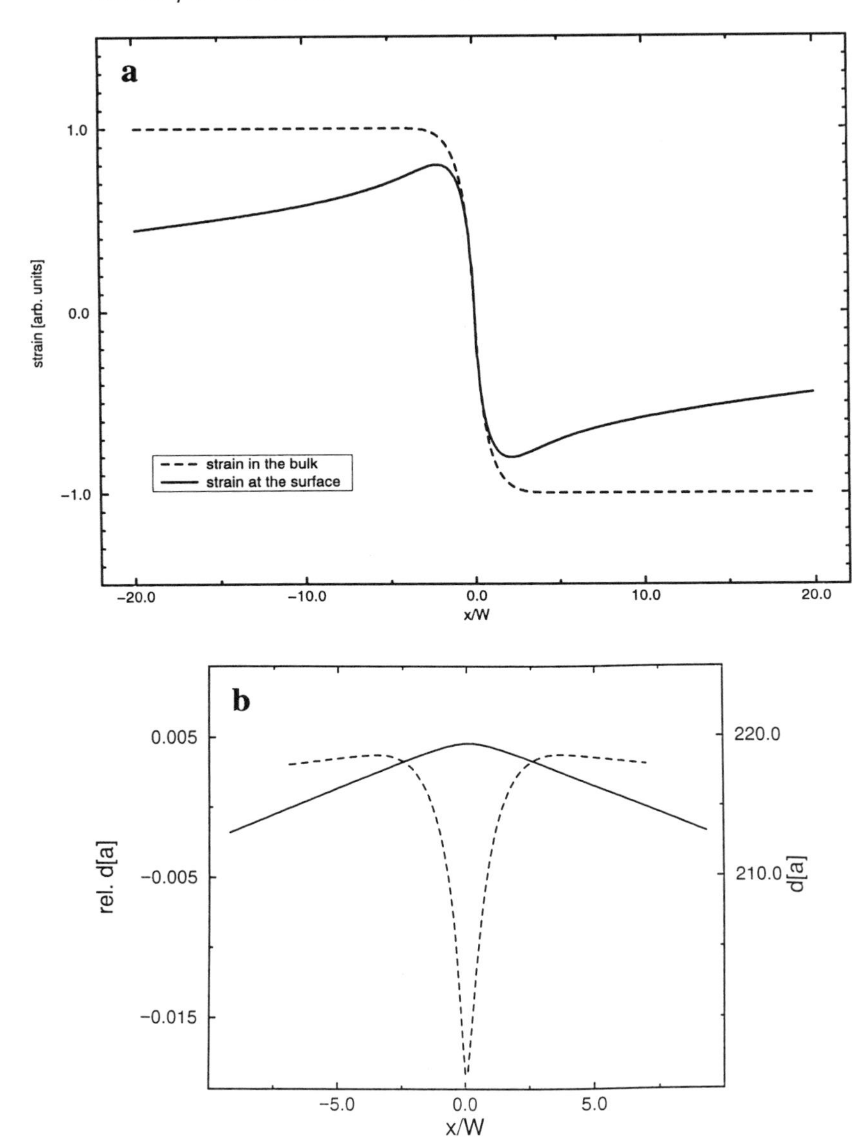

Figure 8. (a) Q/Q_o (solid line, proportional to the strain at the surface) and Q/Q_o (dashed line, proportional to the strain in the bulk). The widths of the twin domain wall in the bulk, W, and at the surface, W_s, are the same. (b) The square of the order parameter at the surface, Q_s^2, related to the chemical potential, and indicative of the areas of maximum decoration, where the distortion of the lattice is at a maximum.

relaxation is larger by about three orders of magnitude than that due to the surface relaxation (Fig. 7).

In an attempt to predict the possible experimental results of AFM investigations of the surface structure of the twin domain wall, the effect that the tip at the end of an AFM cantilever has on the surface of the material was simulated. This was done by displacing each particle in the surface layer by $10^{-8}\alpha$. The resulting lateral force distribution shows a dependence similar to that of the order parameter Q_s (Fig. 8). The normal force distribution has a profile similar to that of the square of the order parameter, Q_s^2, with ridges on both sides of a groove (Fig. 8).

The change in the sign of the order parameter at the surface has been observed (for ferroelectricty) by using a model of imaging developed for the detection of static surface charge (Saurenbach and Terris 1990). For ferroelastics, this corresponds to the profile of the lateral reactive force. The SFM non-contract dynamic mode images (Lüthi et al 1993) would correspond to the distribution of the normal reactive force. The divergence of the lateral force distribution away from the centre of the wall can be attributed to the simulated infinite extension of the lattice. In the simulated array, the lateral component of the force reached a finite value between two adjacent domain walls.

The results can be used as a guide for the future experimental work. In order to determine the twin domain wall width W in the bulk, one only needs to determine the characteristic width W_s of the surface structure of the domain wall. Previously, these features of the twinning materials were investigated using mainly X-ray techniques. In fact, the theoretical work leads to the conclusion that the only necessary information for the determination of the twin domain wall width W are the real space positions of the particles in the surface layer.

REFERENCES

Aird A, Salje EKH (1998) Sheet superconductivity in twin walls: experimental evidence of WO_{3-x}. J Phys: Condens Matter 10:L377-L380

Aird A, Salje, EKH (2000) Enhanced reactivity of domain walls in WO_3 with sodium. Eur Phys J B15:205-210

Aird A, Domeneghetti MC, Mazzi F, Tazzoli V, Salje EKH (1998) Sheet superconductivity in WO_{3-x} : crystal structure of the tetragonal matrix. J Phys: Condens Matter 10:L569-574

Andrews SR, Cowley RA (1986) X-ray scattering from critical fluctuations and domain walls in KDP and DKDP. J Phys C 19:615-635

Ashcroft NW, Mermin D (1976) Solid State Physics. Saunders College, Philadelphia, Pennsylvania, 826 p

Barsch GR, Krumhansl JA (1988) Nonlinear and nonlocal continuum model of transformation precursors in martensites. Met Trans 19:761-775

Bismayer U, Salje EKH, Joffrin C (1982) Reinvestigation of the stepwise character of the ferroelastic phase transition in lead phosphate-arsenate $Pb_3(PO_4)_2 - Pb_3(AsO_4)_2$. J Phys 43:1379-1388

Bismayer U, Hensler J, Salje EKH, Güttler B (1994) Renormalization phenomena in Ba-diluted ferroelastic lead phosphate, $(Pb_{1-x}Ba_x)_3(PO_4)_2$. Phase Trans 48:149-168

Blackburn J, Salje EKH (1999a) Formation of needle shaped twin domains in cordierite: A computer simulation study. J Appl Phys 85:2414-2422

Blackburn J, Salje EKH (1999b) Time evolution of twin domains in cordierite: a computer simulation study. Phys Chem Minerals 26:275-296

Blackburn J, Salje EKH (1999c) Sandwich domain walls in cordierite: a computer simulation study. J Phys: Condens Matter 11:4747-4766

Bleser T, Berge B, Bismayer U, Salje EKH (1994) The possibility that the optical second-harmonic generation in lead phosphate $Pb_3(PO_4)_2$ is related to structural imperfections. J Phys: Condens Matter 6:2093-2099

Boulesteix C, Yangui B, Nihoul G, Bourret A (1983) High-resolution and conventional electronic microscopy studies of repeated wedge microtwins in monoclinic rate-earth sesquioxides. J Microsc 129:315-326

Bratkovsky AM, Salje EKH, Marais SC and Heine V (1994a) Theory and computer simulation of tweed texture. Phase Trans 48:1-13

Bratkovsky AM, Salje EKH and Heine V (1994b) Overview of the origin of tweed texture. Phase Trans 53:77-84

Bratkovsky AM, Marais SC, Heine V, Salje EKH (1994c) The theory of fluctuations and texture embryos in structural phase transitions mediated by strain. J Phys: Condens Matter 6:3679-3696

Bueble S, Knorr K, Brecht E, Schmahl WW (1998) Influence of the ferroelastic twindomain structure on the {100} surface morphology of $LaAlO_3$ HTSC substrates. Surf. Sci 400:345-355.
Burger WG (1945) The genesis of twin crystals. Am Mineral 30:469-482
Bulgaevskii LN (1964) Thermodynamic theory of domain walls in ferroelectric materials with perovskite structure. Sov Phy Sol St 5:2329-2332
Camara F, Doukhan JC, Salje EKH (2000) Twin walls in anorthoclase are enriched in alkali and depleted in Ca and Al. Phase Trans 71:227-242
Cao W, Barsch GR (1990) Quasi-one-dimensional solutions for domain walls and their constraints in improper ferroelastics. Phys Rev B 42:6396-6401
Carpenter M A, Salje EKH, Graeme-Barber A (1998) Spontaneous strain as a determinant of thermodynamic properties for phase transitions in minerals. Eur J Mineral 10:621-691
Chen CH, Cheong SW, Werder DJ, Cooper AS, Rupp LW (1991) Low temperature microstructure and phase transition in $La_{2-x}Sr_xCuO_4$ and $La_{2-x}Ba_xCuO_4$. Physica C 175:301-309
Chrosch J, Salje EKH (1994) Thin domain walls in $YBa_2Cu_3O_{7-d}$ and their rocking curves: An X-ray study. Physica C 225:111-116
Chrosch J, Salje EKH (1997) High-resolution X-ray diffraction from microstructures. Ferroelectrics 194:149-159
Chrosch J, Salje EKH (1999) The temperature dependence of the domain wall width in $LaAlO_3$. J Appl Phys 85:722-727.
Cowley RA (1976) Acoustic phonon instabilities and structural phase transitions. Phys Rev B 13:4877-4885
de Fontaine D (1979) Configurational thermodynamics of solid solutions. Sol St Phys 34:73-274
Ducastelle F (1991) Order and Phase Stability in Alloys. North-Holland, Amsterdam, 511 p
Folk R, Iro H, Schwabl F (1976) Critical statics of elastic phase transitions. Z Phys B 25:69-81
Hayward SA, Salje EKH (1998) Low temperature phase diagrams: non-linearities due to quantum mechanical saturation of order parameters. J Phys Condens Matter 10:1421-1430
Hayward SA, Salje EKH (2000) Twin memory and twin amnesia in anorthoclase. Mineral Mag 64:195-200
Hayward SA, Chrosch J, Carpenter M, Salje EKH (1996) Thickness of pericline twin walls in anorthoclase: an X-ray diffraction study. Eur J Mineral 8:1301-1310
Houchmonzadeh B, Lajzerowicz J, Salje EKH (1992) Interfaces and ripple states in ferroelastic crystals—a simple model. Phase Trans 38:77-78
Hochella MF, Banfield JF (1995) Chemical weathering of silicates in nature: a microscopic perspective with theoretical considerations. Rev Mineral 31:353-406
Jamnik J, Maier J (1997) Transport across boundary layers in ionic crystals. Chem Phys 101:23-40
Khachaturyan AG (1983) Theory of Structural Transformations in Solids. Wiley, New York, 574 p
Krivoglaz MA (1969) Theory of X-ray and thermal neutron scattering by real crystals. Plenum, New York, 405 p
Lee MR, Hodson ME, Parsons I (1998) The role of intragranular microtextures and microstructures in chemical and mechanical weathering: direct comparisons of experimentally and naturally weathered alkali feldspars. Geochim Cosmochim Acta 62:2771-2788
Locherer K, Hayward S, Hirst P, Chrosch J, Yeadon M, Abell JS, Salje EKH (1996) X-ray analysis of mesoscopic twin structures. Phil Trans Roy Soc Lond A 354:2815-2845
Locherer, K, Chrosch J, Salje EKH (1998) Diffuse X-ray scattering in WO_3. Phase Trans 67:51-63
Lüthi R, Haefke H, Meyer KP, Meyer E, Howald L, Günthesrodt HJ (1993) Surface and domain structures of ferroelastic crystals studied with scanning force microscopy. J Appl Phys 74:7461-7471
Marais SC, Heine V, Nex CMM, Salje EKH (1991) Phenomena due to strain coupling in phase transitions. Phys Rev Lett 66:2480-2483
Novak J, Salje EKH (1998a) Simulated mesoscopic structures of a domain wall in ferroelastic lattices. Eur Phys J B 4:279-284
Novak J, Salje EKH (1998b) Surface structure of domain walls. J Phys: Cond Matter 10:L359-L366
Parlinski K, Salje EKH and Heine V (1993a) Annealing of tweed microstructure in high-T_c superconductors studied by a computer simulation. Acta Met 41:839-847
Parlinski K, Heine V, Salje EKH (1993b) Origin of tweed texture in the simulation of a cuprate superconductor. J Phys: Condens Matter 5:497-518
Rabe KM, Waghmare UV (1996) Ferroelastic phase transitions from first principles. J Phys Chem Sol 57:1397-1403
Raterron P, Carpenter M, Doukhan JC (1999) Sillimanite mullitization: a TEM investigation and point defect model. Phase Trans 68:481-500
Roucau C, Tanaka M, Torres J, Ayroles R (1979) Etude en microscopie electronique de la structure liee aux proprietes ferroelastique du phosphat de plomb, $Pb_3(PO_4)_2$. J Micros Spectrosc Electron 4:603-612

Salje EKH (1985) Thermodynamics of sodium feldspar I: order parameter treatment and strain induced coupling effects. Phys Chem Min 12:93-98

Salje EKH, (1992) Application of Landau theory for the analysis of phase transitions in minerals. Phys Rep 215:49-99

Salje EKH (1993) Phase Transitions in Ferroelastic and Co-elastic Crystals, Student Edition. Cambridge University Press, Cambridge, UK, 276 p

Salje EKH (1995) A novel 7-circle diffractometer for the rapid analysis of extended defects in thin films, single crystals and ceramics. Phase Trans 55:37-56

Salje EKH, Devarajan V (1981) Potts model and phase transitions in lead phosphate $Pb_3(PO_4)_2$. J Phys C 14:L1029-L1035

Salje EKH, Hoppmann, G (1976) Direct observation of ferroelasticity in $Pb_3(PO_4)_2$- $Pb_3(VO_4)_2$. Mater Res Bull 11:1545-1550

Salje EKH, Ishibashi Y (1996) Mesoscopic structures in ferroelastic crystals: needletwins and right-angled domains. J Phys: Condens Matter 8:1-19

Salje EKH, Parlinski K (1991) Microstructures in the high-T_c superconductor. Superconductor. Science Tech 4:93-97

Salje EKH, Wruck B (1983) Specific-heat measurements and critical exponents of the ferroelastic phase transiton in $Pb_3(PO_4)_2$ and $Pb_3(P_{1-x}As_xO_4)_2$. Phys Rev B 28:6510-6518

Salje EKH, Devarajan V, Bismayer, U, Guimaraes DMC (1983) Phase transitions in $Pb_3(P_{1-x}As_xO_4)_2$: influence of the central peak and flip mode on the Raman scattering of hard modes. J Phys C 16:5233-52343

Salje EKH, Kuscholke B, Wruck B, Kroll H (1985a) Thermodynamics of sodium feldspar II: experimental results and numerical calculations. Phys Chem Min 12:99-107

Salje EKH, Kuscholke B, Wruck B, Kroll H (1985b) Domain wall formation in minerals: I. Theory of twin boundary shapes in Na-feldspar. Phys Chem Min 12:132-140

Salje EKH, Gallardo MC, Jimenez J, Romero FJ, del Cerro J (1998a) The cubic-tetragonal phase transition in strontium titanate: excess specific heat measurements and evidence for a near-tricritical, mean field type transition mechanism. J Phys: Condens Matter 25:5535-5545

Salje EKH, Buckley A, Van Tendeloo G, Ishibashi Y (1998b) Needle twins and right-angled twins in minerals: comparison between experiment and theory. Am Mineral 83:811-822

Saurenbach F, Terris BD (1990) Imaging of ferroelastic domain walls by atomic force microscopy. Appl Phys Lett 56:1703-1705

Shmytko, IM, Shekhtman VS, Ossipyan YA, Afonikova NS (1989) Twin structure and structure of twin boundaries in 1-2-3-$O_{7-\delta}$. Ferroelectrics 97:151-170

Smith JK, Mclaren AC, O'Donell RG (1987) Optical and electron microscopeinvestigation of temperature-dependent microstructures in anorthoclase. Can J Earth Sci 24:528-543

Stemmer S, Streiffer SK, Ernst F, Rühle M (1995) Atomistic structure of 90-degrees domain walls in ferroeleastic $PbTiO_3$ thin films. Phil Mag A71:713-742

Tsunekawa S, Hara K, Nishitani R, Kasuya A, Fukuda T (1995) Observation of ferroelastic domains in $LaNbO_4$ by atomic- force microscopy. Mat Trans JIM 36:1188-1191

Tung Hsu, Cowley JM (1994) Study of twinning with reflection electron-microscopy (REM). Ultra-microscopy 55:302-307

Tsatskis I, Salje EKH (1996) Time evolution of pericline twin domains in alkali feldspars: A computer-simulation study. Am Mineral 81:800-810

Wruck B, Salje EKH, Zhang M, Abraham T, Bismayer U (1994) On the thickness of ferroelastic twin walls in lead phosphate $Pb_3(PO_4)_2$: an X-ray diffraction study. Phase Trans 48:1-13

Yamamoto N, Yagi K, Honjo G (1977) Electron microscopic studies of ferroelastic and ferroelastic $Gd_2(MoO_4)_3$. Phys Status Solidi A41:523-527

4 High-Pressure Structural Phase Transitions

R. J. Angel

Bayerisches Geoinstitut
Universität Bayreuth
D95440 Bayreuth, Germany

INTRODUCTION

The study of structural phase transitions that occur as the result of the application of pressure is still in its long-drawn-out infancy. This is a direct result of the non-quenchability of structural phase transitions; their characterisation requires measurements of the material to be made *in situ* at high pressures. Although structural phase transitions could be detected by simple macroscopic compression measurements in piston-cylinder apparatus when the volume change arising from the transition was sufficiently large (e.g. calcite by Bridgman 1939, and spodumene by Vaidya et al. 1973) the limitations on sample access precluded their proper microscopic characterisation. The development in the 1970s of *in situ* diffraction methods, specifically the diamond-anvil cell (DAC) for X-ray diffraction and various clamp and gas-pressure cells for neutron diffraction, allowed both the structure of high-pressure phases and the evolution of the unit-cell parameters of both the high- and low-symmetry phases involved in a phase transition to be determined. Several classic studies of phase transitions at relatively low pressures were performed by high-pressure neutron diffraction methods in the 1970s and 1980s (e.g. Yelon et al. 1974, Decker et al. 1979), but the instrument time required for high-pressure studies (a result of limited sample access through, and attenuation by, the high-pressure cells as well as small sample sizes) limited the number of studies performed. In addition, until the advent of the Paris-Edinburgh pressure cell which can achieve pressures of up to 25 GPa (Besson et al. 1992, Klotz et al. 1998), pressure cells for neutron diffraction were limited to maximum pressures of the order of 2 to 4 GPa. In the past two decades the precision and accuracy of high-pressure X-ray diffraction methods has advanced considerably (see Hazen 2000). Single-crystal X-ray diffraction, which can be performed in the laboratory, is routine to pressures of 10 GPa and with care can be extended to 30 GPa or more (e.g. Li and Ashbahs 1998). Powder diffraction methods using synchrotron radiation can be performed to pressures in excess of 100 GPa (1 Megabar) at ambient temperature; for example the structural phase transition in stishovite near 50 GPa has been successfully characterised (Andrault et al. 1998). Much of this development has been driven by the need to measure equations of state of minerals, and to understand the mechanisms of compression of solids. But the experimental methodologies are equally suited to the study of structural phase transitions at high pressures; they therefore provide a rich opportunity for furthering our general understanding of the mechanisms of such transitions.

PRESSURE AND TEMPERATURE

The lattice parameter changes that occur on passing through a phase transition are subject to the same symmetry constraints independent of whether the transition occurs as a result of an isobaric change in temperature, or an isothermal change in pressure. Therefore the Landau approach of expanding the excess free-energy of the low-symmetry phase as a power series in the order parameter Q should be equally applicable to high-pressure phase transitions. The normal Landau expansion for the variation of the excess free energy with temperature in terms of a single order parameter is:

1529-6466/00/0039-0004$05.00

$$G_{ex} = \frac{a}{2}(T - T_{co})Q^2 + \frac{b}{4}Q^4 + \frac{c}{6}Q^6 + \lambda V_{ex}Q^2 + \frac{1}{2}KV_{ex}^2 \tag{1}$$

in which T_{co} is the transition temperature at room pressure. In addition to the direct terms in Q, the term $\lambda V_{ex}Q^2$ represents the lowest allowed direct coupling between the order parameter and the excess volume arising from the phase transition. The term $1/2\,KV_{ex}^2$ is an expression for the excess elastic energy of the low-symmetry phase arising from the volume changes associated with the spontaneous strains, and contains the bulk modulus K of the high-symmetry phase. This term is more correctly written as a summation over the individual elastic constants and components of the strain tensor (see Carpenter et al. 1998, Eqn. 18, for details), but the current form is suitable for purposes of illustration. For stability, the value of G_{ex} must be a minimum with respect to all of the quantities on the right-hand side of Equation (1). Differentiating (1) with respect to V_{ex}, we obtain the result that $V_{ex} = \frac{-\lambda}{K}Q^2$. At constant pressure the bulk modulus K remains approximately constant, and therefore V_{ex} remains proportional to Q^2.

Equation (1) strictly applies to a measurement at zero pressure. At elevated pressures there is an additional contribution to the excess free energy from the excess volume of $PV_{ex} = -\frac{\lambda P}{K}Q^2$. Introduction of these pressure terms into Equation (1) yields the expression for G_{ex} in an isobaric experiment performed at elevated pressure (see Fig. 1):

$$G_{ex} = \frac{a}{2}\left(T - \left(T_{co} + \frac{2\lambda P}{aK}\right)\right)Q^2 + \left(\frac{b}{4} - \frac{\lambda^2}{2K}\right)Q^4 + \frac{c}{6}Q^6 \tag{2}$$

Thus the effect of applying pressure is expected to be two-fold. First, the transition temperature is changed by $2\lambda P/aK$. Therefore the slope of the phase transition boundary $\partial T_c/\partial P_c$ is $2\lambda/aK$, and Equation (2) can be re-written:

$$G_{ex} = \frac{a}{2}\left(T - \left(T_{co} + \frac{\partial T_c}{\partial P_c}P\right)\right)Q^2 + \left(\frac{b}{4} - \frac{\lambda^2}{2K}\right)Q^4 + \frac{c}{6}Q^6 \tag{3}$$

Second, the character of the transition may change as a result of the additional contribution to the coefficient of Q^4. Note that this contribution is negative, so the application of pressure will generally drive transitions towards first-order behavior. Whether the renormalisation becomes stronger or weaker as pressure is increased depends on the balance between the strength of the coupling constant λ and the bulk modulus, both

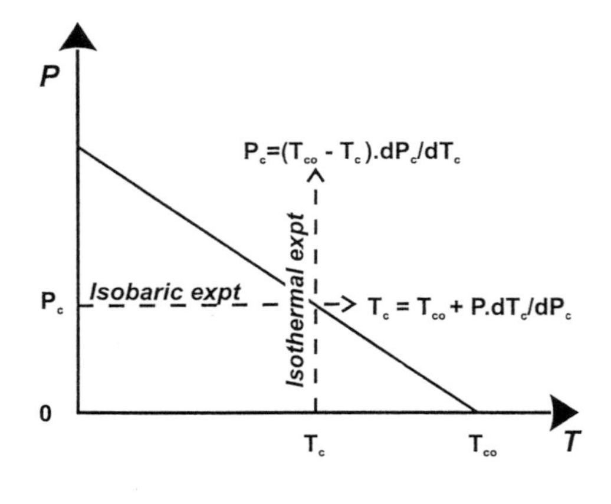

Figure 1. Schematic representation of isobaric and isothermal measurements of a phase transition with a negative slope, $\partial T_c/\partial P_c$, for the transition boundary. The same equations apply to transitions with positive slopes of the boundary.

of which are expected to increase with increasing pressure. The evolution with pressure of the free energy in an isothermal experiment at some temperature T (see Fig. 1) can now be derived from Equation (3) by noting that the transition pressure will be $P_c = (T_{co} - T)\partial P_c/\partial T_c$, thus:

$$G_{ex} = \frac{a}{2}\left(\frac{\partial T_c}{\partial P_c}\right)(P - P_c)Q^2 + \left(\frac{b}{4} - \frac{\lambda^2}{2K}\right)Q^4 + \frac{c}{6}Q^6 \qquad (4)$$

This simple derivation, ignoring the complexities introduced by considering the coupling of individual strain components with the elastic moduli, variations in elastic moduli with pressure and possible curvature of the phase boundary, shows that the evolution of the order parameter with pressure in an isothermal compression experiment should exhibit similar behavior to that observed as temperature varies. Therefore the experimentally observable quantities of spontaneous strain, super-lattice intensities and elasticity can also be expected to evolve in a manner similar to that found in temperature-dependent studies. The one proviso is that one expects some renormalisation of the coefficients of the excess free energy expansion as a result of the application of pressure, and this could result in a change in transition character. Equation (4) also shows that the elasticity and therefore the spontaneous strain not only provide a way of characterising a structural phase transition at high pressure, but their variation with pressure is intimately involved in determining the character of the transition. Therefore, in the rest of the chapter emphasis is placed upon the experimental measurement of these quantities, especially where the experimental methods differ significantly from those applied to studies at high temperatures and ambient pressures.

SPONTANEOUS STRAIN

Experimental methods

When the spontaneous strains are sufficiently large it is possible to measure them directly by measurement of the macroscopic deformation. For example, Battlog et al. (1984) measured the spontaneous strain accompanying the phase transition at 0.5 GPa in ReO_3 with a set of strain gauges mounted directly on the faces of a single-crystal sample within a large-volume high-pressure apparatus. Optical interferometry could, in principle, similarly be employed to measure the changes in dimensions of a single-crystal held in a DAC. Normally, however, spontaneous strains are obtained in a two-step process. First the unit-cell parameters of the phases on either side of the phase transition are determined by diffraction methods. Then the high-symmetry unit-cell parameters are extrapolated to the pressures at which cell parameters were measured for the low-symmetry phase, and the strains calculated.

Unit-cell parameters can be determined by a variety of diffraction methods, each of which have their advantages and disadvantages. Single-crystal diffraction has the advantage that the relative orientation of the unit-cells of the two phases can be unambiguously determined, provided one assumes that the crystal is not physically rotated during the phase transition. By contrast, as discussed by Palmer and Finger (1994), the relative orientations can only be inferred from powder diffraction data. On the other hand, phase transitions that give rise to twinning are best studied by powder diffraction, as are those strongly first-order transitions during which the volume change destroys single crystals.

Single-crystal X-ray diffraction with laboratory sources can yield the most precise high-pressure unit-cell parameter data, and modern DACs allow the routine attainment of pressures of 10 GPa, the hydrostatic limit of the 4:1 methanol:ethanol mixture commonly used as a pressure medium because of its ease of use. With other pressure media structural and lattice parameter data can be obtained to at least 30 GPa (Li and Ashbahs 1998) with laboratory sources. The higher pressure limit is not imposed by the DAC, but by the decreasing size of the sample chamber with increasing pressure, which leads to weaker diffraction signals in higher-pressure experiments. This difficulty will be overcome by the development of single-crystal high-pressure diffraction facilities on synchrotron beamlines.

Synchrotrons have been the X-ray source of choice for some time for high-pressure powder diffraction, which can be performed, with care, to pressures in excess of 100 GPa. Early work employed energy-dispersive diffraction, which is limited in resolution because of the restricted energy resolution of the solid-state detectors involved. The advent of image-plate detectors and improved diamond-anvil cell designs with larger opening angles to allow X-ray access to the sample has resulted in angle-dispersive diffraction becoming the standard method. The data quality from image plates is also greatly increased because the entire diffraction cone can be collected and therefore effects due to, for example, sample texture can be readily identified before integration of the data into a conventional 1-dimensional Intensity vs. 2θ data-set used for refinement. Currently, data quality is such that reliable unit-cell parameters can be obtained from high-pressure powder diffraction as well as structural data in more simple systems. The recent introduction of *in situ* read-out from image plates that allows data to be collected and processed on a ~1 minute cycle makes these detectors competitive with other area detectors such as CCD-based systems for real-time studies of phase transitions.

Neutron diffraction is the method of choice for studies of materials containing both light and heavy atoms. For precise studies up to 0.5 GPa there are a wide variety of gas-pressure cells suitable for both angle-dispersive and time-of-flight diffraction; the latter is especially suited for studies of phase transitions because the resolution is essentially independent of *d*-spacing (David 1992). For slightly higher pressures there are a variety of clamp cells, the latest developments of which can reach pressures of 3.5 GPa and temperatures in excess of 800 K (e.g. Knorr et al. 1997, 1999; for a general review see Miletich et al. 2000), and have been used successfully in studies of phase transitions (e.g. Rios et al. 1999). Scaled-up opposed-anvil cells equipped with sapphire anvils have been used to pressures of at least 3 GPa (e.g. Kuhs et al. 1996). For higher pressures there is the Paris-Edinburgh cell which is capable of developing pressures of up to 25 GPa (Besson et al. 1992, Klotz et al. 1998).

Because structural phase transitions are often ferroelastic or coelastic in character it is essential to have a well-defined stress applied to the crystal at high pressures. In effect, this means that a hydrostatic pressure medium must be used to enclose the crystal. A 4:1 mixture by volume of methanol:ethanol remains hydrostatic to just over 10 GPa (Eggert et al. 1992) and is convenient and suitable for many studies. If the sample dissolves in alcohols, then a mixture of pentane and iso-pentane which remains hydrostatic to ~6 GPa (Nomura et al. 1982), or a solidified gas such as N_2, He, or Ar can be employed. Water appears to remain hydrostatic to about 2.5 GPa at room temperature, just above the phase transition from ice-VI to ice-VII (Angel, unpublished data). The solid pressure media such as NaCl or KCl favoured by spectroscopists are very non-hydrostatic even at pressures below 1 GPa and have been shown to displace phase transitions by at least several kbar (e.g. Sowerby and Ross 1996). Similarly, the "fluorinert" material used in many neutron diffraction experiments because of its low neutron scattering power becomes significantly non-hydrostatic at ~1.3 GPa. Decker et al. (1979) showed that the ferroelastic phase transition that occurs at 1.8 GPa in lead phosphate under hydrostatic conditions is not observed up to 3.6 GPa when fluorinert was used as the pressure medium. At pressures in excess of the hydrostatic limit of the solidified gas and fluid pressure media, the non-hydrostatic stresses can be relaxed after each change in pressure by annealing the sample chamber, either by laser-heating or an external resistance furnace. For example, heating a cell in which the ethanol:methanol mixture is the pressure fluid to 150-200°C for about 1 hour is sufficient to relax the non-hydrostatic stresses developed above 10 GPa (Sinogeikin and Bass 1999). Such procedures not only remove the possibility of the transition being driven (or prevented) by the non-hydrostatic stresses, but also improves the signal-to-noise ratio in diffraction patterns as the diffraction maxima become sharper as the strain broadening is eliminated.

An important consideration in a high-pressure study of a structural phase transition is the method of pressure measurement. The ruby fluorescence method is commonly used to determine pressure in diamond-anvil cell measurements. It is based upon the observation that a pair of electronic transitions in the Cr^{3+} dopant atoms in Al_2O_3 change in energy as the Al_2O_3 lattice is compressed. The fluorescence in the red area of the optical spectrum is strong and easily excited by blue/green laser light, and the shift is quite large, approximately 3.6 Å/GPa. Unfortunately, the fluorescence wavelength is also very sensitive both to temperature, such that a 5° temperature change gives rise to a shift equivalent to 0.1 GPa (Barnett et al. 1973, Vos and Schouten 1991), and to the c/a ratio of the Al_2O_3 host lattice (Sharma and Gupta 1991). As a result, non-hydrostatic stresses increase the observed shift of the stronger R_1 component of the doublet, and can yield an apparent pressure that is higher than the true pressure (Gupta and Shen 1991, Chai and Brown 1996). Other fluorescence sensors have also been employed; for reviews see Holzapfel (1997) and Miletich et al. (2000). Measurement of optical fluorescence is relatively fast, and is extremely useful for setting the approximate pressure in a diamond-anvil cell prior to a diffraction measurement. With the proper precautions it can yield pressures as precise as 0.01 GPa, provided temperature fluctuations are completely excluded. In reality, these and other factors often mean that 0.03 GPa is a more realistic estimate of the precision. For more precise pressure determination internal diffraction standards can be used in DACs, while this is essential for completely enclosed cells, such as the Paris-Edinburgh cell. The pressure is then determined from the unit-cell volume of the standard and its equation of state (EoS). The precision in pressure then depends upon the precision of the volume measurement and the bulk modulus of the material; the softer the standard the more precise the pressure determination. Materials in common use as standards at pressures up to 10 GPa include NaCl (Brown 1999), quartz (Angel et al. 1997) and fluorite (Hazen and Finger 1981, Angel 1993), while metals such as gold (e.g. Heinz and Jeanloz 1984) have been used at higher pressures. It is important to note that there is no absolute pressure standard measurement above 2.5 GPa, so all EoS and all pressure scales are provisional and subject to revision in the light of improved calibrations. As an example, the pressure scale based upon the EoS of NaCl which was introduced by Decker (1971) and developed by Birch (1986) was recently shown to be in significant error by Brown (1999).

Fitting high-pressure lattice parameters

As noted by Carpenter et al. (1998) the non-linear variation of unit-cell parameters and volume (e.g. Fig. 2) with pressure makes the determination of the spontaneous strain a more complex and demanding process than for high-temperature phase transitions where the cell parameter variation with temperature can often be taken as linear. This non-linear behavior under pressure reflects the fact that as the volume of the solid becomes smaller the inter-atomic forces opposing further compression become stronger. The "stiffness" of a solid is characterised by the bulk modulus, defined as $K = -V\partial P/\partial V$ which will generally increase with increasing pressure. Different assumptions can then be made about how K varies with P, or how V varies with P leading to relationships between P and V known as

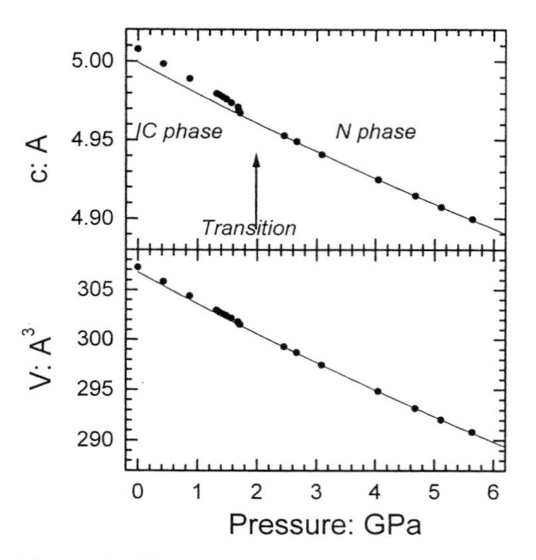

Figure 2. The evolution of the *c*-unit-cell parameter and the unit-cell volume of åkermanite through the incommensurate to normal phase transition at ~1.8 GPa (data from McConnell et al. 2000). Both the cell parameter and volume of the normal phase have been fitted by a Murnaghan EoS, indicated by the solid line. Symbol sizes significantly exceed experimental uncertainties.

"Equations of State" (see Anderson 1995 and Angel 2000 for general reviews). If a linear relationship is used to fit the volume against pressure, then the calculated strains will be incorrect (see Appendix).

For most studies of structural phase transitions at high pressures the simplest available EoS, introduced by Murnaghan (1937), provides a sufficiently accurate representation of the volume variation with pressure. It can be derived by assuming that the bulk modulus varies linearly with pressure, $K = K_0 + K_0'P$ with K_0' being a constant. Integration yields the P-V relationship:

$$V = V_0\left(1 + \frac{K_0'P}{K_0}\right)^{-1/K_0'} \tag{5}$$

in which V_0 is the zero pressure volume. The Murnaghan EoS reproduces both P-V data and yields correct values of the room pressure bulk modulus for compressions up to about 10% (i.e. $V/V_0 > 0.9$), and has the advantage of algebraic simplicity over other formulations such as the Vinet or Birch-Murnaghan EoSs (e.g. Anderson 1995, Angel 2000) which should be used if the range of compression is greater than 10%.

A complete practical guide to fitting both the Murnaghan and other EoS formulations by the method of least-squares is provided by Angel (2000). In addition to the cautions given there, attention must also be paid to data points close to the phase transition. If these display the effects of elastic softening, then their inclusion in the least-squares refinement will bias the resulting EoS parameters. If the high-symmetry phase is the high-pressure phase then the P-V data-set will not include a measurement of the room pressure volume, leading to strong correlations between the EoS parameters V_0, K_0 and K' which can cause instability in the least-squares refinement. This can be avoided by employing the self-similarity of all EoS which allows the pressure scale of the data to be changed by subtracting off a constant reference pressure, say P_{ref}, from all of the pressure values. Fitting the resulting $(P-P_{ref})$ vs. V data-set then yields as parameters the values of V, K and K' at the reference pressure, which can then be transformed to the true zero-pressure values if desired (see Appendix).

As for volume variations with pressure, there is no fundamental thermodynamic basis for specifying the form of cell parameter variations with pressure. It is therefore not unusual to find in the literature cell parameter variations with pressure fitted with a polynomial expression such as $a = a_0 + a_1P + a_2P^2$, even when the P-V data have been fitted with a proper EoS function. Use of polynomials in P is not only inconsistent, it is also unphysical in that a linear expression implies that the material does not become stiffer under pressure, while a quadratic form will have a positive coefficient for P^2, implying that at sufficiently high pressures the material will expand with increasing pressure. A consistent alternative is provided by using the same EoS as that used to fit the P-V data, but substituting the cube of the lattice parameter for the volume in the EoS. Thus for the Murnaghan EoS one obtains:

$$a = a_0\left(1+\frac{K_0'P}{K_0}\right)^{-1/3K'} \tag{6}$$

Note that the value of "*linear-*K_0" obtained from fitting the cell parameters in this way is related to the zero-pressure compressibility β_0 of the axis by $-1/3K_0 = \beta_0 = a_0^{-1}(\partial a/\partial P)_{P=0}$ in which a_0 is the length of the unit-cell axis at zero pressure. For crystals with higher than monoclinic symmetry the definition of the axial compressibilities in this way fully describes the evolution of the unit-cell with pressure because the tensor describing the strain arising from compression is constrained by symmetry from rotating. In the monoclinic system, however, one unit-cell angle may change, and in triclinic crystals all three unit-cell angles may change. The full description of the change in unit-cell shape in these cases must therefore include the full definition of the strain tensor resulting from compression. Fortunately, in monoclinic systems the strain tensor often does not rotate significantly with pressure. Then it may be appropriate to fit quantities such as $a\sin\beta$ against pressure with an EoS function such as Equation (6), or the β angle separately as a polynomial function of pressure (e.g. Angel et al. 1999). The important criterion for the purposes of studying phase transitions is that the resulting expressions provide not only a good fit to the data, but are reliable in extrapolation to the pressures at which data were obtained from the low-symmetry phase. The reliability of these extrapolations can always be tested by parallel calculations with different functions (e.g. Boffa-Ballaran et al. 2000). A further internal check on the robustness of the extrapolations can be obtained by comparing the unit-cell volumes obtained from the extrapolated lattice parameters with those predicted by the EoS function fitted to the unit-cell volume. Robustness is also maximised in systems in which a change in crystal system occurs at the phase transition by fitting the high-symmetry unit-cell parameters and performing the extrapolation *before* subsequently transforming the extrapolated values into the low-symmetry unit-cell setting (e.g. Angel and Bismayer 1999). A final problem is provided in systems in which the high-symmetry phase is the low-pressure phase and the phase transition is at low pressure. This restricts the precision with which the EoS parameters can be determined and often prevents independent determination of K_0 and K' for the volume and cell parameters of the high-symmetry phase. In such cases, K' might be fixed either at a value obtained from the EoS of a similar phase that is stable over a larger pressure range (e.g. Arlt and Angel 2000) or to the value obtained for the high-pressure phase if it has a very similar structure (e.g. Boffa-Ballaran et al. 2000). Either approach must be used with care and the resulting values of strains treated with caution, as fixing the EoS parameters to inappropriate values will obviously bias the resulting values of the spontaneous strain components. Nonetheless such procedures are to be preferred to resorting to linear fits of either cell-parameter or volume variation with pressure.

Calculating strains

Once the pressure dependencies of the unit-cell parameters of the high-symmetry phase have been determined, the values at the pressures of the data points collected from the low symmetry phase are obtained by extrapolation. In this context, note that negative values of pressure can be used in most EoS formalisms (see Appendix) if the high-symmetry data have been fitted with a shifted pressure scale. The components of the spontaneous strain tensor, and their separation into symmetry-breaking and non-symmetry-breaking components, can then be calculated from the extrapolated cell parameters of the high-symmetry phase and those of the low-symmetry phase in the same way as for transitions that occur at high temperatures. Additional examples are provided by Palmer and Finger (1994) and Angel et al. (1999).

What is more difficult is the estimation of the uncertainties of the resulting values of the strain components. These arise from three sources, the uncertainties in the measurement of the low-symmetry cell parameters, the uncertainty in the pressure of that measurement, and the uncertainties in the values of the extrapolated cell parameters of the high-symmetry phase. The variance $\mathbf{V}_{\mathbf{V,V}}$ (= square of the estimated uncertainty) of the extrapolated volume V_r of the high-symmetry phase is given by (see Angel 2000, Eqn. 15):

$$\mathrm{V_{V,V}} = \mathrm{V_{V_0,V_0}}\left(\frac{\partial V_r}{\partial V_0}\right)^2 + \mathrm{V_{K_0,K_0}}\left(\frac{\partial V_r}{\partial K_0}\right)^2 + \mathrm{V_{K_0',K_0'}}\left(\frac{\partial V_r}{\partial K_0'}\right)^2$$
$$+2\mathrm{V_{V_0,K_0}}\left(\frac{\partial V_r}{\partial V_0}\right)\left(\frac{\partial V_r}{\partial K_0}\right) + 2\mathrm{V_{V_0,K_0'}}\left(\frac{\partial V_r}{\partial V_0}\right)\left(\frac{\partial V_r}{\partial K_0'}\right) + 2\mathrm{V_{K_0,K_0'}}\left(\frac{\partial V_r}{\partial K_0}\right)\left(\frac{\partial V_r}{\partial K_0'}\right) \quad (7)$$

in which the other **V** are the components of the variance-covariance matrix of the least-squares fit of the EoS. The derivatives in Equation (7) are calculated from the equation of state. For the Murnaghan EoS:

$$\frac{\partial V_r}{\partial V_0} = \left(\frac{V_r}{V_0}\right), \quad \frac{\partial V_r}{\partial K_0} = \frac{PV_r}{K_0^2}\left(\frac{V_r}{V_0}\right)^{K_0'}, \quad \frac{\partial V_r}{\partial K_0'} = \frac{-V_r}{K_0'}\left(\ln\left(\frac{V_r}{V_0}\right) + \frac{P}{K_0}\left(\frac{V_r}{V_0}\right)^{K_0'}\right) \quad (8)$$

The evolution of $\mathrm{V_{V,V}^{1/2}}$, which is the uncertainty in the volume of the reference state, normally shows a minimum within the pressure range of the data collected from the high-symmetry phase, and increases rapidly outside of this pressure range (Fig. 3a).

The uncertainty in the pressure of the low-pressure datum can be considered as contributing an extra uncertainty to the value of the extrapolated volume of the high-pressure phase, as $\sigma' = \sigma_P(\partial V_r/\partial P) = \sigma_P V_r/K$. The effective uncertainty in the extrapolated volume then becomes $\sqrt{\sigma'^2 + \mathrm{V_{V,V}}}$, and the final uncertainty in the volume strain $V_S = V/V_r - 1$ is :

$$\sigma(V_S) = (1+V_S)\sqrt{\left(\frac{\sigma(V)}{V}\right)^2 + \frac{(\sigma'^2 + \mathrm{V_{V,V}})}{V_r^2}} \quad (9)$$

The calculation of the uncertainties of individual strain components follows the same procedures, except that the details of the error propagation through the equations defining the strain components may differ and, if Equation (5) is used to fit the data, V_r is replaced in Equation (8) with the cube of the lattice parameter. A worked example is provided in the Appendix. Consideration of the form of Equations (7), (8), and (9) suggests that in general the absolute uncertainties in the calculated strains will be smallest at the pressures closest to the phase transition because both the strains and $\mathbf{V}_{\mathbf{V,V}}$ will be smaller than at points further away (Fig. 3b).

This error propagation method can also be applied to the data points of the high-symmetry form for which the calculated strains should be zero. Measurement errors and imperfect fits of EoS functions will, in practise, contribute to small non-zero values being calculated for the strains. Deviations of more than 1 or 2 esd's from zero often indicates that the original high-symmetry unit-cell parameters were not fitted correctly

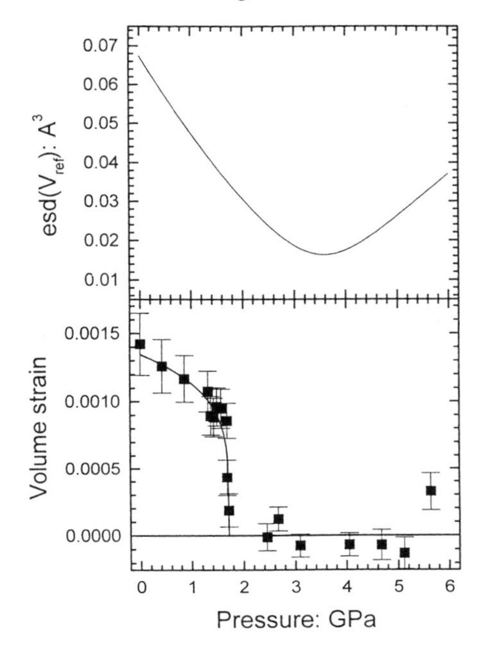

Figure 3. (a: top) The estimated standard deviation of the unit-cell volume of the N-phase of åkermanite calculated from the variance-covariance of the least-squares fit of the *P-V* data through Equation (7). (b: bottom) The evolution with pressure of the spontaneous volume strain in åkermanite. Error bars were calculated through Equation (9).

ELASTICITY

The relationship between the evolution of the elastic moduli of a material undergoing a structural phase transition and the order parameter is much more complex than for strains (e.g. Carpenter and Salje 1998). Nonetheless, as for spontaneous strains, the symmetry rules that govern the behavior of the components of the elastic stiffness tensor at high-pressure transitions remain the same as those for transitions occurring with temperature. Therefore the same experimental methods are, in principle, applicable to the measurement of elastic moduli at high pressures. In practice, the necessity for encapsulation of the sample in a high-pressure environment prevents the general application of measurement techniques, such as dynamical mechanical analysis (e.g. Schranz 1997) that rely on physical deformation of the sample. Similarly, resonance ultrasound spectroscopy, in which the elastic moduli are determined from the physical resonances of the sample at ultrasonic frequencies is limited to maximum pressures of a few hundred bars. Even at these pressures there is significant interaction between the sample and the surrounding gas which has to be accounted for in subsequent data analysis (e.g. Isaak et al. 1998, Oda and Suzuki 1999, Sorbello et al. 2000).

Spectroscopic techniques to measure elastic moduli are readily adaptable to use with diamond-anvil pressure cells with optical access to the sample provided through the diamond anvils themselves. Thus both Brillouin spectroscopy (e.g. Zha et al. 1998, Sinogeikin and Bass 1999) and Impulse Stimulated Phonon Spectroscopy (e.g. Brown et al. 1989, Abramson et al. 1997) have been used to measure elastic constants of single-crystals at high pressures, although no studies of structural phase transitions are known to the author. The major restrictions on transferring these techniques to the diamond-anvil cell is the same as for diffraction - that of access. But with careful experiment design and choice of crystal orientation, high-quality data can be collected. For lower-symmetry crystals it may be necessary to collect data from more than one crystal, each crystal of different orientation being loaded, and measured, separately in the DAC.

Ultrasonic interferometry, in which the travel time of high-frequency elastic waves through a sample is measured, also yields elastic moduli. Because it is a physical property measurement, rather than an optical spectroscopy, it can be used equally well on polycrystalline samples as single-crystals, although polycrystalline measurements only yield the bulk elastic properties, bulk modulus and shear modulus, G. High-pressure ultrasonic interferometry techniques were initially developed in the piston cylinder apparatus (e.g.

Niesler and Jackson 1989). They have been subsequently transferred to the multi-anvil press in which measurements can be made to pressures in excess of 10 GPa, both at room temperature and at elevated temperatures (for a review see Liebermann and Li 1998). Pressure determination is a major potential source of uncertainty in these measurements, because the response of a multi-anvil assembly to the imposed load is not elastic at room temperature, and varies greatly with increasing temperature. Pressure is therefore best determined by a combination of careful and repetitive calibrations against a combination of "fixed points" and measurements of the elasticity of a standard material at room temperature. For high-temperatures, pressure measurement is best performed by diffraction from a standard material such as NaCl included in the sample assembly. This requires the location of the multi-anvil press on a synchrotron beamline.

The diamond-anvil cell has a much smaller sample chamber than the cell assembly of a multi-anvil press—typically 100 μm thick and 200 μm diameter compared to 2-3 mm in the multi-anvil press. The wavelengths of the elastic waves induced in the sample by the frequencies of around 30 MHz employed in conventional ultrasonic measurements are therefore too long for a sample in a DAC. This difficulty has recently been overcome by the development of transducers and equipment that operate in the GHz frequency regime, and can be interfaced with a DAC (Spetzler et al. 1996). Successful measurements of *p*-wave velocities on single-crystals of MgO to pressures in excess of 5 GPa at room temperature have been demonstrated (Reichmann et al. 1998) and extension to high temperatures (Shen et al. 1998) as well as *s*-wave measurements is underway (Spetzler et al. 1999). One disadvantage of GHz ultrasonic interferometry is that the higher frequencies require a far higher quality of sample preparation; surfaces have to be polished such that they bond perfectly, without adhesive, to the surface of the diamond anvil. The advantage of using the DAC for ultrasonic interferometry is that all of the pressure measurement techniques conventionally used for DACs such as diffraction and optical fluorescence can be employed.

All of the fore-going techniques to measure the elasticity of materials actually determine the elastic or phonon wave velocities, both of compressional (V_p for *p*-waves) and transverse waves (V_s for *s*-waves) in the sample. The bulk and shear moduli of a polycrystalline sample are given by:

$$K = \rho\left(V_p^2 - \frac{4}{3}V_s^2\right) \quad \text{and} \quad G = \rho V_s^2 \tag{10}$$

while for a single-crystal, they are determined through the Cristoffel relation:

$$\det\left|c_{ijkl}l_i l_j - \rho V \delta_{ik}\right| = 0 \tag{11}$$

in which the l_i are the direction cosines of the direction in which the velocity is measured, and c_{ijkl} is the elastic stiffness tensor. In order to determine the components of c_{ijkl} it is therefore necessary to know the density, ρ, of the sample at the pressure of the measurement. For ultrasonic measurements, the direct experimental measurement is of travel times, so calculation of the velocities also requires the length of the sample to be known. Both can either be determined directly by diffraction or through a self-consistent calculation known as Cook's method (Cook 1957, and discussion in Kung et al. 2000). However, the application of this method normally involves describing the evolution of the density, or equivalently volume, of the sample by an equation of state. It is therefore *not* generally valid in the neighbourhood of a phase transition when elastic softening occurs, although careful application of the analysis to closely-spaced data could yield correct results. For phase transitions it is probably therefore safer to rely on direct density measurements by diffraction in order to obtain elastic moduli from wave velocity measurements.

Diffraction at high pressure also provides an opportunity to measure some combinations of elastic moduli directly, because the pressure is a stress which results in a strain that is expressed as a change in the unit cell parameters. The compressibility of any direction in the crystal is directly related to the components of the elastic compliance tensor by:

$$\frac{1}{a_i}\left(\frac{\partial a_i}{\partial P}\right) = \beta_l = s_{ijkk} l_i l_j \tag{13}$$

in which l_i are the direction cosines of the axis of interest (see Nye 1957 for details of this derivation). Expansion of the terms on the right-hand side of Equation (13) yields the relationships between the changes in unit-cell parameters with pressure and the elastic compliances. Thus, for orthorhombic crystals, the direction cosines l_i are zero and unity for the unit-cell edges and the compressibilities become (from Nye 1957):

$$\beta_a = (s_{11} + s_{12} + s_{13}), \quad \beta_b = (s_{12} + s_{22} + s_{23}), \quad \beta_c = (s_{13} + s_{23} + s_{33}) \tag{14}$$

and for uniaxial crystal systems:

$$\beta_a = (s_{11} + s_{12} + s_{13}), \quad \beta_c = (2s_{13} + s_{33}) \tag{15}$$

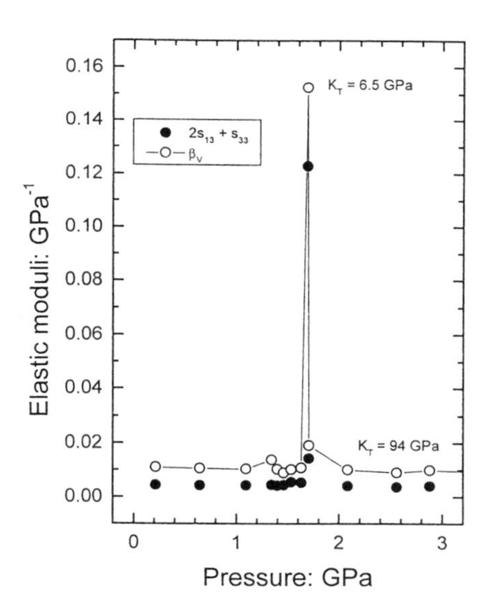

Figure 4. Evolution of the combined elastic modulus $(2s_{13} + s_{33})$ and bulk compressibility of åkermanite, showing significant static softening close to the phase transition. Values of $(2s_{13} + s_{33})$ were calculated from consecutive data points as $c^{-1}(\partial c/\partial P) = 2(c_{i+1} + c_i)^{-1}(c_{i+1} - c_i)(P_{i+1} - P_i)^{-1}$, and values of β in an analogous fashion from the volume data.

The linear compressibility of a cubic crystal is independent of direction in the crystal and equal to $(s_{11} + 2s_{12})$. If lattice parameter measurements are made at sufficiently closely-spaced intervals in pressure, then it is possible to demonstrate elastic softening in the neighborhood of a high-pressure phase transition simply by calculating the derivative in Equation (13) as the difference between consecutive data points (Fig. 4). The same method applied to the unit-cell volume will yield the local value of the bulk modulus as $K = -V\,\partial P/\partial V$. Note that this method provides the *static* and therefore *isothermal* elastic compliances, whereas the other measurement methods discussed previously operate at sufficiently high frequencies that they yield the *adiabatic* compliances. While the relationship between the two quantities is well defined, for example for the bulk modulus $K_S = (1 + \alpha\gamma T)K_T$, in the neighborhood of a structural phase transition both the thermal expansion coefficient α and the Gruneisen parameter γ may diverge from the "background" values due to either intrinsic frequency dispersion or to a finite response time of the order parameter to the applied stresses (e.g. Carpenter and Salje 1998). The net result can be strong frequency dispersion of the elastic constants close to a phase transition as pressure phase transition in KSCN by Schranz and Havlik (1994), with the static and low-frequency measurements exhibiting more elastic softening than those performed with high-frequency techniques such as ultrasonic interferometry.

OTHER TECHNIQUES

A large number of other experimental techniques can be, and have been, applied to the study of phase transitions at high pressure. For optical spectroscopies, including infra-red, Raman and optical absorption, the problems of access to a sample held within a DAC are far less severe than for diffraction. The significant experimental problems include reproducibility of the positioning of the cell within the spectrometer as many DACs require removal in order to change pressure. Secondly, any quantitative analysis of spectra collected from DACs which requires correction for either thickness of the sample or the optical properties of the diamond anvils, can be problematic. The spacing between the diamond anvils can be obtained by observing the interference fringes that arise from multiple reflections between the culet faces (e.g. Osland 1985). Strain-induced birefringence is observable in almost all diamond anvils under load, and is especially a problem for high-pressure optical birefringence studies. The solution adopted by Wood et al. (1980) for optical birefringence measurements of the high-pressure phase transition in $BiVO_4$ is probably the best approach. They first pressurised the DAC and then collected data while the temperature of the DAC was scanned. When used in this way the strain birefringence of the anvils does not change rapidly, and the background signal from this source does not affect the determination of the optical birefringence from the sample.

Phase transitions can also be directly characterised by following the evolution of the structure of the sample as it approaches the phase transition. This normally requires full structural data obtained through collection of intensity datasets by either X-ray or neutron diffraction. The experimental methodologies and the relative advantages and disadvantages of the various techniques available for high-pressure diffraction have recently been thoroughly reviewed (Hazen 2000) and therefore do not require further presentation here. It is sufficient to state here that the difficulties of access to the sample mean that the quality of the intensity data obtained from high-pressure diffraction experiments, and therefore the resulting structure refinements, are usually of lower quality than those obtained through the same diffraction technique at ambient pressures. However, by careful experiment design, useful details of the evolution of a structure towards a phase transition can be obtained. One classic example is provided by the high-pressure transition in ReO_3 in which the corner-linked ReO_6 octahedra tilt. A neutron powder diffraction study by Jorgensen et al (1986) showed that octahedra remained essentially rigid in the low-symmetry phase, and that their angle of tilt evolved as $\phi \propto (P - P_c)^{0.322(5)}$, even up to large tilt angles in excess of 14°. Combination of this direct measurement of the order parameter of the transition with the observation by Battlog et al. (1984) that the excess volume associated with the transition evolves as $V_{ex} \propto (P - P_c)^{2/3}$ provides an experimental confirmation of the general relationship $V_{ex} \propto Q^2$.

ACKNOWLEDGMENTS

I thank Nancy Ross, Ulrich Bismayer, Tiziana Boffa-Ballaran, Jennifer Kung and Ronald Miletich for various collaborations, all of which contributed to this chapter. John Loveday kindly provided material on neutron diffraction at high pressures, and Don Isaak material on RUS. In addition Tiziana Boffa-Ballaran, Michael Carpenter, Simon Redfern, and Ekhard Salje made helpful comments and suggestions for improvement of the original manuscript.

APPENDIX

The process of data reduction to obtain spontaneous strains from measurements of unit-cell parameters is illustrated with the example of the phase transition that occurs in lead phosphate, $Pb_3(PO_4)_2$ at a pressure of ~1.8 GPa and room temperature (Decker et al. 1979, Angel and Bismayer 1999). The unit-cell parameter data used in this appendix are taken from Angel and Bismayer (1999) and are reproduced in Table A1 and illustrated in Figure A1. The high-pressure phase has trigonal symmetry and the low-pressure phase has monoclinic symmetry.

Table A1. Cell parameters of lead phosphate from Angel and Bismayer (1999).

	P, GPa	**a**, Å	**b**, Å	**c**, Å	β, °	**V**, Å^3
P0	0	13.80639(53)	5.69462(21)	9.42700(29)	102.366(3)	723.976(44)
P1	0.146(5)	13.80328(165)	5.67590(63)	9.42867(80)	102.437(10)	721.363(129)
P2	0.482(4)	13.79174(70)	5.63025(31)	9.43315(37)	102.646(4)	714.725(61)
P3	0.819(5)	13.78379(52)	5.58805(21)	9.44176(27)	102.820(3)	709.119(43)
P4	1.031(4)	13.77606(81)	5.56093(34)	9.44648(39)	102.916(5)	705.365(67)
P5	1.154(4)	13.77218(98)	5.54663(38)	9.44877(46)	102.960(6)	703.399(77)
P6	1.249(5)	13.77005(93)	5.53621(36)	9.45117(42)	103.003(5)	702.025(73)
P7	1.543(4)	13.76780(197)	5.49955(64)	9.46639(84)	103.165(1)	697.926(136)
P9	1.816(4)	5.46756(23)		20.09397(75)		520.216(48)
P10	3.021(6)	5.43218(17)		20.00802(57)		511.308(36)
P11	4.086(5)	5.40706(21)		19.94177(70)		504.914(43)
P12	4.710(7)	5.39361(25)		19.90820(89)		501.599(52)
P13	5.981(7)	5.36840(26)		19.84549(84)		495.315(52)

Note: Numbers in parentheses are estimated standard deviations in the last digit(s).

Fitting the high-symmetry data

A fit of the five volume data from the high-symmetry phase of lead phosphate with the Murnaghan EoS (Eqn. 5) yields the parameters listed in Table A2. In this and subsequent fits weights were assigned to each data point based on the estimated uncertainties in *both* V and P, according to the effective variance method (Orear 1982, and Eqn. 12 of Angel 2000). To illustrate the self-similarity of the Murnaghan EoS, the data can also be fitted after 1.7 GPa has been subtracted from each pressure datum. This second fit yields refined parameters V(1.7 GPa) = 521.14(11) Å^3, K(1.7 GPa) = 62.8(1.2) GPa and K' (1.7 GPa) = 11.12(65). These are the same values that would be predicted at this pressure by the EoS based on fitting the data on the original pressure scale, thus:

$$V(1.7\text{ GPa}) = 538.19\left(1+\frac{11.12\text{x}1.7}{43.92}\right)^{-1/11.12} = 521.14\text{A}^3$$

$$K(1.7\text{GPa}) = 43.92 + 11.12\text{x}1.7 = 62.8\text{GPa} \qquad \text{(A1)}$$

In the Murnaghan EoS, K' remains invariant with pressure. Note that a linear fit to the data yields a smaller V_0 ~ 530.8 Å^3 than the Murnaghan EoS and would, in this case, result in an overestimate of the volume strain in the low-symmetry phase.

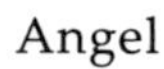

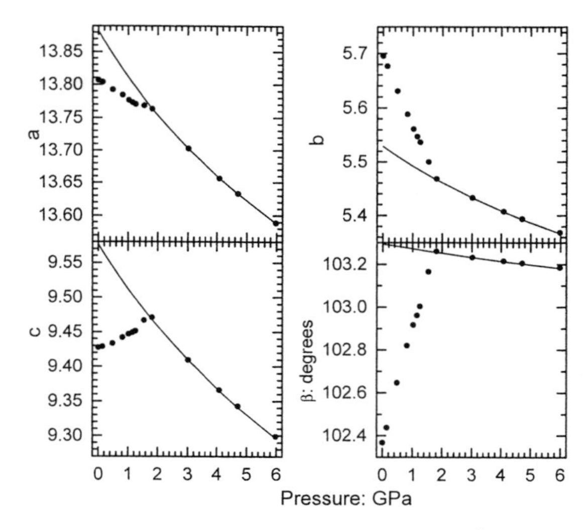

Figure A1. Evolution of the unit-cell parameters of lead phosphate with pressure, redrawn from Angel and Bismayer (1999). The trigonal unit-cell parameters of the high-pressure phase have been transformed to the monoclinic setting through Equation (A3). The lines are the Murnaghan EoS (Table A2) fitted to the trigonal unit-cell parameters and transformed to the monoclinic cell setting.

Table A2. Parameters of the Murnaghan EoS fitted to the trigonal unit-cell data..

	a_0 or V_0	K_0	K'		V_0	K_0	K'
a	5.5380(35)	38.56(2.69)	10.15(78)	V_0	0.10513	-.86085	0.23763
				K_0		7.22491	-2.05079
				K'			0.60338
c	20.2655(54)	59.34(2.58)	14.00(73)	V_0	44.9057	-17.0599	4.66817
				K_0		6.64621	-1.86947
				K'			0.54454
V	538.19(68)	43.93(2.26)	11.12(65)	V_0	0.45574	-1.50957	0.41587
				K_0		5.12896	-1.45300
				K'			0.42631

Note: The arrays on the right-hand side are the components of the variance-covariance matrix from each least-squares refinement. For the two cell parameters, the matrix entries are for the refinement of the cell parameter *cubed.*

For other EoS the expressions for K and K' at high pressures are more complex. In addition, some published expressions for these parameters in the Birch-Murnaghan EoS are either incorrect, or truncated at too low an order in the finite strain. The complete expressions for a 3rd-order Birch-Murnaghan EoS are:

$$f_E = \left[(V_0/V)^{2/3} - 1\right]/2$$

$$P = 3K_0 f_E (1+2f_E)^{5/2}\left(1+\frac{3}{2}(K_0'-4)f_E\right)$$

$$K = K_0\left(1+2f_E\right)^{5/2}\left(1+\left(3K_0'-5\right)f_E+\frac{27}{2}\left(K_0'-4\right)f_E^2\right)$$

$$K' = \frac{K_0}{K}\left(1+2f_E\right)^{5/2}\left(K_0'+\left(16K_0'-\frac{143}{3}\right)f_E+\frac{81}{2}\left(K_0'-4\right)f_E^2\right) \quad \text{(A2)}$$

These are equivalent to the expressions given by Birch (1986) in his Appendix 1, and by Anderson (1995) in his Equations (6.52) to (6.55), except for a typographical error of K' for K'' in his Equation (6.53). The expressions given by Stacey et al. (1981), for example, are correct except that for K', which is truncated at f_E rather than after the f_E^2 which is required for the expression to be exact. Expressions for the 2nd-order Birch-Murnaghan EoS can be obtained by setting $K_0' = 4$ in all of the above.

To obtain the components of the spontaneous strain tensor of lead phosphate it is also necessary to extrapolate the unit-cell parameters of the high-symmetry phase. The *a* and *c* parameters can either be fitted with Equation (6), or be first cubed and fitted with Equation (5). The latter is chosen in order to make the subsequent calculations of esd's more straightforward. The resulting parameters are also listed in Table A2.

Table A3. Calculated trigonal unit-cell parameters for lead phosphate.

	P, GPa	**a**, Å	**c**, Å	**V**, Å^3
P0	0	5.5380(35)	20.2655(54)	538.19(67)
P1	0.146(5)	5.5312(31)	20.2491(48)	536.48(59)
P2	0.482(4)	5.5163(22)	20.2136(35)	532.65(42)
P3	0.819(5)	5.5026(16)	20.1804(25)	529.15(30)
P4	1.031(4)	5.4945(12)	20.1607(20)	527.08(24)
P5	1.154(4)	5.4900(11)	20.1496(18)	525.93(21)
P6	1.249(5)	5.4866(10)	20.1412(16)	525.06(19)
P7	1.543(4)	5.4764(7)	20.1162(11)	522.46(13)
P9	1.816(4)	5.4674(5)	20.0941(9)	520.19(10)
P10	3.021(6)	5.4326(4)	20.0075(6)	511.39(7)
P11	4.086(5)	5.4068(3)	19.9423(6)	504.89(7)
P12	4.710(7)	5.3933(3)	19.9081(6)	501.50(6)
P13	5.981(7)	5.3687(5)	19.8452(9)	495.37(10)

Strain calculation

The cell parameters of the trigonal phase can now be calculated at each pressure for which the monoclinic unit-cell was measured. The estimated uncertainties given in Table A3 were obtained from the components of the variance-covariance matrices of the fits (Table A2) through Equations (7) and (8). Note that for the unit-cell parameters, the variance-covariance matrix used in Equation (7) is that of the fit of the *cubes* of the unit-cell parameters, yielding estimates of the uncertainty of the *cubes* of extrapolated unit- cell parameters. The uncertainty in a cell parameter is then given by

$$\sigma(a) = \sigma\left(a^3\right)/3a^2 .$$

The principles of calculation for other EoS formulations are the same, but the expressions for the derivatives used in Equation (7) will be more complex.

The next step is to define the relationship between the unit-cells of the trigonal and monoclinic phases. By inspection of Figure A2 it can be seen that for lead phosphate, in the absence of symmetry-breaking strains, the unit-cell parameters of the monoclinic phase are:

$$a_m = \sqrt{\frac{1}{3}a_t^2 + \frac{4}{9}c_t^2}, \qquad b_m = a_t, \qquad c_m = a_t\sqrt{3}, \qquad \sin\beta_m = \frac{2c_t}{3a_m} \tag{A3}$$

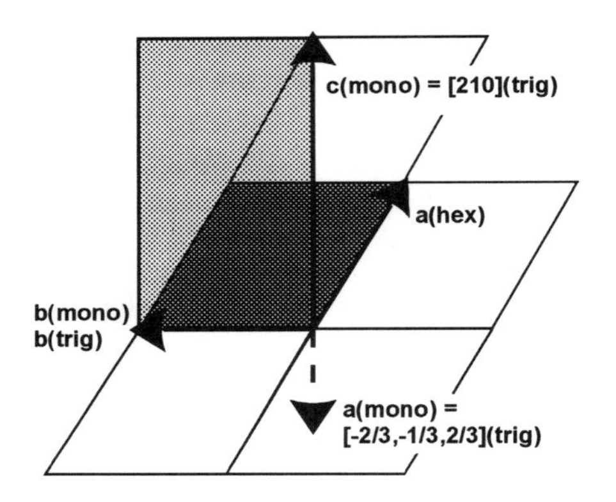

Figure A2. The relationship between the trigonal unit-cell (darker shading) and the monoclinic unit-cell (paler shading) projected down the trigonal *c*-axis.

For calculation of the spontaneous strain components, a Cartesian co-ordinate system must be defined. For convenience we choose the reference system of Guimares (1979) with *i* parallel to the trigonal *c*-axis, *j* parallel to the trigonal *b*-axis, and *k* parallel to the [210] direction in the trigonal unit-cell (Fig. A2). From Equations (43)-(48) of Carpenter et al. (1988), we can now write down the expressions for the spontaneous strain components in terms of the observed monoclinic unit-cell parameters (here without subscript) and the extrapolated trigonal unit-cell parameters transformed into the monoclinic setting (here subscripted with *o* to indicate their use as the reference state):

$$\varepsilon_{11} = a\sin\beta/a_o\sin\beta_o - 1 \qquad \varepsilon_{22} = b/b_o - 1 \qquad \varepsilon_{33} = c/c_o - 1$$

$$\varepsilon_{13} = \frac{1}{2}\left[\frac{a\cos\beta}{a_o\sin\beta_o} - \frac{c\cos\beta_o}{c_o\sin\beta_o}\right] \qquad \varepsilon_{12} = \varepsilon_{23} = 0 \tag{A4}$$

Note the exchange of the expressions for ε_{11} and ε_{33} from Carpenter et al. (1998) arising from our different choice of Cartesian system. The strain component ε_{11} is an expansion of the trigonal *c*-axis and is therefore a purely non-symmetry-breaking strain. In the directions perpendicular to this axis the strain components ε_{22} and ε_{33} can contain both symmetry-breaking and non-symmetry-breaking contributions. The 3-fold symmetry axis of the trigonal cell means the non-symmetry-breaking components must be equal, thus $\varepsilon_{22,nsb} = \varepsilon_{33,nsb}$, and therefore $\varepsilon_{22,sb} = -\varepsilon_{33,sb}$. The ε_{13} component, being a shear, is purely symmetry-breaking. Introduction of these relationships, and the substitution of the expressions in Equation (A3) then leads to the final expressions for the non-zero spontaneous strain components:

$$\varepsilon_{11,nsb} = 3a\sin\beta/2c_t - 1 \qquad \varepsilon_{22,nsb} = \varepsilon_{33,nsb} = \left(b + c/\sqrt{3}\right)/2a_t - 1$$

$$\varepsilon_{13,sb} = \left(c + 3a\cos\beta\right)/4c_t \qquad \varepsilon_{22,sb} = -\varepsilon_{33,sb} = \left(b - c/\sqrt{3}\right)/2a_t \tag{A5}$$

Table A4. Spontaneous strain components of lead phosphate.

	P, GPa	ε_{11}	$\varepsilon_{22,sb}$	$\varepsilon_{22,nsb}$	ε_{13}
P0	0	-0.00180(28)	0.02275(3)	0.02275(64)	0.00687(1)
P1	0.146(5)	-0.00148(30)	0.02100(7)	0.02100(56)	0.00630(3)
P2	0.482(4)	-0.00138(19)	0.01668(3)	0.01668(40)	0.00464(1)
P3	0.819(5)	-0.00100(14)	0.01243(2)	0.01243929)	0.00330(1)
P4	1.031(4)	-0.00096(14)	0.00974(4)	0.00974(23)	0.00259(2)
P5	1.154(4)	-0.00087(14)	0.00832(4)	0.00832(20)	0.00227(2)
P6	1.249(5)	-0.00078(13)	0.00725(4)	0.00725(18)	0.00194(2)
P7	1.543(4)	-0.00036(23)	0.00312(7)	0.00312(15)	0.00074(4)
P9	1.816(4)	-0.00000(6)		0.00003(10)	
P10	3.021(6)	0.00003(4)		-0.00008(7)	
P11	4.086(5)	-0.00003(4)		0.00005(7)	
P12	4.710(7)	0.00000(5)		0.00006(7)	
P13	5.981(7)	0.00001(6)		-0.00005(11)	

Calculated values are given in Table A4. The uncertainty estimates are obtained by standard error propagation through these equations of the uncertainties in the monoclinic unit-cell parameters and the uncertainties in the extrapolated values of the trigonal unit-cell parameters (Table A3). Thus, for example,

$$\sigma(\varepsilon_{22,nsb}) = (1+\varepsilon_{22,nsb})\sqrt{\left(\frac{\sigma(a_t)}{a_t}\right)^2 + \frac{\sigma^2(b)+\sigma^2(c)/3}{\left(b-c/\sqrt{3}\right)^2}}$$

$$\sigma(\varepsilon_{22,sb}) = (\varepsilon_{22,sb})\sqrt{\left(\frac{\sigma(a_t)}{a_t}\right)^2 + \frac{\sigma^2(b)+\sigma^2(c)/3}{\left(b+c/\sqrt{3}\right)^2}} \qquad \text{(A6)}$$

Note that the estimated uncertainties of the non-symmetry-breaking strain components are larger than those of the symmetry-breaking components, largely because of the different pre-multipliers in Equation (A6). A further indication of the uncertainties associated with the strain calculations can be obtained by calculating the strain components in the stability field of the high-symmetry phase. In this case, there are no symmetry-breaking strains, and the non-symmetry-breaking strain components become simply $\varepsilon_{11} = c_{obs}/c_{cal} - 1$ and $\varepsilon_{33} = a_{obs}/a_{cal} - 1$ from Equation (A5) in which the c_{obs} is the measured value and the c_{cal} is that calculated from the EoS. The calculated values for lead phosphate are all less than the associated estimated uncertainties (Table A4) indicating that the EoS fits and the error estimates are probably valid.

Once the components of the spontaneous strain tensor and their uncertainties have been calculated, they can be analysed by standard methods. In the example of lead phosphate used here the strain components all vary linearly with pressure (Fig. A3) indicating that the transition is second order, and the temperature-dependent behavior has been renormalised by the application of high pressure (Angel and Bismayer 1999).

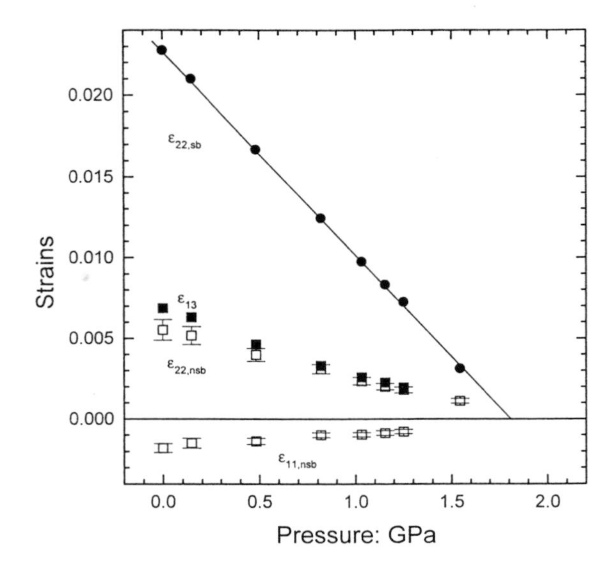

Figure A3. Variation of the spontaneous strain components with pressure. Symbol sizes for the symmetry-breaking components (solid symbols) are larger than their esd's (Table A4).

REFERENCES

Abramson EH, Brown JM, Slutsky LJ, Zaug J (1997) The elastic constants of San Carlos olivine to 17 GPa. J Geophys Res 102:12253-12263

Anderson OL (1995) Equations of State of Solids for Geophysics and Ceramic Science. Oxford University Press, Oxford

Andrault D, Fiquet G, Guyot F, Hanfland M (1998) Pressure-induced Landau-type transition in stishovite. Science 282:720-724

Angel RJ (1993) The high-pressure, high-temperature equation of state of calcium fluoride, CaF_2. J Phys: Cond Matter 5:L141-L144

Angel RJ (2000) Equations of state. *In* Hazen RM (ed) High-Temperature–High-Pressure Crystal Chemistry. Reviews in Mineralogy, Vol 40 (in press)

Angel RJ, Bismayer U (1999) Renormalization of the phase transition in lead phosphate, $Pb_3(PO_4)_2$, by high-pressure: lattice parameters and spontaneous strain. Acta Crystallogr B55:896-901

Angel RJ, Allan DR, Miletich R, Finger LW (1997) The use of quartz as an internal pressure standard in high-pressure crystallography. J Appl Crystallogr 30:461-466

Angel RJ, Kunz M, Miletich R, Woodland AB, Koch M, Xirouchakis D (1999) High-pressure phase transition in $CaTiOSiO_4$ titanite. Phase Trans 68:533-543

Arlt T, Angel RJ (2000) Displacive phase transitions in C-centred clinopyroxenes: Spodumene, $LiScSi_2O_6$ and $ZnSiO_3$. Phys Chem Minerals (in press)

Barnett JD, Block S, Piermarini GJ (1973) An optical fluorescence system for quantitative pressure measurement in the diamond-anvil cell. Rev Sci Instrum 44:1-9

Batlogg B, Maines RG, Greenblatt M, DiGregorio S (1984) Novel p-V relationship in ReO_3 under pressure. Phys Rev B 29:3762-3764

Besson JM, Nelmes RJ, Hamel G, Loveday JS, Weill G, Hull S (1992) Neutron diffraction above 10 GPa. Physica B 180 & 181:907-910

Birch F (1986) Equation of state and thermodynamic parameters of NaCl to 300 kbar in the high-temperature domain. J Geophys Res 91:4949-4954

Boffa-Ballaran T, Angel RJ, Carpenter MA (2000) High-pressure transformation behaviour of the cummingtonite-grunerite solid solution. Eur J Mineral (in press)

Bridgman PJ (1939) The high pressure behavior of miscellaneous minerals. Am J Sci 237:7-18

Brown, JM (1999) The NaCl pressure standard. J Appl Phys 86:5801-5808

Brown JM, Slutsky LJ, Nelson KA, Cheng L-T (1989) Single-crystal elastic constants for San Carlos Peridot: An application of impulsive stimulated scattering. J Geophys Res B94:9485-9492

Carpenter MA, Salje EKH, Graeme-Barber A (1998) Spontaneous strain as a determinant of thermodynamic properties for phase transitions in minerals. Eur J Mineral 10:621-692

Carpenter MA, Salje EKH, (1998) Elastic anomalies in minerals due to structural transitions. Eur J Mineral 10:693-812
Chai M, Brown JM (1996) Effects of non-hydrostatic stress on the R lines of ruby single crystals. Geophys Res Letts 23:3539-3542
Cook RK (1957) Variation of elastic constants and static strains with hydrostatic pressure: a method for calculation from ultrasonic measurements. J Acoust Soc Am 29:445-449
David WIF (1992) Transformations in neutron powder diffraction. Physica B 180 & 181:567-574
Decker DL (1971) High-pressure equations of state for NaCl, KCl, and CsCl. J Appl Phys 42:3239-3244
Decker DL, Petersen S, Debray D, Lambert M (1979) Pressure-induced ferroelastic phase transition in $Pb_3(PO_4)_2$: A neutron diffraction study. Phys Rev B19:3552-3555
Eggert JH, Xu L-W, Che R-Z, Chen L-C, Wang J-F (1992) High-pressure refractive index measurements of 4:1 methanol:ethanol. J Appl Phys 72:2453-2461
Guimaraes DMC (1979) Temperature dependence of lattice parameters and spontaneous strain in $Pb_3(PO_4)_2$. Phase Trans 1:143-154
Gupta YM, Shen XA (1991) Potential use of the ruby R_2 line shift for static high-pressure calibration. Appl Phys Letts 58:583-585
Hazen RM, editor (2000) High-Temperature–High-Pressure Crystal Chemistry. Reviews in Mineralogy, Vol 40 (in press)
Hazen RM, Finger LW (1981) Calcium fluoride as an internal pressure standard in high-pressure/high-temperature crystallography. J Appl Crystallogr 14:234-236
Heinz DL, Jeanloz R (1984) The equation of state of the gold calibration standard. J Appl Phys 55:885-893
Holzapfel WB (1997) Pressure determination. *In* High-Pressure Techniques in Chemistry and Physics. WB Holzapfel, NS Isaacs (eds) Oxford University Press, Oxford, p 47-55
Isaak DG, Carnes JD, Anderson OL, Oda H (1998) Elasticity of fused silica spheres under pressure using resonant ultrasound spectroscopy. J Acoust Soc Am 104:2200-2206
Jorgensen J-E, Jorgensen JD, Batlogg B, Remeika JP, Axe JD (1986) Order parameter and critical exponent for the pressure-induced phase transitions in ReO_3. Phys Rev B 33:4793-4798
Klotz S, Besson JM, Hamel G, Nelmes RJ, Marshall WG, Loveday JS, Braden M (1998) Rev High Pressure Sci Technol 7:217-220
Knorr K, Fütterer K, Annighöfer B, Depmeier W (1997) A heatable large volume high pressure cell for neutron powder diffraction: The Kiel-Berlin Cell I. Rev Sci Instrum 68:3817-3822
Knorr K, Annighöfer B, Depmeier W (1999) A heatable large volume high pressure cell for neutron powder diffraction: The Kiel-Berlin Cell II. Rev Sci Instrum 70:1501-1504
Kuhs W, Bauer FC, Hausmann R, Ahsbahs H, Dorwarth R, Hölzer K (1996) Single crystal diffraction with X-rays and neutrons: High quality at high pressure? High Press Res 14:341-352
Kung J, Angel RJ, Ross NL (2000) Elasticity of $CaSnO_3$ perovskite. Phys Chem Minerals (accepted)
Li Z, Ahsbahs H (1998) New pressure domain in single-crystal X-ray diffraction using a sealed source. Rev High Pressure Sci Technol 7:145-147
Liebermann RC, Li B (1998) Elasticity at high pressures and temperatures. Rev Mineral 37:459-524
McConnell JDC, McCammon CA, Angel RJ, Seifert F (2000) The nature of the incommensurate structure in åkermanite, $Ca_2MgSi_2O_7$, and the character of its transformation from the normal structure. Z Kristallogr (accepted)
Miletich R, Allan DR, Kuhs WF (2000) High-pressure single-crystal techniques. *In* Hazen RM (ed) High-Temperature–High-Pressure Crystal Chemistry. Reviews in Mineralogy, Vol 40 (in press)
Murnaghan FD (1937) Finite deformations of an elastic solid. Am J Math 49:235-260
Niesler H, Jackson I (1989) Pressure derivatives of elastic wave velocities from ultrasonic interferometric measurements on jacketed polycrystals. J Acoust Soc America 86:1573-1585
Nomura M, Nishizaka T, Hirata Y, Nakagiri N, Fujiwara H (1982) Measurement of the resistance of Manganin under liquid pressure to 100 kbar and its application to the measurement of the transition pressures of Bi and Sn. Jap J Appl Phys 21:936-939
Nye JF (1957) Physical Properties of Crystals. Oxford University Press, Oxford
Oda H, Suzuki I (1999) Normal mode oscillation of a sphere with solid-gas-solid structure. J Acoust Soc Am 105:693-699
Osland RCJ (1985) Principles and Practices of Infra-red Spectroscopy. (report) Pye Unicam Ltd, Cambridge, UK
Orear J (1982) Least squares when both variables have uncertainties. Am J Phys 50:912-916
Palmer DC, Finger LW (1994) Pressure-induced phase transition in cristobalite: An X-ray powder diffraction study to 4.4 GPa. Am Mineral 79:1-8
Reichmann H-J, Angel RJ, Spetzler H, Bassett WA (1998) Ultrasonic interferometry and X-ray measurements on MgO in a new diamond-anvil cell. Am Mineral 83:1357-1360

Ríos S, Quilichini M, Knorr K, André G (1999) Study of the (P,T) phase diagram in TlD_2PO_4. Physica B 266:290-299

Schranz W, Havlik (1994) Heat diffusion central peak in the elastic susceptability of KSCN. Phys Rev Letts 73:2575-2578

Schranz W (1997) Dynamical mechanical analysis—a powerful tool for the study of phase transitions. Phase Trans 64:103-114

Sharma SM, Gupta YM (1991) Theoretical analysis of R-line shifts of ruby subjected to different deformation conditions. Phys Rev B 43:879-893

Shen AH, Reichmann H-J, Chen G, Angel RJ, Bassett WA, Spetzler H (1998) GHz ultrasonic interferometry in a diamond anvil cell: P-wave velocities in periclase to 4.4 GPa and 207°C. *In* Manghanni MH, Yagi T (eds) Properties of Earth and Planetary Materials at High Pressure and Temperature. Am Geophys Union, Washington, DC, p 71-77

Sinogeikin SV, Bass JD (1999) Single-crystal elasticity of MgO at high pressure. Phys Rev B 59:R14141-R14144

Sorbello RS, Feller J, Levy M, Isaak DG, Carnes JD, Anderson OL (2000) The effect of gas loading on the RUS spectra of spheres. J Acoust Soc Am 107:808-818

Sowerby J, Ross NL (1996) High-pressure mid-infrared spectra of the $FeSiO_3$ polymorphs. Terra Nova 8(suppl 1):61

Spetzler H, Shen A, Chen G, Hermannsdorfer G, Shulze H, Weigel R (1996) Ultrasonic measurements in a diamond anvil cell. Phys Earth Planet Int 98:93-99

Spetzler H, Jacobsen S, Reichmann H-J, Shulze H, Müller K, Ohlmeyer H (1999) A GHz shear wave generator by P to S conversion. (Absract) EoS Trans Am Geophys Union, F937

Stacey FD, Brennan BJ, Irvine RD (1981) Finite strain theories and comparisons with seismological data. Geophys Surveys 4:189-232

Vaidya SN, Bailey S, Pasternack T, Kennedy GC (1973) Compressibility of fifteen minerals to 45 kilobars. J Geophys Res 78:6893-6898

Vos WL, Schouten JA (1991) On the temperature correction to the ruby pressure scale. J Appl Phys 69:6744-6746

Wood IG, Welber B, David WIF, Glazer AM (1980) Ferroelastic phase transition in $BiVO_4$. II. Birefringence at simultaneous high pressure and temperature. J Appl Crystallogr 13:224-229

Yelon WB, Cox DE, Kortman PJ, Daniels WB (1974) Neutron diffraction study of ND_4Cl in the tricritical region. Phys Rev B 9:4843-4856

Zha C-S, Duffy TS, Downs RT, Mao H-K, Hemley RJ (1998) Brillouin scattering and X-ray diffraction of San Carlos olivine: direct pressure determination to 32 GPa. Earth Planet Sci Letts 159:25-33

5 Order-Disorder Phase Transitions

Simon A. T. Redfern

Department of Earth Sciences
University of Cambridge
Downing Street
Cambridge CB2 3EQ U.K.

INTRODUCTION

Cation ordering is often one of the most efficient ways a mineral can adapt to changing temperature or chemical composition. Disorder of distinct species across different crystallographic sites at high temperature provides significant entropic stabilisation of mineral phases relative to low-temperature ordered structures. For example, the calcite-aragonite phase boundary shows a significant curvature at high temperature due to disorder of CO_3 groups within the calcite structure, associated with an orientation order-disorder phase transition (Redfern et al. 1989a). This leads to an increased stability of calcite with respect to aragonite over that predicted by a simple Clausius-Clapeyron extrapolation of the low pressure-temperature thermochemical data (Fig. 1). Similarly, the pressure-temperature boundary of the reaction albite ↔ jadeite + quartz curves significantly at high temperature due to the entropic stabilisation of albite related to the high-low albite Al/Si order-disorder process. The energy changes associated with cation order-disorder phase transitions in a number of materials have been observed to be as great as the associated melting transitions (see Parsonage and Staveley 1978 for an earlier review). It is unsurprising, therefore, that there has been much interest in recent years in examining and modelling the processes of cation order-disorder in minerals. Computational studies of cation order-disorder have advanced together with experimental investigations and theoretical explanatory frameworks, and the three are increasingly being combined to provide interpretative descriptions of this process.

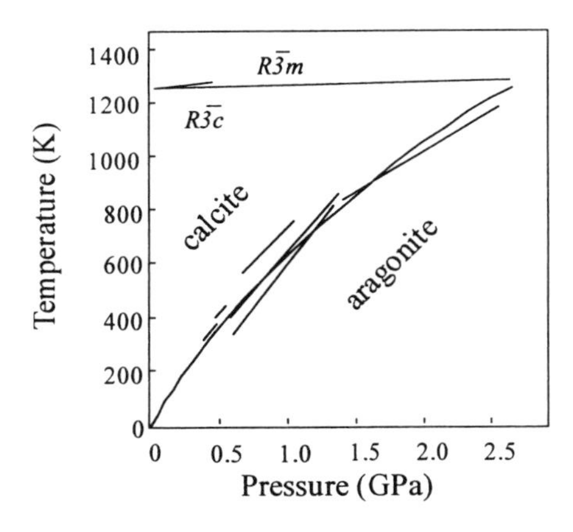

Figure 1. The calcite-aragonite phase boundary in P/T space is curved, due to the temperature-dependence of the entropy of calcite. The entropic stabilisation of calcite at high-T arises from the order-disorder phase transition from orientationally ordered $R\bar{3}c$ calcite at low temperatures to the orientationally disordered $R\bar{3}m$ structure, stable above 1260 K. Figure after Redfern et al. (1989a).

An order-disorder phase transition occurs when the low-temperature phase of a system shows a regular (alternating) pattern of atoms with long-range correlations, but the high-temperature phase has atoms arranged randomly with no long-range correlation. Experimentally, the distinction between the two can often be characterised straight-

1529-6466/00/0039-0005$05.00

forwardly using diffraction techniques, since diffraction measures the long-range correlation of structure. Usually, a minimum enthalpy is achieved by an ordered distribution (for example, in low-albite the preference of Al for one of the four symmetrically distinct tetrahedral sites), but configurational entropy (and often the coupled vibrational entropy) above 0 K results in a situation in which the free energy is a minimum for partially disordered distributions.

Ordering in minerals may occur over a variety of length scales. Most commonly only the long-range order is considered, because this is what is observed by structural diffraction methods, either through the direct measurement of scattering amplitudes at crystallographic sites or bond-lengths in the solid, or less directly through the measurement of coupled strains which may arise through the elastic interplay between the degree of order and the shape and size of the unit cell. Ordering over short length scales can also be measured, however, through experimental probes such as NMR and IR spectroscopy (see the chapters by Bismayer and Phillips in this volume). Often some of the high-frequency infrared vibrations are sensitive to the variations in structure associated with local ordering effects, but because this is an indirect measure of order calibration of some sort is usually required (Salje et al. 2000). NMR can give direct measurements of local site configurations, however (Phillips and Kirkpatrick 1994). Recently, computational methods have been employed successfully to elucidate and illuminate experimental observations of ordering and to begin to separate and compare short- and long-range ordering effects (Meyers et al. 1998, Warren et al. 2000a,b; Harrison et al. 2000b).

Cation ordering in minerals may or may not involve a change in the symmetry of the crystal. This distinction was outlined by Thompson (1969), who defined the two cases as convergent and non-convergent ordering. In convergent ordering two or more crystallographic sites become symmetrically equivalent when their average occupancy becomes identical, and the order-disorder process is associated with a symmetry change at a discrete phase transition. This usually occurs (as a function of temperature) at a fixed temperature, the transition temperature (T_c), or on a phase diagram at a fixed composition in a solid solution, defined by the relative free energies of the two phases. In non-convergent ordering the atomic sites over which disordering occurs never become symmetrically equivalent, even when the occupancies are identical on each. It follows that no symmetry change occurs on disordering, and no phase transition exists. An example of convergent ordering is that of cordierite, where a transition from hexagonal symmetry to orthorhombic occurs on ordering of Al and Si across the tetrahedral sites. In the disordered hexagonal phase particular tetrahedra become equi-valent to one another, both in terms of their Al/Si occupancy and in terms of their crystallographic relationship to one another (Fig. 2). An example of non-convergent ordering is found in spinel, where disordering

Figure 2. The crystal structure of orthorhombic (ordered) cordierite. MgO_6 octahedra are shown heavily shaded. Al and Si order within the rings in the (001) plane as well as along the chains parallel to [001]. In natural cordierites, H_2O, Na^+ and K^+ may be found in the channels running parallel to the *z*-axis.

of Mg and Al between octahedral and tetrahedral sites occurs on heating, but in which an octahedron is always an octahedron, a tetrahedron always a tetrahedron, and the two are never symmetrically equivalent, even when the occupancies of atoms in the two types of site are identical. In fact, the usual behaviour at a non-convergent disordering process is for the degree of order to approach zero asymptotically on heating. This is because the sites over which disordering is occurring are symmetrically different, and therefore usually also have different chemical potentials. They therefore rarely show identical (completely random) occupancies unless the system happens to be at a point where it shows a crossover from order to anti-order. In this chapter the factors controlling the thermodynamics of order-disorder will be reviewed. Their relationship to the kinetics of ordering will also be explored.

The time-temperature dependence of cation ordering and disordering in minerals has considerable petrological importance. Not only does such order/disorder behaviour have significant consequences for the thermodynamic stabilities of the phases in which it occurs, it can also play a significant role in controlling activity-composition relations for components, hence influencing inter-mineral major-element partitioning. Furthermore, since time-temperature pathways affect the final intra-mineral partitioning of (typically) cations within the structure of minerals, inverse modelling may be employed to infer the thermal histories of minerals from measured site occupancies. A quantitative knowledge of the temperatures and pressures of mineral assemblage formation in the crust and mantle is, therefore, fundamental to understanding the thermal evolution of the Earth, and to the development of well constrained petrological and geophysical models. For some time, geothermometric and geobarometric deductions have been based on the compositional variations of coexisting rock-forming minerals (e.g. cation partitioning between orthopyroxene/clinopyroxene, orthopyroxene/garnet, magnetite/ilmenite). Information on cooling rates (geospeedometry) is also potentially available from knowledge of intracrystalline cation partitioning. The convergent ordering of (for example) Al and Si on tetrahedral sites in feldspars has been used in this way as a thermometric indicator and marker of petrogenesis (Kroll and Knitter 1991), as has the non-convergent ordering of Mg and Fe on the M-sites of pyroxenes, which has been shown to be useful in the interpretation of the petrological history of the host rock (Carpenter and Salje 1994a).

EQUILIBRIUM AND NON-EQUILIBRIUM THERMODYNAMICS

Because the change in long-range order can be used to characterise an order-disorder phase transition, it is useful to define a long-range order parameter, Q, as a measure of the long-range correlation. This is usually normalised such that it is unity for a maximally ordered state and zero for a totally disordered state. This can be understood most easily in the case of an AB alloy, in which the structure can be divided into two sublattices α and β (Fig. 3). In such a bipartite framework all A atoms reside on α sites and all B atoms reside on β sites, in the ordered ground state. If the fraction of α sites occupied by A atoms is given as x_A^α then this will be the same as x_B^β for an alloy of 50:50 composition, and the order parameter, Q, is given by:

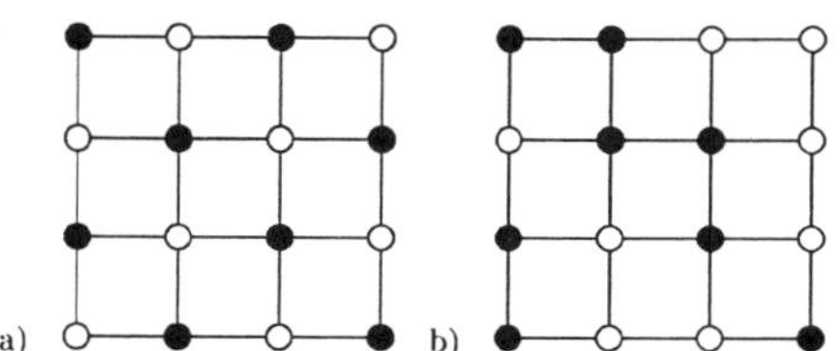

Figure 3. (a) The ordered ground state of an AB alloy on a square lattice, with white and black circles representing the two types of atom. (b) A disordered, high-temperature, arrangement of the same alloy.

$$Q = x_A^\alpha - x_A^\beta = x_B^\beta - x_B^\alpha = 2x_A^\alpha - 1 \quad (1)$$

Those theories of ordering that assume that the thermodynamics of ordering can be explained by Q alone are termed mean-field theories. Here two such theories are explored and compared: that of Bragg and Williams (1934) and that of Landau (1937). While some attempt has been made to consider the pressure-dependence of order-disorder phenomena in minerals (Hazen and Navrotsky 1996), here I shall limit the discussion of these phenomena to their temperature-dependence alone.

The Bragg-Williams model

The Bragg-Williams (1934) model has been comprehensivelydescribed in many texts including Parsonage and Staveley (1978), Ziman (1979), Yeomans (1992) and Putnis (1992). Examples of its use in mineralogy are given by Burton (1987) and by Davidson and Burton (1987). It has several deficiencies, and is fundamentally flawed as an accurate real model for mineral behaviour, but it is used so commonly within mineral sciences and petrology (it is related to the regular solution model) that it is worthwhile spending some time here considering its origins and features, as well as the origins of its failings. Its application to Al/Si order-disorder in framework aluminosilicates was recently critically assessed by Dove et al. (1996), where many of the deficiencies were outlined.

The model separates the free energy of ordering into two parts. The entropy is defined by the configurational entropy, given by the standard Boltzmann formula as $2k_B\ln W$ per molecule or $2R\ln W$ per mole, where W is the number of possible arrangements of $x_A^\alpha N$atoms of type A on N α sites. This is given by

$$W = N!/\left[\left(x_A^\alpha N\right)! \times \left(x_B^\alpha N\right)!\right].$$

By combining Stirling's approximation to this expression with Equation (1) one obtains

$$S = 2Nk_B \ln 2 - Nk_B\left((1+Q)\ln(1+Q) + (1-Q)\ln(1-Q)\right)$$

for the entropy for N α and β sites. The first term in this expression is independent of the degree of order. At phase transitions we are interested in differences and excess quantities. The part that is relevant to the stabilisation of one phase with respect to the other is the excess entropy of ordering, which is

$$S = -Nk_B\left((1+Q)\ln(1+Q) + (1-Q)\ln(1-Q)\right) \quad (2)$$

The enthalpy of ordering is given, in the Bragg-Williams model, by considering the energies of interaction between different pairs of nearest interacting neighbours and the probabilities of those configurations, as given by the long range order parameter. For example, in framework aluminosilicates this is, to a first approximation, dominated by those energies that give rise to the well-known Al-avoidance principle: the enthalpic disadvantage of have two Al atoms in tetrahedral sites next to one another, which drives Al/Si ordering in a host of crustal rock-forming minerals. More generally, the Bragg-Williams enthalpy is derived by considering an AB alloy. If the energy of an A-A nearest neighbour interaction is E_{AA}, then the number of such A-A interactions is the product of the number of a sites (N), the occupancy of these sites by A (x_A^α), the number of nearest neighbour β sites (z) and the occupancy of these β sites by A ($x_A^\beta = 1 - x_A^\alpha$) and the total number of A-A interactions is given by

$$N_{AA} = N.z.x_A^\alpha.\left(1 - x_A^\alpha\right) = \frac{Nz}{4}\left(1 - Q^2\right) \quad (3)$$

The number of B-B interactions is the same, and by similar reasoning the number of A-B interactions is given by considering the number of occurrences of an A in an α site next to a B in a β site and an A in a β site next to a B in an α site:

$$N_{AB} = N.z.x_A^{\alpha 2} + N.z.(1-x_A^{\alpha})^2 = \frac{Nz}{2}(1+Q^2) \tag{4}$$

Hence the internal energy can be expressed as a function of Q as:

$$E = \frac{Nz}{4}\left[(1-Q^2)(E_{AA}+E_{BB}) + 2(1+Q^2)E_{AB}\right] = -\frac{Nz}{4}JQ^2 + E_0 \tag{5}$$

where $E_0 = \frac{Nz}{4}(E_{AA}+E_{BB}+2E_{AB})$ is a constant energy (and hence ignored in the calculation of the energy differences between ordered and disordered states) and $J = E_{AA}+E_{BB}-2E_{AB}$ is the energy of exchange, the energy required to replace two A-B configurations with an A-A and a B-B configuration. Modelling this exchange energy is an essential step in computational studies of cation ordering in silicate, oxide and sulfide minerals.

In the Bragg-Williams model the expressions for entropy and enthalpy (Eqns. 2 and 5) have been obtained by averaging over all configurations with a particular value of Q, giving all such configurations equal weight. This ignores the fact that two configurations with the same degree of long-range order, Q, may have very different values of short-range order and internal energy, and they will not, therefore, occur with equal probability. This is a significant approximation, and is the real downfall of the model. None the less, pursuing this approximation, the free energy of ordering follows from the combination of Equations (5) and (2) (ignoring constant terms):

$$G(Q) = E - TS = -\frac{Nz}{4}JQ^2 + Nk_BT\left((1+Q)\ln(1+Q) + (1-Q)\ln(1-Q)\right) \tag{6}$$

The form of the free energy as a function of degree of order is shown in Figure 4, where it can be seen that the high-temperature minimum state is disordered ($Q = 0$), and that ordering occurs at low temperature to a value of $\pm Q$. At any temperature the equilibrium value of degree of order is given by the solution to $\partial G/\partial Q = 0$ and $\partial^2 G/\partial Q^2 > 0$. Indeed, at the phase transition from disordered to ordered structure the point $Q = 0$ switches from being a minimum to a maximum, and the transition temperature can be determined from the condition $\partial^2 G/\partial Q^2 = 0$:

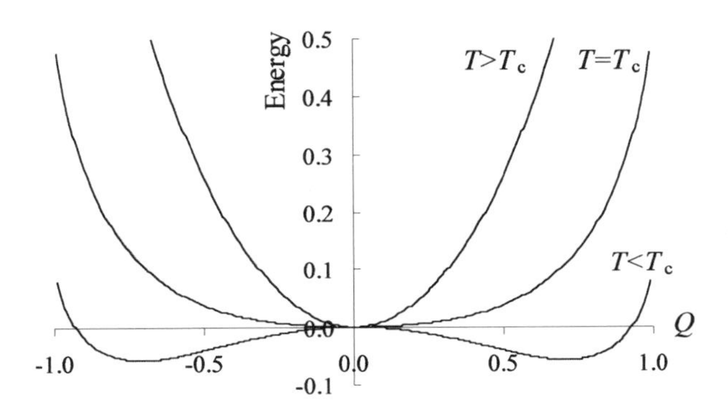

Figure 4. The dependence of the free energy on the degree of order, Q, given by the Bragg-Williams model, at three temperatures (above T_c, below T_c and at T_c).

$$\frac{\partial G}{\partial Q} = -\frac{NzJ}{2}Q + Nk_BT\left(\ln(1+Q) - \ln(1-Q)\right)$$
$$\frac{\partial^2 G}{\partial Q^2} = -\frac{NzJ}{2} + Nk_BT\left(\frac{1}{1+Q} + \frac{1}{1-Q}\right) \qquad (7)$$
$$\left(\frac{\partial^2 G}{\partial Q^2}\right)_{Q=0,T=T_c} = -\frac{NzJ}{2} + 2Nk_BT_c = 0 \ \Rightarrow \ T_c = \frac{zJ}{4k_B}$$

This shows that the Bragg-Williams model predicts that the transition temperature is a direct function of the exchange energy for the order-disorder process. Furthermore, the equilibrium condition $\partial G/\partial Q = 0$ gives the solution

$$Q = \tanh\left(\frac{T_c}{T}Q\right) \qquad (8)$$

which gives the approximate result $Q \propto \sqrt{T_c - T}$ for T close to T_c. Such dependence is typical for a mean field model.

Several authors have pointed the difficulties associated with applying the Bragg-Williams model to order-disorder in minerals: we have already noted that only short-range order that is determined by the long-range order is taken account of. Dove et al. (1996) pointed out that fluctuations in short range order could, however, be taken into account in an adjusted Bragg-Williams model in which the transition temperature is scaled by a factor dependent upon the coordination number of relevant neighbours. For the diamond lattice, for example, the value of T_c is adjusted to a temperature which is only 67.6% of the unadjusted figure.

There are a number of ways that the basic Bragg-Williams model may be extended to more complex situations, and these are discussed in more detail by Dove et al. (1996). One effect that can be considered is dilution. For example, if the number of α and β sites remains the same, but the fraction of atoms A in total is x, less than 0.5, then T_c is modified to $x(1\text{-}x)zJ/k_B$. If the number of α sites is, for example, less than the number of β sites, so that the ratio of α sites to β sites is x, and the ratio of A to B cations is also x (which is less than 0.5), then the transition temperature becomes modified to x^2zJ/k_B. Thus it can be seen that, even if we ignore the adjustments associated with nearest neighbour coordination geometry, the transition temperature for disordering of Al and Si in an aluminosilicate with a ratio of Al to Si of, say, 1 to 3 with 1 α site to 3 β sites (as in albite) would be one quarter that of a framework with a ratio of Al to Si and sites of 2:2 (as in anorthite). In fact, in the case of a dilute system (such as albite) true long range disorder can be achieved while maintaining a high degree of short range order. Indeed, on a square lattice it would be possible to have complete long range disorder and yet have no Al-Al nearest neighbours. However, this deduction ignores the influence of interactions beyond the nearest neighbours, and it turns out that these are, in fact, significant.

The Bragg-Williams model may be used in one of three ways. Given that $Q = 1$ at $T = 0$, the temperature dependence of Q is entirely determined by the value of T_c. This is turn is determined by the value of J, the energy required to form a nearest neighbour like bond. The commonest approach, therefore, is to assume that J is an adjustable parameter This allows one to fit T_c (or Q as a function of T) to the observed behaviour for a system, and to use the model as a phenomenological descriptor of the temperature dependence of the degree of order (and hence entropy and enthalpy and free energy) below an order-disorder phase transition. This assumes that the model is a correct descriptor for the phase transition, which it often is not both because of the failure to address short-range order and also since the enthalpy term (Eqn. 5) is truncated at the Q^2 term, and hence the effects of lattice relaxation and phonon enthalpies are ignored. The model is correct in two aspects however, it satisfies the criteria that $\partial Q/\partial T$ is 0 at $T = 0$ K and ∞ at $T = T_c$.

The second way that the Bragg-Williams model could be used is as a predictive tool for understanding order-disorder transitions. In this case one might obtain the value of J from experimental measurements, and then use this to predict the temperature-dependent ordering behaviour. To obtain J one needs to first measure the enthalpy of ordering from calorimetric measurements of samples with different degrees of order, obtained by equilibrating at different temperatures, for example. Then one needs to measure the number of nearest neighbour interactions as a function of equilibration temperature, typically using NMR spectroscopy or a similar technique. One system for which this process has been carried out is cordierite, $Mg_2Al_4Si_5O_{18}$, which shows an Al/Si order-disorder phase transition from a disordered hexagonal structure to an ordered orthorhombic structure. The transition temperature determined from experimental measurements is estimated as 1720 ± 50 K (Putnis et al. 1987). The number of Al-O-Al linkages annealed for different times was determined by NMR spectroscopy by Putnis and Angel (1985), and Carpenter et al. (1983) measured the enthalpy of ordering on the identical samples. Both quantities were found to vary logarithmically with time, giving a ratio between the enthalpy and the number of Al-O-Al arrangements that was constant, giving the effective interaction energy per Al-O-Al bond. Thayaparam et al. (1996) analysed these results, and show that they provide an effective interaction energy of $J = 0.75$ eV. If this number is used to calculated the Bragg-Williams T_c for this order-disorder phase transition it can be seen that the adjusted Bragg-Williams model (with an atomic Al concentration of 4/9) predicts much stronger ordering than is actually observed. It suggests a transition temperature more than four times higher than that found experimentally. The Bragg-Williams model is fundamentally flawed: experimental measurements of the real interaction energies, J, show that the transitions that Bragg-Williams predicts are all too high in temperature by many orders. This is because Al/Si ordering is often dominated by short-range adjustments. For example, cordierite can lower its entropy substantially by increasing its short-range order (with complete short range order and no Al-O-Al configurations for neighbouring tetrahedra) without developing any long-range order at all.

The third approach to considering the Bragg-Williams model has been to calculate the energies of ordering interactions using computer simulation, swapping atoms over various sites within the computer and then calculating the corresponding lattice energies. Both *ab initio* quantum mechanical calculations (De Vita et al. 1994) and empirical static lattice energy calculations (Bertram et al. 1990, Thayaparam et al. 1994 1996, Dove and Heine 1996) have proved most successful in this regard. The general procedure has been to generate a large number of configurations of ordering atoms within the structure, and for each configuration to determine the numbers of types of linkage (Al-O-Al, Al-O-Si and Si-O-Si linkages, for example). The consequent database of energies can then be used to fit first neighbour interaction energies, and indeed to obtain the energies of interaction to more distant shells of coordinating ordering cations. This has been carried out for Al/Si ordering in a number of minerals, including leucite (Dove et al. 1993), giving $J = 0.65$ eV and a Bragg-Williams T_c of 3340 K (compared with 938 K actually observed), gehlenite (Thayparam et al. 1994), sillimanite (Bertram et al. 1990) and cordierite (Thayparam et al. 1996). From the values of J so obtained, the same picture has emerged as that given by experimental determinations of J, that the Bragg-Williams model predicts transition temperatures that are far too high, and suggests that systems should be more ordered than they really are. There is no indication that the values of J are incorrect by such large factors, rather that the Bragg-Williams model needs to be improved upon. This has spurred recent work into the theory of ordering transitions in minerals that have adopted two alternative approaches. One is the cluster variation method (CVM), and the other employs Monte-Carlo methods. Both succeed in taking proper account of short range ordering, when correctly adopted.

Landau theory

The application of Landau's (1937) theory of symmetry-changing phase transitions to order-disorder in silicates has been described very clearly by Carpenter (1985 1988) and is further discussed in a number of textbooks and seminal papers (Salje 1990, Putnis 1992). The essential feature behind the model is that the excess Gibbs energy can be described by an expansion of the order parameter of the type:

$$\Delta G(Q) = \frac{a}{2}(T - T_c)Q^2 + \frac{b}{4}Q^4 + \frac{c}{6}Q^6 + \ldots \tag{9}$$

where a, b, and c are material-dependent parameters related to the phase transition temperature, T_c, and the scaling of the excess entropy. The form of G as a temperature-dependent function of Q is qualitatively similar to that given by the Bragg-Williams model (Fig. 4) although the details differ. For example, the classical Landau expansion given in Equation (9) does not satisfy the condition that $\partial Q/\partial T = 0$ at $T = 0$ K, and adjustments must be made to incorporate the effects of low-temperature quantum saturation. It turns out that varying the relative sizes of a, b, and c not only allows one to modify the transition temperature, it also modifies the temperature dependence of the order parameter, described in terms of $Q \propto |T_c - T|^\beta$. Thus, classical second-order behaviour ($\beta = 0.5$) is found if $c = 0$ while a and b are > 0. First-order behaviour can be represented by a Landau potential with a and $c > 0$ while $b < 0$. Tricritical behaviour ($\beta = 0.25$), the limiting case between first-order and second-order behaviour, can be described by a Landau expansion with a and $c > 0$ while $b = 0$. As can be seen from Figure 5, any order-parameter behaviour that corresponds to $Q \propto |T_c - T|^\beta$ with $0.25 < \beta < 0.5$ can be accommodated in the model with a Landau potential that has $a, b, c > 0$ but with varying b/c ratio (Ginzburg et al. 1987, Redfern 1992).

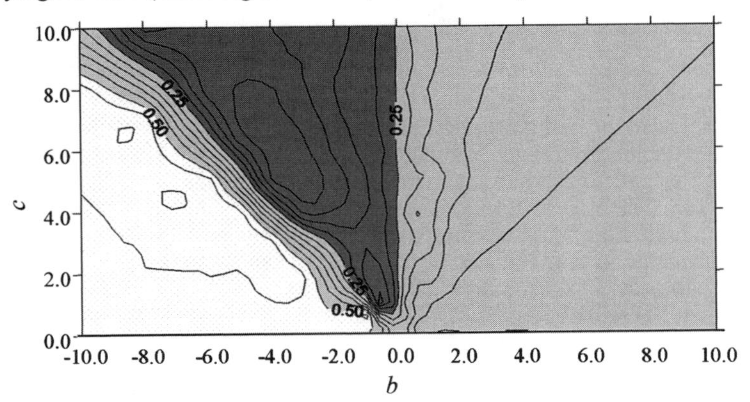

Figure 5. The dependence of the critical exponent β on the Landau coefficients. The critical exponents were calculated by least squares fitting of an equilibrium order parameter evolution from a Landau potential. The transition temperature was kept constant, while the b and c coefficients were varied; thus changing the value of the a coefficient. The dark grey regions show critical exponents of less than 0.25, approaching a first order type of behaviour. The light grey regions indicate a zone were the critical exponent is between 0.25 and 0.5, and the transition is continuous.

The principal differences between the Landau model and the Bragg-Williams models for phase transitions lie in the way they partition the free energy between the enthalpy and the entropy. It can be seen from Equation (9) that within the Landau model the entropy is simply proportional to Q^2, while the enthalpy is a function of Q^2 and Q^4 (and other higher-order terms). Thus, for phase transitions in which the entropy is dominated by vibrational effects (which scale as Q^2), such as displacive phase transitions, the Landau model provides an excellent description. However, for phase transitions in which configurational entropy plays a major role, as in most order-disorder phase transitions, the simplest Landau expansion given in Equation (9) must be augmented by adding a configurational component. It is, therefore, a straightforward matter to incorporate both higher order entropy and higher order enthalpy effects into the Landau potential.

The Landau model for phase transitions is typically applied in a phenomenological manner, with experimental or other data providing a means by which to scale the relative terms in the expansion and fix the parameters a, b, c, etc. The expression given in Equation (9) is usually terminated to the lowest feasible number of terms. Hence both a second-order phase transition and a tricritical transition can be described adequately by a two term expansion, the former as a "2-4" potential and the latter as a "2-6" potential, these figures referring to those exponents in Q present.

Order parameter coupling. The interplay between cation ordering and displacive phase transitions is a very important control on the high-temperature behaviour of many minerals. For example, in a transition such as the cubic to tetragonal transition in garnet (Hatch and Ghose 1989) there is an order parameter describing Mg/Si ordering and anther describing a displacive transition. These two order parameters bring about the same symmetry change and can therefore interact with one another. Physically the interactions must be via some common process, and the usual candidate is a strain, which is common to both order parameters. Framework aluminosilicates, in particular, often display ordering of Al and Si within the tetrahedral sites of the framework as well as elastic instabilities of the framework itself. These processes may couple directly, when allowed by symmetry, through a strain interaction associated with ordering which arises from the difference in size of the Al-O and Si-O bond lengths. Even if symmetry prohibits direct coupling between Al/Si order (whose order parameter we shall term Q_{od}) and elastic instabilities (with associated order parameter Q), higher-order coupling terms are always possible. The manner in which ordering and framework distortions couple depends upon the symmetry relations of these processes. The beauty of the Landau approach to describing phase transitions is the ease with which coupled phenomena may be incorporated into the expression for the free energy of ordering. Thus, for example, it is relatively straightforward to describe the coupling between order-disorder phase transitions and displacive phase transitions within a single phase using the one Landau expansion. This was carried out by Salje et al. (1985) in one of the very first applications of this method in mineralogy: the description of the equilibrium cooling behaviour of albite, which incorporates the two coupled processes of an Al/Si order-disorder phase transition and the displacive elastic transition from monalbite to high-albite.

A theory of order parameter coupling was developed for minerals by Salje and Devarajan (1986). Its application may be illustrated with one simple example, which gives a flavour of the further possibilities that may arise in the case of coupled processes. If two order parameters Q_1 and Q_2 have different symmetries, only coupling in the form $Q_1^2 Q_2^2$ is allowed. This situation is referred to as biquadratic coupling. If, however, two order parameters have the same symmetry (i.e. bringing about the same symmetry breaking transition), other couplings are allowed in the form $Q_1 Q_2$, $Q_1^3 Q_2$, $Q_1^2 Q_2^2$ and $Q_1 Q_2^3$; the first of these is often the dominant term, and this situation is called bilinear coupling.

The other terms are usually assumed to be negligible, and their inclusion is generally not warranted. Other situations may also be envisaged where one phase transition is a zone boundary transition, and the other a zone centre. This can lead to linear-quadratic coupling.

The case of bilinear coupling is important in the description of the high temperature behaviour of albite, and is considered in more detail here. For the case of biquadratic coupling the reader is referred to Salje and Devarajan (1986) and Salje (1990). If we consider two order parameters, Q_1 and Q_2, which are coupled together by a single spontaneous strain, e, the Landau expansion is:

$$\begin{aligned} G = {} & \frac{a_1}{2}\left(T - T_{c1}\right)Q_1^2 + \frac{b_1}{4}Q_1^4 + \frac{c_1}{6}Q_1^2 \\ & + \frac{a2}{2}\left(T - T_{c2}\right)Q_2^2 + \frac{b_2}{4}Q_2^4 + \frac{c_2}{6}Q_2^2 \\ & + deQ_1 + feQ_2 + \frac{1}{2}Ce^2 \end{aligned} \tag{10}$$

This assumes that only linear coupling between strain and order parameters exists, and that the strain obeys Hooke's law. The Landau expansion can then be renormalized as:

$$\begin{aligned} G = {} & \frac{a_1}{2}\left(T - T_{c1}^*\right)Q_1^2 + \frac{b_1}{4}Q_1^4 + \frac{c_1}{6}Q_1^2 \\ & + \frac{a2}{2}\left(T - T_{c2}^*\right)Q_2^2 + \frac{b_2}{4}Q_2^4 + \frac{c_2}{6}Q_2^2 \\ & + ?Q_1Q_2 \end{aligned} \tag{11}$$

where the renormalized temperatures are $T_{c1}^* = T_{c1} + \frac{d}{a_1C}$ and $T_{c2}^* = T_{c2} + \frac{f}{a_2C}$, and the bilinear coupling between the two order parameters is $\lambda = \frac{df}{C}$.

This type of coupling produces three interesting results. The first, and probably most significant, is that there is only one phase transition instead of two distinct transitions, since at equilibrium when $Q_1 = 0$, the second order (Q_2) parameter must also be zero. There is no equilibrium condition where one order parameter equals zero and the other does not. It may be possible for metastable states to occur when one order parameter is zero, and the other not, however.

The second point is that the evolution of the individual order parameters is different to the behaviour without coupling, as is shown in Figure 6, where two order parameters with different transition temperatures are plotted against each other. In one representation the coupling is set at zero. Comparing this to the coupled examples, it can be seen that as the coupling strengthens the order parameter evolution becomes less extreme.

The third point is that the transition temperature is different from the renormalized transition temperatures. This can be seen in Figure 7, where the renormalized transition temperatures are set at constant values. The behaviour of the non-driving order parameter is strongly affected by the coupling, but the actual transition temperature, when $Q_1 = Q_2 = 0$, tends to be similar to that of the driving order parameter, except when the coupling is very strong.

It is also interesting to note that when $T_{c1}^* \neq T_{c2}^*$, there will be a single driving order parameter; this will be the order parameter of the transition with the highest renormalized transition temperature. On cooling in equilibrium, there is a cross-over in the order

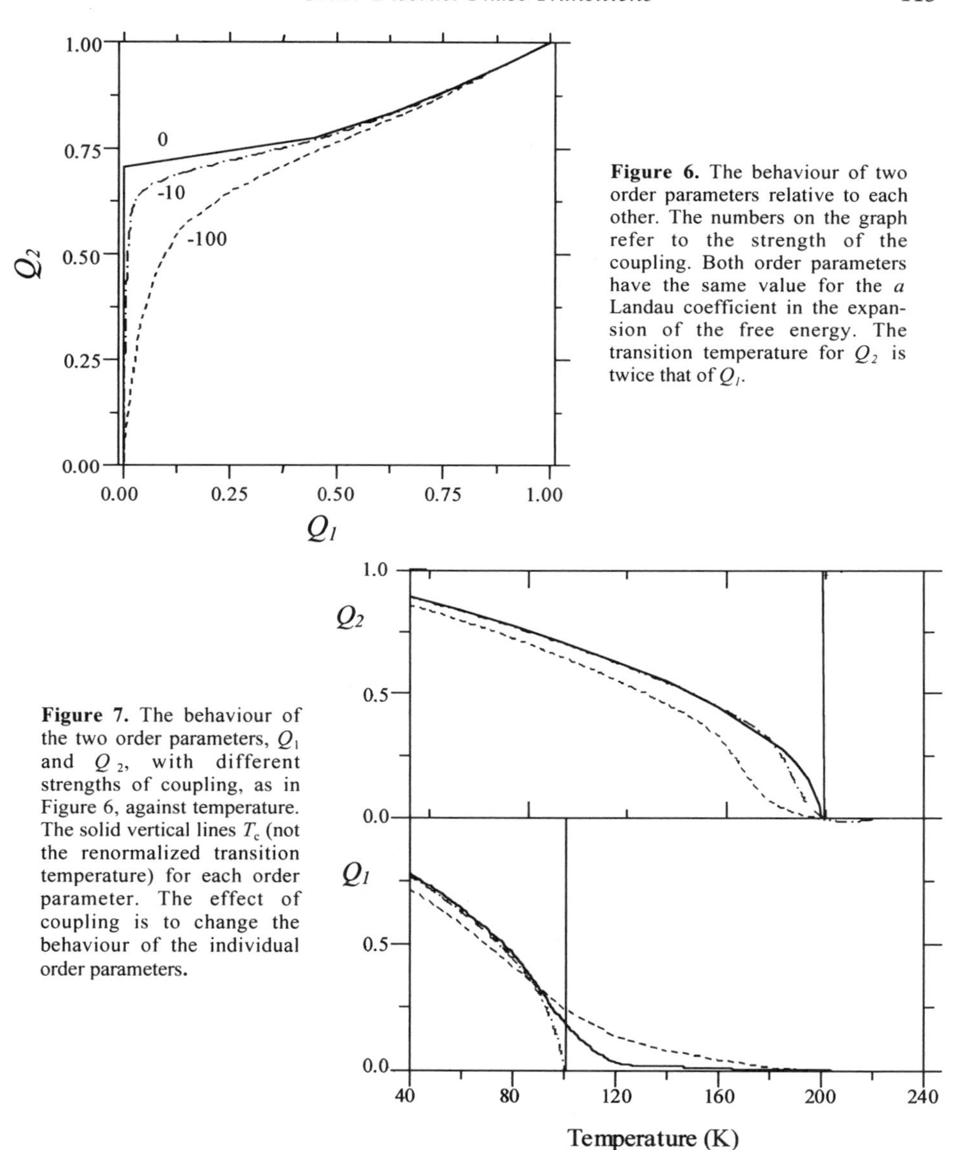

Figure 6. The behaviour of two order parameters relative to each other. The numbers on the graph refer to the strength of the coupling. Both order parameters have the same value for the *a* Landau coefficient in the expansion of the free energy. The transition temperature for Q_2 is twice that of Q_1.

Figure 7. The behaviour of the two order parameters, Q_1 and Q_2, with different strengths of coupling, as in Figure 6, against temperature. The solid vertical lines T_c (not the renormalized transition temperature) for each order parameter. The effect of coupling is to change the behaviour of the individual order parameters.

parameters near the second transition temperature, as shown in Figure 7. The cross-over occurs close to the renormalized transition temperature of the non-driving (slave) order parameter. The term "cross-over" was introduced by Salje and Devarajan (1986). It is a useful term as such, but it does not refer to a specific temperature, rather to a range of temperatures. It also is more difficult to define when the coupling becomes stronger. In actual fact it may be better to regard the cross-over as being the point where the rate of change dQ_1/dQ_2 with respect to one of the order parameters is a maximum, i.e. where $d^2Q_1/dQ_2^2 = 0$. The cross-over then does not have to occur close to either of the renormalized transition temperatures, but between them.

Non-equilibrium description of order-disorder. As well as order parameter coupling, the kinetics of order-disorder have been very successfully described phenomenologically within the Ginzburg-Landau adaptation of the Landau equilibrium model. Time-dependent Landau theory for order-disorder processes in minerals was developed by Carpenter and Salje (1989), and readers are directed to this seminal paper on the subject. The essence of the model is that the rate of change of ordering, dQ/dt is dependent upon the probability of a jump towards the equilibrium state, and the rate of change of the free energy with order at the particular non-equilibrium state that the crystal finds itself in. In the Ginzburg-Landau case the rate of change of order is given by:

$$\frac{dQ}{dt} = -\frac{\gamma \exp(-\Delta G^* / RT)}{2RT} \frac{\partial G}{\partial Q} \tag{10}$$

where γ is the characteristic effective jump frequency of the migrating atom and ΔG^* is the free energy of activation for the jump. The evolution of the order parameter can be understood in terms of its locus across a free energy surface as a function of time (Fig. 8).

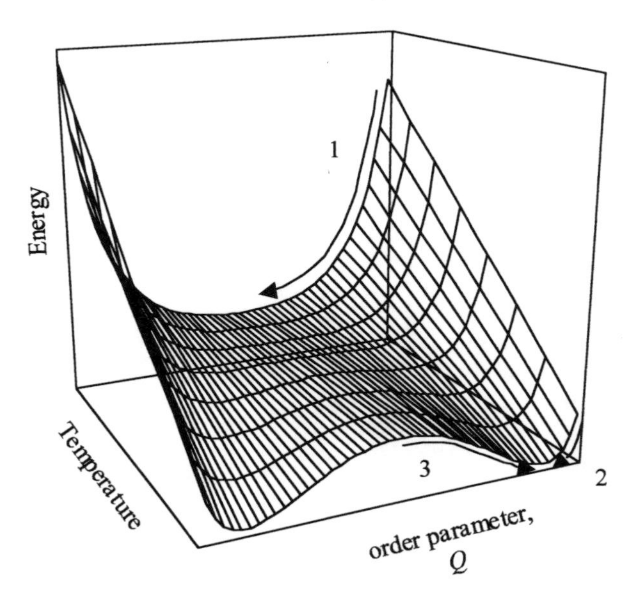

Figure 8. Free energy surface in order parameter–temperature space, showing the pathways taken on annealing an ordered solid at high temperature to generate a disordered state (1), disordering a highly ordered sample at temperature below T_c (2), and ordering a disordered sample below T_c (3).

Integrating Equation (10) under isothermal conditions one obtains:

$$t - t_0 = \int_{Q_0}^{Q} \frac{-2RT}{\gamma \exp(-\Delta G^*/RT)} \left(\frac{\partial \Delta G}{\partial Q} \right)^{-1} dQ \tag{11}$$

where Q_0 is the initial value of Q at time t_0. This equation allows the time taken for a given change in Q to be calculated. Alternatively one may calculate the change in Q for a given annealing time by evaluating Equation (11) numerically, and varying the upper limit of integration in an iterative procedure until the correct annealing time is obtained. The evolution of Q versus t during heating or cooling at a constant rate can be determined by approximating the constant heating or cooling rate by a series of discrete isothermal annealing steps separated by an instantaneous temperature change. The spatial distribution of the order parameter within the sample also plays an important part in the time-dependent evolution of the degree of order, and is itself a time-dependent feature of the material. Its description requires a phenomenological approach that takes account of partial conservation of order parameters, since the growth and change of microstructure involves an alternative method to modifying the time-dependent entropy that is not described by Equation (10). Malecherek et al. (1997) recently described a method by which the kinetic behaviour of the macroscopic order parameter may be understood in systems which are not homogeneous. The effects of microstructure and their relationship to non-equilibrium transformation processes are described in more detail in the chapter by Salje within this volume.

Non-convergent ordering

The extension of the Landau theory for symmetry-breaking phase transitions to non-symmetry breaking non-convergent ordering is disarmingly simple. A magnetic system may be prevented from attaining complete magnetic disorder by applying an external magnetic field to it. In the same way, a cation order-disorder transition will fail to take place if there is a field present, but in this case the field takes the form of the chemical differences between sites that will distinguish them at all temperatures. This can be described as an energy term, $-hQ$, that is linear in the order parameter. It can be seen that by adding such a term to the Landau expansion, one prevents Q = 0 from being a solution to $\partial G/\partial Q = 0$:

$$\Delta G(Q) = -hQ + \frac{a}{2}(T - T_c)Q^2 + \frac{b}{4}Q^4 + \frac{c}{6}Q^6 + \ldots \tag{12}$$

This approach to the description of non-convergent ordering is discussed in some detail by Carpenter et al. (1994). Carpenter and Salje (1994a,b) also describe its application to the non-convergent disordering behaviour of spinels, orthopyroxenes, and potassium feldspar. Kroll et al. (1994) describe the relationship of this Landau formalism of the free energy of non-convergent order-disorder to the description of the same phenomenon that Thompson (1969) suggested, in his early working of a non-convergent Bragg-Williams type phenomenological model. As noted for the case of convergent ordering, the difference lies in the way that the entropy and enthalpy are described. Carpenter et al. (1994) point out that although the entropy in the unadjusted Landau model is truncated at the Q^2 term, this is in fact the first term in the series expansion of the entropy described by Equation (2), above, and that the Landau entropy expression remains a reasonable approximation to the configurational entropy for $Q < 0.9$. The main advantage in applying the Landau model to the description of this phenomenon is that the approach is general, and a fairly comprehensive phenomenological description can be provided with relatively few free parameters. The use of the Ginzburg-Landau expression also allows the modelling of the kinetics of non-convergent order-disorder, in exactly the same way as is done for the kinetics of convergent ordering. In fact, the approach described by Equations (10) and (11) is independent of the method by which the free energy is described, and is also valid for systems described within a Bragg-Williams related model.

Computer modelling of cation ordering

Much progress has been made in recent years in developing effective and accurate models of the energies of exchange that drive cation ordering in minerals. One approach has been to use the exchange energies derived from empirical or quantum mechanical simulations of the atomic configurations to parametrise the energy in terms of the configurations, or more properly the number of interactions present. Formally this is equivalent to a spin model, and the reader is referred to Ross (1990) for an introduction to such models. Dove (1999) derives the energy of an ordering system in terms of the spin variable, σ_j (for site j) with a value of +1 if it is occupied by cation of type A (e.g. Al) and −1 if it is occupied by a cation of type B (e.g. Si). Then the energy is represented by the Hamiltonian:

$$H=\frac{1}{4}\sum_{<j,k>} J_{jk}\sigma_j\sigma_k+\sum_j \mu_j\sigma_j \tag{13}$$

where J_{jk} is the exchange energy and μ_j is the chemical potential. Monte Carlo simulations can then be used to determine the temperature dependence of the average order parameter or average energy. Myers et al. (1998) have explained how one can derive thermodynamic quantities such as entropy and free energy from thermodynamic integration, by splitting the Hamiltonian for the model into one part that can be calculated exactly (usually that of the completely disordered crystal) and one part that can be determined by simulations of the system over a range of states and integrating between limits.

Another method that can be used to estimate the temperature dependence of the order parameter is the Cluster Variation Method (Vinograd and Putnis 1999). In this approximation clusters of sites are considered, together with the proportions of these clusters which are occupied by each possible combination of, for example, Al and Si atoms. Thus, the pair CVM or "quasichemical" approximation is concerned with determining the probability that each nearest-neighbour bond is either Al-O-Al, or Al-O-Si or Si-O-Si. Knowing how many configurations there are with each given set of bond probabilities, it is possible write down an expression for the free energy and minimise it in order to obtain equilibrium bond probabilities. It is possible to derive expression for the number of ways to arrange all $zN/2$ bonds (where N is the number of atoms), but this does not give the correct number of configurations because most of these arrangements are not physically possible. The CVM approximation involves placing z atoms at each of the atomic sites, one atom at the end of each bond, and then deriving the number of ways of distributing these zN atoms, as well as the number of these configurations which have just one type of atom on each site (i.e. the number of ways of distributing the original N atoms with one atom per site). From the configurational information that the model gives it is then possible to determine the temperature dependence of the short and long range ordering. Furthermore, the CVM allows one to explore states of order that are out of equilibrium, providing a route to investigating the kinetics of short range order computationally. The accuracy of CVM calculations can be improved by using larger clusters (beyond simple pairs) and considering the probabilities that they are occupied by each possible combination of atoms. However, the number of variables required to do this increases exponentially with the maximum cluster size. Vinograd and Putnis (1999) have recently discovered an efficient method of carrying out CVM calculations with large clusters on aluminosilicate frameworks, which gives results that agree well with Monte Carlo simulations. However, as yet this method is limited to considering nearest neighbour exchange energies alone. A comparison of the calculated CVM, Bragg-Williams and Monte Carlo results for cordierite and feldspar Al/Si ordering is shown in Figure 9.

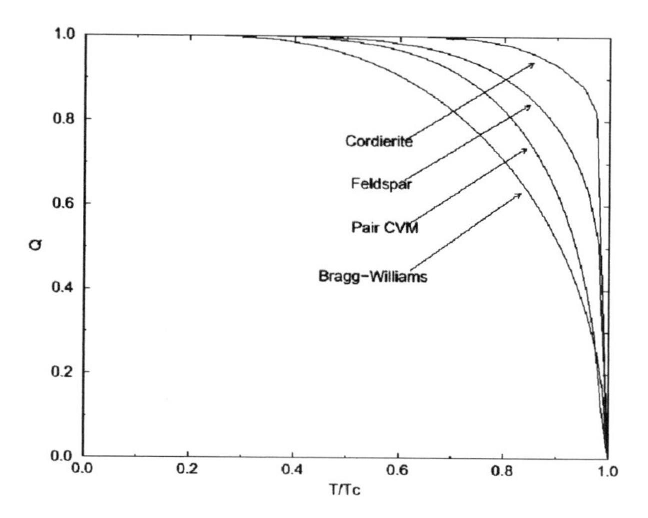

Figure 9. The results of the Monte Carlo simulations of ordering of Al and Si in the cordierite framework (from Thayaparam et al. 1995) and feldspar (Meyers et al. 1998), compared with the equilibrium order parameter predicted by the Bragg-Williams model and the pair CVM model, shown on a rescaled temperature axis.

EXAMPLES OF REAL SYSTEMS

Cation ordering in ilmenite-hematite

Members of the $(FeTiO_3)_x(Fe_2O_3)_{1-x}$ solid solution have large saturation magnetizations and contribute significantly to the palaeomagnetic record. Often such material is observed to acquire self-reversed remnant magnetization. In all cases, the high-temperature $R\bar{3}c$ to $R\bar{3}$ cation ordering transition plays a crucial role in determining the thermodynamic and magnetic properties. This transition involves the partitioning of Ti and Fe cations between alternating (001) layers of the hexagonal-close-packed oxygen sublattice (Fig. 10). Above the transition temperature (T_c) the cations are distributed randomly over all (001) layers. Below T_c the cations order to form Fe-rich A-layers and Ti-rich B-layers. Harrison et al. (2000a) recently carried out an *in situ* time-of-flight

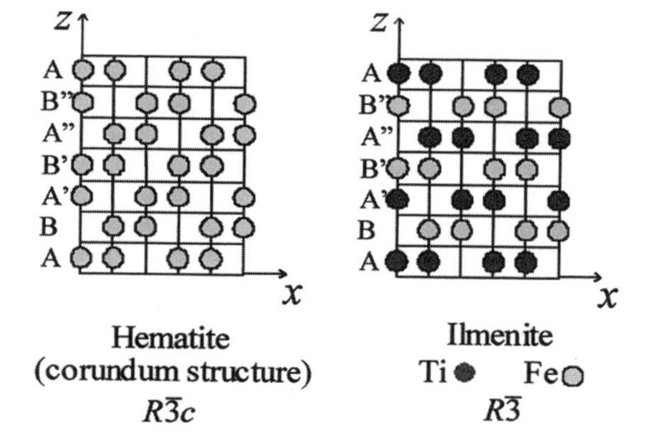

Figure 10. Schematic representation of the ordering scheme of low-temperature ilmenite compared to hematite.

neutron powder diffraction study of synthetic samples of the $(FeTiO_3)_x(Fe_2O_3)_{1-x}$ solid solution with compositions x = 0.7, 0.8, 0.9 and 1.0 (termed ilm70, ilm80, ilm90 and ilm100), which provides an excellent illustrative case study to augment the discussions of models presented above. Harrison et al. (2000a) obtained cation distributions in members of the solid solution at high temperatures directly from measurements of the site occupancies using Rietveld refinement of neutron powder diffraction data. This proved especially powerful in this case because of the very large neutron scattering contrast between Ti and Fe. The measurements offered the first insight into the equilibrium cation ordering behaviour of this system over this compositional range and allow the simultaneous observation of the changes in degree of order, spontaneous strain and the cation-cation distances as a function of temperature. A qualitative interpretation of the observations was provided in terms of the various long- and short-range ordering processes which operate.

In discussing the changes in cation distribution which occur as a function of temperature, T, and composition, x, it is useful to define a long-range interlayer order parameter, Q, as $(X_{Ti}^{b} - X_{Ti}^{a})/(X_{Ti}^{b} + X_{Ti}^{a})$. According to this definition, the order parameter takes a value of $Q = 0$ in the fully disordered state (with Fe and Ti statistically distributed between the A- and B-layers) and a value of $Q = 1$ in the fully ordered state (with the A-layer fully occupied by Fe and all available Ti on the B-layer). Values of Q are shown in Figure 11. In all cases the estimated standard deviation in Q is smaller than the size of the symbols. The value of Q measured at room temperature represents the degree of order maintained after quenching the starting material from the synthesis temperature of 1300°C. In ilm80, ilm90 and ilm100 the quenched starting material is almost fully ordered, with $Q = 0.98$ in all three cases. This apparently low value of Q in ilm70 may be due to the presence of chemical heterogeneities that develop on heating the sample below the solvus that exists at intermediate composition.

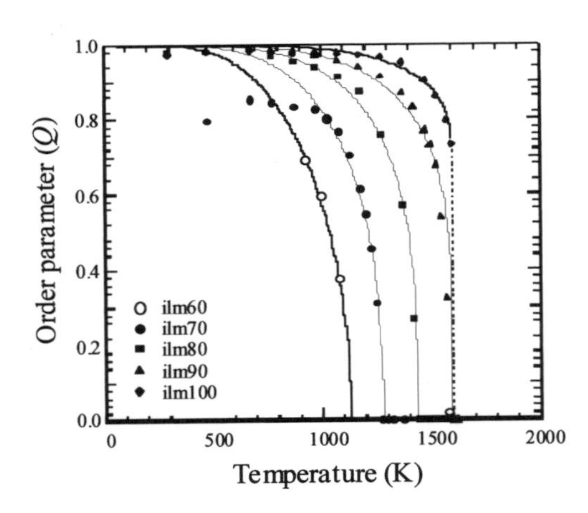

Figure 11. T dependence of Q for members of the ilmenite-hematite solid solution, determined from neutron powder diffraction (solid symbol: Harrison et al. 2000a) and quench magnetization (open symbols: Brown et al. 1993). Solid lines are fits using a modified Bragg-Williams model.

The ordering behaviour in ilm80, ilm90 and ilm100 appears to be fully reversible, but the data close to T_c can only be fitted with a critical exponent for the order parameter, β, which is of the order of 0.1, which does not correspond to any classical mean field Landau-type model. Instead, a modified Bragg-Williams model is required, that describes the free energy phenomenologically in terms of a configurational entropy alongside an enthalpy that contains terms up to Q^4:

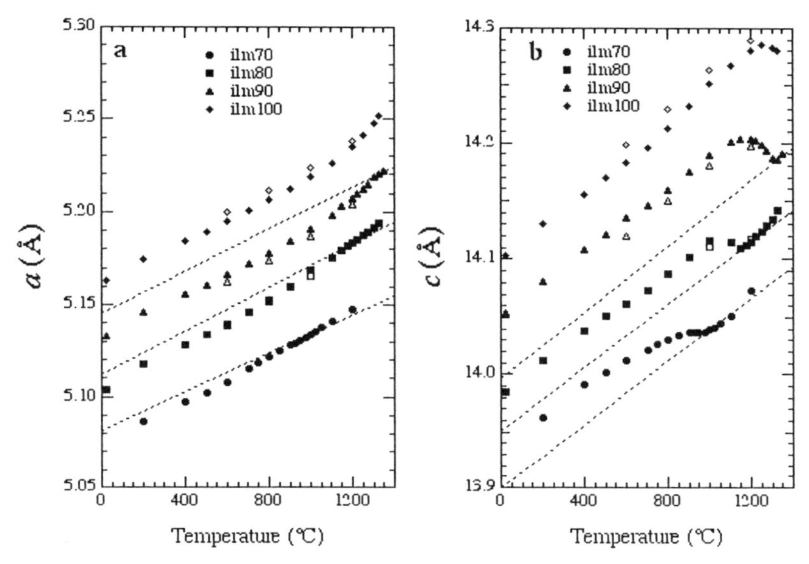

Figure 12. Variation in the cell parameters (a) a, and (b) c as a function of temperature, from Harrison et al. (2000a). Dashed lines are the estimated variation in a_0, and c_0 as a function of temperature.

$$\Delta G = RT \ln W + \frac{1}{2}aQ^2 + \frac{1}{4}bQ^4 \tag{14}$$

The hexagonal unit cell parameters, a and c, and the cell volume, V, are plotted in Figure 12 for all temperatures and compositions measured by Harrison et al. (2000a). There are significant changes in both the a and c cell parameters correlated with the phase transition. Such changes are usually described by the spontaneous strain tensor, ε_{ij}. In the case of the $R\bar{3}c$ to $R\bar{3}$ transition, where there is no change in crystal system, the only non-zero components are $\varepsilon_{11} = \varepsilon_{22}$ and ε_{33}. The estimated variation in a_0 and c_0, the paraphase cell parameters, as a function of temperature is shown by the dashed lines in Figure 12. From Figure 12a one sees that ε_{11} is negative, and that its magnitude increases with increasing Ti-content. The changes in a occur smoothly over a large temperature range and there is no sharp change in trend at $T = T_c$ in any of the samples. In contrast ε_{33} is positive. In ilm70, it is relatively small and c varies smoothly through the transition. In ilm80 and ilm90, ε_{33} is larger and the decrease in c occurs very abruptly at the phase transition. It should be noted that the magnitude of all spontaneous strains associated with ordering in these samples is relatively small. This is consistent with the observations of Nord and Lawson (1989), who studied the twin-domain microstructure associated with the order-disorder transition. The twin boundaries have wavy surfaces, as is expected if there is no strain control over them. Furthermore, the fact that the spontaneous strain on ordering is small provides the first hint that the length scale of the ordering interactions may not be very long-range. Generally, systems that display large strains on ordering tend to behave according to mean field models, as the strain mediates long-range correlations, whereas systems with weak strain interactions tend to show bigger deviations from mean-field behaviour.

The spontaneous strain for long-range ordering in ilmenite is approximately a pure shear (with ε_{11} and ε_{33} having opposite sign and $\varepsilon_V \approx 0$). It seems reasonable to assume that short-range ordering, which is often an important feature of such transitions, will play a significant role in determining the structural changes in the vicinity of the transition temperature in ilmenite. Here, short-range order may be defined by a parameter, σ, which is a measure of the degree of self-avoidance of the more dilute atom (e.g. Al-Al avoidance in aluminosilicates such as feldspars, or Ti-Ti avoidance in the case of ilmenite). The short-range order parameter, σ, is 1 for a structure with no alike nearest neighbours, and 0 for a totally random structure. For example, in ilmenite it may be defined as:

$$\sigma = 1 - \left(\frac{\textit{proportion of Ti} - O - \textit{Ti bonds}}{\textit{proportion of Ti} - O - \textit{Ti bonds in random sample}} \right) \tag{15}$$

Below T_c, σ includes a component due to long-range order. Myers *et al.* (1998) therefore defined a modified short-range order parameter, σ', that excludes short-range order arising from long-range order: $\sigma' = (\sigma - Q^2)/(1 - Q^2)$. Short-range order will become important at temperatures close to and above T_c, where there is mixing of Fe and Ti on both the A- and B-layers. In addition, one expects that short-range ordering above T_c will be more important at compositions close to ilm100, where the Fe:Ti ratio approaches 1:1. Evidence of both these effects can be seen in the cell parameter variation as a function of temperature and composition, as illustrated schematically in Figure 13.

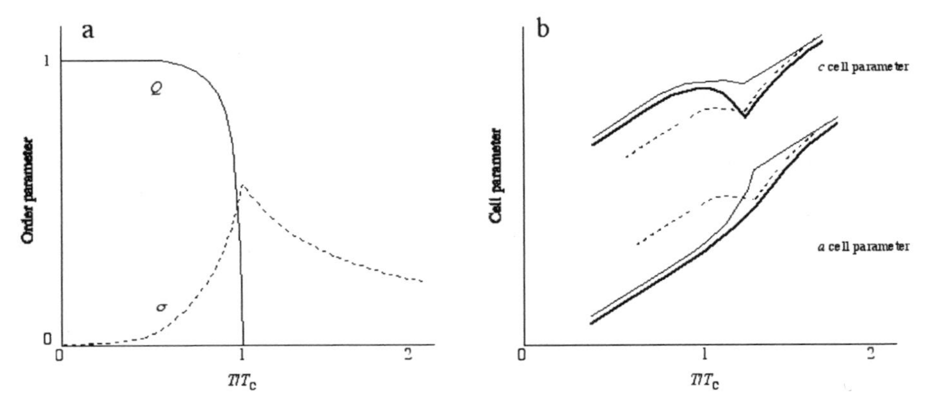

Figure 13. (a) Variation in long-range order, Q, and short-range order , σ, as a function of T. (b) Effect of competing long- and short-range order on a and c parameters. Thin solid lines show long-range ordering effects, dashed lines show short-range effects. The thick lines give their sum (from Harrison et al. 2000a).

There is a rapid increase in the degree of short-range order at temperatures approaching T_c, which correlates with the rapid decrease in long-range order. Above T_c, σ decreases slowly, driven by the increase in configurational entropy at higher temperatures. The thin solid lines in Figure 13b show the effect of long-range ordering on the a and c cell parameters, the dashed lines show the effect of short-range ordering. The thick solid line shows the sum of the long- and short-range effects. In the case of the a cell parameter, the strains due to decreasing Q and increasing σ compensate each other as the transition temperature is approached. This leads to a rather smooth variation in a as function of T, with no sharp change in a at $T = T_c$. In the case of the c cell parameter, the two strain components reinforce each other, leading to a large and abrupt change in c at $T = T_c$, as is observed in Figure 13. According to the arguments above, one expects this effect to be more obvious for bulk compositions close to ilm100, as indeed can be seen. Recent Monte Carlo simulations of this system (Harrison et al. 2000b) confirm this interpretation of the strain effects, with the short-range order parameters due to nearest and next-nearest neighbour cation interactions behaving much as is shown schematically in Figure 13a.

Thermodynamics and kinetics of non-convergent disordering in olivine

The temperature dependence of non-convergent cation exchange between the M1 and M2 octahedral sites of olivine has been the subject of a number of recent neutron diffraction studies, from the single crystal studies of members of the forsterite-fayalite solid solution (Artioli et al. 1995, Rinaldi and Wilson 1996) to powder diffraction studies of the same system (Redfern et al. 2000) as well as the Fe-Mn, Mg-Mn, and Mg-Ni systems (Henderson et al. 1996, Redfern et al. 1996, 1997a, 1998). The high-temperature behaviour of Fe-Mg order-disorder appears to be complicated by crystal field effects, which influence the site preference of Fe^{2+} for M1 and M2, but the cation exchange of the Fe-Mn, Mg-Mn, and Mg-Ni olivines is dominantly controlled by size effects: the larger M2 site accommodating the larger of the two cations in each pair (Mn or Ni, in these cases). In all these experiments, the use of time-of-flight neutron powder diffraction allowed the measurement of states of order at temperatures in excess of 1000°C under controlled oxygen fugacities (especially important given the variable oxidation states that many of the transition metal cations of interest can adopt). Diffraction patterns were collected on the POLARIS diffractometer at the ISIS spallation source (Hull et al. 1992). Structural data were then obtained by Rietveld refinement of the whole patterns giving errors in the cell parameters of about 1 part in 70,000 and estimated errors in the site occupancies of ~0.5% or less. The low errors in refined occupancies result principally from the fact that the contrast between Mn (with a negative scattering length) and the other cations is very strong for neutrons.

All experiments showed the same underlying behaviour of the order parameter. This can be modelled according to a Landau expansion for the free energy of ordering, of the type given in Equation (12). In each case studied (Fig. 14) the order parameter remains constant at the start of the heating experiment, then increases to a maximum before following a steady decline with T to the highest temperatures. This general behaviour reflects both the kinetics and thermodynamics of the systems under study: at low temperatures the samples are not in equilibrium and reflect the kinetics of order-disorder, at high temperatures the states of order are equilibrium states, reflecting the thermodynamic drive towards high-temperature disorder. The initial increase in order results from the starting value being lower than equilibrium, and as soon as the temperature is high enough for thermally activated exchange to commence (on the time scale of the experiments), the occupancies of each site begin to converge towards the equilibrium order-disorder line. Using Ginzburg-Landau theory, which relates the driving force for ordering to the rate of change of order, one can obtain a kinetic and thermodynamic description of the non-convergent disordering process from a single experiment (e.g. Redfern et al. 1997a).

These studies of cation ordering in olivines have shown that, in most cases, the degree of M-site order measured at room temperature is an indication of the cooling rate of a sample, rather than the temperature from which it has cooled. Calculated Q-T cooling pathways for a Fe-Mn olivine are shown in Figure 15, where it is shown that variations in cooling rate over 13 decades might be ascertained from the degree of order locked in to room temperature. This indicates that M-site ordering in olivine might be employed as a practical geospeedometer. However, it is clear that more complex cooling paths may reset the degree of order on, say, re-heating. It would then be difficult to trace back a complex cooling history from a single measurement of Q at room temperature. In the case of these more complex cooling histories, M-site order measurements would have to be carried out in conjunction with other measurements using additional speedometers. The *in situ*

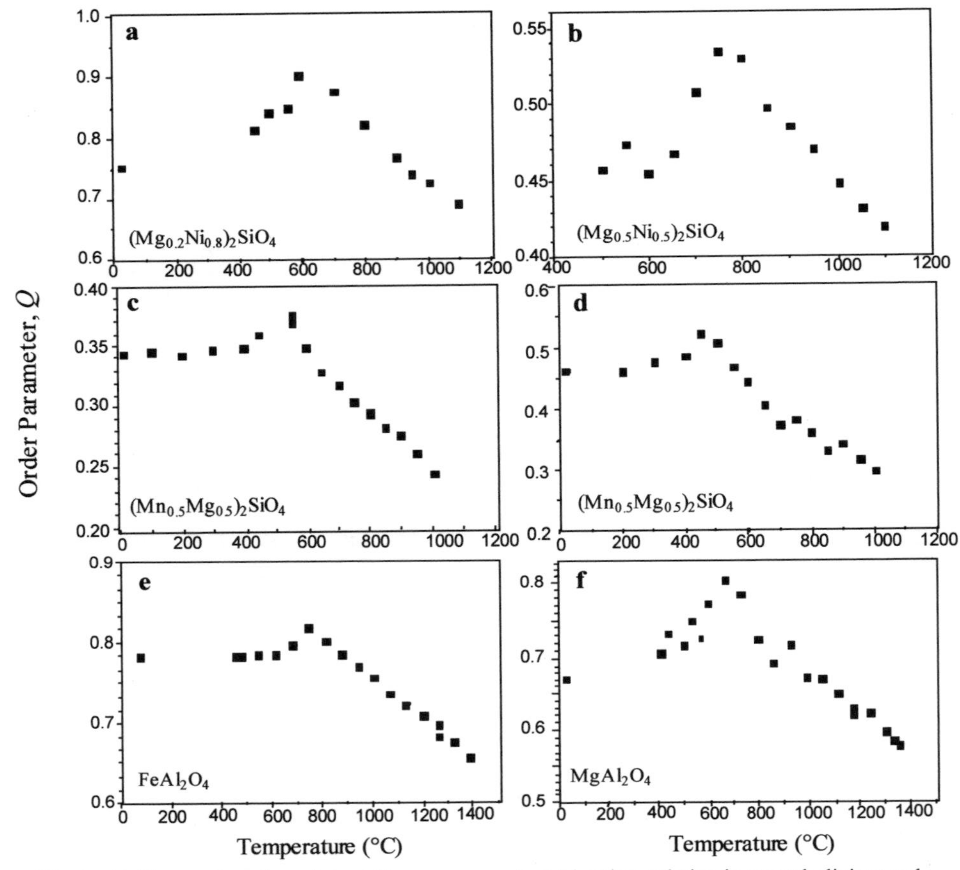

Figure 14. Temperature dependence of non-convergent metal cation-ordering in several olivines and spinels, all measured by Rietveld refinement of neutron powder diffraction data.

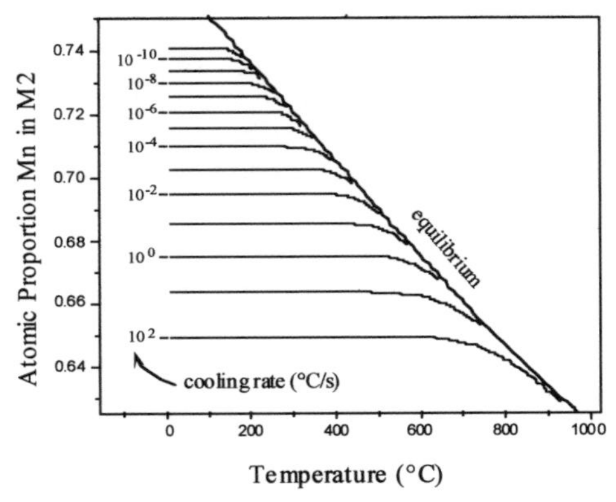

Figure 15. Calculated cooling paths over 13 decades of cooling rates showing dependence of the low-temperature site occupancies of $FeMnSiO_4$ on cooling rate. The room-temperature site occupancy, or degree of non-convergent order, is a direct measure of the cooling rate of the sample.

studies that have been performed in recent years at Rutherford Appleton Laboratory have allowed the temperature dependence of this ordering to be determined accurately to high *T*. In these cases *in situ* study has been essential, since high-temperature disordered states are generally non-quenchable, due to the fast kinetics of cation exchange in olivines and spinels, and the unavoidable re-equilibration of samples on quenching from annealing conditions. Thus, neutron diffraction techniques are invaluable for directly determining the long range ordering characteristics of these important rock-forming minerals.

Modelling non-convergent order-disorder in spinel

Magnesium aluminate spinel ($MgAl_2O_4$) typifies the process of non-convergent ordering-disorder, with Mg^{2+} and Al^{3+} cations ordering over both tetrahedral and octahedral sites without a change in symmetry (Redfern et al. 1999). Warren et al. (2000a,b) used *ab initio* electronic structure calculation methods to calculate the interaction energies that drive the ordering. These methods were required because the ordering over both tetrahedral and octahedral sites necessitates the proper calculation of chemical potential terms, which cannot be estimated from empirical computational models. The energies of interaction were modelled using a Hamiltonian that incorporated three-atom interactions, and these were fed into a Monte Carlo simulation of the long range ordering. The temperature-dependence of the ordering is shown in Figure 16, where it is compared with the experimental results. While the agreement is not perfect, it is remarkably good given that the calculations were made with no input from the experimental data.

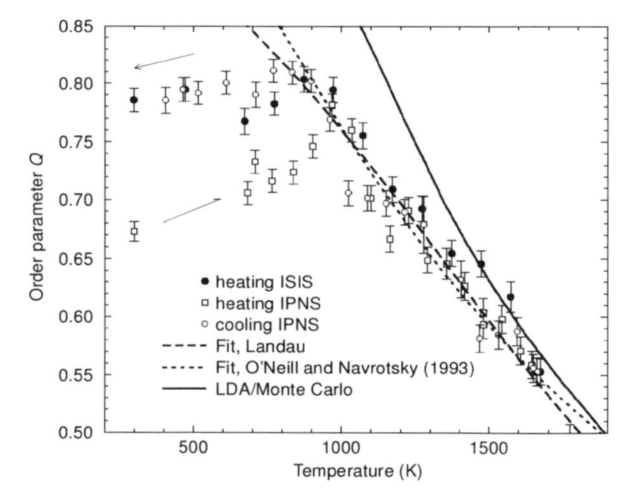

Figure 16. The temperature dependence of the order parameter for spinel calculated in the Monte Carlo simulations of Warren et al. (2000a,b). Experimental data points from Redfern et al. (1999) are shown for comparison.

Bilinear coupling of Q and Q_{od} in albite

Having discussed some of the background to the models that have been used to underpin studies of order-disorder in minerals, let us finally turn to a few of the main results from studies of strain–order-disorder coupling that have been reported for a number of phases. It is worth first considering the case of the high-temperature behaviour of albite ($NaAlSi_3O_8$), as this mineral provides, in many respects, the archetypal example of order parameter coupling in framework silicates. Above around 1260 K albite is monoclinic, $C2/m$, monalbite. On cooling two processes give rise to a transition to $C\bar{1}$ symmetry, ordering of Al and Si in the framework, and a displacive collapse of the framework. Until the mid 1980s these two aspects of the high-temperature behaviour of albite had been treated separately, but each generates a triclinic spontaneous strain which behaves as the active representation for the transition, B_g. Thus, the rate of Al/Si ordering in albite can conveniently be measured in quenched samples by simple powder diffraction methods, as was first demonstrated by Mackenzie in 1957. The displacive transition has also been measured in a number of high-temperature studies. It was not until 1985 that the nature and effects of the bilinear coupling between these two processes was fully elucidated (Salje et al. 1985), and the phase transition was described successfully in terms of a 2-4-6 Landau potential in two order parameters. One is termed Q_{od} and represents A/Si ordering on the tetrahedral sites, the other is termed Q and refers to the displacive transition of the entire framework structure, related to an elastic instability of the structure.

While both processes behave as the active representation, it is found that Q_{od} has a greater influence on the shear strain corresponding to cosγ, while Q dominates the behaviour of cosα^*. This separation of the two order parameters between two shears ε_4 and ε_6 is a common feature of feldspars. The essential result for albite is that, in equilibrium, the development of ordering in the $C\bar{1}$ case is enhanced at high temperatures by the onset of non zero Q, which in turn increases more rapidly at the point (crossover) where Q_{od} would develop in the absence of a displacive transition. It is clear that a transition to $C2/m$ cannot be observed unless both Q and Q_{od} go to zero. Since the kinetics of Al/Si disordering in albites is relatively sluggish, the condition that Q_{od} goes to zero may not be easily attained in well-ordered albites under typical heating rates. The dramatic effect that Al/Si ordering has on the displacive phase transition in albite can be seen from Figure 17, where it is shown that the critical elastic softening at the monoclinic-triclinic displacive transition as a function of temperature is hardened by the onset of ordering. This result follows from computational studies of the displacive transition using mixed Al/Si and Na-K potentials to represent long range order variations on the tetrahedral sites and mixing on the alkali cation site respectively (Redfern et al. 1997b), and is derived from a Landau model fit to those data.

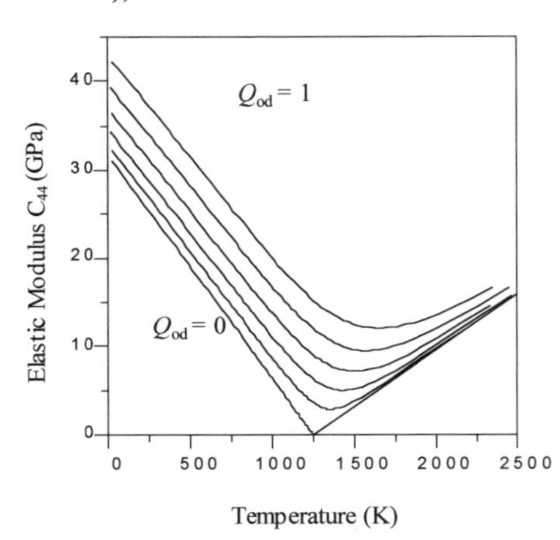

Figure 17. The evolution of the computed C_{44} elastic constant of albite with temperature and differing degrees of Al/Si order, Q_{od}. The C_{44} elastic constant in analbite acts almost as the critical elastic constant as it almost softens to zero. Increasing Al/Si order stiffens the elastic constant (after Wood 1998).

A further consequence of the interplay between the kinetically controlled Q_{od} and spontaneous strain in albite is that, in an ordering or disordering experiment out of equilibrium, a distribution of Q_{od} states exists which can be observed in the evolution of diffraction peaks susceptible to changes in the γ angle (Carpenter and Salje 1989). Furthermore, in kinetically disordered crystals there is a tendency to form spatially non-uniform distributions of the order parameter, and modulated "tweed" microstructures typically develop. Albite has the distinction of being one of the first described by a Landau potential involving bilinearly coupled Q and Q_{od}, and also being one of the first examples of the application of the kinetic extension of Landau theory to minerals (Wruck et al. 1991). Other studies (for example on cordierite, see below) have shown that the incorporation of chemical compositional variation is likely to lead to the further stabilisation of modulated microstructures in kinetically disordered Na-rich plagioclases. In a number of ways, therefore, we see that albite has been a test bed for ideas about the relationship between order-disorder and ferroelastic phenomena in minerals. We next consider how such ideas apply in other mineral systems and in framework structures more generally.

The *P6/mcc–Cccm* transition in pure and K-bearing cordierite: influence of chemical variation.

Cordierite ($Mg_2Al_4Si_5O_{18}$) is one of the few other framework aluminosilicates for which the kinetic order-disorder behaviour has been studied in some detail. Under equilibrium conditions it is orthorhombic below 1450°C, a hexagonal polymorph being stable above that temperature (Schreyer and Schairer 1961, Putnis 1980). Monte Carlo simulations of the equilibrium ordering give the Q-T dependence shown in Figure 9. The transition between the two structures is associated with changes in Al/Si order on the tetrahedral sites (Fig. 2): the hexagonal form cannot accommodate any long-range order of the Al and Si atoms, whereas the orthorhombic structure can attain complete Al/Si order. Glasses, annealed below T_c, crystallise the hexagonal form initially, and this then transforms to the stable orthorhombic polymorph via a modulated intermediate (Putnis et al 1987). The symmetry relations of the high and low-forms of cordierite require that the transformation be first-order in thermodynamic character, but it seems that Al/Si order develops somewhat continuously as a function of time in annealed samples. Nonetheless, the development of Q_{od} does not have a straightforward effect on Q in the manner found in albite, and a first-order step in macroscopic spontaneous strain is observed in annealed samples.

The structure of cordierite (Fig. 2) accommodates additional elements within the channels running parallel to z. In most cases H_2O molecules are present, but Na and K are also incorporate in natural crystals. Redfern et al. (1989b), therefore investigated the influence of K-incorporation on the *P6/mcc–Cccm* transition. Such incorporation directly affects the Al/Si ordering behaviour since the substitution of K^+ into the channels is accompanied by the exchange of Al^{3+} for Si^{4+} in the framework to maintain charge balance. On a macroscopic length scale, the phase transition appears to be triggered by a critical degree of Q_{od}, causing a sudden distortion of the structure from hexagonal to orthorhombic symmetry. This distortion has traditionally, and erroneously, been used as a measure of Q_{od}, the

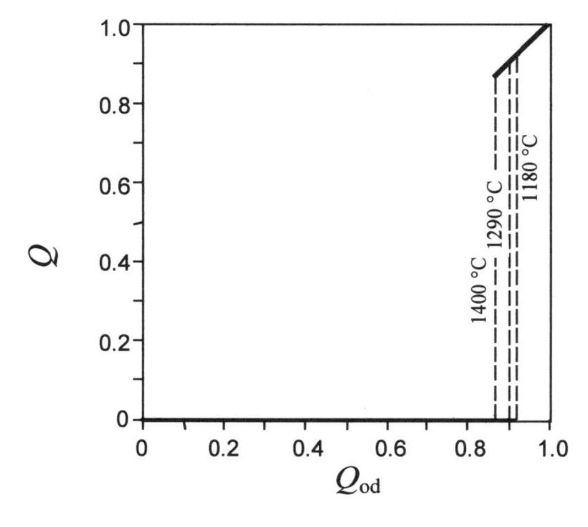

Figure 18. Order parameter vector space for cordierite. Significant ordering takes place on annealing without a macroscopic orthorhombic distortion. The development of such a strain occurs in a stepwise fashion at a critical value of the degree of Al/Si order, Q_{od}, which varies with annealing temperature (after Putnis et al. 1987).

fallacy of which is apparent from the form of the Q-Q_{od} relations found in pure Mg-cordierite by Putnis et al. (1987). As can be seen from Figure 18 (above), there is only a simple relationship between Q_{od} and Q *after* the sample has transformed to orthorhombic symmetry. Furthermore, the critical degree of Q_{od} necessary to trigger the development of a macroscopic orthorhombic strain appears to depend upon the temperature of annealing.

At the shortest annealing times a single Gaussian hexagonal (211) peak is observed in synchrotron X-ray diffraction patterns of K-cordiertite. With longer annealing this peak becomes broadened, corresponding to the onset of a modulated phase, analogous with that observed in pure Mg-cordierite (Redfern et al. 1989b). The width of the (211) peak corresponding to the modulated phase remains constant with further annealing, and precursors the development of the orthorhombic triplet. As soon as this triplet appears, it is already almost completely distorted with respect to the hexagonal cell. The distortion index in K-bearing cordierites does not, therefore, undergo a continuous change with annealing, but undergoes a sudden transformation from zero to 0.155 on transformation to *Cccm*, with a maximum value of $\Delta = 0.17$. These Δ values for K-bearing cordierites are significantly lower than those observed in pure Mg-cordierite, and it seems that the incorporation acts a conjugate field against the order parameter Q.

Cordierite accommodates a large degree of short range Al/Si order (as large as $Q_{od} = 0.9$) while still remaining macroscopically hexagonal, and it is clear that the microscopic strain modulation is one mechanism by which it achieves this, ordering in a macroscopically strain-free manner. The K-containing cordierite must attain a lower degree of Al/Si order than this, but in addition the maximum spontaneous strain developed in the orthorhombic phase is lowered. The effect of doping with K can be thought of in terms of the development of a homogeneous field due to the combined local stress fields of all the individual K^+ ions. Within the Landau formalism, this corresponds to a conjugate field to the order parameter, and the Landau potential for K-bearing cordierite can be expressed as:

$$\Delta G(Q_{od},Q) = \frac{1}{2}a_{od}Q_{od}^2 + \frac{1}{3}b_{od}Q_{od}^3 + \frac{1}{4}c_{od}Q_{od}^4 + \frac{1}{2}aQ^2 + \frac{1}{3}bQ^3 + \frac{1}{4}cQ^4 + \lambda Q_{od}Q + HQ_{od} + hQ + \frac{1}{2}\gamma_{od}(\nabla Q_{od})^2 + \frac{1}{2}\gamma(\nabla Q)^2 \quad (16)$$

where the last four terms represent the coupling with the conjugate fields H and h and the fluctuational Ginzburg terms. The relative stabilities and kinetic behaviour of cordierite depends upon the homogeneity of H and h. Spatial variation of these fields, due to chemical inhomogeneity of incorporated K^+, will lead to the relative stabilisation of the modulated form, as has been pointed out by Michel (1984). The interval of annealing times over which modulated cordierites are found is greater for K-bearing samples than for pure K-free crystals. The introduction of defects such as K^+ within the cordierite structure has a significant influence on the stability and kinetic behaviour of cordierite therefore. Since this is a general feature of phase transitions where a defect stress field acts as a conjugate order parameter, it follows that this is a significant effect in minerals which show solid solution, order disorder, and elastic deformation, for example plagioclase and alkali feldspars.

Ferroelasticity and order/disorder in leucite-related frameworks

Finally, let us consider some aspects of ordering that have been noted for leucites. The leucite framework structure represents an extremely stable topological arrangement. Recent studies have shown that, although comparatively insignificant in nature, the known chemical extent of the leucite family of $X^{I}_{2}(Y^{II}_{x}Z^{III}_{1-2x})Si_{2+x}O_6$ $(0 < x < 0.5)$ compounds is expanding rapidly. Leucite-related compounds have, for example, recently been investigated with an eye to their catalytic properties, especially those with transition metals substituting for Al and Si in the aluminosilicate tetrahedral framework (Heinrich and Baerlocher 1991). Investigations of synthetic leucite analogues with X = K, Rb, Cs; Y = Mg, Zn, Cu, Cd; Z = Al, Fe have revealed a range of symmetrically distinct structures with the same leucite topology (Bell et al. 1994a,b; Bell and Henderson 1994). These studies have highlighted the fact that within anhydrous leucite and its related compounds, three structural phenomena may occur: (1) instabilities of the tetrahedral framework may lead to displacive transitions, (2) ordering of tetrahedral cations may take place on the T-sites, and (3) the size and dynamic behaviour of the alkali cation in the 'W'-site may influence either of the above processes.

Of these, the role of Al/Si order/disorder in natural leucite, and its relation to displacive distortions, has been the subject of much attention. In contrast to other framework aluminosilicates, where strong coupling is often observed between displacive instabilities and Al/Si ordering (e.g. in feldspars), it now seems clear that long-range Al/Si order is only weakly coupled (if at all) to the displacive cubic-tetragonal phase transition in $KAlSi_2O_6$ leucite (Dove et al. 1993). Dove et al. (1996) attribute this to a low ordering temperature for leucite, which they associate with dilution of Al in the tetrahedral network (compared to an Al:Si ratio of 1:1). From calculations of the exchange interaction for ordering, J_1 (equal to the energy difference between an Al-O-Al linkage plus Si-O-Si as against two Al-O-Si linkages), and estimates of the second-nearest-neighbour interaction, J_2, they use a modified Bragg-Williams approach to arrive at an order-disorder transition for leucite of 300°C. They argue that the sluggishness of Al/Si ordering kinetics below such a low T_c renders the process insignificant in leucite, and explains why long-range Al/Si ordering is not found in this mineral.

Tetrahedral cation ordering *is* observed in certain leucite analogues, however. For example, while it appears that Al shows little ordering over the three T-sites of natural $I4_1/a$ $KAlSi_2O_6$, studies of synthetic Fe-leucites ($KFeSi_2O_6$) show that Fe^{3+} tends to order preferentially on T_3 and T_2 rather than T_1 (Bell and Henderson 1994, Brown et al. 1987). More dramatic, however, is the behaviour of tetrahedral cations in certain $K_2Y^{2+}Si_5O_{12}$ leucites (Y^{2+} = Mg, Zn, Cd; Redfern 1994). When synthesised hydro-thermally at relatively low temperatures each of these compounds crystallises as a well-ordered low-symmetry leucite framework, while high-temperature dry synthesis from oxides yields disordered structures isomorphous with the $Ia3d$ high-temperature structure of natural $KAlSi_2O_6$. Coupling between tetrahedral ordering and macroscopic strain in these leucite-analogues is very strong indeed, as is evidenced by the fact that the ordered polymorphs crystallise in low-symmetry monoclinic and orthorhombic structures with strains of a few percent compared to the disordered $Ia3d$ aristotype. Why should their behaviour appear to be so different from that of natural K-Al leucite?

The arguments of Dove et al. (1996) point to dilution as the reason for the low ordering temperature of $KAlSi_2O_6$. It seems paradoxical, therefore, that the $K_2Y^{2+}Si_5O_{12}$ ordered leucites have Y:Si ratios corresponding to even greater dilution, yet they seem to have higher temperature order/disorder transitions. In order to shed further light on the nature and role of order/disorder and its coupling to elastic transitions in these framework structures, Redfern and Henderson (1996) have carried out a study of the high-temperature behaviour of $K_2MgSi_5O_{12}$ leucite which is related to natural leucite by the coupled substitution: $2Al^{3+}$ $Mg^{2+} + Si^{4+}$. The hydrothermally-synthesised monoclinic polymorph has twelve symmetrically distinct tetrahedral sites, while the cubic structure has just one. It remains to be seen why $K_2MgSi_5O_{12}$ can order its tetrahedral cations but $KAlSi_2O_6$ cannot, despite there being a higher dilution of the ordering cation (Mg) in the former. The answer is expected to lie in the relative magnitudes of the J_1 and J_2 (first- and second-nearest neighbour) exchange interaction terms. The size of the MgO_4 tetrahedron is considerably larger than the AlO_4 tetrahedron, thus there will be an enhanced "avoidance rule" in $K_2MgSi_5O_{12}$, compared with $KAlSi_2O_6$, and this appears to be the most important control on T_c for ordering (and hence kinetic accessibility of ordered and disordered states) in this suite of materials. This illustrates once more the importance of the relationship between ordering and elastic interactions in framework minerals.

CONCLUSIONS

The examples of the high-temperature behaviour of albite, cordierite and leucite illustrate some of the most important features of order parameter coupling in framework minerals. We have noted that chemical inhomogenieity can enhance the stability of modulated microstructures in kinetically controlled ordering or disordering experiments, and that K-doped cordierite provides an example of this effect. The coupling of order-disorder with macroscopic strain is a useful phenomenon for the experimentalist, as it can allow techniques such as X-ray diffraction and infrared spectroscopy to be employed to chart the progress of transformations in minerals. If the coupling between Q and Q_{od} is not simple, however, such an approach may have pitfalls. This is seen in the case of cordierite, where the macroscopic strain is a very poor indicator of the development of short-range Q_{od} in the modulated and hexagonal phases. In contrast to the behaviour of albite and cordierite, the elastic cubic-tetragonal phase transition in leucite is independent of Al/Si ordering in the framework. If the exchange interaction for ordering is modified by chemical substitution of Al by larger cations, however, the local strain associated with low temperature disorder is enhanced and ordered polymorphs are stabilised.

We have also seen the benefits of combining accurate experimental observations of order-disorder transitions with realistic theories and models in developing a thorough physical picture of the origins of these processes. In this chapter there has not been the space to review the plethora of studies of order-disorder transitions in minerals that have been published. We have not touched at all on ordering in chain silicates, sheet silicates, and the many other framework and orthosilicates which show numerous order-disorder transitions, both convergent and non-convergent. I hope, however, that I have provided the reader with a starting point for the further exploration of these phenomena, which are all related and united by the over-riding thermodynamic and kinetic controls that dominate their characteristics and the properties of the materials in which they occur.

ACKNOWLEDGEMENTS

I gratefully and freely acknowledge the fruitful collaborations that have illuminated my studies of order-disorder in minerals. In particular I would like to thank Martin Dove, Ekhard Salje, Richard Harrison, Michael Carpenter and Michael Henderson.

REFERENCES

Artioli, G., Rinaldi, R., Wilson, C.C., Zanazzi, P.F. (1995). High temperature Fe-Mg cation partitioning in olivine: *In situ* single-crystal neutron diffraction study. Am Mineral 80:197-200

Bell, A.M.T., Henderson, C.M.B. (1994) Rietveld refinement of the structures of dry-synthesized $MFe^{III}Si_2O_6$ leucites (M = K, Rb, Cs) by synchrotron X-ray powder diffraction. Acta Crystallogr C50:1531-1536

Bell, A.M.T., Henderson, C.M.B., Redfern, S.A.T., Cernik, R.J., Champness, P.E., Fitch, A.N., Kohn, S.C. (1994a) Structures of synthetic $K_2MgSi_5O_{12}$ leucites by integrated X-ray powder diffraction, electron diffraction and ^{29}Si MAS NMR methods. Acta Crystallogr B50:31-41

Bell, A.M.T., Redfern, S.A.T., Henderson, C.M.B., Kohn, S.C. (1994b) Structural relations and tetrahedral ordering pattern of synthetic orthorhombic $Cs_2CdSi_5O_{12}$ leucite: a combined synchrotron X-ray powder diffraction and multinuclear MAS NMR study. Acta Crystallogr B50:560-566

Bertram, U.C., Heine, V., Jones, I.L., Price, G.D. (1990) Computer modelling of Al/Si ordering in sillimanite. Phys Chem Minerals 17:326-333

Bragg, W.L., Williams, E.J. (1934) The effect of thermal agitation on atomic arrangement in alloys. Proc Royal Soc A 145:699-729

Brown, I.W.M., Cardile, C.M., MacKenzie, K.J.D., Ryan, M.J., Meinhold, R.H. (1987) Natural and synthetic leucites studied by solid state 29-Si and 27-Al NMR and 57-Fe Mössbauer spectroscopy. Phys Chem Minerals 15:78-83

Brown, N.E., Navrotsky, A., Nord, G.L., Banerjee, S.K. (1993) Hematite (Fe_2O_3)–ilmenite ($FeTiO_3$) solid solutions: Determinations of FeTi order from magnetic properties. Am Mineral 78:941-951

Burton, B.P. (1987) Theoretical analysis of cation ordering in binary rhombohedral carbonate systems. Am Mineral 72:329-336

Carpenter, M.A. (1985) Order-disorder transformations in mineral solid solutions. *In* Keiffer, S.W., Navrotsky, A. (eds) Microscopic to Macroscopic: Atomic Environments to Mineral Thermodynamics. Rev Mineral 14:187-223

Carpenter, M.A. (1988) Thermochemistry of aluminium/silicon ordering in feldspar minerals. *In* Salje, E.K.H. (ed) Physical Properties and Thermodynamic Behaviour of Minerals. NATO ASI Series C 225:265-323. Reidel, Dordecht, The Netherlands

Carpenter, M.A., Salje, E.K.H. (1989) Time-dependent Landau theory for order/disorder processes in minerals. Mineral Mag 53:483-504

Carpenter, M.A., Salje, E.K.H. (1994a) Thermodynamics of non-convergent caion ordering in minerals: II Spinels and orthopyroxene solid solution. Am Mineral 79:770-776

Carpenter, M.A., Salje, E.K.H. (1994b) Thermodynamics of non-convergent caion ordering in minerals: III. Order parameter coupling in potassium feldspar. Am Mineral 79:1084-1098

Carpenter, M.A., Putnis, A., Navrotsky, A., McConnell, J.D.C. (1983) Enthalpy effects associated with Al,Si ordering in anhydrous Mg-cordierite. Geochim Cosmochim Acta 47:899-906

Carpenter, M.A., Powell, R., Salje, E.K.H. (1994) Thermodynamics of nonconvergent cation ordering in minerals: I. An alternative approach. Am Mineral 79:1053-1067

Davidson, P.M., Burton, B.P. (1987) Order-disorder in omphacite pyroxenes: A model for coupled substitution in the point approximation. Am Mineral 72:337-344

De Vita, A., Heine, V., McConnell, J.D.C. (1994) A first-principles investigation of Al/Si ordering. *In* Putnis, A. (ed) Proceedings of a Workshop on Kinetics of Cation Ordering (Kinetic Processes in Minerals and Ceramics). European Science Foundation, Strasbourg, France, p 34-43

Dove, M.T. (1999) Order/disorder phenomena in minerals: ordering phase transitions and solid solutions. *In* Catow, C.R.A., Wright, K.A. (eds) Microscopic Processes in Minerals. NATO ASI Series, 451-475. Kluwer, Amsterdam, Netherlands

Dove, M.T., Heine, V. (1996) The use of Monte Carlo methods to determine the distribution of Al and Si cations in framework aluminosilicates from ^{29}Si MAS-NMR data. Am Mineral 81:39-44

Dove, M.T., Cool, T., Palmer, D.C., Putnis, A., Salje, E.K.H., Winkler, B. (1993) On the role of Al-Si ordering in the cubic-tetragonal phase transition of leucite. Am Mineral 78:486-492

Dove, M.T., Thayaparam, S., Heine, V., Hammonds, K.D. (1996) The phenomenon of low Al-Si ordering temperatures in aluminosilicate framework structures. Am Mineral 81:349-362

Ginzburg, V.L., Levanyuk, A.P., Sobyanin, A.A. (1987) Comments on the applicability of the Landau theory for structural phase transitions. Ferroelectrics 73:171-182

Harrison, R.J., Redfern, S.A.T., Smith, R.I. (2000a) *In situ* study of the the $R\bar{3}$ to $R\bar{3}c$ transition in the ilmenite-hematite solid solution using time-of-flight neutron powder diffraction. Am Mineral 85:194-205

Harrison, R.J., Becker, U., Redfern, S.A.T. (2000b) Thermodynamics of the $R\bar{3}$ to $R\bar{3}c$ transition in the ilmenite-hematite solid solution. Am Mineral (in press)

Hatch, D.M., Ghose, S. (1989) Symmetry analysis of the phase transition and twinning in $MgSiO_3$ garnet: Implications for mantle mineralogy. Am Mineral 74:1221-1224

Hazen, R.M., Navrotsky, A. (1996) Effects of pressure on order-disorder reations. Am Mineral 81:1021-1035

Heinrich, A.R., Baerlocher, C. (1991) X-ray Rietveld structure determination of $Cs_2CuSi_5O_{12}$ A pollucite analogue. Acta Crystallogr C47:237-241

Henderson, C.M.B., Knight, K.S., Redfern, S.A.T., Wood, B.J. (1996) High-temperature study of cation exchange in olivine by neutron powder diffraction. Science 271:1713-1715

Hull, S., Smith, R.I., David, W.I.F., Hannon, A.C., Mayers, J. Cywinski, R. (1992) The POLARIS powder diffractometer at ISIS. Physica B80:1000-1002

Kroll, H., Knitter, R. (1991) Al, Si exchange kinetics in sanidine and anorthoclase and modeling of rock cooling paths. Am Mineral 76:928-941

Kroll, H., Schlenz, H., Phillips, M.W. (1994) Thermodynamic modelling of non-convergent ordering in orthopyroxenes: a comparison of classical and Landau approaches. Phys Chem Minerals 21:555-560

Landau, L.D. (1937) On the theory of phase transitions, part I. Sov Phys JETP 7:19ff

Mackenzie, W.S. (1957) The crystalline modifications of $NaAlSi_3O_8$. Am J Sci 255:481-516

Malcherek, T., Salje, E.K.H., Kroll, H. (1997) A phenomenological approach to ordering kinetics for partially conserved order parameters. J Phys: Condens Matter 9:8075-8084

Meyers, E.R., Heine, V., Dove, M. (1998) Thermodynamics of Al/Al avoidance in the ordering of Al/Si tetrahedral framework structures. Phys Chem Minerals 25:457-464

Michel, K.H. (1984) Phase transitions in strongly anharmonic and orientationally disordered crystals. Z Physik B 54:129-137

Nord, G.L., Lawson, C.A. (1989) Order-disorder transition-induced twin domains and magnetic properties in ilmenite-hematite. Am Mineral 74:160-176

O'Neill, H.St.C., Navrotsky, A. (1993) Simple spinels: Crystallographic parameters, cation radii, lattice energies, and cation distributions. Am Mineral 68:181-194

Parsonage, N.G., Staveley, L.A.K. (1978) Disorder in Crystals. Clarendon Press, Oxford

Phillips, B.L., Kirkpatrick, R.J. (1994) Short-range Si-Al order in leucite and analcime: determination of the configurational entropy from ^{27}Al and variable-temperature ^{29}Si NMR spectroscopy of leucite, its Cs- and Rb-exchanged derivatives and analcime. Am Mineral 79:125-1031

Putnis, A. (1980) The distortion index in anhydrous Mg-cordierite. Contrib Mineral Petrol 74:135-141

Putnis, A. (1992) Introduction to mineral sciences. Cambridge University Press Cambridge, UK

Putnis, A., Angel, R.J. (1985) Al,Si ordering in cordierite using "magic angle spinning" NMR: II. Models of Al,Si order from NMR data. Phys Chem Minerals 12:217-222

Putnis, A., Salje, E., Redfern, S.A.T., Fyfe, C.A., Stroble, H. (1987) Structural states of Mg-cordierite I: order parameters from synchrotron X-ray and NMR data. Phys Chem Minerals 14:446-454

Redfern, S.A.T. (1994) Cation ordering patterns in leucite-related compounds. *In* Putnis, A. (ed) Proceedings of a Workshop on Kinetics of Cation Ordering (Kinetic Processes in Minerals and Ceramics) European Science Foundation, Strasbourg, France

Redfern, S.A.T. (1992) The effect of Al/Si disorder on the $I\bar{1}-P\bar{1}$ co-elastic phase transition in Ca-rich plagioclase. Phys Chem Minerals 19:246-254

Redfern, S.A.T., Henderson, C.M.B. (1996) Monoclinic-orthorhombic phase transition in $KMg_{0.5}Si_{2.5}O_6$ leucite. Am Mineral 81:369-374

Redfern, S.A.T., Salje, E., Navrotsky, A. (1989a) High-temperature enthalpy at the orientational order-disorder transition in calcite: implications for the calcite/aragonite phase equilibrium. Contrib Mineral Petrol 101:479-484

Redfern, S.A.T., Salje, E., Maresch, W., Schreyer, W. (1989b) X-ray powder diffraction and infrared study of the hexagonal to orthorhombic phase transition in K-bearing cordierite. Am Mineral 74:1293-1299

Redfern, S.A.T., Henderson, C.M.B., Wood, B.J., Harrison, R.J., Knight, K.S. (1996) Determination of olivine cooling rates from metal-cation ordering. Nature 381:407-409

Redfern, S.A.T., Henderson, C.M.B., Knight, K.S., Wood, B.J. (1997a) High-temperature order-disorder in $(Fe_{0.5}Mn_{0.5})_2SiO_4$ and $(Mg_{0.5}Mn_{0.5})_2SiO_4$ olivines: an *in situ* neutron diffraction study. Eur J Mineral 9:287-300

Redfern, S.A.T., Dove, M.T., Wood, D.R.R. (1997b) Static lattice simulation of feldspar solid solutions: ferroelastic instabilities and order/disorder. Phase Trans 61:173-194

Redfern, S.A.T., Knight, K.S., Henderson, C.M.B., Wood, B.J. (1998) Fe-Mn cation ordering in fayalite-tephroite $(Fe_xMn_{1-x})_2SiO_4$ olivines: a neutron diffraction study. Mineral Mag 62:607-615

Redfern, S.A.T., Harrison, R.J., O'Neill, H.St.C., Wood, D.R.R. (1999) Thermodynamics and kinetics of cation ordering in $MgAl_2O_4$ spinel up to 1600°C from *in situ* neutron diffraction. Am Mineral 84:299-310

Redfern, S.A.T., Artioli, G., Rinaldi, R., Henderson, C.M.B., Knight, K.S., Wood, B.J. (2000) Octahedral cation ordering in olivine at high temperature. II: An *in situ* neutron powder diffraction study on synthetic $MgFeSiO_4$ (Fa50). Phys Chem Minerals (in press)

Rinaldi, R., Wilson, C.C. (1996). Crystal dynamics by neutron time-of-flight Laue diffraction in olivine up to 1573K using single frame methods. Solid State Commun 97:395-400

Ross, C.R., II (1990) Ising models and geological applications. *In* Ganuly, J (ed) Diffusion Atomic Ordering and Mass Transport. Advances in Physical Geochemistry 8:51-90, Springer Verlag, Berlin, Germany

Salje, E.K.H. (1990) Phase Transitions in Ferroelastic and Co-elastic Crystals. Cambridge University Press Cambridge, UK, 366 p

Salje, E., Devarajan, V. (1986) Phase transitions in systems with strain-induced coupling between two order parameters. Phase Trans 6:235-248

Salje, E., Kuscholke, B., Wruck, B., Kroll, H. (1985) Thermodynamics of sodium feldspar II: experimental results and numerical calculations. Phys Chem Minerals 12:99-107

Salje, E.K.H., Carpenter, M.A., Malcherek, T., Boffa, Ballaran, T. (2000) Autocorrelation analysis of infrared spactra from minerals. Eur J Mineral 12:503-520

Schreyer, W., Schairer, J.F. (1961) Compositions and structural states of anhydrous Mg-cordierites: a reinvestigation of the central part of the system $MgO-Al_2-SiO_2$. J Petrol 2:324-406

Thayaparam, S., Dove, M.T., Heine, V. (1994) A computer simulation study of Al/Si ordering in gehlenite and the paradox of low transition temperature. Phys Chem Minerals 21:110-116

Thayaparam, S., Heine, V., Dove, M.T., Hammonds, K.D. (1996) A computational study of Al/Si ordering in cordierite. Phys Chem Minerals 23:127-139

Thompson, J.B. Jr. (1969) Chemical reactions in crystals. Am Mineral 54:341-375

Vinograd, V.L., Putnis, A. (1999) The description of Al,Si ordering in aluminosilicates using the cluster variation method. Am Mineral 84:311-324

Warren, M.C., Dove, M.T., Redfern, S.A.T. (2000a) *Ab initio* simulation of cation ordering in oxides: application to spinel. J Phys: Condensed Matter 12:L43-L48

Warren, M.C., Dove, M.T., Redfern, S.A.T. (2000b) Disordering of $MgAl_2O_4$ spinel from first principles. Mineral Mag 64:311-317

Wood, D.R.R. (1998) Aluminium/Silicon Ordering in Na-feldspars. PhD Dissertation, University of Manchester, UK

Wruck, B., Salje, E.K.H., Graeme-Barber, A. (1991) Kinetic rate laws derived from order parameter theory IV: kinetics of Al, Si disordering in Na feldspars. Phys Chem Minerals 17:700-710

Yeomans, J.M. (1992) Statistical Mechanics of Phase Transitions. Clarendon Press, Oxford, UK

Ziman, J.M. (1979) Models of Disorder. Cambridge University Press Cambridge, UK

6 Phase Transformations Induced by Solid Solution

Peter J. Heaney

Department of Geosciences
Pennsylvania State University
University Park, Pennsylvania 16802

INTRODUCTION

Small concentrations of impurities can create profound differences in the thermodynamic stability and the physical behavior of crystalline materials. The dramatic changes produced by chemical substitutions are perhaps best illustrated by the discovery of high-T_c superconducting oxides by Bednorz and Müller in 1986. As a pure endmember, the cuprate that revolutionized solid state physics is an insulator. However, when small amounts of Ba^{2+} or Sr^{2+} replace La^{3+} in La_2CuO_4, the doping induces a series of surprising transformations. At low levels of substitution, $La_{2-x}(Sr,Ba)_xCuO_{4-y}$ remains insulating, but the length scale of antiferromagnetic ordering drops precipitously. With slightly higher concentrations (~0.10 < x < ~0.18), the compound becomes a super-conductor, with critical temperatures as high as 40 K for x_{Sr} = 0.15 (Tarascon et al. 1987). In the four years that followed the announcement of Bednorz and Müller's discovery, materials scientists pushed the superconducting critical temperature in doped cuprates beyond 100 K and published more than 18,000 papers in the process (Batlogg 1991).

The systematic exploration of compositional diversity and its influence on crystal structure has long been a pursuit of materials scientists who attempt to tailor substances for specific technological applications. Geologists, by comparison, have expended much less energy on the study of impurities and their influence on transition behavior in minerals. This inattention is surprising in view of the fact that few natural materials conform exactly to their idealized compositions, and the transitional properties of "dirty" minerals may depart considerably from those of the pure endmember. For example, interstitial and/or substitutional atoms in a mineral structure can: (1) change the stability fields of polymorphs relative to each other; (2) alter the energetics of transformation between polymorphs; (3) stabilize incommensurate phases; (4) decrease the characteristic length scales of twin domains and twin walls; and (5) induce structural transformations isothermally.

Studies of transition behavior in compositionally impure minerals thus offer insights into real rather than idealized mineral transformations. These effects are important among the low-density silicates that constitute the bulk of the Earth's crust, where substitutions of tunnel and cavity ions can occur comparatively freely within the open frameworks of these structures. In addition, substitution reactions are significant in mantle minerals, where Fe-Mg exchange is especially important and can control the transition behavior of a host of Fe-Mg silicates and oxides (reviewed in Fei 1998). Likewise, the iron in the inner core is alloyed with a light element whose identity remains a matter of much debate (Sherman 1995, Alfé et al. 1999), but whose presence may influence the crystalline structure of the core.

The effects of substitutional ions on transition behavior are also important for what they tell us about mineral systems that cannot be observed directly, either because the minerals are thermodynamically unstable or because they cannot be retrieved. Scientists have recognized for some time that substitutions of large atoms for smaller ones in crystal

1529-6466/00/0039-0006$05.00

structures can mimic changes induced by temperature or pressure; V.M. Goldschmidt (1929) was probably the first to observe that "the isomorphic tolerance and the thermal tolerance of crystals appear to be closely related." Consequently, controlled doping of ions in certain mineral structures has allowed geochemists to infer the crystallographic attributes of minerals that are not directly accessible either by field work or by high-pressure laboratory investigations. For example, studies of germanates as analogs to silicates continue to provide insights into the crystal chemistry of high-pressure mantle minerals, such as olivine, spinel, garnet, and perovskite (e.g. Kazey et al. 1982, Gnatchenko et al. 1986, Durben et al. 1991, Liu et al. 1991, Rigden and Jackson 1991, Andreault et al. 1996, Petit et al. 1996).

Although nature has in some instances provided suites of minerals that allow a careful comparison of the variation of structure and transition properties with changing composition (e.g. Carpenter et al. 1985), in many cases natural minerals contain a multiplicity of cationic substitutions that considerably complicate the structural response. Consequently, some of the best-constrained investigations of solid solution effects on phase transitions involve synthetic materials with tightly controlled compositional sequences. These synthetic solids may not boast exact counterparts in nature, but the behaviors observed in these compounds are similar to those seen in natural systems, and they can suggest transition properties that have gone unnoticed in rock-forming minerals. For example, ferroelectric perovskites are uncommon in crustal rocks, but their industrial importance has inspired an enormous body of research from materials scientists. This chapter will review a very small part of the voluminous work that materials scientists have assembled in their characterizations of doped perovskites and perovskite-like compounds, and it will also discuss a number of the more systematic explorations of chemically induced transitions in minerals and mineral analogs.

CONCEPTS OF MORPHOTROPISM

A brief historical background

That solids with different compositions can adopt identical crystal shapes was documented in 1819 by Mitscherlich, who called the phenomenon *isomorphism* (Mitscherlich 1819, Melhado 1980). Isomorphism can describe phases with similar atomic architectures but unlike constituents, such as NaCl and PbS, and it also can refer to members of a continuous solid solution series, such as the olivine group with formula $(Mg,Fe)_2SiO_4$. Three years later, Mitscherlich documented the complementary property of *polymorphism*, whereby phases with identical compositions occur as different structures (Mitscherlich 1822). Although mineralogists of the nineteenth century recognized the important inter-relationship between crystal structure and composition, the crystallographic probes available for structure determination did not keep pace with advances in wet chemical analysis. Consequently, understanding the effects that chemical modifications exert on crystal structures could be revealed only by careful measurements of subtle variations in habit.

The leading figure in this mostly ill-fated effort was Paul von Groth, who coined the term *morphotropism* to describe the changes in crystal form that are induced by chemical substitution (Groth 1870). In one striking success, morphological crystallography revealed that the addition of olivine-like chemical units to the mineral norbergite [$Mg_2SiO_4{\cdot}Mg(F,OH)_2$] generates the suite of minerals that compose the humite series. This behavior is evidenced by a regular variation in one axial parameter due to the addition of olivine-like layers while the other two axes remain constant (Penfield and Howe 1894, Bragg 1929). This construction of composite crystal structures from stoichiometrically distinct subunits now is known as *polysomatism* (Thompson 1978). Polysomatic series can be conceptualized as structural solid solutions between endmember polysomes, such as olivine and norbergite or mica and pyroxene. As in the humite class, intermediate minerals within these polysomatic solutions may be readily discriminated from adjacent members of the series by macroscopic measures if crystals are sufficiently large. By contrast, neighboring members of ionic solid solutions typically are distinguished only by techniques that are sensitive to slight variations in atomic arrangement. It is this latter style of solid solution that is the focus of this chapter, and readers interested in the transitional modes of polysomatic mineral series are directed to Veblen's (1991) excellent review.

Von Laue's discovery of X-ray diffraction by crystals in 1912 provided the first direct means of unraveling the relationship between crystal structure and crystal chemistry at the atomic scale. As the structures of simple salts emerged, crystallographers were able to calculate ionic radii for the major elements, and Linus Pauling and V.M. Goldschmidt led the effort to quantify the influence of ionic size on structure type (Pauling 1927, 1929; Goldschmidt 1926, 1927, 1929). Goldschmidt formulated the well-known radius-ratio rules for predicting cationic coordination numbers, and he developed the notion of tolerance factors to explain the preference of certain compositions for specific structure types. As an example, he argued that perovskite-type structures with formula ABX_3 are most stable when unit cells are cubic or nearly so; this constraint limits the allowable compositions that can adopt perovskite isotypes to ion assemblages whose tolerance factors t lie between 0.7 and 1.2, where $t = (R_A + R_X)/\sqrt{2}\,(R_B + R_X)$ and R represents the radii of the A, B, and X ions. This simple relation continues to be useful for considerations of perovskite-structure stability for specific compositions under mantle conditions (e.g. Leinenweber et al. 1991, Linton et al. 1999).

Analogies between morphotropism and polymorphism

Reconstructive and displacive transitions. Goldschmidt redefined morphotropism as a structural transition "effected by means of chemical substitution and representing a discontinous alteration, surpassing the limits of homogeneous deformation" (Goldschmidt 1929). In other words, when a specific ion is replaced with one so different in size that tolerance factors or radius ratio rules are violated, then alternative structure types are adopted. For example, the replacement of Ca by Sr or Ba in carbonates induces a morphotropic transition from the calcite to the aragonite structure type. The analogous transition in polymorphic systems is the *reconstructive* transformation, in which changes in temperature or pressure induce a breaking of primary bonds and a rearrangement of anionic frameworks (Buerger 1951).

The analogy between polymorphic and morphotropic transitions is sufficiently robust that Buerger's notion of *displacive* transformations also may be applied to chemically induced structural changes. Displacive transitions involve the bending but not the breaking of primary bonds with a concomitant distortion of the anionic framework. Typically they involve densification through framework collapse in response to increased pressure or decreased temperature. Similarly, the substitution of cations that fall within the tolerance factors for a given structure type may preserve the network topology, but small differences in the sizes of the substitutional ions and those they replace may result in framework distortions. These deformations typically follow modes that also can be activated by temperature or pressure.

K-Rb-Cs leucite. The much-studied feldspathoid leucite ($KAlSi_2O_6$) provides an example of the close correspondence between polymorphic and morphotropic displacive transitions. Leucite readily accommodates substitutions of its alkali cation, and

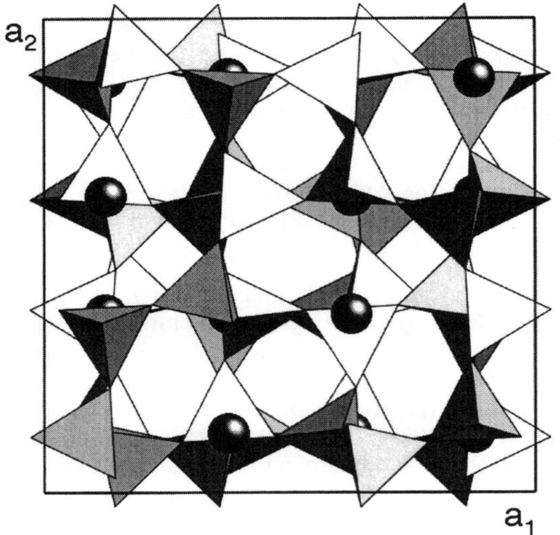

Figure 1. A projection of the structure of cubic leucite along the *a* axis. Spheres are K cations.

replacement cations affect the structure in a way that closely parallels changes in temperature (Taylor and Henderson 1968, Martin and Lagache 1975, Hirao et al. 1976, Henderson 1981). The structure of leucite (Fig. 1) consists of corner-sharing tetrahedra that reticulate into 4-membered rings normal to the *c* axis and 6-membered rings normal to [111]. This network creates large cavities that host alkali cations in 12 coordination. At high temperatures, the space group for leucite with complete Al-Si disorder is $Ia3d$ (Peacor 1968). When cooled below ~665°C, the tetrahedral framework collapses and the structure ultimately adopts the space group $I4_1/a$. The transition may involve an intermediate phase with space group $I4_1/acd$, and the complexity of the inversion has inspired investigations by a variety of techniques (Lange et al. 1986, Heaney and Veblen 1990, Dove et al. 1993, Palmer et al. 1989, 1990; Palmer and Salje 1990).

Martin and Lagache (1975) synthesized solid solutions within the K-Rb-Cs leucite system and examined changes in lattice parameters as a function of composition at room temperature (Fig. 2). These results may be compared with the variations of lattice parameters as a function of temperature for end-member K-, Rb-, and Cs-leucite (Fig. 3) as obtained by Palmer et al. (1997). The similarities in the response of the structure to decreased temperature and to decreased substitutional cation size is apparent in the matching ferroelastic transitions from cubic to tetrag-onal crystal systems. Moreover, Palmer et al. (1997) report that the style of framework distortion associated with decreased temperature is identical to that produced by the replacement of Cs by Rb and K (Fig. 4); both cooling and small-cation substitution provoke a twisting of tetragonal prisms about [001]. Because the larger cations prop open the framework and limit the

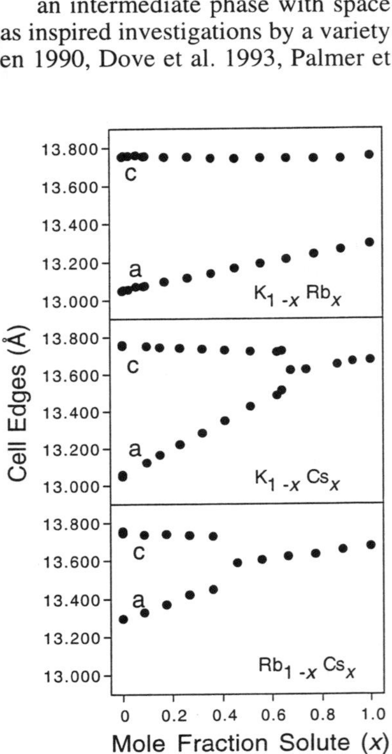

Figure 2. Dependence of lattice parameters with solute concentration in: (*top*) $K_{1-x}Rb_xAlSi_2O_6$; (*middle*) $K_{1-x}Cs_xAlSi_2O_6$; and (*bottom*) $Rb_{1-x}Cs_xAlSi_2O_6$. Data from Martin and Lagache (1975).

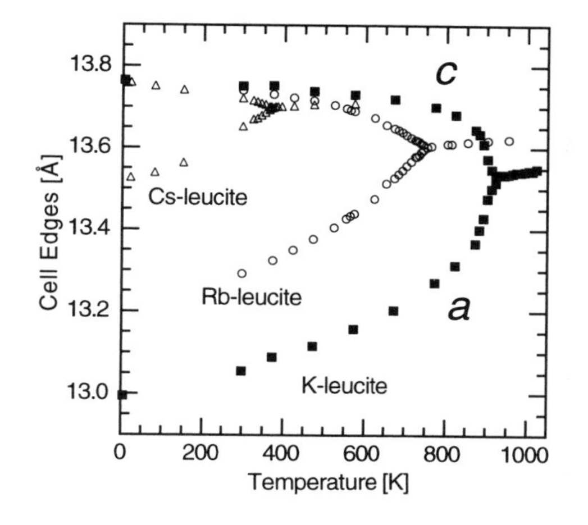

Figure 3. Variation of lattice parameters with temperature in endmember $KAlSi_2O_6$, $RbAlSi_2O_6$, and $CsAlSi_2O_6$. From Figure 5 in Palmer et al. (1997).

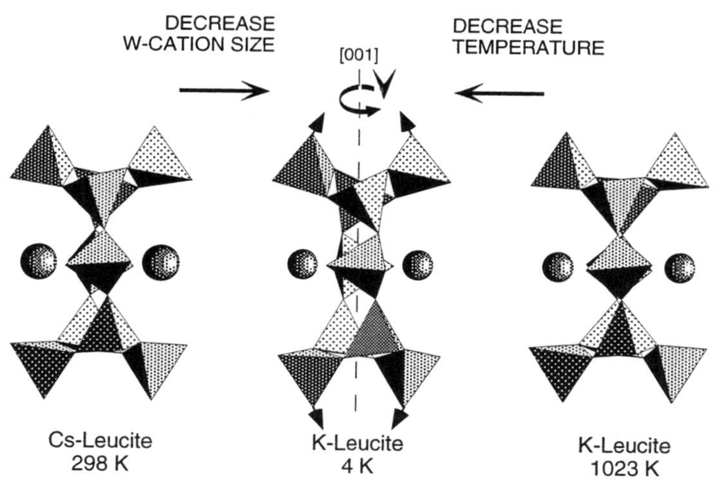

Figure 4. Distortion of the leucite structure as induced by temperature or composition, viewed along [110]. Smaller channel cations effect a twisting distortion about the *c* axis, resulting in axial elongation along *c* and radial compression along *a*. From Figure 9 in Palmer et al. (1997).

degree of tetrahedral tilting, they enlarge the stability field of the cubic phase. Accordingly, the transition temperatures decrease from 665°C (K-leucite) to 475°C (Rb-leucite) to 97°C (Cs-leucite). (Discrepancies in transition temperatures in the studies of Martin and Lagache (1975) and Palmer et al. (1997) may be attributed to differences in sample synthesis.) In addition, Palmer et al. demonstrate that the changes in the mean distance between alkali ions and oxygen as a function of temperature are quite similar when normalized by unit cell volumes.

Nevertheless, the analogies between morphotropic and polymorphic transitions in leucite are not exact. Palmer et al. (1989, 1997) calculate a total spontaneous strain (ε_{tot}) produced by the transition, and they subdivide this total strain into a nonsymmetry-breaking volume strain (ε_a) and a symmetry-breaking ferroelastic strain (ε_e) using the relations:

$$\varepsilon_{tot} = \sqrt{\left(\frac{c - a_0}{a_0}\right)^2 + 2\left(\frac{a - a_0}{a_0}\right)^2} \quad (1)$$

$$\varepsilon_a = \sqrt{3}\left(\frac{a_0 - (c+2a)/3}{a_0}\right) \quad (2) \qquad \varepsilon_e = \varepsilon_{tot} - \varepsilon_a \quad (3)$$

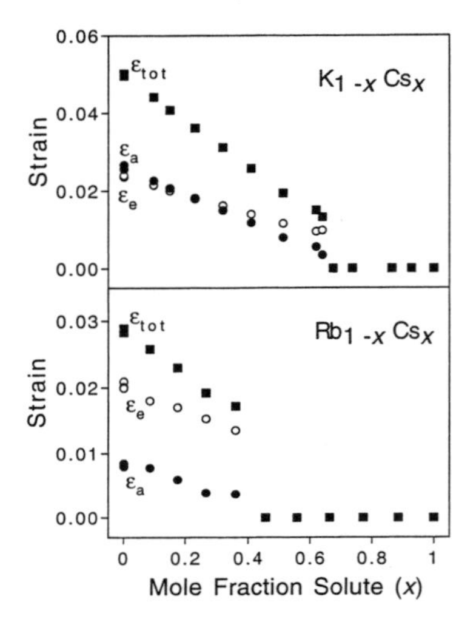

Figure 5. Evolution of spontaneous strain components ε_a (filled circles), ε_e (open circles), and total strain ε_{tot} (filled squares) in substituted leucite as a function of solute content. (*Top*) $K_{1-x}Cs_xAlSi_2O_6$; (*Bottom*) $Rb_{1-x}Cs_xAlSi_2O_6$.

where the subscript 0 denotes the paraelastic value as extrapolated from the observed behavior of the high-temperature phase to the low-temperature phase field. For endmember K-leucite, they find that the transition is continuous, that the variation of the square of ε_e with temperature is linear up to ~500°C, and that the ferroelastic transition therefore is consistent with a second-order Landau free energy expansion.

One can apply the relations of Palmer et al. (1997) to the solid solution data of Martin and Lagache (1975) along the two substitutional joins that exhibit transitions, namely $K_{1-x}Cs_xAlSi_2O_6$ and $Rb_{1-x}Cs_xAlSi_2O_6$. In this analysis, the spontaneous strain is dependent not on temperature but on the concentration of the dopant Cs. This treatment demonstrates that ε_{tot}, ε_a, and ε_e (and not the squares of these values) are linear with composition (Fig. 5). In addition, both the volumetric and the ferroelastic strains appear to be discontinuous at the critical compositions, which occur at $x_{Cs} = 0.67$ for K-Cs leucite and $x_{Cs} = 0.45$ for Rb-Cs leucite. Consequently, the behavior of the compositionally induced transition in leucite is clearly first-order in both the volumetric and the ferroelastic components of the strain. This behavior contrasts with the second-order behavior associated with the thermally induced transition of K-leucite.

In short, K-leucite and its Rb- and Cs-derivatives nicely demonstrate a general rule concerning polymorphic and morphotropic transitions for a given mineral system: The similarities are striking, but observations based on one transition mechanism are not automatically transferrable to the other.

PRINCIPLES OF MORPHOTROPIC TRANSITIONS

Types of atomic substitutions

It seems fair to argue that mineralogists are far from having the ability to predict the detailed response of a mineral structure to ionic substitutions. To begin, the phenomenon of solid solution is itself tremendously complex. As is well known, isomorphism of compounds does not guarantee the existence of a solid solution between them. Calcite ($CaCO_3$) and smithsonite ($ZnCO_3$) are isostructural and share the same space group ($R\bar{3}c$), but they exhibit virtually no solid solution. Conversely, Fe can substitute more than 50 mol % in sphalerite (ZnS), even though sphalerite is not isomorphous with the FeS endmember, troilite (Berry and Mason 1959, Hutchison and Scott 1983). Moreover, the same kind of cation exchange can induce a structural transformation in one system and not in another. For example, solid solution in olivine ($Mg_{2-x}Fe_xSiO_4$) involves cation exchange of octahedral Fe and Mg, but no transition has been detected within this series (Redfern et al. 1998). On the other hand, complete solid solution in Fe and Mg has been recorded in the anthophyllite-grunerite amphiboles ($Mg_{7-x}Fe_xSi_8O_{22}(OH)_2$), but when x exceeds ~2, the stable structure is no longer orthorhombic (S.G. *Pnma*) but monoclinic (S.G. *C2/m*) (Gilbert et al. 1982).

In addition, chemical substitutions come in a variety of styles. The simplest involves the exchange of one cation for another, as occurs in magnesiowüstite ($Mg_{1-x}Fe_xO$). When cation substitutions are not isovalent, then the exchange may be accompanied by other ions to maintain charge balance. For instance, in the stuffed derivatives of silica, Si^{4+} is replaced by Al^{3+}, and an M^+ cation is sited interstitially within a tunnel or cavity to ensure electrostatic neutrality (Buerger 1954, Palmer 1994). Alternatively, the substitution of a cation with one having a different valence can create an anionic vacancy, as very commonly occurs in the perovskite-like superconductors, such as $La_{2-x}Sr_xCuO_{4-\delta}$ with $\delta > 0$. In addition, substitutions may appear as ionic pairs as in the Tchermak's exchange, $Mg^{VI}Si^{IV} \leftrightarrow Al^{VI}Al^{IV}$, which relates muscovite ($KAl_2Si_3AlO_{10}(OH)_2$) and celadonite ($KAlMgSi_4O_{10}(OH)_2$).

In addition, the substitutions themselves may be inextricably tied to other processes. For example, in the plagioclase series ($NaAlSi_3O_8$–$CaAl_2Si_2O_8$), the coupled exchange $(CaAl)^{5+} \rightarrow (NaSi)^{5+}$ not only involves a substitution of cavity ions with different sizes and valences, it also introduces additional Al into tetrahedral sites in a fashion that may promote a transformation from a disordered to an ordered arrangement. A number of recent papers have explored the nature of isothermal order/disorder reactions induced by Al-Si substitution and Lowenstein avoidance considerations (Dove et al. 1996, Myers et al. 1998, Vinograd and Putnis 1999). Teasing apart the structural changes that are actuated purely by cation size or valence from those that are incited by concomitant cation ordering requires models that involve the coupling of multiple order parameters (e.g. Salje 1987, Holland and Powell 1996, Phillips et al. 1997).

Finally, it should be noted that in the seminal study of defects and phase transitions by Halperin and Varma (1976), the authors distinguish among substituents that violate the symmetry of the high-temperature phase (so-called Type A defects) and those that do not (Type B defects). Type A defects can couple linearly to the order parameter (Q) and locally induce non-zero values of Q, even above the critical temperature. Type B defects are not expected to couple with the order parameter when present in low concentrations. Although replacement cations often are of Type A and interstitials of Type B, this generalization does not always hold. Darlington and Cernik (1993) observe that Li cations in the perovskite $(K_{1-x}Li_x)TaO_3$ are Type A impurities because they do not statically occupy a single site but hop between two equivalent symmetry-breaking sites. By contrast, Nb in $K(Ta_{1-x}Nb_x)O_3$ is a category B impurity. As acknowledged by Halperin and Varma (1976), such situations require an understanding of the time-dependent properties of the defect in conjunction with their spatial distributions.

Linear dependence of T_c on composition

A number of studies have treated the effects of impurities on phase transitions from a theoretical perspective (Halperin and Varma 1976, Höck et al. 1979, Levanyuk et al. 1979, Weyrich and Siems 1981, Lebedev et al. 1983, Bulenda et al. 1996, Schwabl and Täuber 1996). By and large, however, theoreticians have focused on the way in which local interactions between defect fields and the order parameter produce an anomalous central peak in neutron scattering cross-sections of impure ferroelectrics up to 65°C above the critical temperature (Shirane and Axe 1971, Shapiro et al. 1972, Müller 1979).

Moreover, many of these studies concern themselves only with very low concentrations of defects (N) such that the volume per defect ($1/N$) is very much larger than the volume over which individual defect fields interact ($4\pi r_c^3/3$, where r_c is the correlation length).

When concentrations of substitutional atoms are high, the presumption generally has been that transition temperatures will vary in a simple linear fashion with dopant concentration. This supposition can be rationalized by a Landau-Ginzburg excess Gibbs free-energy expansion (Salje et al. 1991). A simple second order phase transition for phase A with the regular free energy expression

$$\Delta G^A = \frac{1}{2}A(T-T_c)Q^2 + \frac{1}{4}BQ^4 + \cdots \tag{4}$$

must be modified to incorporate the effects of solid solution with phase B if the solutes are sufficiently abundant to couple with the order parameter. If the defect fields overlap uniformly over the whole structure, then the transition is convergent and the coupling occurs with even powers of Q. The revised energy for the solid solution A-B can be written as

$$\Delta G^{A\text{-}B} = \frac{1}{2}A(T-T_c)Q^2 + \frac{1}{4}BQ^4 + \cdots + \xi_1 X_B Q^2 + \xi_2 X_B Q^4 + \cdots \tag{5}$$

where X_B is the mole fraction of the B component and ξ_i is the coupling strength. If the fourth and higher order terms are insignificant, it is clear that the defect interaction generates a renormalized critical temperature T_c^* that is equal to the critical temperature of pure A as linearly modified by the dopant B:

$$T_c^* = T_c - \xi_1 X_B . \tag{6}$$

This linear relationship between the effective critical temperature and the composition is observed for a number of systems, including the paraelectric-ferroelectric transition in a host of perovskites, such as $Pb(Zr_{1-x}Ti_x)O_3$ (Rossetti and Navrotsky 1999, Oh and Jang 1999) and $(Pb_xBa_{1-x})TiO_3$ (Subrahmanyham and Goo 1998). In mineral and mineral analog systems, linearity in the dependence of T_c and composition is observed in the $C2/m$–$C\bar{1}$ transition in alkali feldspars ($Na_{1-x}K_xAlSi_3O_8$) (Zhang et al. 1996), the $I\bar{1}$–$I2/c$ transition in Sr-doped anorthite ($Ca_{1-x}Sr_xAl_2Si_2O_8$) (Bambauer and Nager 1981, Tribaudino et al. 1993), the $P3_221$–$P6_222$ transition in Li,Al-doped quartz ($Li_{1-x}Al_{1-x}Si_{1+x}O_4$) (Xu et al. 2000), the $F\bar{4}3c$–$Pca2_1$ transition in boracite-congolite ($Mg_{1-x}Fe_xB_7O_{13}Cl$) (Burns and Carpenter 1996), the $R\bar{3}m$–$C2/c$ transitions for As-rich portions of $Pb_3(P_{1-x}As_xO_4)_2$ (Bismayer et al. 1986) and the Ba-rich regions of $(Pb_{1-x}Ba_x)_3(PO_4)_2$ (Bismayer et al. 1994).

Morphotropic phase diagrams (MPDs)

If morphotropic phase boundaries (MPBs) are planar or nearly planar surfaces in T-P-X diagrams, one can describe five classes of displacive transition profiles in systems with solid solution. In the simplest type of phase diagram, the MPB consists of a single sloping surface that intersects non-zero transition temperatures and pressures at either end of the join. Denoted as Type Ia in Figure 6a, this relationship describes the morphotropic phase relations in the K-Rb-Cs leucite system (inferred from Martin and Lagache 1975, Palmer et al. 1997, though more T_c measurements are needed for intermediate compositions) and in the boracite-congolite system (Burns and Carpenter 1996). A slight variant to this class of phase diagram can be called Type Ib, in which the coupling strength ξ_1 (Eqn. 6) is sufficiently great that the temperature is renormalized to 0 K for compositions within the binary join (Fig. 6b). The quartz–β-eucryptite (Xu et al. 2000)

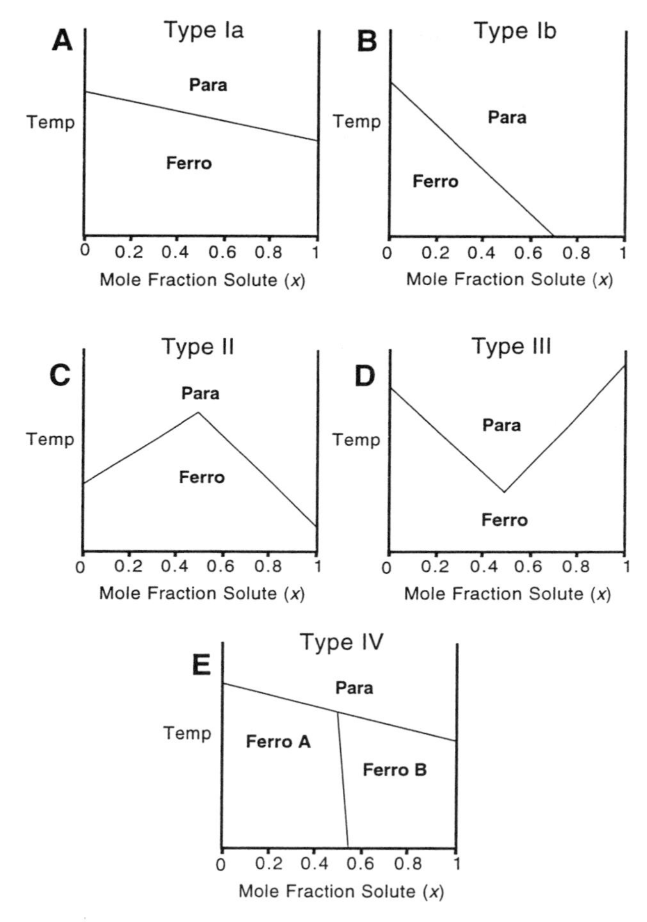

Figure 6. Schematic representations of morphotropic phase diagrams. (*a*) Type Ia; (*b*) Type Ib; (*c*) Type II; (*d*) Type III; and (*e*) Type IV.

and the Ca-anorthite–Sr-anorthite (Bambauer and Nager 1981) joins exemplify this behavior.

In addition, two morphotrophic phase boundaries may intersect within the solid solution system so as to produce a maximum critical temperature (Type II MPD in Fig. 6c) or a minimum critical temperature (Type III MPD in Fig. 6d). The Type II MPD is similar to that of the diopside–jadeite ($CaMgSi_2O_6$–$NaAlSi_2O_6$) solid solution (Carpenter 1980, Holland 1990, Holland and Powell 1996). Each endmember has a monoclinic structure with space group *C*2/*c*, but the midpoint of the solid solution, omphacite, has a reduced symmetry (S.G. *P*2/*n*) due to an ordering transition. Salje et al. (1991) argue that this midpoint composition actually represents the "pure" component of the omphacite join. A Type III MPD is observed in the lead phosphate-arsenate system (Bismayer et al. 1986).

A more complicated style of morphotropism includes two intersecting MPBs so as to create 3 phase fields. The so-called Type IV morphotropic phase diagram (Fig. 6e) is

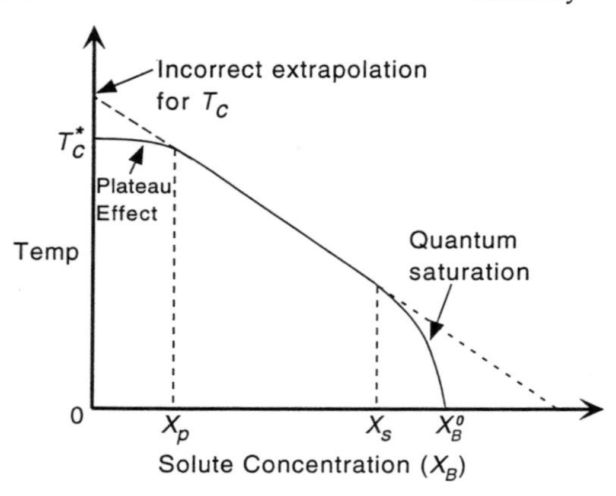

Figure 7. Schematic view of the dependence of the critical temperature T_C on the concentration of solute, X_B. Departures from linearity occur for concentrations below X_p, which defines the upper limit for the plateau effect, and above X_s, which marks the onset of quantum saturation. Modified from Figure 1 in Salje (1995).

characteristic of the important ferroelectric $Pb(Zr_{1-x}Ti_x)O_3$ (reviewed in Cross 1993) and of the tungsten bronze $Pb_{1-x}Ba_xNb_2O_6$ (Randall et al. 1991). The group theoretical relations within the Type IV MPD distinguishes it from the other classes. In the Type IV system, each of the two lower temperature morphotrophs is a subgroup of the high-temperature aristotype. However, the lower temperature morphotrophs do not share a subgroup-supergroup association with each other. Consequently, in Type IV MPDs, increasing levels of solute do not ultimately yield a substituted structural analog to the high-temperature aristotype. By contrast, the MPB in the Type I system does separate structures that share a subgroup-supergroup relationship. Thus, in Type I MPDs, increasing concentrations of dopant lead to a transition (or a series of transitions for multiple phase boundaries) that closely parallel displacive transitions induced by heating.

Quantum saturation, the plateau effect, and defect tails

Salje and collaborators (Salje et al. 1991, Salje 1995) emphasize that the variation of critical temperature with dopant concentration typically is not linear over the entire substitutional sequence (Fig. 7). Specifically, departures from linearity are noticeable over two regimes:

(1) For Type Ib systems, a given phase displacively transforms at lower temperatures with higher impurity contents (i.e. $\xi_I > 0$), and the renormalized critical temperatures equal 0 K when X_B is sufficiently large. If we call X_B^0 the composition at which $T_c^* =$ 0 K, then only the high-temperature structure has a field of stability for $X_B > X_B^0$. Near absolute zero, the dependence of T_c^* on X_B is expected to strengthen as quantum effects play a larger role, and the T_c^*-X_B curve will steepen. This phenomenon is described as *quantum saturation.*

(2) When concentrations of solute atoms are low (i.e. X_B less than ~0.01 mol % to ~2 mol %, depending on the mineral), the fields generated by the defects will not overlap, and the dependence of T_c^* on composition will be very weak. Consequently, the T_c^*-X_B curve flattens as X_B approaches 0 to create a *plateau effect.*

Salje (1995) models the excess energies associated with the plateau effect for three classes of substitutions: (1) Solutes that generate random fields that interact with the host; (2) Solutes that cannot generate fields but locally modify the transition temperature; and (3) Solutes that are annealed in disequilibrium configurations at temperatures below T_c. For solutes of the first category, the excess free energy expression must include gradients in Q associated with the local fields about each dopant, with g as the coupling constant. In addition, the solutes are assumed to generate a field h that conjugates to Q rather than

Q^2, rendering the transition non-convergent. The resulting Landau-Ginzburg expression for small X_B is

$$\Delta G^{A\text{-}B} = \frac{1}{2}A(T - T_c)Q^2 + \frac{1}{4}BQ^4 + \cdots + \frac{1}{2}g(\nabla Q)^2 - hQ \qquad (7)$$

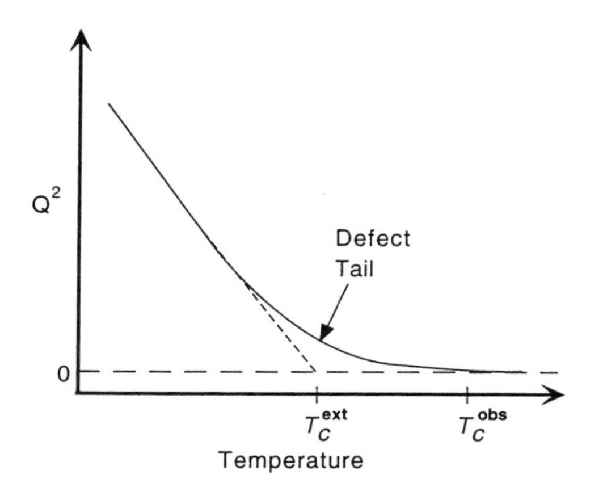

Figure 8. Representation of a defect tail. The order parameter Q goes to 0 at a lower temperature (T_c^{ext}) when extrapolated from low temperature data than is observed experimentally (T_c^{obs}).

This expression leads to a temperature dependence for Q that is distinctly different from that implicit in Equation (5) above. For small values of h, Q decays steeply near the critical temperature and then levels off, creating a characteristic *defect tail* (Fig. 8). These defect tails are commonly observed in the temperature evolution of properties in impure solids that are proportional to the square of the order parameter, such as birefringence and spontaneous strain. To the extent that the sudden strengthening in the dependence of Q with temperature near T_c diverges from the dependence at lower temperatures, two transition temperatures must be differentiated: the critical temperature extrapolated from lower temperatures (T_c^{ext}), and the observed T_c as measured experimentally (T_c^{obs}).

The field h in Equation (7) may be subdivided into a uniform component (h_0), which is associated with the host matrix, and a random field (h^s), which represents the fields associated with a random spatial distribution of solute atoms. The total field $h(r)$ is the sum of these two: $h(r) = h_0 + h^s\delta(r)$. In the immediate vicinity of a single defect, h_0 can be ignored, and invoking the assumption of thermodynamic stability ($\delta G/\delta Q = 0$), then

$$A(T - T_c)Q + g(\nabla Q) = h^s\delta(r) \qquad (8)$$

The solution to this equation (Salje 1995) has the form

$$Q(r) = \frac{h^s}{4\pi g|r|} e^{-|r|/r_c} \qquad (9)$$

with the correlation radius $r_c = \sqrt{g/A(T - T_c)}$. Equation (9) is an Ornstein-Zernike function and it represents the variation in the order parameter near a solute atom. The dependence of Q on the radial distance r reveals a steeply dipping profile close to the defect but a long-ranging tail that can modify the order parameter over large distances (Fig. 9). With an atomic field strength $h_{at} = 4\pi g^{3/2}/B^{1/2}$, it can be shown that Equation (9) leads to an expression for the dependence of the critical temperature on the density of solute atoms (N_B), which is directly proportional to the molar concentration X_B:

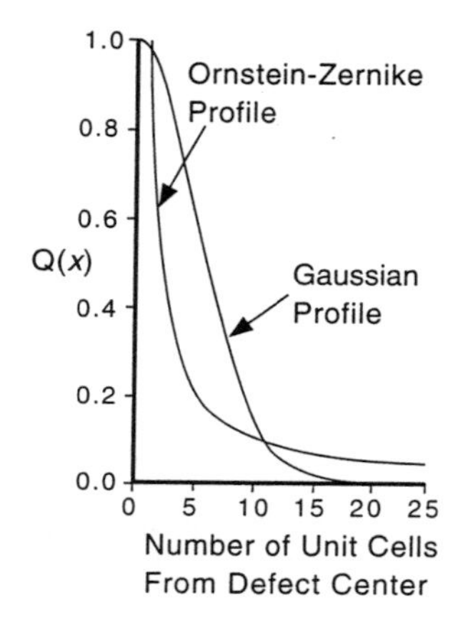

Figure 9. Profiles of the order parameter Q near a defect. Ornstein-Zernike profiles exhibit longer-ranging tails than Gaussian profiles, implying a more extensive modification of the order parameter of the host structure. Modified from Figure 5 in Salje (1995).

$$\frac{T_c^0 - T_c}{T_c} = \frac{9}{2AT_c}(6\pi)^{2/3}\left(\frac{h^s}{h_{at}}\right)^{4/3} N_B^{2/3} \tag{10}$$

The term T_c^0 refers to the critical temperature for the pure phase ($X_B = 0$), and the result demonstrates that $T_c(X_B)$ is proportional to $X_B^{2/3}$ rather than to X_B, as is the case when concentrations of solute are high. The resulting profile of the variation of the critical temperature with composition actually is concave up and represents an "inverse plateau." When solutes are not randomly distributed but present as clusters, Salje (1995) argues that the proportionality of T_c goes as $X_B^{4/3}$. This profile is concave downward and more closely resembles the plateau effects measured experimentally.

These departures from linearity are most prominent at the boundaries of the T-P-X_B phase space, but Salje (1995) points out that ramifications of non-linearity must be considered for several reasons. Most obviously, transition temperatures for pure endmembers will be incorrect if they are determined by linear extrapolation from members within the solid solution (Fig. 7). This practice can lead to errors in T_c at $X_B = 0$ by as much as 15°C. In addition, as noted in the preceding paragraph, critical temperatures for slightly impure phases may vary greatly depending upon the degree of randomness with which defects are distributed, and the disposition of impurities in turn is controlled by annealing temperatures and times for a mineral. Salje (1993) suggests that metamorphic minerals heated for long times below the critical temperature will exhibit less extensive plateau effects than those heated above T_c such that positional randomization of defects occurs.

Impurity-induced twinning

Memory effects. Twin walls provide hospitable environments for point defects. As the order parameter changes sign on crossing a twin boundary, within the wall itself Q may equal 0 and the structure will adopt the configuration of the high-temperature aristotype. Consequently, solutes that induce local distortions will impart a lower degree of strain when situated within a twin wall, and point defects within walls are energetically inhibited from further migration. Indeed, the attraction of twin walls and point defects to each other is so strong that it leads in many instances to memory effects with respect to twin wall positions as minerals are cycled above and below the critical temperature (Heaney and Veblen 1991b, Carpenter 1994, Xu and Heaney 1997). In quartz, for example, point defect diffusion at high temperature actually will control twin wall migration during cycling episodes.

If the spacing between twin walls is on the same order as the distance between defects in a mineral, the twin walls will attract the impurities with high efficiency. This behavior is especially strong outside the plateau region, where the order parameter is strongly dependent on composition, so spatial variations in Q are large and gradients in chemical potential are steep (Salje 1995). Once impurities have diffused to twin walls, they diminish wall mobility and can increase the effective critical temperature. In fact, minerals that have persisted below the critical temperature for long periods may actually display anomalous first-order behavior when first heated above T_c because of structural relaxation about impurity defects. Therefore, it is always important to cycle a natural mineral above and below T_c to exercise the structure and disperse defects before performing calorimetry or structural analysis near the phase transition.

Tweed twinning. When impurity concentrations are sufficiently large, twin walls are stabilized and thus twin wall densities can be extremely high. As a result, mean twin domain sizes can decrease dramatically for relatively small concentrations of solute. In some instances, the twin walls adopt a characteristic checkerboard-like microstructure known as a tweed pattern with orthogonal modulations of ~100-200 Å. The formation and energetics of tweed twins are discussed in Chapter 3 of this volume by Salje and also in Putnis and Salje (1994) and Salje (1999). Examples of compositionally induced tweed textures include the doped superconductor $YBa_2(Cu_{1-x}M_x)_3O_{7-\delta}$ for M = Co and Fe (Van Tendeloo et al 1987, Schmahl et al 1989, Xu et al. 1989) as well as As-doped lead phosphate (Bismayer et al. 1995) and Sr-doped anorthite (Tribaudino et al. 1995). Diffraction patterns of twin tweeds exhibit characteristic cross-shaped intensities superimposed on the primary spots.

Nanoscale domains and titanite. When the strains associated with a phase transition are small, domain boundaries are not crystallographically controlled and tweed patterns tend not to form. Rather, twin boundaries will be irregularly oriented. This behavior is common (though not universal) for merohedral twins, which involve transitions within the same crystal system, and antiphase domains or APDs, which form by the loss of a translational symmetry operation (Nord 1992). Mean sizes of these irregular domains often fall below the correlation length of X-ray scattering techniques, and the presence of these domains is discernible only by probes that are sensitive to symmetry violations over unit-cell scales.

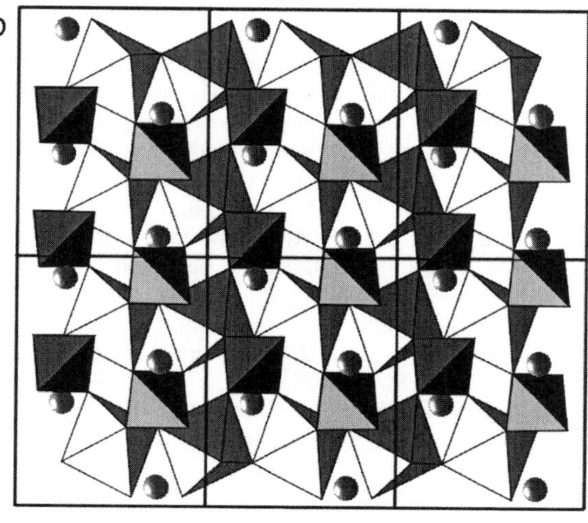

Figure 10. Projection of the structure of titanite ($CaTiSiO_5$) on the (001) plane. Ti cations within octahedral chains parallel to *a* are interconnected via silica tetrahedra. Spheres are Ca cations.

Titanite ($CaTiSiO_5$) is an excellent example of a mineral in which small amounts of impurities will precipitate high densities of antiphase boundaries. Titanite contains parallel kinked chains of TiO_6 octahedra that are interconnected by SiO_4 tetrahedra, with Ca cations in irregular 7-coordination (Fig. 10). At room temperature, the Ti cations are displaced from the geometric centers of the coordination octahedra by 0.88 Å along the chain direction. Although the sense of displacement is the same for all Ti cations within a single chain, in adjacent octahedral chains the sense of Ti displacement occurs in the opposite direction (Speer and Gibbs 1976). When titanite is heated to 235°C, however, each Ti cation shifts to the center of its coordinating octahedron, and the space group symmetry transforms from $P2_1/a$ to $A2/a$ (Taylor and Brown 1976). In X-ray diffraction experiments, this symmetry change is accompanied by the disappearance of all spots for which $k+l$ is odd due to the A-centering.

In natural titanite, Al and Fe commonly replace Ti according to the following coupled exchange scheme: $(Al,Fe)^{3+} + (OH,F)^{1-} \leftrightarrow Ti^{4+} + O^{2-}$. The monovalent anion substitutes for the corner-sharing O^{2-} atoms along the octahedral chain, and the $(Al,Fe)^{3+}$ cations interrupt the displacement sequence of the Ti cations in low titanite. These two effects promote the formation of APBs in $P2_1/a$ titanite (Taylor and Brown 1976). When the substitution of Ti exceeds 3 mol %, the size of the antiphase domains approximates the correlation length of X-rays, and the superlattice diffractions ($k+l$ odd) grow diffuse. When substitution exceeds 15 mol %, the $k+l$ odd diffractions are no longer visible (Higgins and Ribbe 1976). Thus, the apparent symmetry for titanite becomes $A2/a$ when substitution levels are sufficiently high. Indeed, Hollabaugh and Foit (1984) report that a natural titanite sample with 10 mol % $(Al,Fe)^{3+}$ yielded a superior refinement at room temperature in the $A2/a$ rather than the $P2_1/a$ space group, despite the presence of diffuse superlattice diffractions. Dark field TEM imaging of the APBs in highly doped titanite is frustrated by the faintness of these diffractions, but the presence of the diffuse diffraction spots indicates that ordered domains do exist over short length scales. This inference is confirmed by measurements of dielectric loss for a natural crystal of titanite with 15 mol % $(Al,Fe)^{3+}$ substitution that revealed evidence for a broad transition from ~150°C to ~350°C (Heaney et al. 1990).

Incommensurate phases and solid solutions

Even when solid solution fails to change the bulk symmetry of a mineral, sometimes the incorporation of substitutional and/or interstitial cations can induce structural modulations over length scales that are slightly larger than the unit cell. Consequently, solid solutions may be characterized by isothermal transitions between commensurate and incommensurate structures. In some cases, these modulated structures occur in metastable phases. For example, plagioclase ideally is not a solid solution at room temperature; rather, subsolidus plagioclase under true equilibrium conditions should be represented by the co-existence of near endmember $C\bar{1}$ albite and $P\bar{1}$-anorthite (Smith 1983). Nevertheless, virtually all plutonic plagioclase crystals with compositions between An_{20} and An_{70} contain "e"-type superstructures. Many models have been proposed to account for the nature of these incommensurate modulations (e.g. Kitamura and Morimoto 1975, Grove 1977, Kumao et al. 1981). In general, the superperiodicities may be ascribed to oscillations in Na and Ca content produced by spinodal unmixing, with lattice energy minimization controlling the orientations of the interfaces (Fleet 1981). The wavelength of the incommensuration decreases from ~50 Å in An_{70} to ~20 Å in An_{30}, with a discontinuity at An_{50} (Slimming 1976); this boundary separates the so-called "e_2" superstructures observed in An_{20} to An_{50} from the "e_1" superstructures found in An_{50} to An_{70}. In their Landau analysis of plagioclase based on hard mode infrared spectroscopy, Atkinson et al. (1999) observe that no empirical order parameter has been identified for the production of the e_1 and e_2 phases, but that the degree of Al-Si order also changes over these intermediate compositions. The e_1 state in An_{70} has local Al-Si order, but the degree of order decreases with increasing Ab content.

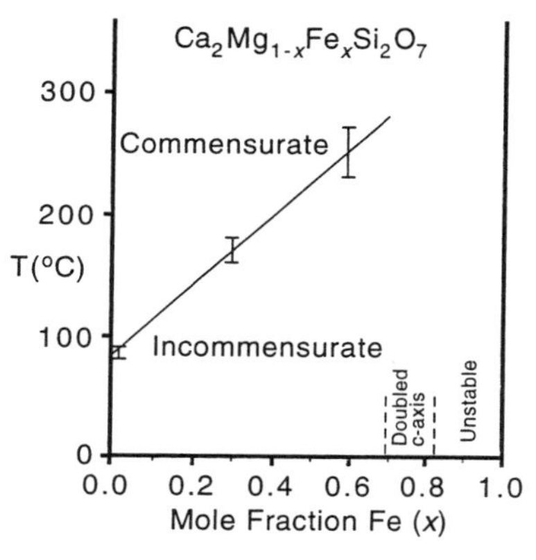

Figure 11. Phase diagram for åkermanite solid solutions ($Ca_2MgSi_2O_7$ – $Ca_2FeSi_2O_7$). Transition temperatures between commensurate and incommensurate phases increase with Fe content. Modified from Figure 12 in Seifert et al. (1987).

Compositionally induced incommensurate-commensurate transitions also were documented by Seifert et al. (1987), who examined synthetic melilites in the åkermanite system ($Ca_2(Mg_{1-x},Fe_x)Si_2O_7$) by transmission electron microscopy. They found that superstructures with modulations of ~19 Å are present in these minerals at room temperature for 0 x_{Fe} 0.7. However, the incommensurate phases disappear on heating, and as Fe content increases, the critical temperatures for the incommensurate-commensurate phases also increase (Fig. 11). Incommensurate phases also have been described for solid solutions between åkermanite and gehlenite ($Ca_2Al_2SiO_7$) (Hemingway et al. 1986, Swainson et al. 1992) and Sr-doped melilite systems, such as $(Ca_{1-x}Sr_x)_2CoSi_2O_7$ (Iishi et al. 1990) and $(Ca_{1-x}Sr_x)_2MgSi_2O_7$ (Jiang et al. 1998). The cause of the superperiodic modulations generally is attributed to a mismatch in the dimensions of the layered tetrahedral network and the divalent cations, and the transition from the incommensurate to the commensurate structure is effected by changes in temperature, pressure, or composition that work to minimize the misfit (Brown et al. 1994, Yang et al. 1997, Riester et al. 2000).

CASE STUDIES OF DISPLACIVE TRANSITIONS INDUCED BY SOLID SOLUTION

Because theoretical treatments of the relation between transition behavior and solid solutions are still in development, phenomenological approaches are necessary to illuminate the ways in which structures respond to increasing solute contents. A discussion of some systematic studies in minerals and mineral analogs follows.

Ferroelectric perovskites

Crystal chemistry. The effect of solid solution on the transition behavior of perovskite (ABX_3) structures has been intensively scrutinized for more than 50 years. These materials have merited continuous attention because of their enormous technological versatility. As multilayer capacitors, piezoelectric transducers, and positive temperature coefficient (PTC) thermistors they generate a market of over $3 billion every year (Newnham 1989, 1997). In addition to ease of fabrication, these compounds exhibit a number of attributes required of ideal actuators: (1) They display very large field-induced strains; (2) They offer quick response times; and (3) Their strain-field hysteresis can be chemically controlled to be very large or negligibly small, depending on the application. Details of their technical applications can be found in Jaffe et al. (1971) and Cross (1993).

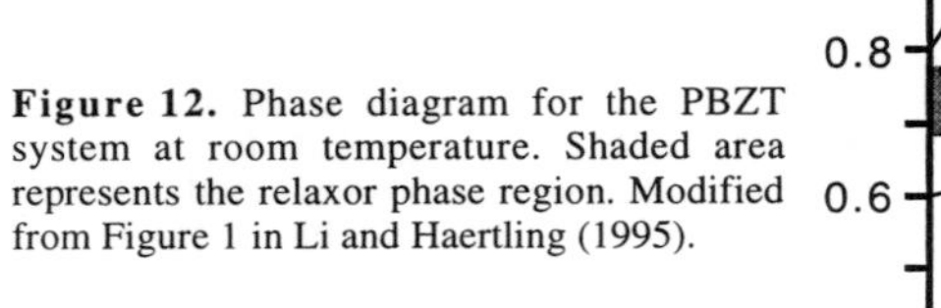

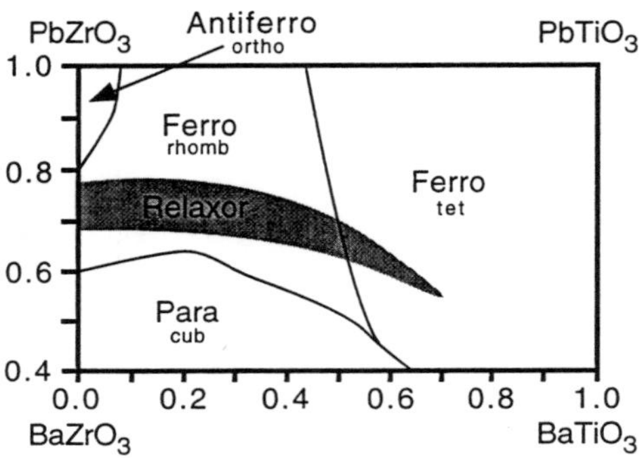

Figure 12. Phase diagram for the PBZT system at room temperature. Shaded area represents the relaxor phase region. Modified from Figure 1 in Li and Haertling (1995).

The compositional perovskite series that has served as the basis for much of this research is the so-called PBZT system. The quadrilateral that joins the endmembers $PbZrO_3$–$BaZrO_3$–$BaTiO_3$–$PbTiO_3$ exhibits virtually complete solid solution (Fig. 12). Nevertheless, this system displays a number of morphotropic phase transitions that involve transformations of several kinds: ferroelastic, ferroelectric, antiferroelectric, and relaxor. Understanding the nature of these isothermal transitions requires a review of the thermal distortions that occur as these perovskites are cooled.

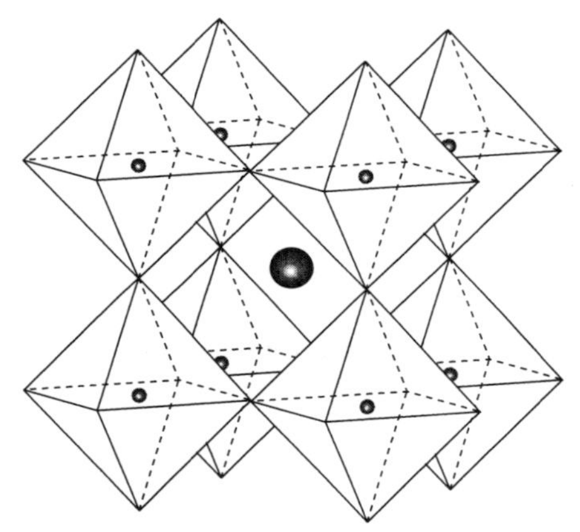

Figure 13. Structure of the cubic aristotype for perovskite compounds.

At high temperatures, these compounds adopt the idealized perovskite aristotype (Fig. 13). Taking $BaTiO_3$ as an example, Ba^{2+} and O^{2-} each have radii of ~1.4 Å and together create a face-centered cubic (or cubic closest packed) array with each Ti^{4+} cation in octahedral coordination with oxygen. As temperatures are lowered, the Ti cations displace from the geometric centers of their coordinating octahedra, and the O 2*p* electrons hybridize with the Ti 3*d* electrons to minimize the short-range repulsions that attend this displacement (Cohen 2000). The directions of Ti displacement are strongly temperature dependent: (1) Below the Curie temperature of 130°C, the Ti atoms shift parallel to [001], violating the cubic symmetry and generating a tetragonal structure; (2) Below 0°C, the Ti atoms displace along [110], generating an orthorhombic structure; and (3) Below -90°C, the Ti atoms move along [111], generating a rhombohedral structure.

Ferroelectricity. Each of these transitions involves a change in crystal system, thereby inducing ferroelastic distortions that create twins (Nord 1992, 1994). In addition, each of these displacements alters the ferroelectric character of the structure. Some earth scientists may be unfamiliar with ferroelectricity, but the principles are straightforward. Polar compounds are easily oriented by electric fields, and the ease of polarizability is measured by the dielectric constant for that substance. Formally, the dielectric constant ε is defined as $1 + \chi$ where χ is the electric susceptibility and serves as the proportionality constant between the polarization $\vec{P}$ of the compound and the electric field intensity $\vec{E}$. This relation is expressed as $\vec{P} = \chi\varepsilon_0 \vec{E}$, where ε_0 is the permittivity of free space and equals 8.85×10^{-12} C^2/Nm^2. Thus, compounds that become highly polarized for a given electric field will have high dielectric constants. The dielectric constant for H_2O at room temperature is 80, whereas that for $Ba(Ti_{1-x}Zr_x)O_3$ ranges up to 50,000.

The high electric susceptibility of PBZT ceramics can be attributed to the easy displacement of the octahedrally coordinated cation. Above the Curie temperature, the cations are centered and exhibit no spontaneous polarization. In this paraelectric state, the compounds obey the Curie-Weiss Law; the polarization varies directly with the strength of the electric field and inversely with the temperature due to thermal randomization. Compounds near $BaZrO_3$ in composition are paraelectric at room temperature (Fig. 12). Below their Curie temperature, the compounds are spontaneously polarized, and the directions of the spontaneous polarization vectors are dictated by the directions of atomic displacement: P_s is parallel to [100] in the tetragonal phase, [110] in the orthorhombic phase, and [111] in the rhombohedral phase. The high degree of spontaneous polarization in the PBZT compounds explains their use as capacitors, as they are capable of storing considerable electric charge. Moreover, their high values of P_s account for their excellence as transducers, since the spontaneously polarized structures are strongly pyroelectric and piezoelectric. Consequently, thermomechanical energy is readily translated into electrical energy, and vice versa.

In the compounds that are close to $PbZrO_3$ in composition, room-temperature structures are orthorhombic (Fig. 12), but the octahedral cations do not uniformly displace parallel to [110]. Rather, for every Zr^{4+} cation that displaces parallel to [110], a neighboring Zr^{4+} shifts parallel to $[\bar{1}\bar{1}0]$. The net polarization therefore is zero, and the material is classified as antiferroelectric. Antiferroelectric materials also exhibit higher-than-average dielectric constants, but they are not so extreme as those observed in ferroelectric compounds.

PZT and the morphotropic phase boundary. Despite its similarity to $BeTiO_3$, $PbTiO_3$ exhibits a simpler transition sequence than $BaTiO_3$. Where $BaTiO_3$ experiences four transitions from high to low temperatures, $PbTiO_3$ undergoes only one transition from cubic to tetragonal symmetry. This disparity is attributable to the electronic structures of Ba^{2+} and Pb^{2+}. Whereas Ba^{2+} is highly ionized and quite spherical, Pb^{2+} has 2 $6s$ lone pair electrons that are easily polarized. This polarization stabilizes the strain associated with the tetragonal phase (Cohen 1992), and transformation to the orthorhombic phase is inhibited by repulsions between the $6s^2$ electrons displaced along [110] and the surrounding O atoms.

As a consequence, the joins for $(Pb_{1-x}Ba_x)TiO_3$ at low temperature and for $Pb(Zr_{1-x}Ti_x)O_3$ at room temperature are interrupted by a morphotropic phase boundary (MPB), which separates tetragonal and rhombohedral phases (Fig. 14). The structural state of the oxides in the vicinity of the MPB is a subject of active inquiry, because many of the physical properties of PBZT ferroelectrics are maximized at the MPB. These include the dielectric constant, the piezoelectric constant, and the electromechanical coupling coefficients (Jaffe 1971, Thomann and Wersing 1982, Heywang and Thomann 1984). For industrial purposes, this behavior is exploited by annealing PBZT ferroelectrics with compositions near the MPB close to the Curie temperature in an

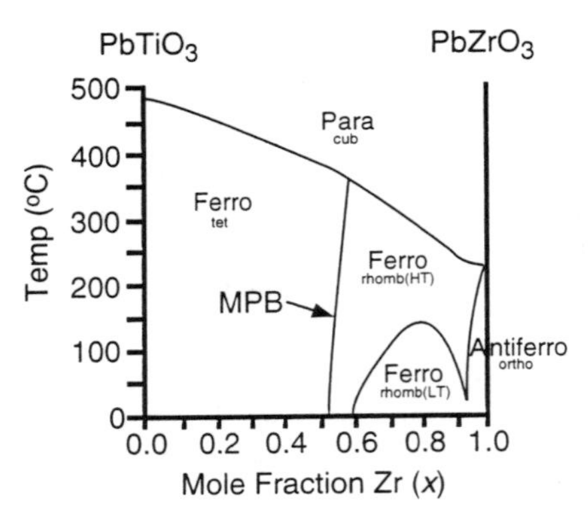

Figure 14. Phase diagram for the PZT ($PbTiO_3$ – $PbZrO_3$) system. Tetragonal and rhombohedral phase fields are separated by a nearly vertical morphotropic phase boundary (MPB). Adapted from Figure 1 in Oh and Jang (1999).

intense dc electric field; this "poling" process significantly increases the bulk polarization.

The MPB allows researchers to search for mechanisms of morphotropic transformation at room temperature, and two significant questions have driven these examinations. What is the structure of the material at the MPB, and what causes the exceptionally high susceptibilities near the MPB?

As seen in Figures 12 and 14, the MPB in the PBZT system occurs quite close to compositions with equal quantities of Zr and Ti, and the position of the boundary has only a small dependence on temperature and pressure (Oh and Jang 1999). Two models have been posited to explain the nature of the transition at the MPB. One scenario treats the MPB as a rough analog to the univariant boundary in a *P*-*T* polymorphic phase diagram. The assumption here is that the MPB represents a single fixed composition for a given temperature and pressure, so that the transition across the MPB is structurally and compositionally continuous (Karl and Hardtl 1971, Kakegawa et al 1982, Lal et al. 1988). In this conception, the octahedral distortions associated with the tetragonal and rhombohedral structures diminish as the MPB is approached from either side. At the MPB itself, structures adopt a strain-free cubic symmetry with $Q = 0$. Alternatively, classical solution chemistry suggests that the MPB represents a zone of co-existing phases (Arigur and Benguigui 1975, Isupov 1975). In this scenario, the MPB comprises a miscibility gap.

This latter hypothesis has been difficult to test; if a miscibility gap exists, it must occur over a very narrow compositional range. However, recent high-resolution X-ray diffraction analyses of $Pb(Zr_xTi_{1-x})O_3$ across the MPB offer strong support for the immiscibility model (Singh et al. 1995, Mishra et al. 1997). These studies reveal phase coexistence for $x_{Zr} = 0.525$, such that phases are tetragonal for x_{Zr} 0.520 and rhombohedral for x_{Zr} 0.530. The inferred compositional gap thus has a magnitude in $\Delta x_{Zr}$of only 0.01. If these results are correct, the morphotropic transition in the PZT system is structurally and compositionally discontinuous.

The traditional explanation for the high susceptibility of ferroelectrics near the MPB relates to the large number of possible directions for polarization near the MPB. The tetragonal phase offers 6 directions for polarization when all of the cubic equivalents of the <100> displacement family are considered, and the rhombohedral phase presents 8 directions for the <111> family. As compositions near the MPB allow the coexistence of tetragonal and rhombohedral phases, this model proposes that polarization domains can adopt 14 orientations for alignment during poling (Heywang 1965). This idea is supported by TEM observations of PZT at the morphotropic boundary by Reaney (1995), who interpreted parallel twin walls along {110} as both tetragonal and rhombohedral twin members.

On the other hand, Cross (1993) suggests that instabilities in the tetragonal phase near the MPB lead to a maximum in the electromechanical coupling maximum, k_p. Since the dielectric constant is proportional to k_p^2, this instability also serves to maximize the electric susceptibility. Measurements of k_p for PZT phases near the MPB by Mishra and Pandey (1997) support this second model. The value for k_p for the rhombohedral phase is lower than that for the tetragonal phase, and accordingly they find that electromechanical coupling is maximized in the tetragonal phase near the MPB and that co-existence of rhombohedral PZT with tetragonal PZT at the MPB actually decreases k_p.

Order of morphotropic transitions. When the coupling strength ξ_l in the Landau-Ginzburg expression for solid solutions (Eqn. 5) is positive, the transition temperature decreases with increasing solute concentration X_B (Eqn. 6). It is generally observed that in such cases where $dT_c/dX_B < 0$, the energy of transition tends to decrease and the order of the transition to increase. The tetragonal to cubic (or ferroelectric to paraelectric) transition in PZT nicely demonstrates this trend. Careful calorimetric measurements of $Pb(Zr_xTi_{1-x})O_3$ by Rossetti and Navrotsky (1999) for $x = 0$, 0.15, 0.30, and 0.40 reveal a first-order character to the transition for 0 x 0.30: Latent heats of transition were detected for these compositions; the hysteresis in transition temperatures on heating and cooling lay outside experimental error; and the dependence of excess entropy ($S^{xs} = \int(\Delta C_p/T)\, dT$) with temperature displayed step-like behaviors (Fig. 15).

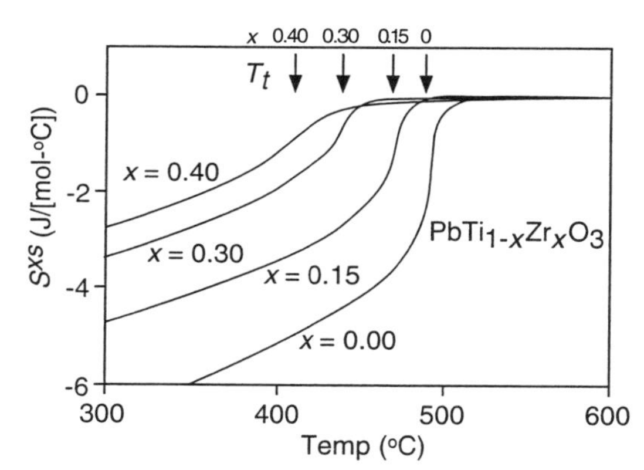

Figure 15. The dependence of excess entropy (S^{xs}) on temperature for compositions $x = 0.0$, 0.15, 0.30, and 0.40 in $PbTi_{1-x}Zr_xO_3$. T_t represents transition temperatures for these compositions. Modified from Figure 5 in Rosetti and Navrotsky (1999).

Nevertheless, the intensity of the transition diminishes with increasing x_{Zr}. The latent heat decreases from 1.93 to 0.96 to 0.39 kJ/mol for $x_{Zr} = 0$, 0.15, and 0.30 respectively. The hysteresis in transition temperature decreases from 12°C at $x = 0$ to 5°C at $x = 0.30$. Likewise, the discontinuity in S^{xs} lessens with increasing x_{Zr}. For compounds with $x_{Zr}=$ 0.40, the transition appears to be second-order. No latent heat is detected, and the hysteresis in T_c falls below experimental error. Consquently, with increasing Zr concentration, the paraelectric to tetragonal inversion transforms from a first-order to a second-order transition. Rossetti and Navrotsky (1999) argue that these effects are analogous to those induced by increased pressure, and they infer from this correspondence that the tricritical point is close to x_{Zr} = 0.38; at this composition, the Goldschmidt tolerance factor is almost exactly equal to 1, so that the PZT perovskite is ideally cubic closest packed. Other authors also have documented an increase in the order of the transition with increasing Zr substitution in PZT, though the details differ. Mishra and Pandey (1997), for instance, set the tricritical composition point very close to the MPB, at x_{Zr} = 0.545. These disparities may arise from different methods of sample preparation.

Stabilized cubic zirconia

Crystal chemistry. Cubic zirconia (CZ) is a popular simulant for diamond, and deservedly so. With a Mohs hardness of 8.25 and and a refractive index of 2.15, it closely matches the scratch resistance and brilliance of authentic diamond, and its dispersion coefficient is higher (Shigley and Moses 1998). (Fortunately for those who like their gems natural, diamond and CZ are easily distinguished by density and by thermal conductivity.) Cubic zirconia also offers important technological uses as a refractory high-temperature ceramic (Meriani and Palmonari 1989).

Pure baddeleyite (ZrO_2) is monoclinic (S.G $P2_1/c$) at room temperature. When heated to ~1100°C, it undergoes a martensitic transformation to a tetragonal modification (S.G. $P4_2/nmc$), and at ~2300°C it inverts displacively to a cubic polymorph (S.G. $Fm3m$) (Smith and Cline 1962, Teufer 1962). The high-temperature aristotype is the fluorite structure, and the symmetry-breaking distortions that occur on cooling involve small displacements of the oxygen anions from their idealized sites (Fig. 16).

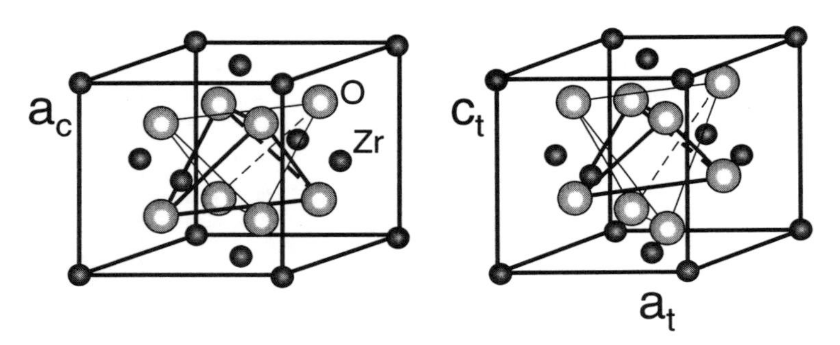

Figure 16. Structures of cubic (*left*) and tetragonal (*right*)) ZrO_2. Positions of Zr cations (small spheres) remain unchanged. O anions (large spheres) are arranged as intersecting tetrahedra that are regular in the cubic structure but distorted in the tetragonal polymorph. Modified from Figure 6 in Heuer et al. (1987).

Morphotropic transitions. Room temperature monoclinic baddeleyite (m-ZrO_2) has a low tolerance for substitutional cations, and miscibility gaps separate nearly pure m-ZrO_2 from dopant oxides. When the dopant oxides are cubic at room temperature, the miscibility gap thus separates monoclinic from cubic phases, and the extent of the cubic stability field is strongly dependent on the dopant. The gap between m-ZrO_2 and cubic CeO_2 is virtually complete at room temperature (Fig. 17), whereas (1-x)ZrO_2–xY_2O_3 has a stable cubic solid solution for x 0.09 (Tani et al. 1983, Zhou et al. 1991). Most cubic zirconia on the gem market is fabricated with small amounts of Ca or Y, because m-ZrO_2 is significantly birefringent; this causes a doubling of facet edges when observed through the gem.

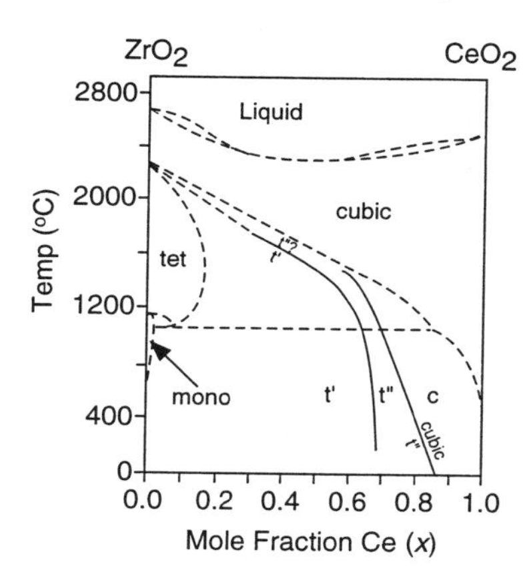

Figure 17. Phase diagram for Ce-doped zirconia. Dashed lines outline stability fields for monoclinic, tetragonal, and cubic phases. Solid lines demarcate metastable phase fields. Modified from Figure 6 in Yashima et al. (1994).

On the other hand, ZrO_2 is very susceptible to metastable solid solution in a range of substitutional systems, such as $(Zr_{1-x}Ce_x)O_2$ (Yashima et al. 1993, 1994), $(Zr_{1-x}U_x)O_2$ (Cohen and Schaner 1963), and $(Zr_{1-x}RE_x)O_2$, with RE = Y, Er, and other rare earth elements (Scott 1975, Heuer et al. 1987, Zhou et al. 1991). As first shown by Lefèvre (1963), when ZrO_2 is alloyed with small amounts of these dopants, the cubic phase can be quenched from high temperatures directly to a metastable tetragonal polymorph, denoted as t'. In $(1-x)ZrO_2–xY_2O_3$, this transition can be induced for x = ~0.03 to ~0.08 (Scott 1975). The metastable structure called t' belongs to space group $P4_2/nmc$, like that of the high-temperature polymorph for pure ZrO_2, and the unit cell is metrically tetragonal ($c/a > 1$).

In $(Zr_{1-x}Ce_x)O_2$, the t' to cubic transition requires higher dopant concentrations than is the case with yttria stabilized zirconia. Although initial X-ray diffraction experiments suggested that the morphotropic transition in $(Zr_{1-x}Ce_x)O_2$ to cubic symmetry occurs at x_{Ce} = 0.65 to 0.70 (Yashima et al. 1993), Raman spectroscopy reveals 6 modes that are active for the tetragonal structure up to x_{Ce} = 0.80 (Yashima et al. 1994). For compositions with x_{Ce} = 0.9 and 1.0, only the 3 Raman modes characteristic of cubic zirconia are observed. Consequently, an intermediate t" phase for $\sim0.65 < x_{Ce} < \sim0.85$ separates the t' and the cubic c fields in $(Zr_{1-x}Ce_x)O_2$. This t" phase is metrically cubic but symmetrically tetragonal. Cerium-stabilized cubic zirconia thus exhibits 2 morphotropic transition boundaries (from monoclinic to tetragonal and from tetragonal to cubic) that are analogous to those induced by temperature.

Lead phosphate analogs to palmierite

Crystal chemistry. Solid solutions of lead phosphate compounds with structures like that of palmierite ($(K,Na)_2Pb(SO_4)_2$) exhibit transition sequences that are well modeled by Landau-Ginzburg excess free energy expressions. Systems that have been especially heavily studied include $Pb_3(P_{1-x}As_xO_4)_2$ (Tolédano et al. 1975, Torres 1975, Bismayer and Salje 1981, Bismayer et al. 1982, 1986; Salje and Wruck 1983) and $(Pb_{1-x}A_x)_3(PO_4)_2$, where A = Sr or Ba (Bismayer et al. 1994, 1995). The phase diagram for the phosphate-arsenate join is not completely mapped, but Bismayer and collaborators have identified a series of transitions for the P-rich and the As-rich fields. The structure consists of two sheets of isolated $(P,As)O_4$-tetrahedra with vertices pointing towards each other and a sheet of Pb atoms in the plane of the apical oxygen atoms. Layers of Pb atoms lie between these tetrahedral sheets (Fig. 18) (Viswanathan and Miehe 1978).

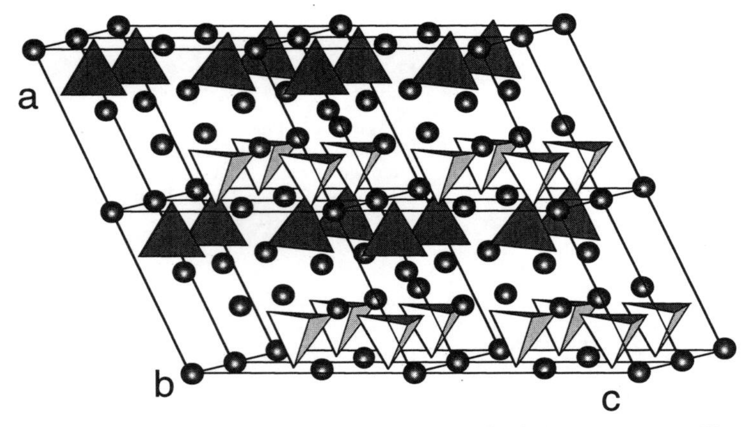

Figure 18. Projection of the palmierite-type structure along the *b* axis. Spheres represent Pb atoms.

Phase transitions. At high temperatures, the space group for the so-called *a* polymorph is $R\bar{3}m$, and on cooling this phase transforms to the monoclinic *b* phase with space group *C2/c*. The ferroelastic transition temperature decreases with increasing solute concentration from either endmember, such that the minimum T_c in $Pb_3(P_{1-x}As_xO_4)_2$ occurs close to $x_{As} = 0.5$ (Fig. 19). Likewise, the order of the transition decreases with increasing solute concentration. The critical exponent β in the relation $\langle Q \rangle^2 \propto |T - T_c|^{2\beta}$ is 0.235 in pure $Pb_3(PO_4)_2$, as is consistent with a tricritical transition. With the substitution of P by As, β equals 0.5, which is indicative of a second-order reaction (Bismayer 1990). The transition from phase *a* to *b* occurs via two steps. First, at temperatures above the ferroelastic *a*-to-*b* transition, the Pb atoms displace normal to the 3-fold inversion axis, generating nanoscale clusters with locally monoclinic symmetry. The presence of these clusters is revealed by diffuse superlattice reflections above T_c (Bismayer et al. 1982, Salje and Wruck 1983), but the bulk structure of this intermediate a_b phase remains metrically trigonal. Second, the displaced Pb atoms adopt a preferential orientation parallel to the 2-fold axis, and a ferroelastic transition to the monoclinic phase occurs. The a_b intermediate phase occurs over the whole of the $Pb_3(P_{1-x}As_xO_4)_2$ join except for endmember $Pb_3(PO_4)_2$, which nevertheless displays dynamical excitations for more than 80°C above the ferroelastic T_c of 180°C (Salje et al. 1983). The excitations of the Pb atoms above T_c can be represented by a Landau potential in $\{Q_3\}$ and by orientational terms with components $\{Q_1, Q_2\}$. Based on a three states Potts model developed by Salje

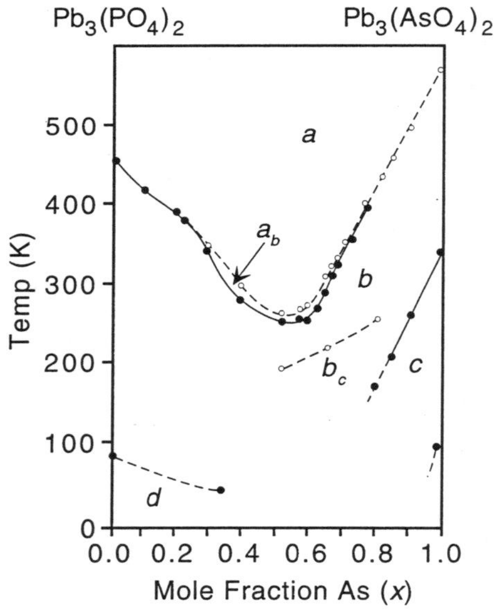

Figure 19. Phase diagram for the system $Pb_3(P_{1-x}As_xO_4)_2$. Modified from Figure 1 in Bismayer et al. (1986).

and Devarajan (1981), Bismayer (1990) presents a Landau-Ginzburg Hamiltonian that accounts for the short-range monoclinic distortions in the paraelastic phase over the P-As solid solution.

In addition, Viswanathan and Miehe (1978) observed a second lower temperature transition at 64°C in $Pb_3(AsO_4)_2$ from the *b* phase to a *c* polymorph with space group $P2_1/c$. Bismayer et al. (1986) have shown that this *b*-to-*c* transition is sharply discontinuous for the endmember arsenate (x_{As} = 1.0). The first-order nature of this transition is confirmed by a temperature hysteresis of ~14°C. However, the transformation becomes continuous with increasing P content, as can be seen from the dependence of birefringence on temperature for different compositions in the $Pb_3(P_{1-x}As_xO_4)_2$ series (Fig. 20).

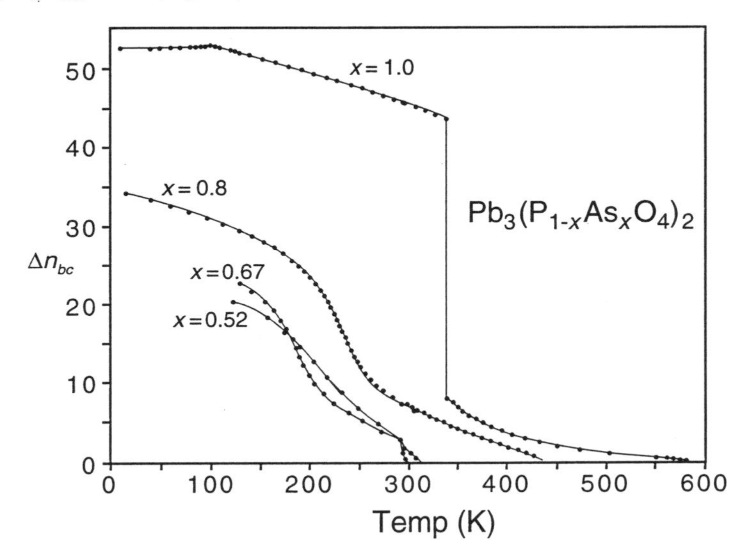

Figure 20. Dependence of birefringence on temperature for compositions x = 0.52, 0.67, 0.8, and 1.0 in the system $Pb_3(P_{1-x}As_xO_4)_2$. Modified from Figure 3 in Bismayer et al. (1986).

Similar damping effects on the transition occur when Pb is replaced by Sr or Ba (Bismayer et al. 1994, 1995). Heat capacity profiles at the $R\bar{3}m$–$C2/c$ transition become broader and less intense with higher concentrations of solute (Fig. 21), and the temperature of the maximum specific heat occurs at lower temperatures. The dramatic renormalization of T_c to lower values as a function of Ba and Sr contents is depicted in Figure 22 on the basis of birefringence data. Bismayer and colleagues report that the sizes of the phase fields for the intermediate a_b structure is enlarged in the Sr- and Ba-doped phosphates, though no local monoclinic signals are detected for x_{Sr} > 0.4 in $(Pb_{1-x}Sr_x)_3(PO_4)_2$. As can be seen in Figure 22, the coupling strength ξ_l (Eqn. 6) for Ba is considerably stronger than that for Sr, and the plateau effect is much greater for Sr than for Ba. For $(Pb_{1-x}Sr_x)_3(PO_4)_2$, the plateau extends to x_{Sr} = ~0.08, whereas in $(Pb_{1-x}Ba_x)_3(PO_4)_2$ the plateau boundary occurs at x_{Ba} = ~0.002. Based on these compositions, Bismayer et al. (1995) determine that the plateau edge for Sr-doped lead phosphate occurs when Sr atoms are separated by ~14 Å on average, whereas the mean distance for Ba atoms at the plateau limit is ~50 Å. Consequently, the characteristic interaction length for Ba in doped lead phosphate is 3.6 times longer than that for Sr.

Cuproscheelite–sanmartinite solid solutions

Cuproscheelite ($CuWO_4$) and sanmartinite ($ZnWO_4$) are topologically identical to the Fe-Mn tungstate wolframite, but at room temperature they exhibit different symmetries from each other (Filipenko et al. 1968, Kihlborg and Gebert 1970). As with wolframite, the sanmartinite structure belongs to space group *P*2/*c*, but that for cuproscheelite is $P\bar{1}$. Nevertheless, these minerals exhibit complete solid solution across the binary join (Schofield and Redfern 1992). Explorations of the transition behavior in this system are motivated by their technological potential. Cuproscheelite is an *n*-type semiconductor with possible uses as a photoanode, and sanmartinite may serve as a high *Z*-number scintillator (Doumerc et al. 1984, Redfern et al. 1995). The structure consists of edge-sharing chains of WO_6 octahedra that are each corner-linked to 4 chains of edge-sharing $(Cu,Zn)O_6$ octahedra; each $(Cu,Zn)O_6$ chain is likewise corner-linked to 4 WO_6 chains (**Fig**. 23).

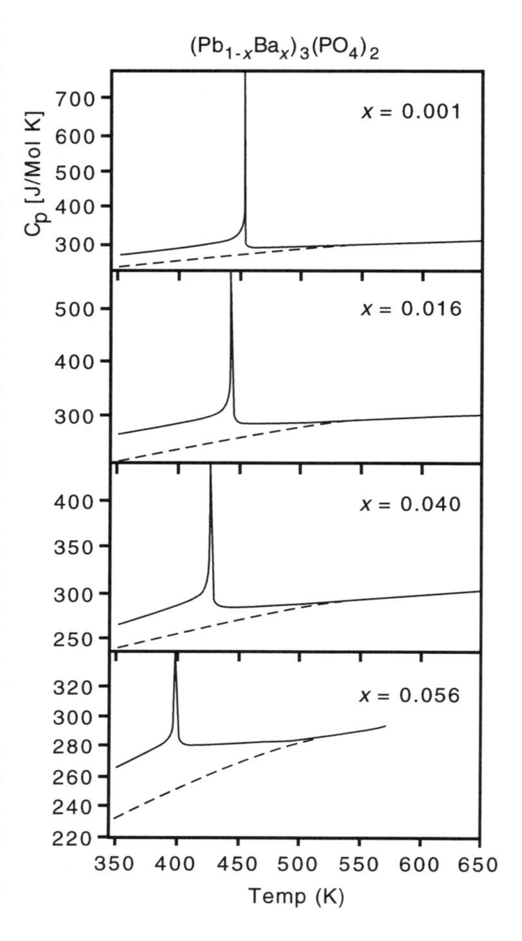

Figure 21. Evolution of specific heat profiles in $(Pb_{1-x}Sr_x)_3(PO_4)_2$ with increasing concentrations of Ba. Note change of C_p scales. Modified from Figure 2 in Bismayer et al. (1995).

The ZnO_6 and the CuO_6 octahedra deviate strikingly from the geometric ideal. In pure $ZnWO_4$, the ZnO_6 octahedra contain 4 long Zn-O bonds (2.11 Å in length) in square planar configuration with 2 Zn-O axial bond lengths of only 1.95 Å (Schofield et al. 1994). Thus, the ZnO_6 octahedra are axially compressed. By contrast, the Cu^{2+} octahedra in endmember $CuWO_4$ exhibit marked Jahn-Teller distortions;

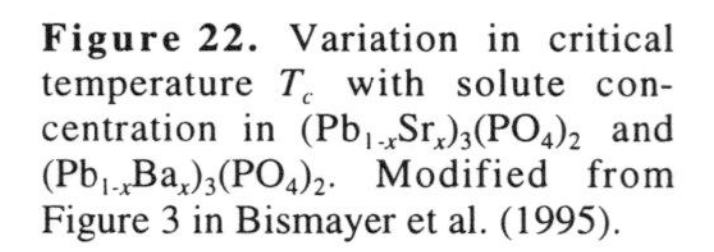

Figure 22. Variation in critical temperature T_c with solute concentration in $(Pb_{1-x}Sr_x)_3(PO_4)_2$ and $(Pb_{1-x}Ba_x)_3(PO_4)_2$. Modified from Figure 3 in Bismayer et al. (1995).

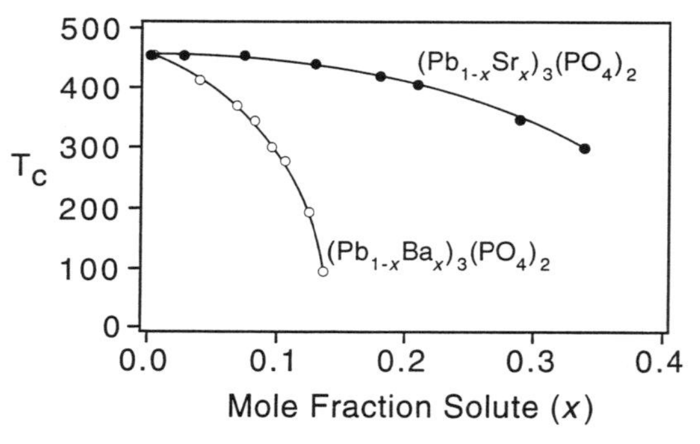

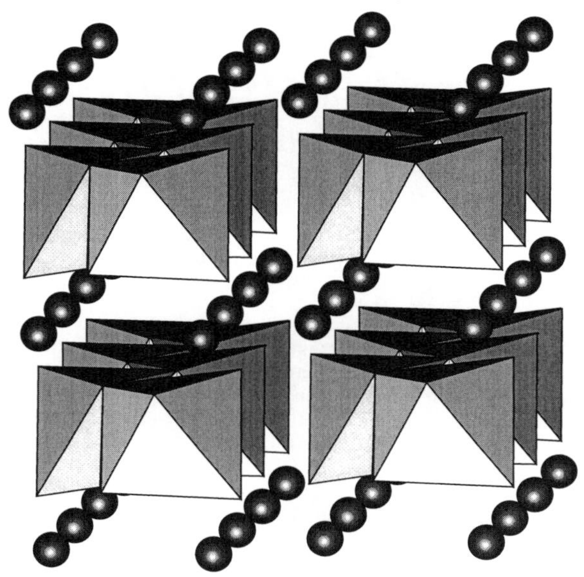

Figure 23. Projection of the structure of sanmartinite ($ZnWO_4$) along the *c* axis. Spheres are W cations.

the two axial Cu-O distances average 2.397 Å for the two axial bonds and 1.975 Å for the four coplanar vertices (Schofield et al. 1997). It is the Jahn-Teller character of the Cu-rich octahedra that reduces the symmetry of the monoclinic aristotype to triclinic.

Using EXAFS, Schofield et al. (1994) systematically examined the local bonding environments for both Cu^{2+} and Zn^{2+} over the entire $Cu_{1-x}Zn_xWO_4$ series to determine the interrelation between the axially elongate Cu octahedra and the axially compressed Zn octahedra. Their results indicate that for small degrees of Zn substitution, most of the Zn cations occupy octahedra that are axially elongate, but a minority reside in compressed coordination octahedra. With increasing Zn concentrations, the ratio of compressed to elongate octahedra increases so that at $x_{Zn} = 0.5$, roughly even proportions of the two sites exist. By $x_{Zn} = 0.78$, the room temperature bulk structure transforms to the *P2/c* symmetry of the sanmartinite endmember, and for 0.8 x_{Zn} 1.0, the Zn *K*-edge EXAFS spectra yield a superior fit for a compressed octahedral coordination. By contrast, the CuO_6 octahedra remain axially elongate over most of the series, even beyond the critical composition X_{Zn}. Consequently, these results demonstrate that on either side of X_{Zn}, the (Zn,Cu) octahedra are geometrically discordant with the bulk symmetry as detected by elastic scattering.

Analyses of the variation of transition temperatures with composition (Fig. 24) reveal an unambiguous departure from linearity (Schofield and Redfern 1993). For example, the plateau region is more extensive than in other oxide systems, ranging from 0 x_{Zn} 0.12. In addition, the dependence of T_c on x_{Zn} steepens near X_c. Consequently, when the square of the spontaneous strain is plotted as a function of composition (Fig. 25), the data depart slightly from ideal second-order behavior. Schofield et al. (1997) attribute this deviation and the broad plateau to the short-range interaction length of the strain fields associated with the Zn cations.

In addition, the non-linearity of the change in transition temperature with

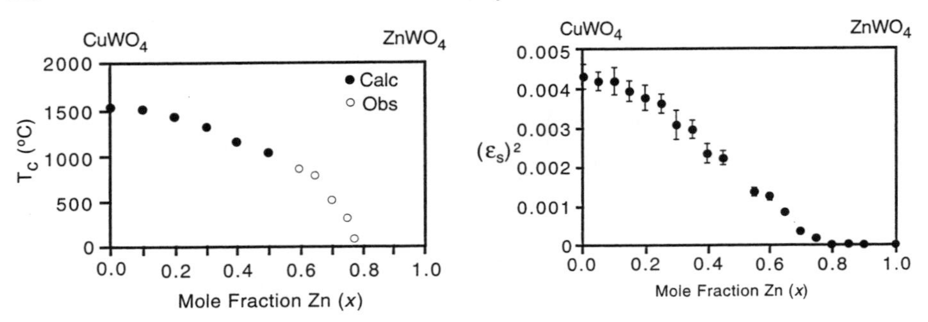

Figure 24 (left). Variation of the critical temperature T_c with composition in $Cu_{1-x}Zn_xWO_4$. Transition values for compositions $x_{Zn} < 0.6$ were calculated on the basis of spontaneous strain systematics. Modified from Figure 6 in Schofield and Redfern (1993).

Figure 25 (right). Dependence of the square of spontaneous strain $[(\varepsilon_s)^2]$ with composition in $Cu_{1-x}Zn_xWO_4$. Non-linearity in the profile is apparent. Modified from Figure 11 in Schofield et al. (1997).

composition suggests that the Landau-Ginzburg excess free energy expression must include Q^4 terms, and Schofield and Redfern (1993) suggest that the solute content X_{Zn} couples with the order parameter in the form $\xi_1 X_{Zn} Q^2 + \xi_2 X_{Zn} Q^4$. Further, these authors observe sharp discontinuities in the evolution of spontaneous strain as a function of temperature for compositions x_{Zn} 0.65 in $Cu_{1-x}Zn_xWO_4$. When x_{Zn} 0.70, strain decreases gradually to 0 as the critical temperature is approached. Thus, as with most other compositionally induced transitions, higher levels of dopant modify the character of the inversion from first- to second-order. The tricritical composition in the $Cu_{1-x}Zn_xWO_4$ system must lie at $0.65 < x_{Zn} < 0.70$.

Substitutions in feldspar frameworks

Transitions in feldspar minerals. The feldspars arguably constitute the most important class of minerals that experience phase transitions in response to solid solution. The feldspar group composes 60% of the Earth's crust according to the estimates of Clarke (1904), and they are abundant if not predominant in most igneous, metamorphic, and sedimentary environments. However, the phase transitions within these framework silicates are among the most complex of all to model, as natural samples exhibit various degrees of cation substitution, Al-Si order, and framework collapse. In addition, large portions of the feldspar system are characterized by high degrees of immiscibility. Perthitic exsolution textures are common in slowly cooled plutonic alkali feldspars, and even the join between albite ($NaAlSi_3O_8$) and anorthite ($CaAl_2Si_2O_8$) exhibits immiscibility at the microscopic scale. Plagioclase minerals display a variety of exsolution features, such as peristerite intergrowths in albitic compositions, Bøggild intergrowths, which give rise to iridescence in labradorite, and Huttenlocher intergrowths in anorthitic feldspar (Champness and Lorimer 1976, Smith and Brown 1988).

Nevertheless, immiscibility and Al-Si ordering are minimized when specimens anneal at high temperatures for long periods and then are rapidly quenched. In natural and synthetic specimens that have experienced such cooling histories, the effects of cationic substitutions can be analyzed independent of other factors. For instance, a number of authors have examined transition temperatures from $C2/m$ to $C\bar{1}$ symmetry in completely disordered alkali feldspars ($Na_{1-x}K_xAlSi_3O_8$), and the results (Fig. 26) indicate that the critical temperature decreases linearly with K content, since the larger K^+ cation inhibits structural collapse (Kroll et al. 1980, 1986; Salje 1985, Harrison and Salje 1994,

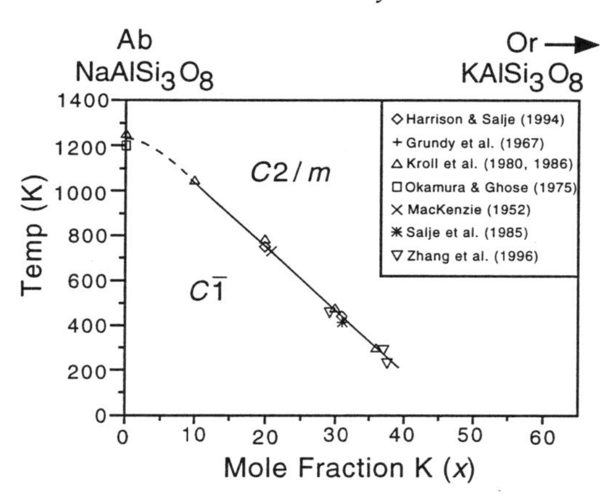

Figure 26. Dependence of critical temperature on composition in completely disordered alkali feldspars ($Na_{1-x}K_xAlSi_3O_8$). Modified from Figure 14 in Zhang et al. (1996).

Zhang et al. 1996). The critical composition x_K at room temperature is ~0.35, and the transition disappears at T_c = 0 K when X_K exceeds ~0.47. Plateau effects extend from X_K = 0 to 0.02, suggesting an interaction range for K^+ substitutions of ~10 Å (Hayward et al. 1998).

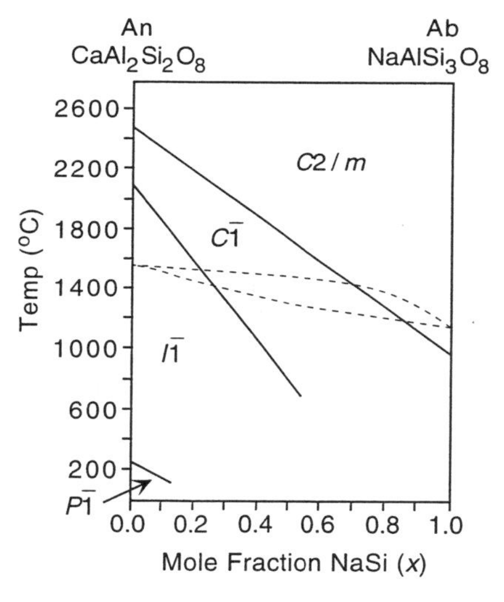

Figure 27. Inferred phase relations in plagioclase feldspars based on Carpenter (1988) and Redfern (1990).

Plagioclase solid solutions exhibit different behaviors. The idealized transition sequences for endmember albite and anorthite are presumed to parallel those experienced by alkali feldspars (Carpenter 1988). At high temperature, the structures belong to space group *C2/m*, and on cooling the symmetries change to $C\bar{1}$ to $I\bar{1}$ to $P\bar{1}$ (Fig. 27). Of course, the observed transition sequences for real plagioclase feldspars deviate strikingly from this model. Endmember albite undergoes a single displacive transition at ~1000°C from *C2/m* to $C\bar{1}$ (Smith and Brown 1988, Carpenter 1994, Atkinson et al. 1999), and the only subsolidus transition exhibited by anorthite is an inversion from $I\bar{1}$ to $P\bar{1}$ (Redfern and Salje 1987, 1992). This transition occurs at ~241°C for pure metamorphic samples, and it involves framework expansion and possibly Ca positional disordering (Laves et al. 1970, Staehli and Brinkmann 1974, Adlhart et al. 1980, Van Tendeloo et al. 1989, Ghose et al. 1993).

The substitution of Na^+ for Ca^{2+} in plagioclase is expected to stabilize the expanded framework, since Na^+ is slightly larger than Ca^{2+} (Shannon 1976), and the difference in valence will favor the collapsed configuration in the anorthitic endmember. This effect is reinforced by the concomitant replacement of Al^{3+} with the smaller Si^{4+}, which decreases the mean tetrahedral size. Consequently, the *c*-reflections that are diagnostic of $P\bar{1}$ symmetry grow extremely diffuse and disappear with the substitution of Na^+ for Ca^{2+} in anorthite, suggesting a morphotropic transition to $I\bar{1}$ symmetry (Adlhart et al. 1980).

Hard mode infrared spectroscopy indicates that $P\bar{1}$ character is absent even at the microscale when Ca contents fall below An_{96} (Atkinson et al. 1999). Moreover, the character of the transition changes in response to an increasing albitic component. Endmember Ca-anorthite transforms tricritically from $P\bar{1}$ to $I\bar{1}$ symmetry, but with an increase in Na+Si, the transition becomes second-order (Redfern 1990). Interestingly, Redfern et al. (1988) and Redfern (1992) present evidence that an increasing albitic component acts to increase critical temperatures of anorthitic solid solutions when samples are equally disordered with respect to Al and Si. Th effects of Al-Si order/disorder are addressed by Redfern in a chapter in this volume.

Studies of Sr-anorthite. Despite the fact that neither anorthite nor albite travels the full transition pathway from $P\bar{1}$ to $C2/m$, the supersolidus transitions that are never directly observed strongly influence mineral behavior at low temperatures (Salje 1990). For instance, the Al-Si order-disorder reaction that underlies the $C\bar{1}$ to $I\bar{1}$ transition occurs above the melting point of anorthite, but this transition controls much of the behavior of subsolidus anorthite. Therefore, understanding the structural response of the feldspars at each of these transitions is highly desirable. As the intrusion of melting temperatures does not allow us to perform the necessary experiments at ambient pressures in the $NaAlSi_3O_8$–$CaAl_2Si_2O_8$ system, instead we must turn to substituted compositions that mimic the behavior of the plagioclase system.

Some of the most instructive studies of this kind have focused on compositionally induced transformations along the $CaAl_2Si_2O_8$–$SrAl_2Si_2O_8$ join. Replacement of Ca with the larger Sr cation drives structural expansion through the displacive triclinic-monoclinic inversion (Bambauer and Nager 1981). However, the transition sequence differs slightly from that of the plagioclase system. With increasing Sr content, the symmetry changes from $P\bar{1}$ to $I\bar{1}$to $I2/c$ (Bruno and Gazzoni 1968, Sirazhiddinov et al. 1971, Bruno and Pentinghaus 1974, Chiari et al. 1975). This transition path is strongly dependent on the degree of Al-Si order. For example, Grundy and Ito (1974) found that a highly disordered, non-stoichiometric Sr-anorthite ($(Sr_{0.84}Na_{0.03}[]_{0.13})Al_{1.7}Si_{2.3}O_8$) yielded a symmetry of $C2/m$, as is typical of the plagioclase feldspars.

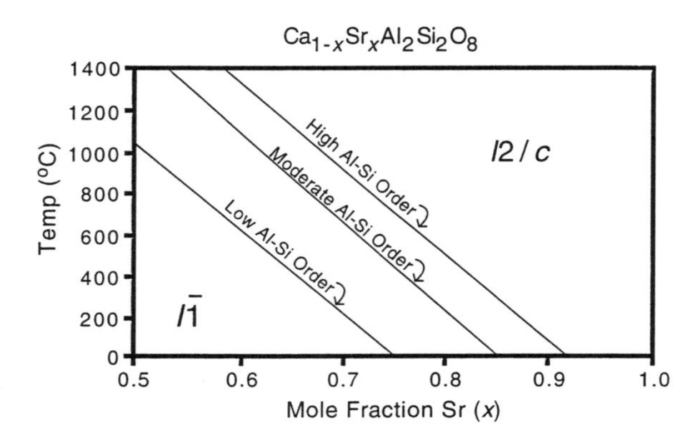

Figure 28. Phase diagram for $Ca_{1-x}Sr_xAl_2Si_2O_8$ with different phase boundaries for various degrees of Al-Si order. Adapted from Figure 3 in Tribaudino et al. (1993).

The critical temperatures for the monoclinic to triclinic transition decrease with increasing Sr content (Fig. 28). As with the transition sequence, the room-temperature critical composition x_{Sr} varies with the degree of Al-Si order. Experimental data (Bruno and Gazzoni 1968, Bambauer and Nager 1981, Tribaudino et al. 1993, 1995) demonstrate that x_{Sr} equals 0.91 for a relatively ordered sample, but that disorder can shift x_{Sr} to 0.75 (Fig. 28). McGuinn and Redfern (1994) synthesized a series of compounds within the $Ca_{1-x}Sr_xAl_2Si_2O_8$ solid solution with long annealing times and found a critical x_{Sr} of 0.91. Phillips et al. (1997) then used ^{29}Si MAS NMR to determine the state of order for these samples and found σ = ~0.92 for compounds near the critical composition, where σ is defined as $1 - 0.5\ (N_{Al\text{-}Al})$, and $N_{Al\text{-}Al}$ is the number of Al-O-Al linkages per formula unit. All of these empirical results indicate that Al-Si disorder significantly stabilizes the monoclinic structure. Indeed, lattice-energy minimization calculations by Dove and Redfern (1997) indicate that a fully ordered phase is triclinic across the entire composition range at 0 K. Their model also reveals that Al-Si disorder decreases the slope of the displacive phase boundary and favors *C*-centered rather than *I*-centered structures.

The symmetry-breaking strains in the monoclinic-to-triclinic transition for feldspars are ε_4 and ε_6, where

$$\varepsilon_4 = \frac{1}{\sin\beta_0^*}\left(\frac{c\cos\alpha}{c_0} + \frac{a\cos\beta_0^*\cos\gamma}{a_0}\right) \tag{11}$$

$$\varepsilon_6 = \frac{a\cos\gamma}{a_0} \tag{12}$$

after Redfern and Salje (1987). It has been shown empirically that ε_4 behaves as the primary order parameter (Q) and that ε_6 is sensitive to the Al-Si order-disorder (Q_{OD}) (Salje et al. 1985). The calculations of Dove and Redfern (1997) reveal that the relationship between ε_4 and ε_6 in plagioclase is quite close to that for the alkali feldspars. In addition, McGuinn and Redfern (1994) demonstrate that the dependence of ε_4^2 on the Sr content (expressed as $|x_{crit} - x_{Sr}|$) is nearly exactly linear, suggesting that β = 1/2 and that the compositionally induced transition is classically second order. However, when ε_4 is plotted as a function of Sr content (Fig. 29), a triclinic strain tail is clearly observed near the critical composition. Consequently, the critical composition as extrapolated through the low-Sr data yields a value for x_c of 0.86, which is lower than the composition at which the strains are measured as zero (x_{Sr} = 0.91). This persistent triclinic strain also is detected in ^{29}Si MAS-NMR peak broadening near the critical composition, and it probably is related to Al-Si ordering processes or local structural fluctuations near the alkali cation (Phillips et al. 1997).

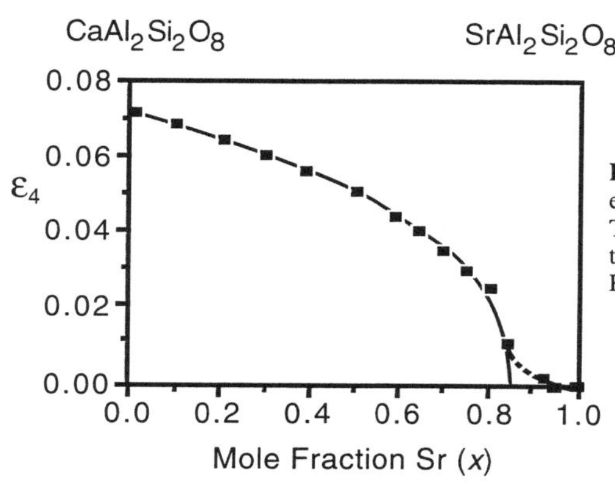

Figure 29. Dependence of the strain parameter ε_4 with composition in $Ca_{1-x}Sr_xAl_2Si_2O_8$. The dashed line represents an observed defect tail. Modified from Figure 7 in McGuinn and Redfern (1994).

Stuffed derivatives of quartz

Impurities and the α–β-quartz transition. The α–β-quartz transition was the basis for one of the earliest systematic investigations of the variation of transition temperatures in response to impurities. Pure α-quartz undergoes a first-order transition to a microtwinned incommensurate structure at 573°C, and this modulated phase transforms to β-quartz at 574.3°C with second-order behavior (Van Tendeloo et al. 1976, Bachheimer 1980, Dolino 1990). Tuttle (1949) and Keith and Tuttle (1952) investigated 250 quartz crystals and observed that T_c for natural samples varied over a 38°C range. In their examination of synthetic specimens, substitution of Ge^{4+} for Si^{4+} raised the critical temperature by as much as 40°C, whereas the coupled exchange of $Al^{3+}+Li^{+} \leftrightarrow Si^{4+}$ depressed T_c by 120°C. They concluded from their analyses that the departure of the α–β-quartz inversion temperature from 573°C could be used to assess the chemical environ-ment and the growth conditions for natural quartz.

Subsequent studies, however, have dampened enthusiasm for using the measured T_c for the α–β-quartz inversion as a universal petrogenetic indicator. As reviewed in Heaney (1994), a number of scientists have explored relationships between the incorporation of impurities (especially Al^{3+}) and the formation of quartz, but promising trends generally have been quashed by subsequent investigation (Perry 1971, Scotford 1975). While it is true that careful studies (e.g. Ghiorso et al. 1979, Smith and Steele 1984) demonstrate a roughly linear response between Al^{3+} concentration and transition temperature, most natural materials contain many kinds of impurities, and the degree of scatter is large. Moreover, as noted by Keith and Tuttle (1952), over 95% of natural quartz crystals invert within a 2.5°C range of 573°C. Consequently, the usefulness of quartz inversion temperatures for the determination of provenance or for geothermometry is limited.

On the other hand, the stuffed derivatives of β-quartz have found extremely widespread technological application because of their unusually low coefficients of thermal expansion (CTEs). In particular, members of the Li_2O–Al_2O_3–SiO_2 (LAS) system are major components of glass ceramic materials that require high thermal stability and thermal shock resistance, such as domestic cookware and jet engine components (Roy et al. 1989, Beall 1994). β-eucryptite ($LiAlSiO_4$) is isostructural with β-quartz, and it can engage in a metastable solid solution with the silica endmember. By a long-standing convention in the ceramics literature, members of this series are represented as $Li_{1-x}Al_{1-x}Si_{1+x}O_4$, even though the Type Ib morphotropic trends displayed by the system suggest that SiO_2 should be considered the solvent and (Li,Al) the solute.

Crystal chemistry of β-eucryptite. The β-quartz structure (S.G. $P6_222$ or $P6_422$) consists of intertwined tetrahedral helices that spiral about the *c*-axis. On cooling below 537°C, the tetrahedra tilt about the *a*-axes and the structure displacively transforms to the α-quartz modification (S.G. $P3_221$ or $P3_121$). Although pure β-quartz is not quenchable, the incorporation of small ions (such as Li^+ and Be^{2+}) in the channels along *c* can prop open the framework and stabilize the β-quartz modification (Buerger 1954, Palmer 1994, Müller 1995). Crystal structure analyses of the endmember β-eucryptite have revealed that despite its topological identity with β-quartz, the translational periodicity of $LiAlSiO_4$ is doubled along the *c*- and *a*-axes relative to β-quartz (Fig. 30) (Winkler 1948, Schulz and Tscherry 1972a,b; Tscherry et al. 1972a,b; Pillars and Peacor 1973). This superstructure arises from the ordering of Al and Si ions in alternate layers normal to *c* with concomitant ordering of Li within two distinct channels.

Recent studies by Xu et al. (1999a,b; 2000) have demonstrated that two structural transitions occur within the $Li_{1-x}Al_{1-x}Si_{1+x}O_4$ system at room temperature. When x_{Si} > ~0.3, the Al^{3+} and Si^{4+} cations disorder over the tetrahedral sites, and when $x_{Si} > 0.65$, the

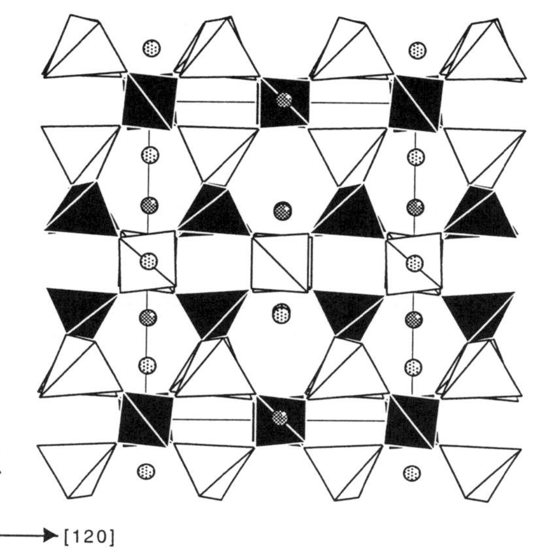

Figure 30. A projection of the structure in ordered β-eucryptite along the *a* axis. Spheres represent Li ions. Si- and Al-tetrahedra are plotted in black and white, respectively. From Figure 4 in Xu et al. (2000).

structure collapses from the expanded β-quartz to the collapsed α-quartz configuration. The studies by Xu et al. reveal that this displacive morphotropic transition is very similar to the thermally induced α-β quartz transition both energetically and mechanistically. In pure quartz, the enthalpy of the α-β transition is negligibly small, on the order of 0.5 to 0.7 kJ/mol (Ghiorso et al. 1979, Gronvold et al. 1989). Similarly, the enthalpy change associated with the morphotropic transition at x_{Si} = 0.65 in $Li_{1-x}Al_{1-x}Si_{1+x}O_4$ fell below detection limits in the drop solution calorimetry experiments of Xu et al. (1999a).

Moreover, high-resolution structural studies demonstrate that those atomic displacement pathways that effect the α-β transition in pure quartz are identical to those activated in the morphotropic transition. Because β-quartz cannot persist below its inversion temperature, it is not possible to measure directly the lattice parameters for β-quartz at 25°C. However, strain-free parameters for quartz may be inferred by two approaches: Values may be extrapolated from the high-temperature lattice parameters for β-quartz as determined by Carpenter et al. (1998); and the lattice constants can be extrapolated through the β-quartz-like phases within the $Li_{1-x}Al_{1-x}Si_{1+x}O_4$ series. Xu et al. (2000) have demonstrated that these two approaches yield identical values for *a*, *c*, and unit cell volume within error (Fig. 31). This coincidence is explained by crystallographic analyses of LAS phases with 0.65 x_{Si} 1.0 that revealed a decrease in tetrahedal tilting about the *a*-axis with increasing Li+Al content. In other words, both temperature and cation substitution expand the structure via tetrahedral rotation, which can serve as the order parameter for both the polymorphic and the morphotrophic transitions.

The similarity in behaviors of thermally and compositionally induced transitions suggests that the dependence of spontaneous strain on Li+Al content may parallel the dependence of strain on temperature. Xu et al. (2000) determine elastic strains ($e_1 = e_2 = a/a_0 - 1$; $e_3 = c/c_0 - 1$) and volume strains ($V_s = V/V_0 - 1$) by referencing the paraelastic cell dimensions to the β-quartz-like phases within the $Li_{1-x}Al_{1-x}Si_{1+x}O_4$ series. They find that the data are consistent with the relation e_1 (or e_3 or V_s) = A $(X - X_c)^{1/2}$, and with $e_l \propto e_3 \propto V_s \propto Q^2$, the morphotropic transition appears to be tricritical. This transition character conforms closely to that observed by Carpenter et al. (1998), who argued that the thermally induced α-β quartz transition is first-order but very close to tricritical. In

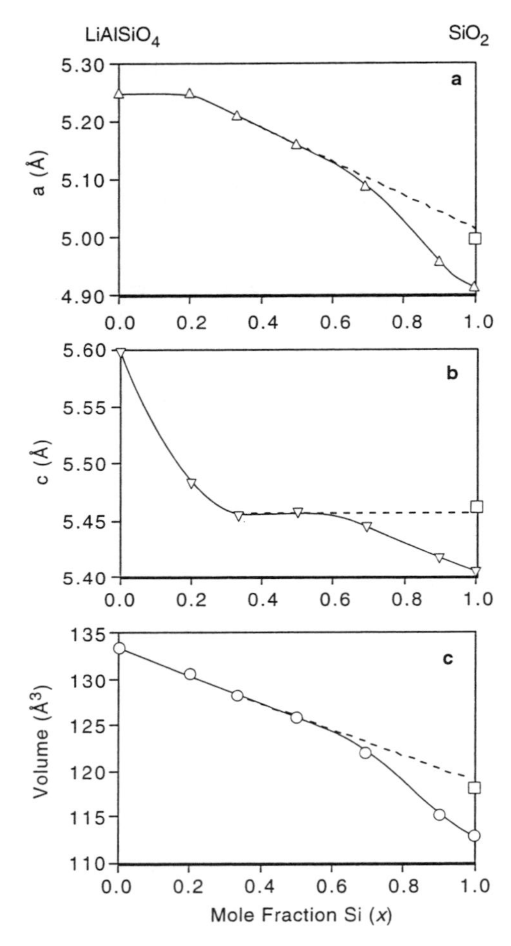

Figure 31. Variation in lattice parameters as a function of composition in the system $(LiAl)_{1-x}Si_{1+x}O_4$. (a) *a* axis; (b) *c* axis; (c) unit cell volume. The hypothetical room-temperature cell parameters of pure β-quartz are also shown for comparison (squares). From Figure 5 in Xu et al. (2000).

fact, additional studies are needed near the critical composition in $Li_{1-x}Al_{1-x}Si_{1+x}O_4$ to determine the precise character of this morphotropic transition.

GENERAL CONCLUSIONS

Pure materials that undergo displacive transitions may exhibit anomalies in heat capacity, birefringence, volume expansion, and other properties at their transition points. When the low-symmetry structures are doped with transition-inducing solutes, however, the structures can experience partial transformation even when impurity levels fall well below the critical composition. Consequently, the anomalies associated with thermally induced transitions are less intense with higher dopant contents. Typically, the critical temperatures, energies of transition and ranges of hysteresis decrease; the heat capacity profile alters from a sharp spike to a broader lambda-like figure; and the character of the transition changes from first-order to tricritical to second-order. These trends are coupled to, and to some extent caused by, a continual decrease in the mean sizes of twin and antiphase domains; the escalating wall volume that attends the diminishing domain size gradually increases the amount of locally paraelastic structure within the ferroelastic host. This phenomenon is not unlike the microtwinning that accompanies thermally induced transitions in many minerals, such as quartz and anorthite.

The similarity in compositionally induced phase transitions and displacive transformations activated by heat or pressure is not completely surprising. In structures that thermally transform via rigid unit modes (see chapter by Dove, this volume), the replacement of small cavity-dwelling ions with larger species leads to framework expansion by polyhedral tilting, and the re-orientation of these polyhedra will occur along the same pathways that are followed during thermal expansion. Structural studies of the quartz and leucite systems have demonstrated mechanistic similarities in the rigid unit rotations induced by both temperature and composition. Alternatively, compounds with distorted, low symmetry polyhedral units (such as are generated by Jahn-Teller deformation) may see diminished distortion when the central polyhedral cations are replaced; the resultant increase in local polyhedral symmetry can lead to an increase in the bulk symmetry of the crystal. This behavior is typified by the replacement of Cu by Zn in cuproscheelite.

As a result of these processes, symmetry analyses of doped materials are particularly susceptible to confusion. The perceived symmetry of any structure depends on the correlation length of the analytical probe. Prolific twinning and the strain fields that surround an impurity atom can generate regional violations of the bulk symmetry. Especially in the vicinity of morphotropic phase boundaries, techniques that measure structures over relatively long length scales may not be sensitive to short-range symmetry-breaking distortions. For example, elastic scattering intensities are sharp only when the interference function includes more than ~10 unit cells. Consequently, diffraction methods may measure a higher overall symmetry than spectroscopic techniques that probe over length scales of a few unit cells. Thus, it is especially important that analyses of morphotropic transitions include a multiplicity of techniques in order to capture the evolution of symmetry over both long and short ranges.

Finally, it is clear that many more solid solutions must be examined for minerals and mineral analogs in order to achieve a fundamental understanding of compositionally induced phase transitions. Currently it is not possible to predict how transition temperatures will change when a particular impurity substitutes in a mineral structure, nor can we predict the interaction length for that impurity in the mineral. Landau-Ginzburg analysis provides an ideal framework for comparing the character of phase transitions that are activated by different variables (temperature, pressure, composition), and future studies of this type will lay an empirical foundation from which the detailed character of morphotropic transitions in minerals may be inferred.

ACKNOWLEDGMENTS

I am grateful to Simon Redfern and Martin Dove for reviewing this chapter and to David Palmer for contributing two figures and for inventing CrystalMaker, which generated many of the structure drawings in this chapter. A. Navrotsky, F. Siefert, U. Bismayer, D. Palmer, and H. Xu kindly gave permission to reproduce or adapt figures for this paper. In addition, I thank Bob Newnham, Ekhard Salje, Michael Carpenter and Hongwu Xu and for many insightful discussions regarding the effects of impurities on phase transitions.

REFERENCES

Adlhart, W, Frey F, Jagodzinski H (1980) X-ray and neutron investigations of the $P\bar{1}$–$I\bar{1}$ transition in anorthite with low albite content. Acta Crystallogr A36:461-470

Alfé D, Price GD, Gillan MJ (1999) Oxygen in the Earth's core: A first-principles study. Phys Earth Plan Int 110:191-210

Andrault D, Itie JP, Farges F (1996) High-temperature structural study of germanate perovskites and pyroxenoids. Am Mineral, 81:822-832

Ari-Gur P, Benguigui L (1975) Direct determination of the co-existence region in the solid solutions $Pb(Zr_xTi_{1-x})O_3$. J Phys D 8:1856-1862

Atkinson AJ, Carpenter MA, Salje EKH (1999) Hard mode infrared spectrosopy of plagioclase feldspars. Eur J Mineral 11:7-21

Bachheimer JP (1980) An anomaly in the β phase near the α-β transition of quartz. J Physique–Lett 41:L559-L561

Batlogg B (1991) Physical properties of high-T_c superconductors. Phys Today 44:44-50

Bambauer HU, Nager HE (1981) Lattice parameters and displacive transformation of synthetic alkaline earth feldspars. I. System $Ca[Al_2Si_2O_8]$–$Sr[Al_2Si_2O_8]$–$Ba[Al_2Si_2O_8]$. Neues Jahrb Mineral Abh 141:225-239

Beall GH (1994) Industrial applications of silica. Rev Mineral 29:468-505

Bednorz JG, Müller KA (1986) Possible high T_c superconductivity in the Ba-La-Cu-O system. Z Phys B 64:189-193

Berry LG, Mason B (1959) Mineralogy: Concepts, Descriptions, Determinations. Freeman, San Francisco. 630 p

Bismayer U (1990) Stepwise ferroelastic phase transitions in $Pb_3(P_{1-x}As_xO_4)_2$. *In* Salje EKH Phase Transitions in Ferroelastic and Co-elastic Crystals. Cambridge University Press, Cambridge, UK, p 253-267

Bismayer U, Hensler J, Salje E, Güttler B (1994) Renormalization phenomena in Ba-diluted ferroelastic lead phosphate, $(Pb_{1-x}Ba_x)_3(PO_4)_2$. Phase Trans 48:149-168

Bismayer U, Röwer RW, Wruck B (1995) Ferroelastic phase transition and renormalization effect in diluted lead phosphate, $(Pb_{1-x}Sr_x)_3(PO_4)_2$ and $(Pb_{1-x}Ba_x)_3(PO_4)_2$. Phase Trans 55:169-179

Bismayer U, Salje E (1981) Ferroelastic phases in $Pb_3(PO_4)_2$–$Pb_3(AsO_4)_2$: x-ray and optical experiments. Acta Crystallogr A37:145-153

Bismayer U, Salje E, Glazer AM, Cosier J (1986) Effect of strain-induced order-parameter coupling on the ferroelastic behaviour of lead phosphate-arsenate. Phase Trans 6:129-151

Bismayer U, Salje E, Joffrin C (1982) Reinvestigation of the stepwise character of the ferroelastic phase transition in lead phosphate-arsenate, $Pb_3(PO_4)_2$–$Pb_3(AsO_4)_2$. J Physique 43:1379-1388

Bragg WL (1929) Atomic arrangement in the silicates. Proc Faraday Soc, London, p 291-314

Brown NE, Ross CR, Webb SL (1994) Atomic displacements in the normal-incommensurate phase-transition in Co-åkermanite ($Ca_2CoSi_2O_7$). Phys Chem Minerals 21:469-480

Bruno E, Gazzoni G (1968) Feldspati sintetici della serie $Ca[Al_2Si_2O_{8X}]$–$Sr[Al_2Si_2O_8]$. Atti della Accademia delle Scienze de Torino 102:881-893

Bruno E, Pentinghaus H (1974) Substitution of cations in natural and synthetic feldspars. *In* MacKenzie WS, Zussmann, J (eds) The Feldspars. Manchester Univ Press, Manchester. p 574-609

Buerger MJ (1951) Crystallographic aspects of phase transformations. *In* Smoluchowski R (ed) Phase Transformations in Solids. Wiley, New York, p 183-211

Buerger MJ (1954) The stuffed derivatives of the silica structures. Am Mineral 39:600-614

Bulenda M, Schwabl F, Täuber UC (1996) Defect-induced condensation and the central peak at elastic phase transitions. Phys Rev B 54:6210-6221

Burns PC, Carpenter MA (1996) Phase transitions in the series boracite-trembathite-congolite: Phase relations. Can Mineral 34:881-892

Carl K, Härdtl KH (1971) On the origin of the maximum in the electromechanical activity in $Pb(Zr_xTi_{1-x})O_3$ ceramics near the morphotropic phase boundary. Phys Stat Sol A8:87-98

Carpenter MA (1980) Mechanisms of ordering and exsolution in omphacites. Contrib Mineral Petrol 71:289-300

Carpenter MA (1988) Thermochemistry of aluminum/silicon ordering in feldspar minerals. *In* Salje EKH (ed) Physical properties and thermodynamic behaviour of minerals. Reidel, Dordrecht, p 265-323

Carpenter MA (1994) Evolution and properties of antiphase boundaries in silicate minerals. Phase Trans 48:189-199

Carpenter MA, McConnell JDC, Navrotsky A (1985) Enthalpies of ordering in the plagioclase feldspar solid solution. Geochim Cosmochim Acta 49:947-966

Carpenter MA, Salje EKH, Graeme-Barber A, Wruck B, Dove MT, Knight KS (1998) Calibration of excess thermodynamic properties and elastic constant variations associated with the $\alpha\leftrightarrow\beta$ phase transition in quartz. Am Mineral 83:2-22

Champness PE, Lorimer GW (1976) Exsolution in silicates. *In* Wenk H-R et al. (eds) Electron Microscopy in Mineralogy. Springer-Verlag, Berlin, p 174-204

Chiari G, Calleri M, Bruno E, Ribbe PH (1975) The structure of partially disordered, synthetic strontium feldspar. Am Mineral 60:111-119

Clarke FW (1904) Analyses of rocks from the laboratory of the United States Geological Survey, 1880-1903. Bull U S Geol Surv 228

Cohen I, Schaner BE (1963) A metallographic and X-ray study of the UO_2–ZrO_2 system. J Nucl Mater 9:18-52

Cohen RE (1992) Origin of ferroelectricity in oxide ferroelectrics and the difference in ferroelectric behavior of $BaTiO_3$ and $PbTiO_3$. Nature 358:136-138

Cohen RE (2000) Theory of ferroelectrics: A vision for the next decade and beyond. J Phys Chem Sol 61:139-146

Cross LE (1993) Ferroelectric Ceramics: Tutorial Reviews, Theory, Processing and Applications. Birkhauser, Basel

Darlington CNW, Cernik RJ (1993) The effects of isovalent and non-isovalent impurities on the ferroelectric phase transition in barium titanate. J Phys: Condens Matter 5:5963-5970

Dolino G (1990) The α-inc-β transitions of quartz: A century of research on displacive phase transitions. Phase Trans 21:59-72

Doumerc JP, Hejtmanek J, Chaminade JP, Pouchard M, Krussanova M (1984) A photochemical study of $CuWO_4$ single crystals. Phys Stat Sol A 82:285-294

Dove MT, Cool T, Palmer DC, Putnis A, Salje EKH, Winkler B (1993) On the role of Al-Si ordering in the cubic-tetragonal phase transition of leucite. Am Mineral 78:486-492

Dove MT, Redfern SAT (1997) Lattice simulation studies of the ferroelastic phase transitions in $(Na,K)AlSi_3O_8$ and $(Sr,Ca)Al_2Si_2O_8$ feldspar solid solutions. Am Mineral 82:8-15

Dove MT, Thayaparam S, Heine V, Hammonds KD (1996) The phenomenon of low Al-Si ordering temperatures in aluminosilicate framework structures. Am Mineral 81:349-362

Durben DJ, Wolf GH, McMillan, PF (1991) Raman scattering study of the high-temperature vibrational properties and stability of $CaGeO_3$ perovskite. Phys Chem Minerals 18:215-223

Fei Y (1998) Solid solutions and element partitioning at high pressures and temperatures. Rev Mineral 37:343-367

Filipenko OS, Pobedimskaya EA, Belov NV (1968) Crystal structure of $ZnWO_4$. Sov Phys Crystallogr 13:127-129

Fleet ME (1981) The intermediate plagioclase structure: An explanation from interface theory. Phys Chem Minerals 7:64-70

Ghiorso MS, Carmichael ISE, Moret LK (1979) Inverted high-temperature quartz. Contrib Mineral Petrol 68:307-323

Ghose S, Van Tendeloo G, Amelinckx S (1988) Dynamics of a second-order phase transition: $P\bar{1}$–$I\bar{1}$ phase transition in anorthite, $CaAl_2Si_2O_8$. Science 242:1539-1541

Gilbert MC, Helz RT, Popp RK, Spear FS (1982) Experimental studies of amphibole stability. Rev Mineral 9B:229-353

Gnatchenko SL, Eremenko SV, Sofroneev NF, Kharchenko M, Devignes M, Feldmann P, LeGall A (1986) Spontaneous phase transitons and optical anisotropy in manganese germanium garnet $(Ca_3Mn_2Ge_3O_{12})$. Sov Phys JEPT 63:102-109

Goldschmidt VM (1926) Geochemische Verteilungsgesetze der Elemente. VII. Die Gesetze der Krystallochemie. Norske Videnskaps-Akademi, Oslo, 117 p

Goldschmidt VM (1927) Geochemische Verteilungsgesetze der Elemente. VIII. Untersuchungen über Bau und Eigenschaften von Krystallen. Norske Videnskaps-Akademi, Oslo, 156 p

Goldschmidt VM (1929) Crystal structure and chemical constitution. Proc Faraday Soc, London, p 253-283

Gronvold F, Stolen S, Svendsen SR (1989) Heat capacity of α quartz from 298.15 to 847.3 K, and of β quartz from 847.3 to 1000 K—Transition behaviour and reevaluation of the thermodynamic properties. Thermochim Acta 139:225-243

Groth PHR von (1870) Über Beziehungen zwischen Krystallform und chemischer Constitution bei einigen organischen Verbindungen. Ber Dtsch Chem Ges 3:449-457

Grove TL (1977) A periodic antiphase structure model for the intermediate plagioclases (An_{33} to An_{75}). Am Mineral 62:932-941

Grundy HD, Brown WL, MacKenzie WS (1967) On the existence of monoclinic $NaAlSi_3O_8$ at elevated temperatures. Mineral Mag 36:83-88

Grundy HD, Ito J (1974) The refinement of the crystal structure of a synthetic non-stoichiometric Sr feldspar. Am Mineral 59:1319-1326

Halperin BI and Varma CM (1976) Defects and the central peak near structural phase transitions. Phys Rev B 14:4030-4044

Harrison RJ, Salje EKH (1994) X-ray diffraction study of the displacive phase transition in anorthoclase, grain-size effects and surface relaxations. Phys Chem Minerals 21:325-329

Hayward SA, Salje EKH, Chrosch J (1998) Local fluctuations in feldspar frameworks. Mineral Mag 62:639-645

Heaney PJ (1994) Structure and chemistry of the low-pressure silica polymorphs. Rev Mineral 29:1-40

Heaney PJ, Salje EKH, Carpenter MA (1991) A dielectric study of the antiferroelectric to paraelectric phase transition in synthetic and natural titanites. (abstr) Trans Am Geophys Union Eos 72:554

Heaney PJ, Veblen DR (1990) A high-temperature study of the low-high leucite phase transition using the transmission electron microscope. Am Mineral 75:464-476

Heaney PJ, Veblen DR (1991b) Observation and kinetic analysis of a memory effect at the α-β quartz transition. Am Mineral 76:1459-1466

Hemingway BS, Evans HT Jr, Nord GL Jr, Haselton HT Jr, Robie RA, McGee JJ (1987) Åkermanite: phase transitions in heat capacity and thermal expansions, and revised thermodynamic data. Can Mineral 24:425-434

Henderson CMB (1981) The tetragonal-cubic inversion in leucite solid solutions. Progress Exp Petrol 5:50-54

Heuer AH, Chaim R, Lanteri V (1987) The displacive cubic → tetragonal transformation in ZrO_2 alloys. Acta Metall 35:661-666

Heywang W (1965) Ferroelektrizität in perowskitischen Systemen und ihre technischen Anwendungen. Zeit Angew Physik 19:473-481

Heywang W, Thomann H (1984) Tailoring of piezoelectric ceramics. Ann Rev Mater Sci 14:27-47

Higgins JB, Ribbe PH (1976) The crystal chemistry and space groups of natural and synthetic titanites. Am Mineral 61:878-888

Hirao K, Soga N, Kunugi M (1976) Thermal expansion and structure of leucite-type compounds. J Phys Chem 80:1612-1616

Höck KH, Schäfer R, Thomas H (1979) Dynamics of a locally distorted impurity in a host crystal with displacive phase transition. Z Phys B 36:151-160

Hollabaugh CL, Foit FF Jr (1984) The crystal structure of an Al-rich titanite from Grisons, Switzerland. Am Mineral 69:725-732

Holland T (1990) Activities in omphacitic solid solutions: An application of Landau theory to mixtures. Contrib Mineral Petrol 105:446-453

Holland T, Powell R (1996) Thermodynamics of order-disorder in minerals: II. Symmetric formalism applied to solid solutions. Am Mineral 81:1425-1437

Hutchison MN, Scott SD (1983) Experimental calibration of the sphalerite cosmobarometer. Geochim Cosmochim Acta 47:101-108

Iishi K, Fujino K, Furukawa Y (1990) Electron microscopy studies of akermanites $(Ca_{1-x}Sr_x)_2CoSiO_7$ with modulated structure. Phys Chem Minerals 17:467-471

Isupov VA (1975) Comments on the paper "X-ray study of the PZT solid solutions near the morphotropic phase transition." Sol Stat Comm 17:1331-1333

Jaffe B, Cook WR, Jaffe H (1971) Piezoelectric Ceramics. Academic Press, London

Jiang JC, Schosnig M, Schaper AK, Ganster K, Rager H, Toth L (1998) Modulations in incommensurate $(Ca_{1-x}Sr_x)_2MgSi_2O_7$ single crystals. Phys Chem Minerals 26:128-134

Kakagawa K, Mohri J, Shirashki S, Takahashi K (1982) Sluggish transition between tetragonal and rhombohedral phases of $Pb(Zr,Ti)O_3$ prepared by application of electric field. J Am Ceram Soc 65:515-519

Kazey ZA, Novak P, Soklov VI (1982) Cooperative Jahn-Teller effect in the garnets. Sov Phys JETP 56:854-864

Keith ML, Tuttle OF (1952) Significance of variation in the high-low inversion of quartz. Am J Sci Bowen Volume, p 203-280

Kihlborg L, Gebert E (1970) $CuWO_4$ a distorted wolframite-type structure. Acta Crystallogr B26:1020-1025

Kitamura M, Morimoto N (1975) The superstructure of intermediate plagioclase. Proc Japan Acad 51:419-424

Kroll H, Bambauer H-U, Schirmer U (1980) The high albite-monalbite and analbite-monalbite transitions. Am Mineral 65:1192-1211

Kroll H, Schmiemann I, von Cölln G (1986) Feldspar solid solutions. Am Mineral 71:1-16

Kumao A, Hashimoto H, Nissen H-U, Endoh H (1981) Ca and Na positions in labradorite feldspar as derived from high-resolution electron microscopy and optical diffraction. Acta Crystallogr A37:229-238

Lal R, Krishnan R, Ramakrishnan P (1988) Transition between tetragonal and rhombohedral phases of PZT ceramics prepared from spray-dried powders. Br Ceram Trans J 87:99-102

Lange RA, Carmichael ISE, Stebbins JF (1986) Phase transitions in leucite $KAlSi_2O_6$, orthorhombic $KAlSiO_4$, and their iron analogues ($KFeSi_2O_6$, $KFeSiO_4$). Am Mineral 71:937-945

Laves F, Czank M, Schulz H (1970) The temperature dependence of the reflection intensities of anorthite ($CaAl_2Si_2O_8$) and the corresponding formation of domains. Schweiz mineral petrogr Mitt 50:519-525

Lebedev NI, Levanyuk AP, Morozov AI, Sigov AS (1983) Defects near phase transition points: approximation of quasi-isolated defects. Sov Phys Sol Stat 25:1716-1718

Lefèvre J (1963) Contribution à l'étude de différentes modifications structurales des phases de type fluorine dans les systèmes à base de zircone ou d'oxyde de hafnium. I. Étude d'une déformation discontinue de la maille fluorine. La transformation allotropique de la zircone. Ann Chim 8:118-124

Leinenweber K, Utsumi W, Tsuchida Y, Yagi T, Kurita K (1991) Unquenchable high-pressure perovskite polymorphs of $MnSnO_3$ and $FeTiO_3$. Phys Chem Minerals 22:251-258

Levanyuk AP, Osipov VV, Sigov AS, Sobyanin AA (1979) Change of defect structure and the resultant anomalies in the properties of substances near phase-transition points. Sov Phys JEPT 49:176-188

Linton JA, Fei Y, Navrotsky A (1999) The $MgTiO_3$–$FeTiO_3$ join at high pressure and temperature. Am Mineral 84:1595-1603

Liu X, Wang Y, Liebermann RC, Maniar PD, Navrotsky A. (1991) Phase transition in $CaGeO_3$ perovskite: evidence from X-ray powder diffraction, thermal expansion and heat capacity. Phys Chem Minerals 18:224-230

MacKenzie WS (1952) The effect of temperature on the symmetry of high-temperature soda-rich feldspars. Am J Sci, Bowen Volume, p 319-342

Martin RF, Lagache M (1975) Cell edges and infrared spectra of synthetic leucites and pollucites in the system $KAlSi_2O_6$–$RbAlSi_2O_6$–$CsAlSi_2O_6$. Can Mineral 13:275-281

McGuinn MD, Redfern SAT (1994) Ferroelastic phase transition along the join $CaAl_2Si_2O_8$–$SrAl_2Si_2O_8$. Am Mineral 79:24-30

Melhado E (1980) Mitscherlich's discovery of isomorphism. Hist Studies Phys Sci 11:87-123

Meriani S, Palmonari C (eds) (1989) Zirconia '88—Advances in Zirconia Science and Technology. Elsevier, New York, 379 p

Mishra SK, Pandey D (1997) Thermodynamic nature of phase transitions in $Pb(Zr_xTi_{1-x})O_3$ ceramics near the morphotropic phase boundary. II. Dielectric and piezoelectric studies. Phil Mag B 76:227-240

Mishra SK, Singh AP, Pandey D (1997) Thermodynamic nature of phase transitions in $Pb(Zr_xTi_{1-x})O_3$ ceramics near the morphotropic phase boundary. I. Structural Studies. Phil Mag B 76:213-226

Mitscherlich E (1819) Über die Kristallisation der Salze, in denen das Metall der Basis mit zwei Proportionen Sauerstof verbunden ist. Abh Königlichen Akad Wiss Berlin 5:427-437

Mitscherlich E (1822) Sur la relation qui existe entre la forme cristalline et les proportions chimiques. I. Mémoire sur les arseniates et les phosphates. Ann Chim Phys 19:350-419

Molodetsky I, Navrotsky A (1998) The energetics of cubic zirconia from solution calorimetry of yttria- and calcia-stabilized zirconia. Z Physikal Chemie 207: 59-65

Müller G (1995) The scientific basis. In Bach H (ed) Low Thermal Expansion Glass Ceramics. Springer-Verlag, Berlin, p 13-49

Müller KA (1979) Intrinsic and extrinsic central-peak properties near structural phase transitions. *In* Enz CP (ed) Dynamical critical phenomena and related topics. Springer, Berlin, p 210-250

Myers ER, Heine V, Dove MT (1998) Thermodynamics of Al/Al avoidance in the ordering of Al/Si tetrahedral framework structures. Phys Chem Minerals 25:457-464

Newnham RE (1989) Structure-property relationships in perovskite electroceramics. *In* Navrotsky A, Weidner DI (eds) Perovskite: a structure of great interest to geophysics and materials science. Geophys Monogr 45:91-98, American Geophysical Union, Washington, DC

Newnham RE (1997) Molecular mechanisms in smart materials. Mat Res Soc Bull 22: 20-34

Nord GL Jr (1992) Imaging transformation-induced microstructures. Rev Mineral 27:455-508

Nord GL Jr (1994) Transformation-induced twin boundaries in minerals. Phase Trans 48:107-134

Oh HS, Jang HM (1999) Ferroelectric phase transitions and three-dimensional phase diagrams of a $Pb(Zr,Ti)O_3$ system under a hydrostatic pressure. J Appl Phys 85:2815-2820

Okamura FO, Ghose S (1975) Analbite → monalbite transition in a heat-treated, twinned Amelia albite. Contrib Mineral Petrol 50:211-216

Palmer DC (1994) Stuffed derivatives of the silica polymorphs. Rev Mineral 29:83-122

Palmer DC, Bismayer U, Salje EKH (1990) Phase transitions in leucite: order parameter behavior and the Landau potential deduced from Raman spectroscopy and birefringence studies. Phys Chem Minerals 17:259-265

Palmer DC, Dove MT, Ibberson RM, Powell BM (1997) Structural behavior, crystal chemistry, and phase transitions in substituted leucite: High-resolution neutron powder diffraction studies. Am Mineral 82:16-29

Palmer DC, Salje EKH (1990) Phase transitions in leucite: Dielectric properties and transition mechanism. Phys Chem Minerals 17:444-452

Palmer DC, Salje EKH, Schmahl WW (1989) Phase transitions in leucite: X-ray diffraction studies. Phys Chem Minerals 16:714-719

Pauling L (1927) The sizes of ions and the structure of ionic crystals. J Am Chem Soc 49:765-790

Pauling L (1929) The principles determining the structure of complex ionic crystals. J Am Chem Soc 51:1010-1026

Peacor DR (1968) A high temperature single crystal diffractometer study of leucite, $(K,Na)AlSi_2O_6$. Z Kristallogr 127:213-224

Penfield SL, Howe WTH. (1894) On the chemical composition of chondrodite, humite, and clinohumite. Am J Sci 47:188-206

Perry EC Jr (1971) Implications for geothermometry of aluminum substitution in quartz from Kings Mountain, North Carolina. Contrib Mineral Petrol 30:125-128

Petit PE, Guyot F, Fiquet G, Itie JP (1996) High-pressure behaviour of germanate olivines studied by X-ray diffraction and X-ray absorption spectroscopy. Phys Chem Minerals 23:173-185

Phillips BL, McGuinn MD, Redfern SAT (1997) Si-Al order and the $I\bar{1}$–$I2/c$ structural phase transition in synthetic $CaAl_2Si_2O_8$–$SrAl_2Al_2Si_2O_8$ feldspar: A ^{29}Si MAS-NMR spectroscopic study. Am Mineral 82:1-7

Pillars WW, Peacor DR (1973) The crystal structure of beta eucryptite as a function of temperature. Am Mineral 85:681-690

Putnis A, Salje E (1994) Tweed microstructures: Experimental observations and some theoretical models. Phase Trans 48:85-105

Randall CA, Guo R, Bhalla AS, Cross LE (1991) Microstructure-property relations in tungsten bronze lead barium niobate, $Pb_{1-x}Ba_xNb_2O_6$. J Mater Res 6:1720-1728

Reaney IM (1995) TEM observations of domains in ferroelectric and nonferroelectric perovskites. Ferroelectrics 172:115-125

Redfern SAT (1990) Strain coupling and changing transition character in Ca-rich plagioclase. *In* Salje EKH, Phase Transitions in Ferroelastic and Co-elastic Crystals. Cambridge University Press, Cambridge, UK, p 268-282

Redfern SAT (1992) The effect of Al/Si disorder on the $I\bar{1}$–$P\bar{1}$ co-elastic phase transition in Ca-rich plagioclase. Phys Chem Minerals 19:246-254

Redfern SAT, Bell AMT, Henderson CMB, Schofield PF (1995) Rietveld study of the structural phase transition in the sanmartinite ($ZnWO_4$)–cuproscheelite ($CuWO_4$) solid solution. Eur J Mineral 7:1019-1028

Redfern SAT, Graeme Barber A, Salje E (1988) Thermodynamics of plagioclase III: Spontaneous strain at the $I\bar{1}$-$P\bar{1}$ phase transition in Ca-rich plagioclase. Phys Chem Minerals 16:157-163

Redfern SAT, Knight KS, Henderson CMB, Wood BJ (1998) Fe-Mn cation ordering in fayalite-tephroite $(Fe_xMn_{1-x})_2SiO_4$ olivines: a neutron diffraction study. Mineral Mag 62:607-615

Redfern SAT, Salje E (1987) Thermodynamics of plagioclase II: Temperature evolution of the spontaneous strain at the $I\bar{1}$–$P\bar{1}$ phase transition in anorthtie. Phys Chem Minerals 14:189-195

Redfern SAT, Salje E (1992) Microscopic dynamic and macroscopic thermodynamic character of the $I\bar{1}$–$P\bar{1}$ phase transition in anorthite. Phys Chem Minerals 18:526-533

Riester M, Böhm H, Petricek V (2000) The commensurately modulated structure of the lock-in phase of synthetic Co-åkermanite, $Ca_2CoSi_2O_7$. Z Kristallogr 215:102-109

Rigden SM, Jackson I (1991) Elasticity of germanate and silicate spinels at high pressure. J Geophys Res B 96:9999-10,006

Rossetti GA Jr, Navrotsky A (1999) Calorimetric investigation of tricritical behavior in tetragonal $Pb(Zr_xTi_{1-x})O_3$. J Sol Stat Chem 144:188-194

Roy R, Agrawal DK, McKinstry HA (1989) Very low thermal expansion coefficient materials. Ann Rev Mat Sci 19:59-81

Salje E (1985) Thermodynamics of sodium feldspar: I. Order parameter treatment and strain induced coupling effects. Phys Chem Minerals 12:93-98

Salje E (1987) Thermodynamics of plagioclase I: Theory of the $I\bar{1}$–$P\bar{1}$ phase transition in anorthite and Ca-rich plagioclase. Phys Chem Minerals 14:181-188

Salje EKH (1990) Phase Transitions in Ferroelastic and Co-elastic Crystals. Cambridge Univ Press, Cambridge, UK

Salje E (1993) Phase Transitions in Ferroelastic and Co-elastic Crystals, Student Edition. Cambridge Univ Press, Cambridge, UK

Salje EKH (1995) Chemical mixing and structural phase transitions: The plateau effect and oscillatory zoning near surfaces and interfaces. Eur J Mineral 7:791-806

Salje EKH (1999) Ferroelastic phase transitions and mesoscopic structures. Ferroelectrics 221:1-7

Salje E, Bismayer U, Wruck B, Hensler J (1991) Influence of lattice imperfections on the transition temperatures of structural phase transitions: The plateau effect. Phase Trans 35:61-74

Salje E, Devarajan V (1981) Potts model and phase transition in lead phosphate $Pb_3(PO_4)_2$. J Phys C 14:L1029-L1035

Salje E, Devarajan V, Bismayer U, Guimaraes DMC (1983) Phase transitions in $Pb_3(P_{1-x}As_xO_4)_2$: Influence of the central peak and flip mode on the Raman scattering of hard modes. J Phys C 16:5233-5243

Salje E, Kuscholke B, Wruck B, Kroll H (1985) Thermodynamics of sodium feldspar II: Experimental results and numerical calculations. Phys Chem Minerals 12:99-107

Salje E, Wruck B (1983) Specific-heat measurements and critical exponents of the ferroelastic phase transition in $Pb_3(PO_4)_2$ and $Pb_3(P_{1-x}As_xO_4)_2$. Phys Rev B28:6510-6518

Schofield PF, Charnock JM, Cressey G, Henderson CMB (1994) An EXAFS study of cation site distortions through the $P2/c$–$P\bar{1}$ phase transition in the synthetic cuproscheelite-sanmartinite solid solution. Mineral Mag 58:185-199

Schofield PF, Knight KS, Redfern SAT, Cressey G (1997) Distortion characteristics across the structural phase transition in $(Cu_{1-x}Zn_x)WO_4$. Acta Crystallogr B53:102-112

Schofield PF, Redfern SAT (1992) Ferroelastic phase transition in the sanmartinite ($ZnWO_4$)–cuproscheelite ($CuWO_4$) solid solution. J Phys Condensed Matter 4:375-388
Schofield PF, Redfern SAT (1993) Temperature- and composition-dependence of the ferroelastic phase transition in $(Cu_xZn_{1-x})WO_4$. J Phys Chem Solids 54:161-170
Schmahl WW, Putnis A, Salje E, Freeman P, Graeme-Barber A, Jones R, Singh KK, Blunt J, Edward PP, Loram J, Mirza K (1989) Twin formation and structural modulations on orthorhombic and tetragonal $YBa_2(Cu_{1-x}Co_x)_3O_{7-\delta}$. Phil Mag Lett 60:241-248
Schulz H, Tscherry V (1972a) Structural relations between the low- and high-temperature forms of β-eucryptite ($LiAlSiO_4$) and low and high quartz. I. Low-temperature form of β-eucryptite and low quartz. Acta Crystallogr B28:2168-2173
Schulz H, Tscherry V (1972b) Structural relations between the low- and high-temperature forms of β-eucryptite ($LiAlSiO_4$) and low and high quartz. II. High-temperature form of β-eucryptite and high quartz. Acta Crystallogr B28:2174-2177
Schwabl F, Täuber WC (1996) Continuous elastic phase transitions in pure and disordered crystals. 354:2847-2873
Scotford DM (1975) A test of aluminum in quartz as a geothermometer. Am Mineral 60:139-142
Scott HG (1975) Phase relationship in the zirconia-yttria system. J Mater Sci 10:1827-1835
Seifert F, Czank M, Simons B, Schmahl W (1987) A commensurate-incommensurate phase transition in iron-bearing åkermanites. Phys Chem Minerals 14:26-35
Shannon RD (1976) Revised effective ionic radii and systematic studies of interatomic distances in halides and chalcogenides. Acta Crystallogr A32:751-767
Shapiro SM, Axe JD, Shirane G, Riste T (1972) Critical neutron scattering in $SrTiO_3$ and $KMnF_3$. Phys Rev B 6:4332-4341
Sherman DM (1995) Stability of possible Fe-FeS and Fe-FeO alloy phases at high pressure and the composition of the Earth's core. Earth Planet Sci Lett 132:87-98
Shigley JE, Moses T (1998) Diamonds as gemstones. *In* Harlow GE (ed) The Nature of Diamonds. Cambridge Univ Press, Cambridge, UK, p 240-254
Shirane G, Axe JD (1971) Acoustic-phonon instability and critical scattering in Nb_3Sn. Phys Rev Lett 27:1803-1806
Singh AP, Mishra SK, Lal R, Pandey D (1995) Coexistence of tetragonal and rhombohedral phases at the morphotropic phase boundary in PZT powders. I. X-ray diffraction studies. Ferroelectrics 163:103-113
Sirazhiddinov NA, Arifov PA, Grebenshichikov RG (1971) Phase diagram of strontium-calcium anorthites. Izvestiia Akad Nauk SSSR, Neorganicheskie Materialy 7:1581-1583
Slimming EH (1976) An electron diffraction study of some intermediate plagioclases. Am Mineral 61:54-59
Smith DK, Cline CF (1962) Verification of existence of cubic zirconia at high temperature. J Am Ceram Soc 45:249-250
Smith JV (1983) Phase equilibria of plagioclase. Rev Mineral 2:223-239
Smith JV, Brown WL (1988) Feldspar Minerals. 1. Crystal Structures, Physical, Chemical and Microtextural Properties (2nd Edition). Springer Verlag, Berlin, 828 p
Smith JV, Steele IM (1984) Chemical substitution in silica polymorphs. Neues Jahrb Mineral Mon 3:137-144
Speer JA, Gibbs GV (1976) The crystal structure of synthetic titanite, $CaTiOSiO_4$, and the domain texture of natural titanites. Am Mineral 61:238-247
Subrahmanyham S, Goo E (1998) Diffuse phase transitions in the $(Pb_xBa_{1-x})TiO_3$ system. J Mat Sci 33:4085-4088
Staehli JL, Brinkmann D (1974) A nuclear magnetic resonance study of the phase transition in anorthite, $CaAl_2Si_2O_8$. Z Kristallogr 140:360-373
Swainson IP, Dove MT, Schmahl WW, Putnis A (1992) Neutron powder diffraction study of the åkermanite-gehlenite solid solution series. Phys Chem Minerals 19:185-195
Tani E, Yoshimura M, Somiya S (1983) The confirmation of phase equilibria in the system ZrO_2–CeO_2 below 1400°C. J Am Ceram Soc 66:506-510
Tarascon JM, Greene LH, McKinnon WR, Hull GW, Geballe TH (1987) Superconductivity at 40 K in the oxygen-defect perovskites $La_{2-x}Sr_xCuO_{4-y}$. Science 235:1373-1376
Taylor M, Brown GE (1976) High-temperature structural study of the $P2_1/a$–$A2/a$ phase transition in synthetic titanite, $CaTiSiO_5$. Am Mineral 61:435-447
Taylor D, Henderson CMB (1968) The thermal expansion of the leucite group of minerals. Am Mineral 53:1476-1489
Taylor M, Brown GE (1976) High-temperature structural study of the $P2_1/a$ to $A2/a$ phase transition in synthetic titanite, $CaTiSiO_5$. Am Mineral 61:435-447
Teufer G (1962) Crystal structure of tetragonal ZrO_2. Acta Crystallogr 15:1187

Thomann H, Wersing W (1982) Principles of piezoelectric ceramics for mechanical filters. Ferroelectrics 40:189-202

Thompson JB Jr (1978) Biopyriboles and polysomatic series. Am Mineral 63:239-249

Tolédano JC, Pateau L, Primot J, Aubrée J, Morin D (1975) Etude dilatométrique de la transition ferroélastique de l'ortho phosphate de plomb monocristallin. Mat Res Bull 10:103-112

Torres J (1975) Symétrie du paramètre d'ordre de la transition de phase ferroélastique du phosphate de plomb. Phys Stat Sol B71:141-150

Tribaudino M, Benna P, Bruno E (1993) $I\bar{1}$-$I2/c$ phase transition in alkaline-earth feldspars along the $CaAl_2Si_2O_8$–$SrAl_2Si_2O_8$ jo*In* Thermodynamic behavior. Phys Chem Minerals 20:221-227

Tribaudino M, Benna P, Bruno E (1995) $I\bar{1}$–$I2/c$ phase transition in alkaline-earth feldspars: Evidence from TEM observations of Sr-rich feldspars along the $CaAl_2Si_2O_8$–$SrAl_2Si_2O_8$ join. Am Mineral 80:907-915

Tscherry V, Schulz H, Laves F (1972a) Average and super structure of β-eucryptite ($LiAlSiO_4$), Part I. average structure. Z Kristallogr 135:161-174

Tscherry V, Schulz H, Laves F (1972b) Average and super structure of β-eucryptite ($LiAlSiO_4$), Part II. super structure. Z Kristallogr 135:175-198

Tuttle OF (1949) The variable inversion temperature of quartz as a possible geologic thermometer. Am Mineral 34:723-730

Van Tendeloo G, Van Landuyt J, Amelinckx S (1976) The α-β phase transition in quartz and $AlPO_4$ as studied by electron microscopy and diffraction. Phys Status Solidi 33:723-735

Van Tendeloo G, Ghose S, Amelinckx S (1989) A dynamical model for the $P\bar{1}$–$I\bar{1}$ phase transition in anorthite, $CaAl_2Si_2O_8$: I. Evidence from electron microscopy. Phys Chem Minerals 16:311-319

Van Tendeloo G, Zandbergen HW, Amelinckx S (1987) Electron diffraction and electron microscopic study of Ba-Y-Cu-O superconducting materials. Sol Stat Comm 63:389-393

Veblen DR (1991) Polysomatism and polysomatic series: A review and applications. Am Mineral 76:801-826

Vinograd VL, Putnis A (1999) The description of Al,Si ordering in aluminosilicates using the cluster variation method. Am Mineral 84:311-324

Viswanathan K, Miehe G (1978) The crystal structure of low temperature $Pb_3(AsO_4)_2$. Z Kristallogr 148:275-280

Weyrich KH, Siems R (1984) Molecular dynamics calculations for systems with a localized "soft-mode." Ferroelectrics 55:333-336

Winkler HGF (1948) Synthese und Kristallstruktur des Eukryptits, $LiAlSiO_4$. Acta Crystallogr 1:27-34

Xu H, Heaney PJ (1997) Memory effects of domain structures during displacive phase transitions: a high-temperature TEM study of quartz and anorthite. Am Mineral 82:99-108

Xu H, Heaney PJ, Yates DM, Von Dreele RB, Bourke MA (1999a) Structural mechanisms underlying near-zero thermal expansion in β-eucryptite: a combined synchrotron X-ray and neutron Rietveld analysis. J Mat Res 14:3138-3151

Xu H, Heaney PJ, Navrotsky A, Topor L, Liu J (1999b) Thermochemistry of stuffed quartz-derivative phases along the join $LiAlSiO_4$–SiO_2. Am Mineral 84:1360-1369

Xu H, Heaney PJ, Beall GH (2000) Phase transitions induced by solid solution in stuffed derivatives of quartz: A powder synchrotron XRD study of the $LiAlSiO_4$–SiO_2 join. Am Mineral 85:971-979

Xu Y, Suenaga M, Tafto J, Sabatini RL, Moodenbaugh AR (1989) Microstructure, lattice parameters and superconductivity of $YBa_2(Cu_{1-x}Fe_x)_3O_{7-\delta}$ for 0 x 0.33. Phys Rev B 39:6667-6680

Yang H, Hazen RM, Downs RT, Finger LW (1997) Structural change associated with the incommensurate-normal phase transition in åkermanite, $Ca_2MgSi_2O_7$ at high pressure. Phys Chem Minerals 24:510-519

Yashima M, Arashi H, Kakihana M, Yoshimura M (1994) Raman scattering study of cubic-tetragonal phase transition in $Zr_{1-x}Ce_xO_2$ solid solution. J Am Ceram Soc 77:1067-1071

Yashima M, Morimoto K, Ishizawa N, Yoshimura M (1993) Diffusionless tetragonal-cubic transformation temperature in zirconia-ceria solid solutions. J Am Ceram Soc 76:2865-2868

Zhang M, Wruck B, Graeme Barber A, Salje EKH, Carpenter MA (1996) Phonon spectra of alkali feldspars: phase transitions and solid solutions. Am Mineral 81:92-104

Zhou Y, Lei TC, Sakuma T (1991) Tetragonal phase in zirconia-yttria ceramics. J Am Ceram Soc 74:633-640

7 Magnetic Transitions in Minerals

Richard J. Harrison

Institut für Mineralogie
Westfälische Wilhelms-Universität Münster
Corrensstrasse 24
48149 Münster, Germany

INTRODUCTION

Magnetic minerals have fascinated man since they were first used as compasses by the Chinese over 4000 years ago. Their scientific study has given rise to the interrelated disciplines of mineral magnetism, rock magnetism, and paleomagnetism, which have contributed to some of the most important scientific discoveries of the last century and continue to be at the forefront of scientific investigation at the beginning of this one.

Rock magnetism is concerned with understanding the processes by which rocks become magnetized in nature and the factors which influence their ability to maintain a faithful record of the Earth's magnetic field over geological time (the reader is referred to Dunlop and Özdemir 1997 for the definitive guide to this subject). Mineral magnetism aims to understand the physical, chemical, and thermodynamic consequences of magnetic ordering in minerals and how their magnetic properties are influenced by structural and microstructural changes associated with phase transformations. This is the subject of this chapter.

A wide range of different transformation processes occur in magnetic minerals, including convergent and non-convergent cation ordering, vacancy and charge ordering, oxidation and reduction, reconstructive or inversion transformations, and subsolvus exsolution. No review of this size could adequately cover all these topics, and so I will describe a small number of examples which illustrate the general principles involved, concentrating on the magnetic properties of oxide minerals. The reader is referred to Coey and Ghose (1987) for a review which covers the magnetic properties of silicate minerals.

MAGNETIC ORDERING

Magnetic ordering describes the transition from a disordered (paramagnetic) arrangement of magnetic moments above the Curie temperature, T_c, to an ordered arrangement of aligned magnetic moments below T_c. Several theories describing the magnetic and thermodynamic consequences of magnetic ordering are in common use today. This section provides a brief description of the different approaches and their application to different aspects of mineral magnetism.

Driving force for magnetic ordering

Alignment of magnetic moments is a consequence of the exchange interaction between unpaired electron spins on neighbouring atoms. The exchange interaction arises because the Coulomb energy of two electrons occupying overlapping orbitals is different when their spins are parallel or antiparallel to each other (a consequence of the Pauli exclusion principle). The exchange interaction energy is expressed as:

$$E = -2\,J_E\,\boldsymbol{S}_i \cdot \boldsymbol{S}_j \qquad (1)$$

where J_E is the exchange integral and $\boldsymbol{S}_i$ and $\boldsymbol{S}_j$ are the spins on neighbouring atoms i and j.

1529-6466/00/0039-0007$05.00

For positive J_E, parallel alignment of adjacent spins is energetically favoured, for negative J_E, antiparallel alignment is favoured. In most oxide minerals, cations are too far apart for there to be significant direct overlap of their 3*d* orbitals. In this case a "superexchange" interaction occurs via overlap with the 2*p* orbitals of intervening oxygen anions. The magnitude and sign of the superexchange interaction is sensitive to the cation-oxygen separation as well as the cation-oxygen-cation bond angle. This provides a fundamental link between the magnetic properties and crystal chemistry of minerals (Blasse 1964, Gorter 1954).

Classification of ordered (collinear) magnetic structures

The magnetic structures encountered in this chapter are summarised Figure 1. The ferromagnetic structure consists of a single magnetic lattice with all spins parallel. The antiferromagnetic structure consists of two magnetic sublattices of equal magnitude, with parallel alignment of spins within each sublattice but antiparallel alignment of one sublattice relative to the other. The ferrimagnetic structure is formed from two antiparallel sublattices of unequal magnitude. The canted antiferromagnetic structure is a result of imperfect antiferromagnetic coupling between two sublattices. The sublattices are rotated by a small angle relative to each other, producing a small "parasitic" ferromagnetic moment in a direction perpendicular to the spin alignment.

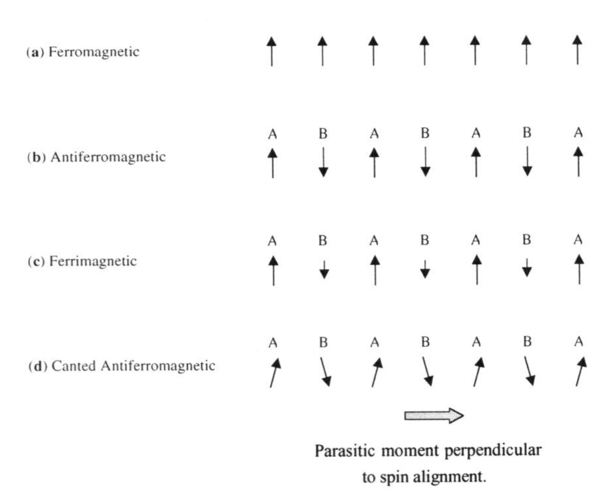

Figure 1. Classification of ordered collinear magnetic structures.

Models of magnetic ordering

Molecular field theory. The process of ferromagnetic ordering was first described by Weiss (1907) using the molecular field theory. Each magnetic moment, μ, experiences an ordering force due to exchange interaction with its nearest neighbours. This ordering force can be described by the presence of an effective magnetic field (the so-called molecular field, $\boldsymbol{H}_m$). Although the molecular field originates from short-range interactions (and is therefore a function of the local environment of each individual atom) one makes the mean field approximation that every atom within a uniformly magnetized domain experiences the same field. The molecular field is then proportional to the overall magnetization of the domain ($\boldsymbol{H}_m = \lambda \boldsymbol{M}$) and the magnetostatic potential energy of each magnetic moment is $E = -\mu_0 \lambda \mu \cdot \boldsymbol{M}$. For a two-state model, in which individual moments are either parallel or antiparallel to the molecular field (i.e. for atoms with an electron angular momentum quantum number $J = 1/2$), the net magnetization of an assembly of N moments is given by the difference between the number of up and down moments via Boltzmann statistics:

$$M = N\frac{\mu\,exp\left(\frac{\mu_0\mu\lambda M}{kT}\right) - \mu\,exp\left(-\frac{\mu_0\mu\lambda M}{kT}\right)}{exp\left(\frac{\mu_0\mu\lambda M}{kT}\right) + exp\left(-\frac{\mu_0\mu\lambda M}{kT}\right)} = N\mu\,tanh\left(\frac{\mu_0\mu\lambda M}{kT}\right). \tag{2}$$

Equation (2) describes a paramagnetic to ferromagnetic phase transition at $T_c = N\mu^2\mu_0\lambda/k$. A similar calculation for atoms with $J > 1/2$ can be performed by summing over each of the $(2J + 1)$ magnetization states. The result is that the tanh function in Equation (2) is replaced by the Brillouin function (see any textbook of solid state physics, e.g. Kittel 1976 or Crangle 1977).

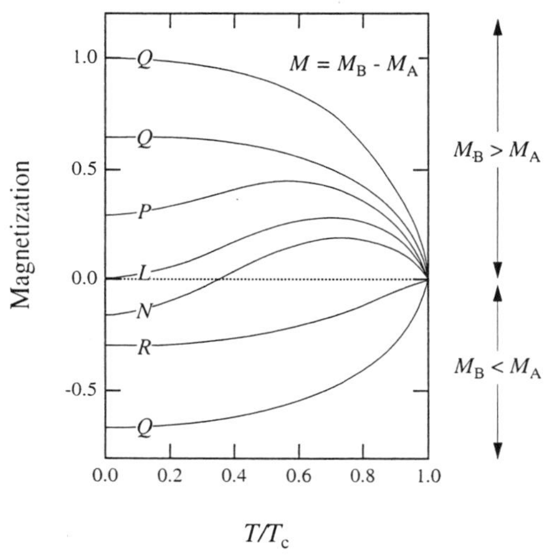

Figure 2. Néel's classification of magnetization curves for ferrimagnetic materials.

Néel (1948) applied the Weiss molecular field theory to describe magnetic ordering in antiferromagnetic and ferrimagnetic materials containing two magnetic sublattices, A and B. This was achieved by calculating separate molecular fields for each sublattice, each field being the sum of contributions from intersublattice (AB) and intrasublattice (AA + BB) interactions. The net magnetization is given by the difference between the A and the B sublattice magnetizations ($M = M_B - M_A$), resulting in different types of M-T curve due to the different possible temperature dependencies of M_A and M_B. These curves are classified as Q-, P-, L-, N-, and R-type, as illustrated in Figure 2. Néel's theory was modified by Stephenson (1972a,b) to account for mixing two types of magnetic species over the two magnetic sublattices. This and similar theories are widely used in the description of natural magnetic minerals (Readman and O'Reilly 1972, Moskowitz 1987).

Macroscopic thermodynamic models. Magnetic ordering has a large effect on the phase diagram topology of mineral and alloy solid solutions (Inden 1981, 1982; Kaufman 1981, Miodownik 1982, Burton and Davidson 1988, Burton 1991). To predict these effects, the mean field model described above is often rewritten as an equivalent macroscopic expression for the excess free energy of magnetic ordering (Meijering 1963). Ghiorso (1997) describes how the magnetic entropy can be calculated for atoms with $J > 1/2$. For example, an Fe^{3+} cation with $J = 5/2$ has 5 unpaired electrons in its half-filled $3d$ orbital and can exist in any one of the following $(2J + 1)$ spin states: ↑↑↑↑↑, ↓↑↑↑↑, ↓↓↑↑↑, ↓↓↓↑↑, ↓↓↓↓↑, and ↓↓↓↓↓. Above T_c each state is occupied with equal probability, yielding a total entropy contribution of R ln 6. For an intermediate degree of magnetic alignment one defines a magnetic order parameter, Q_m, such that $Q_m = 0$ in the disordered state and $Q_m = 1$ in the fully ordered state. The entropy is then given by (Fowler 1936):

$$\Delta S_{\text{mag}} = R\,(\ln p - Q_{\text{m}}\,J \ln \xi) \tag{3}$$

where

$$p = \xi^{J} + \xi^{J-1} + \ldots + \xi^{-J+1} + \xi^{-J} \tag{4}$$

and ξ is obtained as a function of Q_{m} by solving the equation:

$$\xi \frac{dp}{d\xi} = Q_{\text{m}} J p. \tag{5}$$

In many cases it is sufficient to write a simplified expression for the magnetic entropy (Harrison and Putnis 1997, 1999a). In Landau theory, for example, the magnetic entropy is approximated by:

$$\Delta S_{\text{mag}} = -\frac{1}{2} a_{\text{m}} Q_{\text{m}}^{2} \tag{6}$$

where a_{m} is an adjustable coefficient. This approximation is reasonable for values up to $Q_{\text{m}} \approx 0.8\text{-}0.9$. Irrespective of which entropy expression is used, the enthalpy of magnetic ordering is well described by an even polynomial expansion of the magnetic order parameter, which in Landau theory leads to the standard free energy potential (Landau and Lifshitz 1980, Tolédano and Tolédano 1987, Salje 1990):

$$\Delta G_{\text{m}} = \frac{1}{2} a_{m} (T - T_{\text{c}}) Q_{\text{m}}^{2} + \frac{1}{4} b_{\text{m}} Q_{\text{m}}^{4} + \ldots\ldots \frac{1}{n} c_{\text{m}} Q_{\text{m}}^{n} \tag{7}$$

where b_{m} and c_{m} are adjustable coefficients and n is an even integer. The coefficients may be determined by fitting to experimental data such as the temperature dependence of the sublattice magnetization (Riste and Tenzer 1961, Shirane et al. 1962, van der Woude et al. 1968) or the magnetic specific heat capacity anomaly (Ghiorso 1997).

Atomistic models. Macroscopic models fail to describe the real physics of the magnetic ordering process due to the assumption that the interactions driving ordering are long-range and therefore insensitive to the atomistic details of the local crystal structure. This is clearly not the case for magnetic ordering, which is driven by very short-range (often nearest-neighbour only) exchange interactions. A more physically rigorous description is achieved using the Ising model, which describes the energy of the system in terms of an array of spins ($J = 1/2$) with nearest neighbour interactions. The microscopic Hamiltonian for such a model has the form:

$$E = -J_{\text{E}} \sum_{\langle \text{i,j} \rangle} \sigma_{\text{i}} \cdot \sigma_{\text{j}} \tag{8}$$

where σ_{i} and σ_{j} are the spin variables for atoms i and j ($\sigma_{\text{i}} = \pm 1$) and the sum <i,j> is over the nearest neighbours of all lattice sites (see Eqn. 1). Exact solutions to the Ising model are only possible in one and two dimensions (the one dimensional case does not exhibit a magnetic transition, however). Ising models for three dimensional systems can be solved to a good degree of approximation using CVM and Monte Carlo techniques (Burton and Kikuchi 1984, Mouritsen 1984, Burton 1985, Binder and Heermann 1988).

An even closer physical description of magnetic ordering is obtained if one takes account of the fact that atoms may have $J > 1/2$ and their moments can be oriented in any direction (not just up and down). Such effects are accounted for by Heisenberg models, which are described by Hamiltonians of the form (Mouritsen 1984):

$$E = -J_{\text{E}} \sum_{\langle \text{i,j} \rangle} \bar{S}_{\text{i}} \cdot \bar{S}_{\text{j}} + A \sum_{\text{i}} \left(S_{\text{i}}^{z}\right)^{2} - g\mu_{\text{B}} H \sum_{\text{i}} S_{\text{i}}^{x} \tag{9}$$

where $\bar{S}_i$ is the spin vector, $S_i = (S_i^x, S_i^y, S_i^z)$. The first term describes the exchange interaction between adjacent spins. The second term describes the magnetocrystalline anisotropy energy, dictating which crystallographic directions are preferred directions for magnetic alignment. For $A > 0$ the spins prefer to lie in the x-y plane, for $A < 0$ the spins prefer to align parallel to the z-axis. The final term describes the magnetostatic effect of an external magnetic field, H, applied in the x-y plane (g is the Landé factor and μ_B is the Bohr magneton).

Micromagnetic models. In addition to the exchange, anisotropy, and magnetostatic energies described by Equation (9), any finite magnetized grain has a self energy (usually called the demagnetizing energy) due to long-range magnetic dipole-dipole interactions. The magnetic field at a point within a magnetized grain is the vector sum of the magnetic dipole fields acting at that point. The contribution to this internal field from compensated magnetic poles within the grain is small. Uncompensated poles occur at the surface of the grain, creating an internal magnetic field which acts in the opposite direction to the magnetization of the grain (the demagnetizing field). Each magnetic moment has a potential energy due to the presence of this field, which, when integrated over the whole grain, yields the demagnetizing energy. This energy can be greatly reduced by subdividing the grain into a large number of differently oriented magnetic domains (reducing the number of uncompensated surface poles).

Micromagnetic calculations are motivated by the need to understand variations in domain structure as a function of grain size and grain shape, and must, therefore, take proper account of the demagnetizing energy. This is achieved using a coarse-grained approach in which a magnetic grain of finite size (~5 μm or less) is subdivided into a number of small cells (each cell containing approx. 10-20 unit cells of magnetic material along its shortest dimension). The magnetization within a cell is assumed to be uniform, and the direction of magnetization is allowed to vary continuously (although the possible orientations may be constrained to simplify the calculation). The total energy of the grain for a given distribution of magnetization vectors is calculated by summing the exchange, anisotropy, magnetostatic, and demagnetizing energies. There are several techniques for evaluating the demagnetizing energy, but the most popular is derived from the theory of Rhodes and Rowlands (1954), in which the sum of dipole-dipole interactions over the volume of the grain is replaced by summing Coulomb interactions due to magnetic charges at the surface of each cell. The magnetization vectors of all cells are varied in an iterative manner, mapping out the local energy minimum states of the grain. An example of the three-dimensional "vortex" structure in a 0.2-μm cubic grain of magnetite is shown in Figure 3 (Fabian 1998). Each arrow represents the direction of magnetization in one cell. The vortex structure ensures that magnetic flux lines close within the body of the grain, effectively eliminating surface poles and minimizing the demagnetizing energy. An excellent summary of micromagnetic techniques is given by Dunlop and Özdemir (1997). Specific examples are described by Moon and Merrill (1984), Newell et al. (1993), Xu et al. (1994), Fabian et al. (1996), Wright et al. (1997), and Williams and Wright (1998).

CATION ORDERING

Cation ordering affects the magnetic properties of minerals in many different ways, either directly by changing the distribution of cations between the magnetic sublattices, or indirectly through the development of ordering-induced microstructures. This section first describes the cation ordering phenomena which occur in the two most important types of magnetic oxide. The direct magnetic consequences of cation ordering in these systems are then reviewed.

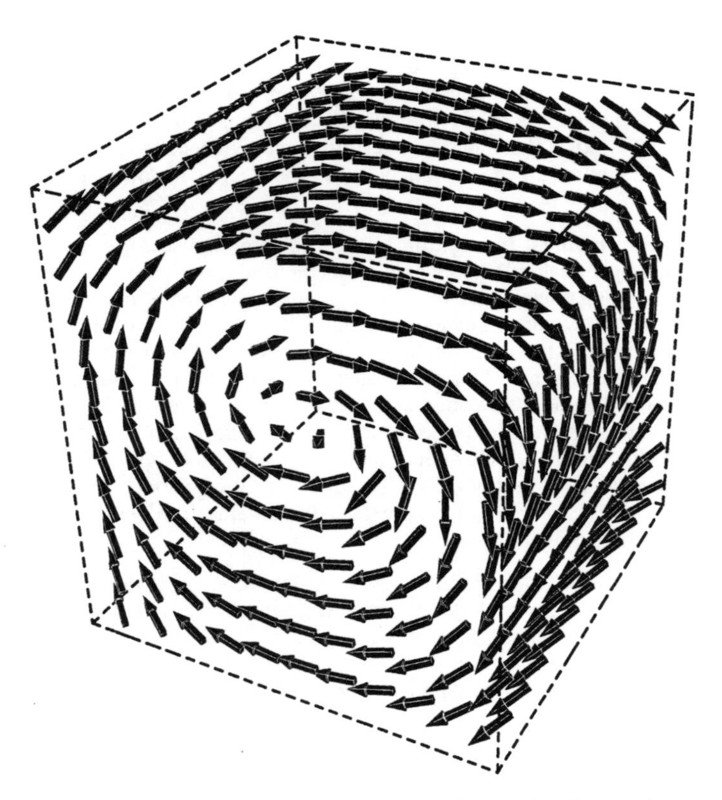

Figure 3. Vortex magnetization state in a 0.2-μm cube of magnetite, determined using 3-dimensional micromagnetic calculations (after Fabian 1998).

Non-convergent cation ordering in oxide spinels

Harrison and Putnis (1999a) have recently reviewed the relationship between the crystal chemistry and magnetic properties of oxide spinels, so this subject will only be dealt with briefly here. Spinels have the general formula XY_2O_4 and cubic space group $Fd\bar{3}m$. The most important natural examples are magnetite (Fe_3O_4) and ulvöspinel (Fe_2TiO_4), which together form the titanomagnetite solid solution, the dominant carrier of paleomagnetic remanence in rocks. The oxygens form an approximately cubic close packed arrangement and the cations occupy one tetrahedral site and two octahedral sites per formula unit. Two ordered configurations exist at zero temperature: the normal configuration $X[Y]_2O_4$ and the inverse configuration $Y[X_{0.5}Y_{0.5}]_2O_4$, where brackets refer to cations on octahedral sites. With increasing temperature the distribution of cations becomes increasingly disordered (tending toward the fully disordered distribution $X_{1/3}Y_{2/3}[X_{1/3}Y_{2/3}]_2O_4$ at infinite temperature). The thermodynamics of this ordering process have been described by many workers (O'Neill and Navrotsky 1983, 1984; Nell and Wood 1989, Sack and Ghiorso 1991, Carpenter and Salje 1994, Holland and Powell 1996a,b; Harrison and Putnis 1999a). Studies pertaining to the temperature dependent cation distribution in titanomagnetites are Stephenson (1969), O'Donovan and O'Reilly (1980), Wu and Mason (1981), Trestman-Matts et al. (1983), and Wißmann et al. (1998).

Verwey transition in magnetite

Magnetite adopts the inverse spinel structure at room temperature, $Fe^{3+}[Fe^{3+}_{0.5}Fe^{2+}_{0.5}]_2O_4$. Below 120 K it undergoes a first-order phase transition (the Verwey transition) to a monoclinic structure due to ordering of Fe^{2+} and Fe^{3+} cations on octahedral sites (Verwey 1939). An Fe^{2+} cation can be considered as an Fe^{3+} cation plus an extra electron. Above the Verwey transition the extra electrons are free to hop from one octahedral site to another, giving magnetite a significant electrical conductivity. Below the Verwey transition the extra electrons stop hopping, producing discrete Fe^{2+} and Fe^{3+} cations in a long-range ordered superstructure (Mizoguchi 1985, Miyamoto and Chikazumi 1988). The precise details of the charge ordering scheme are still not known with any certainty. Diffraction studies show that the unit cell of the ordered phase has space group *Cc* with a monoclinic unit cell corresponding to $\sqrt{2}a \times \sqrt{2}a \times 2a$ of the original cubic spinel structure (Iizumi et al. 1982). Possible charge ordering schemes are proposed by Iida (1980) on the basis of various NMR and Mössbauer spectroscopy observations. Insight into the electronic structure via ab initio calculations is given by Zhang and Satpathy (1991). Many physical properties of magnetite (e.g. electrical conductivity and specific heat capacity) show anomalies at or close to the Verwey transition (Honig 1995).

Convergent cation ordering in rhombohedral oxides

Rhombohedral oxides have the general formula A_2O_3 or ABO_3 and are based on the corundum structure (Waychunas 1991). The most important natural examples are hematite (Fe_2O_3) and ilmenite ($FeTiO_3$), which together form the titanohematite solid solution. This solid solution exhibits many interesting magnetic properties due to the presence of a convergent cation ordering phase transition.

Endmember Fe_2O_3 has space group $R\bar{3}c$. The oxygens form a distorted hexagonal close packed arrangement and the Fe^{3+} cations occupy two thirds of the octahedral sites, forming symmetrically-equivalent layers parallel to (001) (Fig. 4a). End-member $FeTiO_3$ adopts a related structure, in which the Fe^{2+} and Ti cations are partitioned onto alternating (001) layers (Fig. 4b). This partitioning destroys the equivalence of the layers, reducing the symmetry to $R\bar{3}$.

The solid solution is formed via the substitution $2Fe^{3+} = Fe^{2+} + Ti^{4+}$. At high temperatures, Fe^{2+}, Fe^{3+}, and Ti are equally partitioned between the layers and the symmetry of the solid solution is $R\bar{3}c$ (Fig. 5). Below a critical temperature, T_{od}, Fe^{2+} and Ti partition onto alternating (001) layers and there is a phase transition to the $R\bar{3}$ structure. This phase transition was first was first demonstrated by Ishikawa using measurements of saturation magnetization on quenched samples (Ishikawa and Akimoto 1957, Ishikawa 1958, 1962; Ishikawa and Syono 1963). The transition has recently been studied in some detail using *in situ* time-of-flight neutron diffraction and Monte Carlo simulation (Harrison et al. 2000a, 2000b).

Magnetic consequences of cation ordering

Saturation magnetization. The net magnetization of a Néel ferrimagnet is simply given by the difference between the two sublattice magnetizations, and is therefore intrinsically linked to the concentration of magnetic species on each sublattice. Spinels adopt the Néel collinear ferrimagnetic structure, with the A and B magnetic sublattices coinciding with the tetrahedral and octahedral sites respectively (Fig. 1c). This structure is a consequence of the dominant antiferromagnetic superexchange interaction between tetrahedral and octahedral sites (Blasse 1964). Since the cation distribution of any spinel is temperature dependent, the observed magnetization will be a function of the thermal history of the sample. This is the basis of the "quench-magnetization" technique, which is often used as an empirical test of cation distribution and oxidation models in titanomagnetite solid solutions (Akimoto 1954, Néel 1955, O'Reilly and Banerjee 1965, O'Reilly 1984, Kakol et al. 1991).

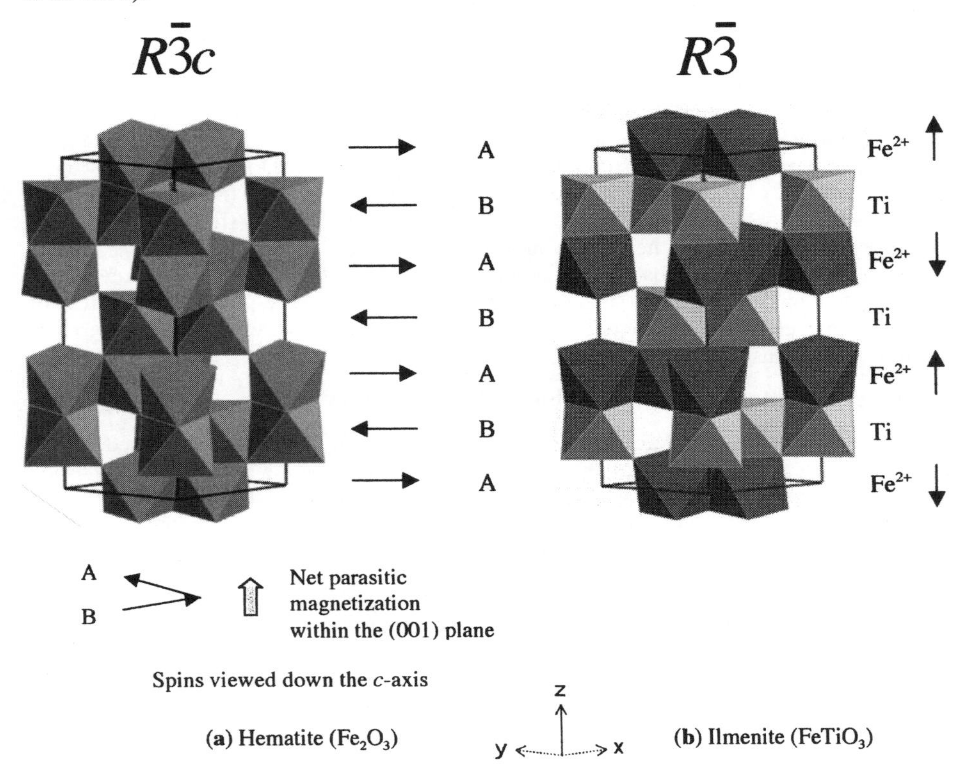

Figure 4. Crystal and magnetic structure of (a) hematite (Fe_2O_3) and (b) ilmenite ($FeTiO_3$).

Similarly, the saturation magnetization of the ilmenite-hematite solid solution is a complex function of bulk composition and degree of cation order (Brown et al. 1993). Endmember hematite has a canted antiferromagnetic structure at temperatures between 675°C (the Néel temperature) and -15°C (the Morin transition). The A and B magnetic sublattices coincide with the alternating (001) cation layers (Fig. 4a). The sublattice spins align perpendicular to [001] (i.e. within the basal plane) but are rotated by a small angle about [001], producing a parasitic magnetic moment within the basal plane, perpendicular to the sublattice magnetizations (Dzyaloshinsky 1958). The disordered solid solution has essentially the same canted antiferromagnetic structure as hematite (labelled CAF in Fig. 5) since Fe is equally distributed over the (001) layers and M_A and M_B have equal magnitudes. The ordered solid solution is ferrimagnetic (labelled FM in Fig. 5) since Fe^{2+} and Ti are partitioned onto alternating (001) layers and M_A and M_B have different magnitudes. The ideal saturation magnetization of the fully ordered solid solution is $4x\ \mu_B$, where x is the mole fraction of $FeTiO_3$ (assuming a spin-only moment of 4 μ_B for Fe^{2+}). Observed values of M_s are compared to the ideal values in Figure 6. The magnetization is zero for $x < 0.4$ due to the lack of long-range cation order in quenched samples of this composition. The

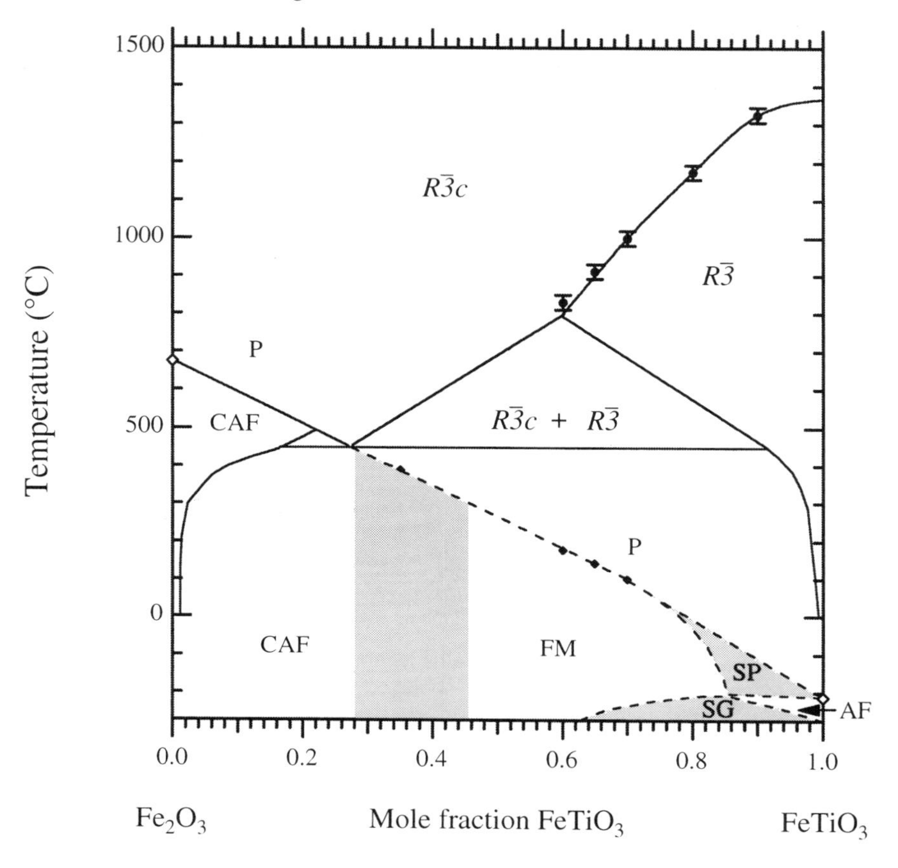

Figure 5. Summary of phase relations in the ilmenite-hematite solid solution: $R\bar{3}c$ = cation disordered; $R\bar{3}$ = cation ordered; P = paramagnetic; CAF = canted antiferromagnetic; FM = ferrimagnetic; AF = antiferromagnetic; SP = superparamagnetic; SG = spin glass. Closed circles are T_{od} for the $R\bar{3}c$ to $R\bar{3}$ phase transition (Harrison et al. 2000a; Harrison and Redfern, in preparation). Closed diamonds show T_c for stoichiometric ilm35, ilm60, ilm65, and ilm70 (Harrison, in preparation). Endmember T_c's and SP, SG, and AF fields from Ishikawa et al. (1985). Upper miscibility gap calculated by Harrison et al. (2000b). Lower miscibility gap is schematic (modified from Burton 1985). Boundary between CAF and FM depends on thermal history (represented by broad shaded region). Other lines are guides to the eye.

magnetization rises rapidly to a maximum around $x = 0.8$ due to presence of the $R\bar{3}c$ to $R\bar{3}$ phase transition. Magnetizations in this range are close to the ideal value (shown by the dashed line). The decrease in magnetization for $x > 0.8$ is caused by the gradual onset of a spin glass transition at low temperatures (see below).

A useful industrial application arising from Figure 6 is the magnetic roasting of ilmenite feedstocks used in the production of TiO_2 pigments (Nell and den Hoed 1997). Pure ilmenite is paramagnetic at room temperature and has a low magnetic susceptibility. In this state, magnetic separation is an unsuitable technique for purifying the feedstock. An increase in magnetic susceptibility can be achieved by partially oxidising the ilmenite (roasting), which, if the conditions are correct, reacts to give a mixture of TiO_2 and titanohematite with $x \approx 0.7$. This material is highly magnetic at room temperature and can be easily separated from less magnetic impurity phases.

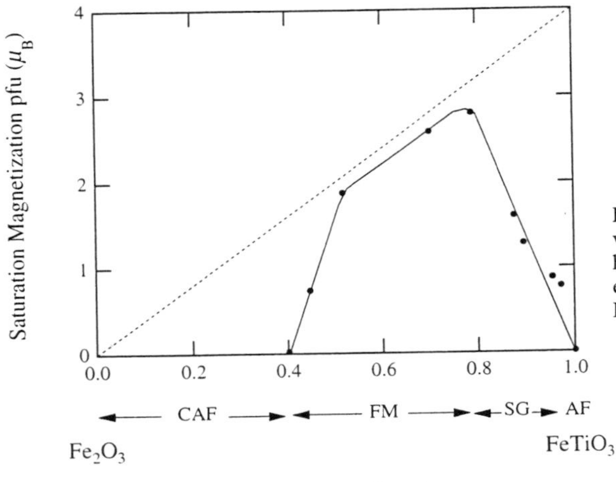

Figure 6. Saturation magnetization versus composition in the ilmenite-hematite solid solution (Ishikawa et al. 1985). Abbreviations as in Figure 5.

In this state, magnetic separation is an unsuitable technique for purifying the feedstock. An increase in magnetic susceptibility can be achieved by partially oxidising the ilmenite (roasting), which, if the conditions are correct, reacts to give a mixture of TiO_2 and titanohematite with $x \approx 0.7$. This material is highly magnetic at room temperature and can be easily separated from less magnetic impurity phases.

Curie temperature. Changing the distribution of magnetic species between the magnetic sublattices changes the relative numbers of AA, BB, and AB superexchange interactions and, consequently, the temperature at which magnetic ordering occurs. A well studied example is the inverse spinel magnesioferrite, $MgFe_2O_4$ (O'Neill et al. 1992, Harrison and Putnis 1997, 1999b). This material shows a strong linear correlation between T_c and the degree of cation order, Q, defined as the difference between the octahedral and tetrahedral Fe site occupancies (Fig. 7a). This relationship can be understood in terms of any of the magnetic ordering models discussed earlier, although Harrison and Putnis (1997) found rather poor agreement between the observed variations in T_c and those predicted using the molecular field model. The reason for this lies in the mean field approximation, which neglects the fact that exchange interactions are short ranged and therefore sensitive to the atomistic structure of the material. For example, the probability of finding a strong Fe-Fe AB interaction differs by only 11% in ordered and disordered $MgFe_2O_4$ if short-range correlations between site occupancies are ignored. This is insufficient to explain the magnitude of the effect seen in Figure 7a. If short-range ordering favours the formation of Mg-Fe pairs over Fe-Fe and Mg-Mg pairs then nearest-neighbour exchange interactions can be greatly reduced. Such effects are accounted for by atomistic models (e.g. Burton 1985) and can be parameterised successfully using macroscopic thermodynamic models by including an energy term describing coupling between the magnetic and cation order parameters (Harrison and Putnis 1997, 1999a; Ghiorso 1997).

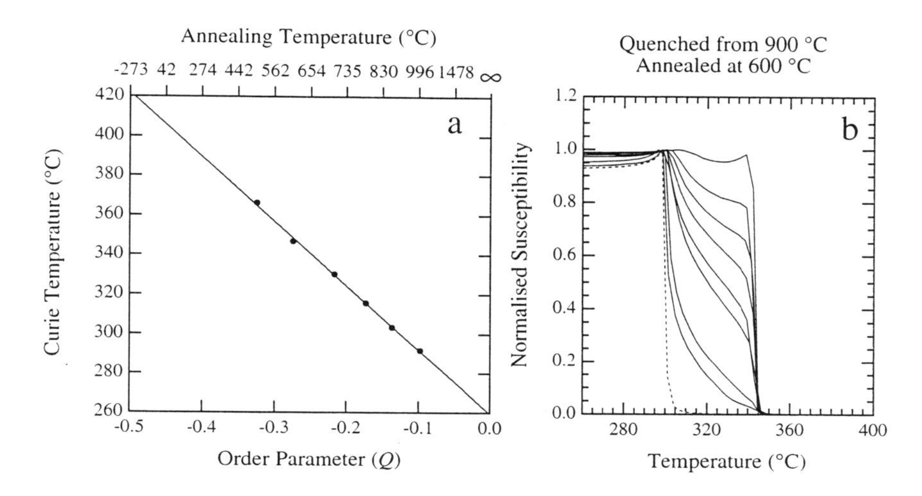

Figure 7. (a) Curie temperature of $MgFe_2O_4$ spinel as a function of the cation order parameter, $Q = X_{Fe}^{oct} - X_{Fe}^{tet}$ (O'Neill et al. 1992). The equilibration temperature for a given value of Q is shown on the top axis. (b) Normalised susceptibility as a function of temperature for $MgFe_2O_4$ quenched from 900 °C (dashed curve) and then annealed at 600°C for (1) 1.4 h, (2) 3.2 h, (3) 19.2 h, (4) 24.4 h, (5) 45.2 h, (6) 88 h, (7) 134 h, and (8) 434 h (Harrison and Putnis 1999b).

The strong correlation between cation distribution and Curie temperature allows magnetic measurements to be used as a tool to study the process of cation ordering in heterogeneous systems. Harrison and Putnis (1999b) used measurements of alternating-field magnetic susceptibility (χ) to study the kinetics of cation ordering in samples of $MgFe_2O_4$ quenched from 900°C and annealed at lower temperatures for various times (Fig. 7b). Susceptibility drops sharply as material is heated through T_c and therefore the χ-T curves reveal the range of T_c's present in the sample. In Figure 7b the starting material (dashed line) shows a single sharp drop in χ at a temperature of 300°C, corresponding to T_c of homogeneous material quenched from 900°C. On subsequent annealing at a lower temperature the cation distribution becomes more ordered (Q becomes more negative) and T_c increases. Ordering does not occur homogeneously throughout the sample. Instead one observes a process similar to nucleation and growth, where some regions of the sample attain their equilibrium degree of order very quickly and then grow slowly into the disordered matrix. Such mechanisms are predicted by rate-law theory (Carpenter and Salje 1989, Malcherek et al. 1997, 2000).

Magnetic structure. The Néel collinear antiferromagnetic/ferrimagnetic structure occurs when strong antiferromagnetic AB interactions dominate the weaker AA and BB interactions. When one of the sublattices is diluted by non-magnetic cations, the probability of AB exchange interaction is reduced and the weaker AA and BB interactions become increasingly important. This can lead to deviations from Néel-type behaviour.

Dilution leads to complex magnetic behaviour in the ilmenite-hematite system for $x > 0.8$ (Ishikawa et al. 1985, Arai et al. 1985a,b; Arai and Ishikawa 1985). In endmember $FeTiO_3$ only every second plane of cations contains Fe and the magnetic exchange interaction must operate over two intervening oxygen layers (Fig. 4b). These next-nearest neighbour (n.n.n.) interactions are approximately a factor of 15 weaker than the nearest-neighbour (n.n.) interactions between adjacent cation layers and produce an antiferromagnetic structure with spins parallel to [001]. The effect of substituting small amounts of Fe_2O_3 into $FeTiO_3$ is illustrated schematically in Figure 8 (Ishikawa et al. 1985). An Fe^{3+} cation in the Ti layer forces nearby spins in the adjacent layers to become parallel to each other due to the large n.n. exchange interaction, forming a ferrimagnetic cluster. If the concentration of Fe^{3+} is large enough, these clusters join together and establish the long-range ordered ferrimagnetic structure (more precisely, if the concentration of Fe^{3+} is greater than the percolation threshold, $x_P \approx$ 2/number of nearest

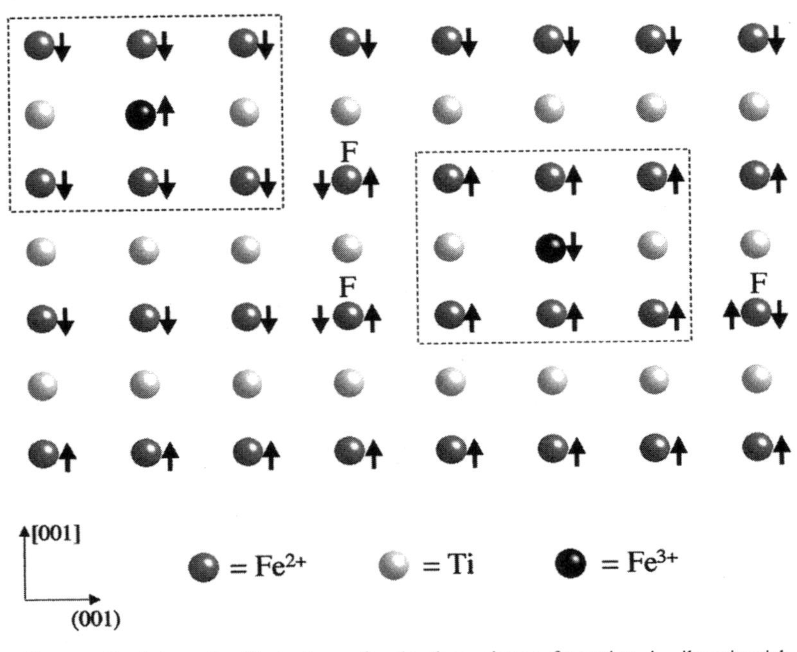

Figure 8. Schematic illustration of spin-glass cluster formation in ilmenite-rich members of the ilmenite-hematite solid solution (after Ishikawa et al. 1985). Dashed lines show ferrimagnetic clusters surrounding Fe^{3+} cations within the (001) Ti layers. Atoms labelled 'F' have frustrated spins.

neighbours = 2/9 = 0.22; Ishikawa et al. 1985). If the concentration of Fe^{3+} on the Ti layer is less than the percolation threshold, the ferrimagnetic clusters remain isolated from one another (weakly coupled via n.n.n. interactions) and behave superparamagnetically at high temperatures (i.e. magnetic order exists within the ferrimagnetic cluster, but the orientation of net magnetization fluctuates due to thermal excitation). This region is labelled SP in Figure 5. Neutron diffraction studies show that the spins within the SP region are predominantly oriented within the basal plane (Arai et al. 1985). At lower temperatures, antiferromagnetic ordering of the type observed in endmember $FeTiO_3$ starts to occur in-between the ferrimagnetic clusters (AF in Fig. 5), creating frustrated spins at the cluster boundaries (atoms labelled F in Fig. 8). Below a critical temperature, T_g, a complex canted spin structure develops. This structure is classified as a spin glass (SG in Fig. 5) due to its lack of long-range order. The spin glass transition is accompanied by a rotation of spins from the basal plane towards the [001] axis.

A similar effect can be observed in spinels. For example, fully ordered normal spinels such as $FeAl_2O_4$, $MnAl_2O_4$ and $CoAl_2O_4$ have magnetic cations exclusively on tetrahedral sites. Tetrahedral-tetrahedral superexchange interaction in these spinels is weak and negative, leading to an antiferromagnetic structure in which nearest-neighbour tetrahedral cations have antiparallel spins. Roth (1964) observed this structure in fully ordered $MnAl_2O_4$ and $CoAl_2O_4$ using neutron diffraction but found no long-range magnetic order in $FeAl_2O_4$. This was attributed to the partially disordered cation distribution in $FeAl_2O_4$ quenched from high temperature, which displaces small amounts of magnetic species onto the octahedral sublattice (Larsson et al. 1994, Harrison et al. 1998). The presence of a small number of strong AB interactions interrupts the long-range antiferromagnetic ordering on the tetrahedral sublattice.

Magnetocrystalline anisotropy. The concept of magnetocrystalline anisotropy was introduced in Equation (9). This term describes the fact that it is energetically favourable for spins to align parallel to some crystallographic directions (the so-called 'easy' axes) and energetically unfavourable for spins to align parallel to others (the 'hard' axes). Anisotropy has two microscopic origins (Dunlop and Özdemir 1997). Single-ion anisotropy is caused by coupling between the spin (S) and orbital (L) contributions to the electron angular momentum (J). Fe^{3+} cations have $L = 0$ and therefore little or no single-ion anisotropy. Fe^{2+} cations have $L = 2$, and although most of this is removed by the influence of the crystal field (a process called 'quenching', Kittel 1976), the residual L-S coupling is large enough to impart significant single-ion anisotropy. Dipolar anisotropy is the result of anisotropic exchange interactions, which have a pseudo-dipolar and pseudo-quadrupolar angular dependence. In simple cubic structures, the dipole component of the anisotropic exchange interaction sums to zero and only the quadrupole component leads to an orientation dependence of the exchange interaction energy. In non-cubic structures, however, the dipole component contributes significantly to the total magnetocrystalline anisotropy.

Magnetocrystalline anisotropy in cubic minerals can be described macroscopically via the energy term:

$$E_K = K_1 V(\alpha_1^2\alpha_2^2 + \alpha_1^2\alpha_3^2 + \alpha_2^2\alpha_3^2) + K_2 V \alpha_1^2\alpha_2^2\alpha_3^2 \tag{10}$$

where K_1 and K_2 are constants, V is volume, and α_i are the direction cosines of the magnetization vector with respect to $\langle 100\rangle$. For $K_1 > 0$, the easy axes are parallel to $<100>$ (minimum E_K) and the hard directions are parallel to $\langle 111\rangle$ (maximum E_K). For $K_1 < 0$ the reverse is true. Lower symmetry minerals may have only one easy axis (uniaxial anisotropy) described by the macroscopic energy term:

$$E_K = K_{u1} V \sin^2\theta + K_{u2} V \sin^4\theta \tag{11}$$

where K_{u1} and K_{u2} are uniaxial anisotropy constants and θ is the angle between the magnetization vector and the axis. For $K_{u1} > 0$ the magnetization lies parallel to the axis. For $K_{u1} < 0$ the magnetization lies in an easy plane perpendicular to the axis.

The magnetocrystalline anisotropy constants are functions of temperature, composition, cation distribution, and crystal structure. For certain combinations of these variables it is possible that the principal anisotropy constant, K_1 or K_{u1}, changes from positive to negative, passing through an 'isotropic point'. This leads to a 'spin-flop' transition, where the alignment of spins changes abruptly to a new easy axis. An example of this phenomenon is the Morin transition in hematite (Lieberman and Banerjee 1971, Kvardakov et al. 1991). The single-ion and dipolar anisotropies in hematite have opposite signs and different temperature dependencies, causing them to cancel each other out at the

Morin transition (T_M = -15°C) (Banerjee 1991). Above T_M, K_{1u} is negative, and the spins lie within the basal plane (Fig. 4a). Below T_M, K_{1u} is positive, and the spins are aligned parallel to [001].

Single-ion anisotropy can vary greatly as a function of cation distribution. For example, disordered $Co_{0.5}Pt_{0.5}$ has low anisotropy and easy axes parallel to ⟨111⟩, whereas ordered $Co_{0.5}Pt_{0.5}$ has large anisotropy and an easy axis parallel to [001] (Razee et al. 1998). A similar phenomenon occurs in magnetite, associated with the ordering of Fe^{2+} and Fe^{3+} cations at the Verwey transition. Below the Verwey transition, discrete Fe^{2+} cations on octahedral sites have a large positive single-ion anisotropy due to the presence of unquenched orbital angular momentum. This dominates other sources of anisotropy so that the net K_1 is positive and the easy axes are parallel to <100>. Above the Verwey transition, electron hopping converts discrete Fe^{2+} and Fe^{3+} cations into average $Fe^{2.5+}$ cations, greatly reducing the single-ion anisotropy. Above 130 K (10 K higher than the Verwey transition) the small positive anisotropy due to octahedral $Fe^{2.5+}$ is outweighed by the negative anisotropy due to tetrahedral Fe^{3+}, K_1 changes sign and the easy axes switch to ⟨111⟩ (Banerjee 1991).

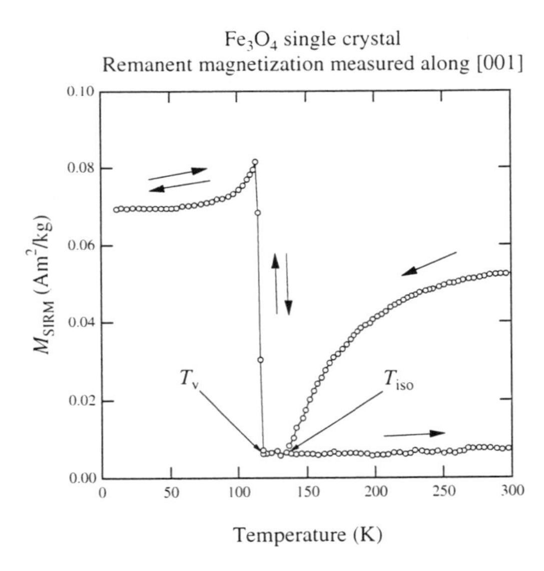

Figure 9. Temperature dependence of saturation remanence M_{SIRM}, produced by a 2.5 T field along [001] at 300 K, during zero-field cooling from 300 K to 10 K and zero field warming back to 300 K (after Özdemir and Dunlop 1999). An irreversible loss of remanence occurs on cooling from room temperature to the isotropic point (T_{iso} = 130 K). A small but stable remanence remains between T_{iso} and the Verwey transition (T_v = 120 K). Below T_v there is a reversible jump in remanence due to the changes in magnetocrystalline anisotropy and microstructure associated with the cubic to monoclinic phase transition.

The magnetic consequences of heating and cooling through the Verwey transition (T_V = 120 K) and isotropic point (T_{iso} = 130 K) are best seen in magnetic measurements on oriented single crystals (Özdemir and Dunlop 1998, 1999). Figure 9 shows the variation in saturation induced remanent magnetization (M_{SIRM}) as a function of temperature in a 1.5 mm diameter single crystal of magnetite as it is cycled through T_V and T_{iso}. M_{SIRM} is the remanent magnetization obtained after saturating the sample at room temperature in a field of 2.5 T applied parallel to [001]. The remanence is due to pinning of magnetic domain walls by crystal defects. M_{SIRM} decreases as the sample is cooled due to two general principles: firstly, the width of a magnetic domain wall is proportional to $(K_{u1})^{-0.5}$ (Dunlop and Özdemir 1997) and therefore walls get broader as the isotropic point is approached; secondly, broad walls are less effectively pinned by crystal defects (Xu and Merrill 1989) and can move under the influence of the demagnetizing field, destroying the remanence. The remanence does not disappear entirely on reaching T_{iso}; a small stable remanence (about

14% of the original M_{SIRM}) remains between T_{iso} and T_V. This hard remanence is carried by walls which are magnetoelastically pinned by strain fields due to dislocations (Özdemir and Dunlop 1999).

On cooling below T_V there is a discontinuous increase in remanence to a value exceeding the original M_{SIRM}. This increase is correlated with the first-order transition from cubic to monoclinic. The monoclinic phase has uniaxial anisotropy, with the easy axis parallel to the *c*-axis (Abe et al. 1976, Matsui et al. 1977). The *c*-axis forms parallel to the <100> axes of the cubic phase, leading to the development of twin domains (Chikazumi et al. 1971, Otsuka and Sato 1986). Within individual twin domains the magnetization is parallel or antiparallel to the *c*-axis and conventional uniaxial magnetic domains structures are observed (Moloni et al. 1996). The magnetization rotates through 90° on crossing a twin domain boundary due to the change in orientation of the *c*-axis. A detailed anlaysis of the magnetic domain structures above and below the Verwey transition has recently been performed using the micromagnetic techniques described earlier (Muxworthy and Williams 1999, Muxworthy and McClelland 2000a,b). Here it is argued that closure domains at the surface of magnetite grains are destroyed below T_V, leading to increased flux leakage and an increase in remanence. Özdemir and Dunop (1999) suggest that twin domain boundaries play a significant role, acting as pinning sites for magnetic domain walls. However, TEM observations demonstrate that twin boundaries in magnetite are extremely mobile, moving rapidly through the crystal in response to small changes in temperature or magnetic field (Otsuka and Sato 1986). In addition, the first-order nature of the Verwey transition implies that the low temperature phase forms by a nucleation and growth mechanism. This was confirmed by Otsuka and Sato (1986), who observed a coherent interface between the low and high temperature phases using *in situ* TEM. The changes in remanence which occur as this interface sweeps through the crystal are likely to be a complex function of the magnetic structures behind and ahead of the interface, as well as the exchange and magnetoelastic coupling between high and low temperature phases at the interface. Further *in situ* TEM work, revealing the relationship between twin domains and magnetic domains in the vicinity of T_V, is needed to resolve this problem.

SELF-REVERSED THERMOREMANENT MAGNETIZATION (SR-TRM)

Thermoremanent magnetization (TRM) is acquired when a mineral is cooled through its blocking temperature (T_B) in the presence of an external magnetic field (Néel 1949, 1955). T_B is the temperature below which thermally-activated rotation of the magnetization away from the easy axes stops and the net magnetization of the grain becomes frozen in. Usually, magnetostatic interaction with the external field ensures that more grains become blocked with their magnetizations parallel to the field than antiparallel to it, leading to a normal TRM. Some materials, however, acquire a TRM which is antiparallel to the applied field. The phenomenon of SR-TRM is a fascinating example of how phase transformations influence magnetic properties, which at the time of its discovery in the 1950s, threatened to discredit the theory that the Earth's magnetic field reverses polarity periodically.

Mechanisms of self reversal

An interesting consequence of the relationship between saturation magnetization and cation distribution discussed earlier is the occurrence of "compensation points", where for certain degrees of cation order (or for certain compositions in a solid solution) the two sublattice magnetizations of a ferrimagnet become equal and opposite (Harrison and Putnis 1995). This provides a possible mechanism of self-reversal (Néel 1955, Verhoogen 1956). Consider a material which displays a compensation point after being annealed and quenched from a temperature T_{comp} (i.e. the net magnetization is positive for material quenched from above T_{comp} and negative for material quenched from below T_{comp}). A sample cooled quickly from above T_{comp} would initially acquire a normal TRM, which would then become reversed as the cation distribution reequilibrated at the lower temperature. This mechanism requires that the kinetics of cation ordering are slow enough to prevent ordering during cooling but fast enough to allow ordering to occur below the Curie temperature. Such constraints are rarely met in nature and no natural examples of this mechanism are known.

Material close to a compensation point often displays an *N*-type magnetization curve (Fig. 2), which is another potential source of SR-TRM. Natural examples occur in low-temperature oxidised titanomagnetites (Schult 1968, 1976). Low temperature oxidation of the titanomagnetite solid solution leads to the development of a metastable defect spinel (titanomaghemite), in which the excess charge due to conversion of Fe^{2+} to Fe^{3+} is balanced by the formation of vacancies on the octahedral sublattice (O'Reilly 1984). This reduces the size of the octahedral sublattice magnetization, which for certain bulk compositions and degrees of oxidation can generate *N*-type magnetization curves and SR-TRM (Stephenson 1972b).

A third mechanism proposed by Néel (1951) involves two phases which interact magnetostatically. The two phases have different Curie temperatures, with the high-T_c phase acquiring a normal TRM on cooling and the low-T_c phase acquiring a reversed TRM due to negative magnetostatic coupling with the first phase (Veitch 1980). This mechanism is not thought to be important in nature since the magnetostatic interaction is very weak and the resulting SR-TRM is relatively unstable.

The fourth, and most important, mechanism of self reversal involves two phases which are coupled via exchange interactions. Exchange interaction is possible when the two phases form a coherent intergrowth, and is governed by the interaction between cations immediately adjacent to the interface. Néel (1951) identified the magnetic characteristics of the two phases required to produce self reversal. The phases must have different Curie temperatures, with the high-T_c phase (often referred to as the '*x*-phase') having an antiferromagnetic structure with a weak parasitic moment and the low-T_c phase having a ferromagnetic or ferrimagnetic structure. Exchange coupling between the phases must be negative, so that their net magnetizations align antiparallel to each other. Many examples of self reversal via exchange interaction have been documented. In each case, understanding the self-reversal mechanism requires the identification of the two phases involved, their magnetic characteristics, and the origin of the negative exchange coupling.

Self-reversal in the ilmenite-hematite solid solution

The first natural occurrence of SR-TRM was discovered by Nagata et al. (1952) in samples from the Haruna dacite and later shown to be carried by intermediate members of the ilmenite-hematite solid solution (Uyeda 1955). Since then there have been many attempts to determine the origin of the self-reversal effect (Uyeda 1957, 1958; Ishikawa 1958, Ishikawa and Syono 1963, Hoffman 1975, 1992; Varea and Robledo 1987, Nord and Lawson 1989, 1992; Hoffmann and Fehr 1996, Bina et al. 1999). Many of the early attempts to interpret experimental observations on self-reversing material were hampered by the lack of information about the equilibrium phase diagram. It is useful, therefore, to reappraise this work in light of recent experimental and theoretical studies which provide stricter constraints on the phase transformation behaviour in this system (Burton 1984, 1985, Ghiorso 1997, Harrison et al. 2000a,b).

Nature of the reversal process. Experimental studies have demonstrated that self reversal in ilmenite-hematite is the result of negative exchange coupling between two phases (Uyeda 1958). The most compelling evidence of this is the very large magnetic fields required to suppress SR-TRM in synthetic materials (up to 1.6 T). Such interaction fields could not be caused by magnetostatic effects. Further evidence is the observation of right-shifted hysteresis loops and $\sin\theta$ torque curves, which are classic indicators of unidirectional anisotropy caused by exchange interaction (Meiklejohn and Carter 1960, Meiklejohn 1962).

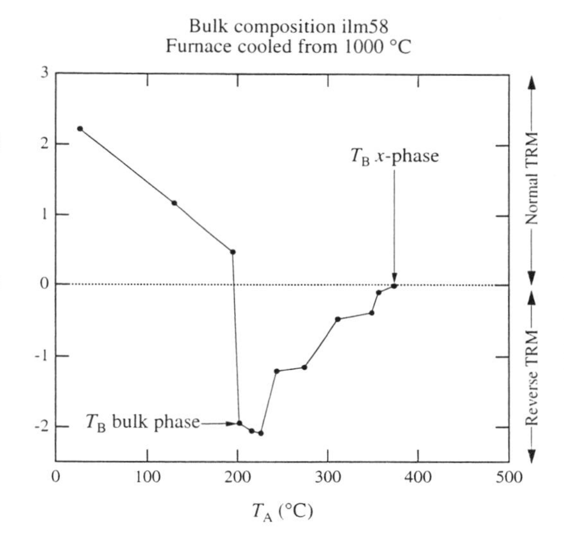

Figure 10. Partial TRM acquisition in cation ordered ilm58 (furnace cooled from 1000°C). TRM was acquired by cooling from 450°C to T_A in a magnetic field and then to room temperature in zero field (Ishikawa and Syono 1963). Arrows show the blocking temperatures, T_B, of the bulk phase (ilm58) and the *x*-phase (approx. ilm40).

Figure 10 shows TRM acquisition measurements on a synthetic sample of ilm58 (58% $FeTiO_3$ 42% Fe_2O_3) annealed at 1000°C and then furnace cooled to room temperature (Ishikawa and Syono 1963). Each data point represents a partial TRM, acquired by cooling the sample in a magnetic field from 450°C to a temperature, T_A, and then cooling to room temperature in zero field. Acquisition of SR-TRM (negative values) begins when $T_A < 350$°C, indicating that the Curie temperature of the high-T_c phase is at least 350°C. This corresponds to an ilmenite-hematite phase with composition around ilm40, i.e. significantly richer in Fe than the bulk composition of the sample (Fig. 5). The intensity of SR-TRM increases with decreasing T_A over the range 200°C < T_A < 350°C, indicating that the high-T_c phase has a range of different compositions or particle sizes (or both). When $T_A < 200$°C the TRM suddenly becomes normal. This temperature corresponds to T_c for the bulk composition ilm58 (Fig. 5). The change from reversed to normal TRM occurs because the sample adopts a multidomain magnetic structure below 200°C and any magnetic domains which are not exchange coupled to the high-T_c phase are free to move in response to the external field, producing a large normal component to the TRM.

Origin of the 'x-phase'. A possible source of compositional heterogeneity in such samples is the twin walls which develop on cooling through the $R\bar{3}c$ to $R\bar{3}$ order-disorder phase transition (Nord and Lawson 1989). Twin domains occur due to the loss of symmetry associated with the transition. In this case, adjacent domains are related to each other by 180° rotation about the *a*-axis. Figure 11a is a dark-field TEM micrograph showing the twin domain microstructure observed in ilm70 annealed within the single-phase $R\bar{3}$ field for 10 hours at 800°C. Since both the ordered and disordered phases belong

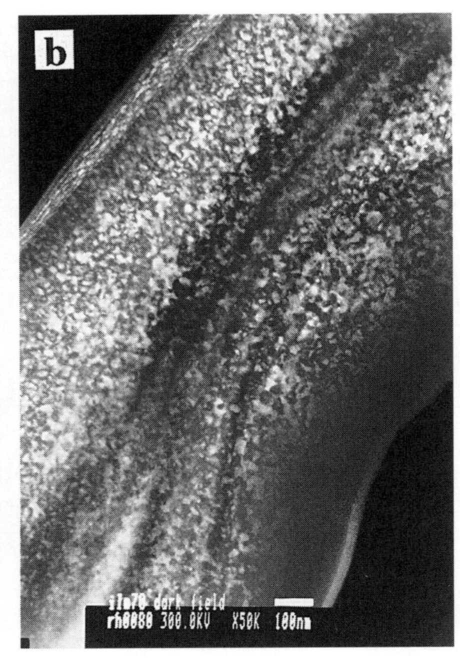

Figure 11. Dark-field TEM micrographs of ilm70 using the (003) reflection. (a) ilm70 quenched from 1300°C and annealed at 80°C for 10 hours. (b) ilm70 as quenched from 1300°C.

to the same crystal system and the magnitude of the spontaneous strain is relatively small (Harrison et al. 2000a), the twin walls are not constrained to a particular crystallographic orientation and meander through the crystal in a manner normally associated with antiphase domains. In fact, the domains do have an antiphase relationship with respect to the partitioning of Fe and Ti between the (001) layers, so that an Fe-rich layer becomes a Ti-rich layer on crossing the twin wall and vice versa.

The size of the twin domains depends on the thermal history of the sample. Domains coarsen rapidly on annealing below T_{od} (Fig. 11a) but can be extremely fine scale in samples quenched from above T_{od} (Fig. 11b). Nord and Lawson (1989, 1992) performed a systematic TEM and magnetic study of ilm70 with a range of twin domain sizes. They found a strong correlation between the magnetic properties and the domain size, with self reversal occurring only when the average domain size was less than 800-1000 Å. This is due to the close relationship between twin domains and magnetic domains in this system (Fig. 12). Each horizontal bar in Figure 12 represents a single (001) sublattice layer, shaded according to its cation occupancy (dark = Fe-rich, light = Ti-rich). Two cation ordered domains are shown, separated by a twin wall. The ordered domains are ferrimagnetic due to the different concentrations of Fe on each layer. The net magnetization reverses across the twin wall due to the antiphase relationship between the cation layers in adjacent domains, creating a magnetic domain wall. These magnetic walls are static (they can only exist where there is a twin wall), and if the twin domains are small enough (<800 Å), the material displays single-domain magnetic behaviour. If the twin domains are large (>6500 Å), then free-standing conventional magnetic domain walls can nucleate and move within the body of the twin domain, destroying the SR-TRM component (Fig. 10). The correlation between SR-TRM acquisition and twin domain size is therefore explained

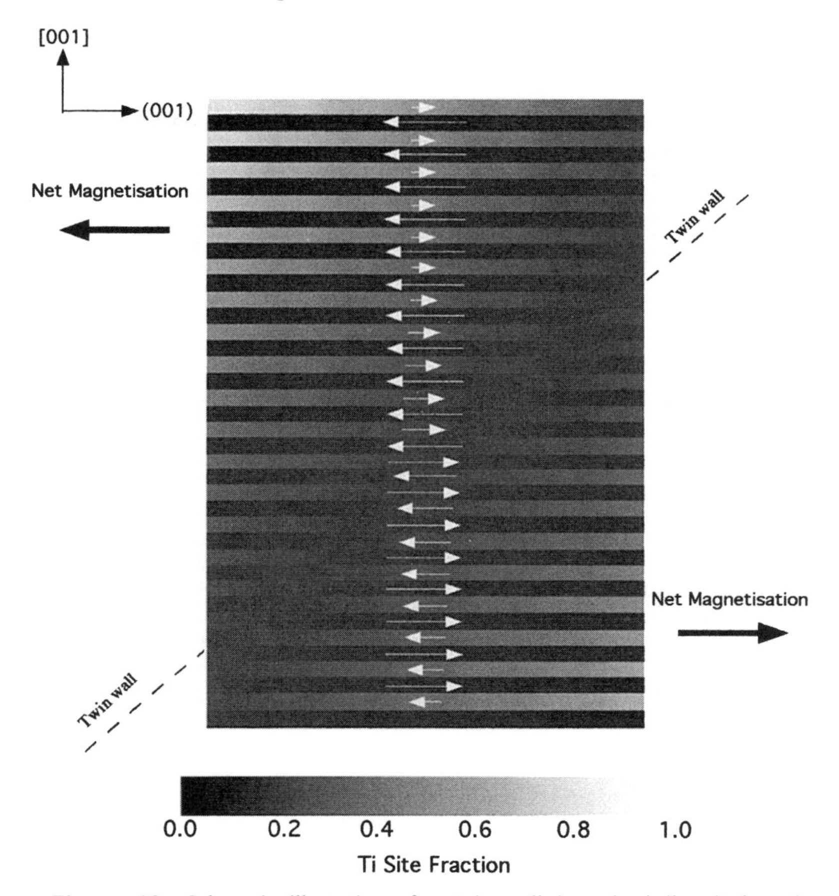

Figure 12. Schematic illustration of a twin wall in ordered ilmenite-hematite. Horizontal bars represent (001) cation layers, shaded according to their Fe-occupancy (Fe = dark, Ti = light). There are two twin domains (upper left and lower right) in antiphase with respect to their cation occupancies. Arrows show the magnitude and direction of the sublattice magnetizations along the central portion of the diagram. The antiferromagnetic coupling is uninterrupted at the twin wall, but the net ferrimagnetic moment reverses, forming a magnetic wall.

by the transition from single-domain to multi-domain behaviour with increasing twin domain size (Hoffman 1992).

Nord and Lawson (1992) and Hoffman (1992) proposed that the twin walls themselves act as the high-T_c 'x-phase' during self reversal. As illustrated in Figure 12, the twin wall is equivalent to a thin ribbon of the disordered $R\bar{3}c$ phase and presumably has a canted antiferromagnetic structure similar to that of endmember hematite. The model requires that the twin walls are Fe-rich and of sufficient thickness to have blocking temperatures more than 100°C higher than domains themselves. Thick twin walls were suggested by Hoffman (1992). However, an Fe-rich wall of finite thickness will experience a coherency strain similar to that experienced by fine-scale exsolution lamellae, forcing it to lie parallel to (001) (Haggerty 1991). This is inconsistent with the meandering twin walls observed using TEM (Fig. 11), which suggest atomically thin walls with little coherency strain. Nord and Lawson (1992) observed twin walls with a zig-zag morphology in samples annealed below the solvus, which would be more consistent with the Fe-enrichment model. Harrison et al. (2000a) argue that thick Fe-rich walls could be produced below the solvus, where an intergrowth of disordered Fe-rich material and ordered Ti-rich material is thermodynamically stable. *In situ* neutron diffraction experiments demonstrate that a highly heterogeneous state of order with a (001) texture develops when material containing a high density of twin domain walls is heated below the solvus (Harrison and Redfern, in preparation). Homogeneous ordering is observed in samples with low densities of twin domain walls. These lines of evidence suggest that Fe-enrichment occurs more readily in samples quenched from above T_{od} and subsequently annealed below or cooled slowly through the solvus.

Recent studies of natural samples from the Nevado del Ruiz and Pinatubo dacitic pumices suggest a completely different origin of the *x*-phase (Haag et al. 1993, Hoffmann 1996, Hoffmann and Fehr 1996, Bina et al. 1999). Ilmenite-hematite grains from these volcanoes are zoned, with the rims being slightly Fe-richer than the cores. The chemical zonation is thought to be caused by the injection of a more basaltic magma into the dacitic magma chamber shortly before erruption. Despite the small difference in composition between the rim and the core (rim = ilm53-57, core = ilm58), their magnetic properties are very different. The rim is antiferromagnetic with a weak parasitic moment, while the core is strongly ferrimagnetic. This suggests that the rim phase grew within the disordered $R\bar{3}c$ stability field and was quenched fast enough on eruption to prevent cation ordering, while the core phase either grew within the ordered $R\bar{3}$ stability field or cooled at a much slower rate, allowing time for cation ordering to take place after eruption. Hoffmann and Fehr (1996) suggest that the disordered Fe-rich rim of these grains acts as the *x*-phase and exchange coupling at the rim-core interface is responsible for the SR-TRM. The possible influence of fine scale microstructures in the ordered core of these natural samples has still to be determined.

Negative exchange coupling. The least well understood aspect of SR-TRM in the ilmenite-hematite system is the origin of the negative exchange coupling between high- and low-T_c phases. Hoffman (1992) has addressed this problem and proposed a spin-alignment model which could lead to self reversal (Fig. 13). The boxes in Figure 13 represent adjacent (001) sublattices viewed down the *c*-axis (one above and one below), shaded according to their cation occupancy as in Figure 12. The model shows two ordered domains (left and right) separated by a disordered twin wall. The twin wall has a canted antiferromagnetic structure and orders with its parasitic moment parallel to the external magnetic field. The domains are ferrimagnetic and order with their net magnetizations perpendicular to the external field and antiparallel to each other (Fig. 13a). At this point, the net magnetization of the sample is due entirely to the twin wall and is parallel to the external field. Hoffman (1992) proposed that SR-TRM develops at lower temperatures due to rotation of spins in the ordered domains (Fig. 13b). Each sublattice moment rotates in the opposite sense to those in the twin wall, so that all ordered domains develop a component of magnetization antiparallel to the twin wall magnetization. Experimental evidence in support of this non-collinear spin model comes from the neutron diffraction study of Shirane et al. (1962). They observed much lower sublattice magnetizations than expected in ordered intermediate members of the ilmenite-hematite solid solution, which was attributed to tilting of a certain fraction of spins. The magnitude and direction in which a spin tilts is a reflection of the local magnetocrystalline anisotropy, which is influenced by the distribution of cations among its nearest neighbours (a function of the composition and degree of short-range order). Such effects can be modelled from first-principles using spin-polarised relativistic density functional theory (e.g. Razee et al. 1998). Application of these methods.

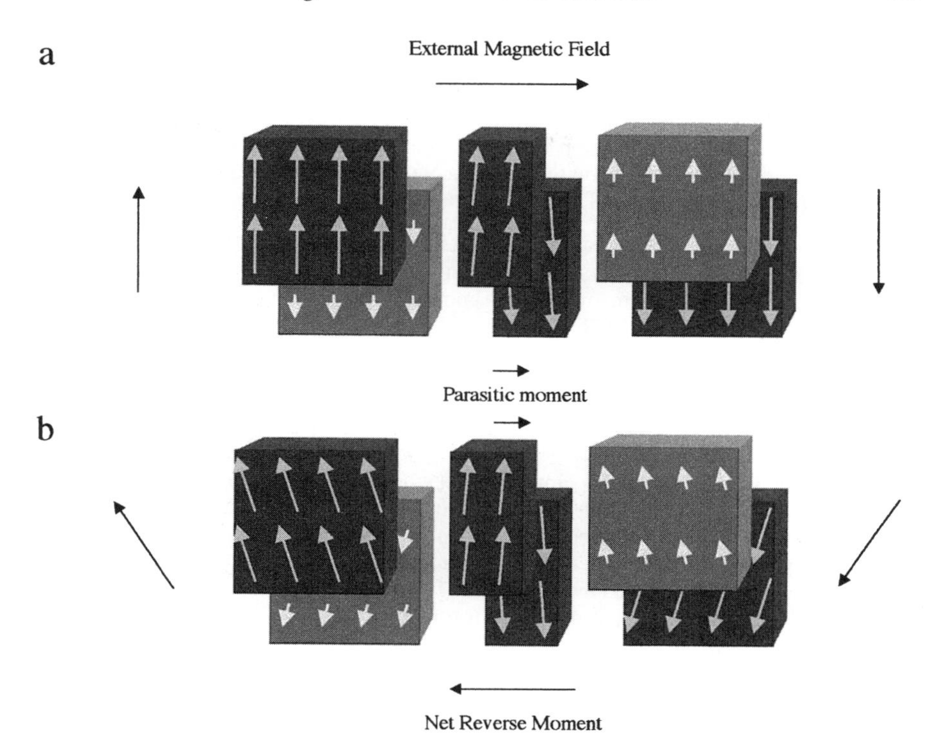

Figure 13. Model of self-reversal in the ilmenite-hematite solid solution (after Hoffman 1992). Boxes represent (001) cation layers (viewed down the *c*-axis), shaded according to their Fe-occupancy (Fe = dark, Ti = light). Two ordered ferrimagnetic domains are shown (left and right) separated by a twin wall (central) with a canted antiferromagnetic structure. (a) The twin wall orders first with its parasitic moment parallel to the external field. The ferrimagnetic domains order perpendicular to external field and antiparallel to each other. (b) The domain moments tilt away from the wall moment at lower temperatures, creating a large reverse component of magnetization.

might provide a way to gain new insight into the true mechanism of the self-reversal process in the ilmenite-hematite system.

CHEMICAL REMANENT MAGNETIZATION (CRM)

Principles of CRM

Chemical remanent magnetization (CRM) is acquired when a magnetic phase forms (or transforms) in the presence of a magnetic field. Formation of new magnetic material may occur via many processes (e.g. nucleation and growth from an aqueous solution, subsolvus exsolution from a solid solution, oxidation or reduction, hydrothermal alteration etc.). A common transformation leading to CRM is the inversion of a metastable parent phase to a stable daughter phase. The reverse process is thought to occur during the magnetization of natural lodestones (essentially pure magnetite), where lightening strikes hitting the sample induce large magnetizations and simultaneously create large numbers of metastable stacking faults (Banfield et al. 1994). These stacking faults pin magnetic domain walls and stabilise the CRM.

The simplest form of CRM occurs when a magnetic grain nucleates with a volume smaller than its blocking volume, V_B, and then grows to a volume greater than V_B (growth CRM). Grains with $V < V_B$ are superparamagnetic and will acquire a large induced magnetization in an applied field but no remanent magnetization. As the grains grow larger than V_B, the induced magnetization becomes frozen in and the material acquires a stable CRM. A good example of growth CRM occurs when ferromagnetic particles of Co precipitate from a paramagnetic Cu-Co alloy in the presence of a magnetic field (Kobayashi 1961).

CRM acquisition is more complex when new magnetic material is formed from a magnetic parent phase. There may be magnetostatic or exchange coupling between the phases and the orientation of CRM may be influenced by a preexisting remanence in the parent phase. This is of considerable importance in paleomagnetism, since the resultant CRM may point in a direction completely unrelated to the Earth's magnetic field direction.

TRANSFORMATION OF γ-FEOOH $\rightarrow$ γ-FE$_2$O$_3$ $\rightarrow$ α-FE$_2$O$_3$

The transformation of lepidocrocite (γ-FeOOH) $\rightarrow$ maghemite (γ-Fe_2O_3) $\rightarrow$ hematite (α-Fe_2O_3) provides an excellent demonstration of how transformation microstructures control CRM acquisition. Lepidocrocite has an orthorhombic crystal structure based on a cubic close-packed arrangement of oxygen anions. It dehydrates at temperatures above 200°C to yield maghemite. Maghemite is the fully-oxidised form of magnetite and has a defect spinel structure represented by the formula $Fe^{3+}\lfloor Fe^{3+}_{5/3}\square_{1/3}\rfloor O_4$ ($\square$ = cation vacancy). Vacancies occur exclusively on octahedral sites, where they order to yield a tetragonal superstructure with $c = 3a$ (Boudeulle et al. 1983, Greaves 1983). Maghemite is metastable and transforms to the stable phase, hematite, at temperatures between 350 and 600°C.

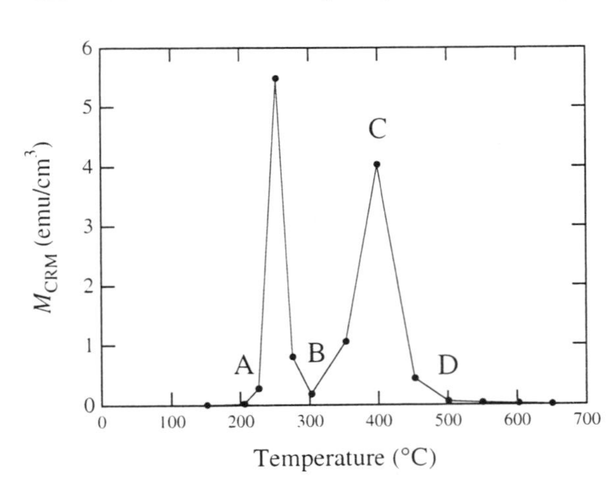

Figure 14. CRM acquisition during the transformation of γ-FeOOH $\rightarrow$ γ-Fe_2O_3 $\rightarrow$ α-Fe_2O_3 (after Özdemir and Dunlop 1999). The dehydration reaction γ-FeOOH $\rightarrow$ γ-Fe_2O_3 begins at A and is complete at B. Significant inversion of γ-Fe_2O_3 $\rightarrow$ α-Fe_2O_3 begins at C. CRM between A and C is parallel to applied field. CRM is dominated by α-Fe_2O_3 from point D onwards and is almost perpendicular to applied field.

Dehydrating lepidocrocite in a magnetic field produces CRM due to nucleation and growth of ferrimagnetic maghemite (Hedley 1968, Özdemir and Dunlop 1988, 1993; McClelland and Goss 1993). Since lepidocrocite is paramagnetic (T_c = 77 K) there is no parent-daughter coupling and the remanence is acquired parallel to the applied field. Figure 14 shows the intensity of CRM in samples of lepidocrocite heated for 2.5 hours at the given temperture in a 50-μT magnetic field (Özdemir and Dunlop 1993). CRM rises rapidly between 200 and 250°C due to the onset of dehydration. The reasons for the sudden decrease in CRM above 250°C are not clear. Özdemir and Dunlop (1993) proposed a model based on the generation of antiphase boundaries (APB's) due to vacancy ordering in the maghemite. They suggest that static magnetic domain walls are produced at the APB's, similar to the magnetic walls which are produced at twin boundaries in the ilmenite-hematite system (Fig. 12). This forces half the antiphase domains to have their magnetization opposed to the applied field, reducing the CRM to very small values. The increase in CRM above 300°C is suggested to be caused by nucleation of hematite on the APB's, which breaks the negative exchange coupling and allows all antiphase domains to align parallel to the field.

The argument for negative coupling at APB's in this system is speculative. Negative exchange at APB's normally occurs when the ordering species (cations or vacancies) partition between both ferrimagnetic sublattices (as in ilmenite-hematite). Vacancy ordering in maghemite occurs on just one magnetic sublattice (the octahedral sublattice), which should lead to positive exchange coupling at the APB.

An alternative explanation for the decrease in CRM above 275°C is that the maghemite grains have blocking temperatures close to 275°C. In this case, maghemite formed below 275°C would be able to acquire a stable CRM, whereas that grown above 275°C would not. The blocking temperature of single-crystal maghemite of the size used in these experiments is estimated to be close to the Curie temperature (675°C), i.e. far too high to explain the sudden decrease in CRM above 275°C. A possible way round this problem is suggested by McClelland and Goss (1993). They observed that maghemite produced by dehydration of lepidocrocite has a porous microstructure consisting of a polycrystalline aggregate of maghemite particles with diameter approximately 40 Å. Isolated crystallites would be superparamagnetic, but when in close contact (as part of a polycrystalline aggregate) they act in unison and behave as a single magnetic domain with $T_B < T_C$. On annealing, the crystallites coarsen and join together to form larger single-crystal grains of maghemite with higher blocking temperatures. The double peak in the CRM curve can then be explained if the maghemite formed during dehydration at low temperatures has a polycrystalline microstructure with T_B < 275°C, whereas maghemite formed during dehydration at high temperature forms single-crystal grains with 275°C < T_B < 675°C.

The decrease in CRM above 400°C is caused by the gradual unblocking of maghemite at higher and higher temperatures as well as the onset of the maghemite to hematite transformation. Hematite has only a weak parasitic magnetic moment, but nevertheless acquires a CRM during its formation. Some fraction of the parent maghemite may still be below T_B as it transforms to hematite, introducing the possibility of magnetic coupling beween the preexisting maghemite CRM and the developing hematite CRM. Overwhelming evidence in support of parent-daughter coupling is provided by the large angle between the high-temperature CRM and the applied field. Özdemir and Dunlop (1993) observed CRM orientations almost orthogonal to the applied field in material annealed between 500 and 600°C. McClellend and Goss (1993) observed self-reversed CRM over a similar temperature range. The nature of this coupling is difficult to predict because of the large number of possible crystallographic relationships between parent and daughter phases ($\{001\}_{hem}//\{111\}_{mag}$, $\langle 100\rangle_{hem}//\langle 110\rangle_{mag}$) and the large number of possible magnetiza-tion directions for each phase (maghemite easy axes are $\langle 111\rangle$, hematite easy plane is (001) with parasitic moment perpendicular to spin alignment). It is clear, however, that such coupling exists and may greatly influence the direction of CRM carried by hematite grains formed from maghemite, providing a cautionary note for any paleomagnetic interpretations based on such data.

CLOSING REMARKS

The aim of this review was to illustrate the diversity of phase transformations effecting magnetic minerals and some of the interesting, unusual, and potentially useful ways in which phase transformations influence their magnetic properties. The intrinsic magnetic and thermodynamic consequences of magnetic and cation ordering are well understood in terms of the theories outlined here and in the other chapters of this volume. These theories provide fundamental insight as well as a quantitative framework for the description of magnetic properties. It is clear, however, that one of the most important factors controlling the magnetic properties of minerals (namely, the exchange interaction between coupled phases) is poorly understood at present. Exchange coupling at interfaces is a crucial aspect of many mineral magnetic problems, yet current theories of exhange coupling are largely based on a mixture of common sense and intuition, and at best offer only a qualitative description of its effects. This is now one of the most pressing areas for future research in this field. Much-needed insight may be gained through the application of polarised-beam neutron diffraction, which has been successful in the study exhange coupling in magnetic mulitlayer materials (Ijiri et al. 1998). The use of first-principles calculations to determine the nature of magnetic interactions at internal interfaces is only just beginning to be explored, but may eventually provide a theoretical basis for some of the models presented here.

ACKNOWLEDGMENTS

I thank Andrew Putnis for his comments on the manuscript and his continued support of mineral magnetic research in Münster. This work was supported by the Marie Curie Fellowship Scheme.

REFERENCES

Abe K, Miyamoto Y, Chikazumi S (1976) Magnetocrystalline anisotropy of the low temperature phase of magnetite. J Phys Soc Japan 41:1894-1902

Akimoto S (1954) Thermomagnetic study of ferromagnetic minerals in igneous rocks. J Geomag Geoelectr 6:1-14

Arai M, Ishikawa Y, Saito N, Takei H (1985a) A new oxide spin glass system of (1-x)$FeTiO_3$-$x$$Fe_2O_3$. II. Neutron scattering studies of a cluster type spin glass of 90$FeTiO_3$-10Fe_2O_3. J Phys Soc Japan 54:781-794

Arai M, Ishkawa Y, Takei H (1985b) A new oxide spin-glass system of (1-x)$FeTiO_3$-$x$$Fe_2O_3$. IV. Neutron scattering studies on a reentrant spin glass of 79 $FeTiO_3$–21 Fe_2O_3 single crystal. J Phys Soc Japan 54:2279-2286

Arai M, Ishikawa Y (1985) A new oxide spin glass system of (1-x)$FeTiO_3$-$x$$Fe_2O_3$. III. Neutron scattering studies of magnetization processes in a cluster type spin glass of 90$FeTiO_3$-10Fe_2O_3. J Phys Soc Japan 54:795-802

Banerjee SK (1991) Magnetic properties of Fe-Ti oxides. Rev Mineral 25:107-128

Banfield JF, Wasilewski PJ, Veblen RR (1994) TEM study of relationships between the microstructures and magnetic properties of strongly magnetized magnetite and maghemite. Am Mineral 79:654-667

Bina M, Tanguy JC, Hoffmann V, Prevot M, Listanco EL, Keller R, Fehr KT, Goguitchaichvili AT, Punongbayan RS (1999) A detailed magnetic and mineralogical study of self-reversed dacitic pumices from the 1991 Pinatubo eruption (Philippines). Geophys J Int 138:159-78

Binder K, Heermann DW (1988) Monte Carlo Simulation in Statistical Physics, An Introduction. Springer-Verlag, New York

Blasse G (1964) Crystal chemistry and some magnetic properties of mixed metal oxides with spinel structure. Philips Res Rep Supp 3:1-139

Boudeulle M, Batis-Landoulsi H, Leclerq CH, Vergnon P (1983) Structure of γ-Fe_2O_3 microcrystals: vacancy distribution and structure. J Solid State Chem 48:21-32

Brown NE, Navrotsky A, Nord GL, Banerjee SK (1993) Hematite (Fe_2O_3)-ilmenite ($FeTiO_3$) solid solutions: Determinations of Fe-Ti order from magnetic properties. Am Mineral 78:941-951

Burton B (1984) Thermodynamic analysis of the system Fe_2O_3-$FeTiO_3$. Phys Chem Minerals 11:132-139

Burton BP (1985) Theoretical analysis of chemical and magnetic ordering in the system Fe_2O_3-$FeTiO_3$. Am Mineral 70:1027-1035

Burton BP (1991) Interplay of chemical and magnetic ordering. Rev Mineral 25:303-321

Burton BP, Davidson PM (1988) Multicritical phase relations in minerals. *In* S Ghose, JMD Coey, E Salje (eds) Structural and Magnetic Phase Transitions in Minerals. Springer-Verlag, Berlin, Heidelberg, New York, Tokyo

Burton BP, Kikuchi R (1984) The antiferromagnetic-paramagnetic transition in α–Fe_2O_3 in the single prism approximation of the cluster variation method. Phys Chem Minerals 11:125-131

Carpenter MA, Salje E (1989) Time-dependent Landau theory for order/disorder processes in minerals. Mineral Mag 53:483-504

Carpenter MA, Salje EKH (1994) Thermodynamics of nonconvergent cation ordering in minerals: II. Spinels and the orthopyroxene solid solution. Am Mineral 79:1068-1083

Chikazumi S, Chiba K, Suzuki K, Yamada T (1971) Electron microscopic observation of the low temperature phase of magnetite. In: Y. Hoshino, S. Iida, M. Sugimoto (eds) Ferrites: Proc Int'l Conf. University of Tokyo Press, Tokyo

Coey JMD, Ghose S (1987) Magnetic ordering and thermodynamics in silicates. *In* EKH Salje (ed) Physical Properties and Thermodynamic Behaviour of Minerals. NATO ASI Series D. Reidel, Dordrecht, The Netherlands

Crangle J (1977) The Magnetic Properties of Solids. Edward Arnold Limited, London

Dunlop DJ, Özdemir Ö (1997) Rock Magnetism: Fundamentals and Frontiers. Cambridge University Press, Cambridge, UK

Dzyaloshinsky I (1958) A thermodynamic theory of "weak" ferromagnetism of antiferromagnetics. J Phys Chem Solids 4:241–255

Fabian K, Kirchner A, Williams W, Heider F, Leibl T, Huber A (1996) Three-dimensional micromagnetic calculations for magnetite using FFT. Geophys J Int 124:89-104

Fabian K (1998) Neue Methoden der Modellrechnung im Gesteinsmagnetismus. PhD Thesis, Münich

Fowler RH (1936) Statistical Mechanics. Cambridge University Press, London

Ghiorso MS (1997) Thermodynamic analysis of the effect of magnetic ordering on miscibility gaps in the FeTi cubic and rhombohedral oxide minerals and the FeTi oxide geothermometer. Phys Chem Minerals 25:28-38

Gorter EW (1954) Saturation magnetization and crystal chemistry of ferrimagnetic oxides. Philips Res Rep 9:295-355

Greaves C (1983) A powder neutron diffraction invenstigation of vacancy ordering and covalence in γ-Fe_2O_3. J Solid State Chem 49:325-333

Haag M, Heller F, Lutz M, Reusser E (1993) Domain observations of the magnetic phases in volcanics with self-reversed magnetization. Geophys Res Lett 20:675-678

Haggerty SE (1991) Oxide textures—a mini-atlas. Rev Mineral 25:129-219

Harrison RJ, Putnis A (1995) Magnetic properties of the magnetite-spinel solid solution: Saturation magnetization and cation distributions. Am Mineral 80:213-221

Harrison RJ, Putnis A (1997) The coupling between magnetic and cation ordering: A macroscopic approach. Eur J Mineral 9:1115-1130

Harrison RJ, Putnis A (1999a) The magnetic properties and crystal chemistry of oxide spinel solid solutions. Surveys Geophys 19:461-520

Harrison RJ, Putnis A (1999b) Determination of the mechanism of cation ordering in magnesioferrite ($MgFe_2O_4$) from the time- and temperature-dependence of magnetic susceptibility. Phys Chem Minerals 26:322-332

Harrison RJ, Becker U, Redfern SAT (2000b) Thermodynamics of the $R\bar{3}$ to $R\bar{3}c$ phase transition in the ilmenite-hematite solid solution. Am Mineral (in press)

Harrison RJ, Redfern SAT, O'Neill HSC (1998) The temperature dependence of the cation distribution in synthetic hercynite ($FeAl_2O_4$) from *in situ* neutron diffraction. Am Mineral 83:1092-1099

Harrison RJ, Redfern SAT, Smith RI (2000a) *In situ* study of the $R\bar{3}$ to $R\bar{3}c$ phase transition in the ilmenite-hematite solid solution using time-of-flight neutron powder diffraction. Am Mineral 85:194-205

Hedley IG (1968) Chemical remanent magnetization of the FeOOH, Fe_2O_3 system. Phys Earth Planet Inter 1:103-121

Hoffman KA (1975) Cation diffusion processes and self-reversal of thermoremanent magnetization in the ilmenite-hematite solid solution series. Geophys J Royal Astro Soc 41:65-80

Hoffman KA (1992) Self-Reversal of thermoremanent magnetization in the ilmenite-hematite system: Order-disorder, symmetry, and spin alignment. J Geophys Res 97:10883-10895

Hoffmann V (1996) Experimenteller Mikromagnetismus zur aufklärung der physikalischen Prozesse des Erwerbs und der Stabilität von Daten des Paläo-Magnetfeldes der Erde. Habilitationsschrift, Univ, Münich

Hoffmann V, Fehr KT (1996) Micromagnetic, rockmagnetic and minerological studies on dacitic pumice from the Pinatubo eruption (1991, Phillipines) showing self-reversed TRM. Geophys Res Lett 23:2835-2838

Holland T, Powell R (1996a) Thermodynamics of order-disorder in minerals: I. Symmetric formalism applied to minerals of fixed composition. Am Mineral 81:1413-1424

Holland T, Powell R (1996b) Thermodynamics of order-disorder in minerals: II. Symmetric formalism applied to solid solutions. Am Mineral 81:1425-1437

Honig JM (1995) Analysis of the Verwey transition in magnetite. J Alloys Compounds 229:24-39

Iida S (1980) Structure of Fe_3O_4 at low temperatures. Philos Mag 42:349-376

Iizumi M, Koetzle TF, Shirane G, Chikazumi S, Matsui M, Todo S (1982) Structure of magnetite (Fe_3O_4) below the Verwey transition temperature. Acta Crystallogr B38:2121-2133

Ijiri Y, Borchers JA, Erwin RW, Lee SH, van der Zaag PJ, Wolf RM (1998) Perpendicular coupling in exchange-biased Fe_3O_4/CoO superlattices. Phys Rev Lett 80:608-611

Inden G (1981) The role of magnetism in the calculation of phase diagrams. Physica 103B:82-100

Inden G (1982) The effect of continuous transformations on phase diagrams. Bull Alloy Phase Diagrams 2:412-422

Ishikawa Y (1958) An order-disorder transformation phenomenon in the $FeTiO_3$-Fe_2O_3 solid solution series. J Phys Soc Japan 13:828-837

Ishikawa Y (1962) Magnetic properties of the ilmenite-hematite system at low temperature. J Phys Soc Japan 17:1835-1844

Ishikawa Y, Akimoto S (1957) Magnetic properties of the $FeTiO_3$–Fe_2O_3 solid solution series. J Phys Soc Japan 12:1083–1098

Ishikawa Y, Syono Y (1963) Order-disorder transformation and reverse thermoremanent magnetization in the $FeTiO_3$–Fe_2O_3 system. J Phys Chem Solids 24:517-528

Ishikawa Y, Saito N, Arai M, Watanabe Y, Takei H (1985) A new oxide spin glass system of $(1-x)FeTiO_3$-xFe_2O_3. I. Magnetic properties. J Phys Soc Japan 54:312-325

Kakol Z, Sabol J, Honig JM (1991) Cation distribution and magnetic properties of titanomagnetites $Fe_{3-x}Ti_xO_4$ ($0 \le x < 1$). Phys Rev B 43:649-654

Kaufman L (1981) J.L. Meijering's contribution to the calculation of phase diagrams—a personal perspective. Physica 103:1-7

Kittel C (1976) Introduction to Solid State Physics. John Wiley, New York

Kobayashi K (1961) An experimental demonstration of the production of chemical remanent magnetization with Cu-Co alloy. J Geomag Geoelectr 12:148–164

Kvardakov VV, Sandonis J, Podurets KM, Shilstein SS, Baruchel J (1991) Study of Morin transition in nearly perfect crystals of hematite by diffraction and topography. Physica B 168:242–250

Landau LD, Lifshitz EM (1980) Statistical Physics. Pergamon Press, Oxford, New York, Seoul, Tokyo

Larsson L, O'Neill HSC, Annersten H (1994) Crystal chemistry of the synthetic hercynite ($FeAl_2O_4$) from XRD structural refinements and Mössbauer spectroscopy. Eur J Mineral 6:39-51

Lieberman RC, Banerjee SK (1971) Magnetoelastic interactions in hematite: implications for geophysics. J Geophys Res 76:2735–2756

Malcherek T, Kroll H, Salje EKH (2000) Al,Ge cation ordering in $BaAl_2Ge_2O_8$-feldspar: Monodomain ordering kinetics. Phys Chem Minerals 27:203-212

Malcherek T, Salje EKH, Kroll H (1997) A phenomenological approach to ordering kinetics for partially conserved order parameters. J Phys Cond Matter 9:8075-8084

Matsui M, Todo S, Chikazumi S (1977) Magnetization of the low temperature phase of Fe_3O_4. J Phys Soc Japan 43:47-52

McClelland E, Goss C (1993) Self reversal of chemical remanent magnetization on the transformation of maghemite to hematite. Geophys J Inter 112:517-532

Meijering JL (1963) Miscibility gaps in ferromagnetic alloy systems. Philips Res Rep 13:318-330

Meiklejohn WH (1962) Exhange anisotropy—a review. J App Phys 33:1328-1335

Meiklejohn WH, Carter RE (1960) Exchange anisotropy in rock magnetism. J App Phys 31:164S-165S

Miodownik AP (1982) The effect of magnetic transformations on phase diagrams. Bull Alloy Phase Diagrams 2:406-412

Miyamoto Y, Chikazumi S (1988) Crystal symmetry of magnetite in low-temperature phase deduced from magneto-electric measurements. J Phys Soc Japan 57:2040–2050

Mizoguchi M (1985) Abrupt change of NMR line shape in the low-temperature phase of Fe_3O_4. J Phys Soc Japan 54:4295–4299

Moloni K, Moskowitz BM, Dahlberg ED (1996) Domain structures in single-crystal magnetite below the Verwey transition as observed with a low-temperature magnetic force microscope. Geophys Res Lett 23:2851-2854

Moon T, Merrill RT (1984) The magnetic moments of non-uniformly magnetized grains. Phys Earth Planet Inter 34:186-194

Moskowitz BM (1987) Towards resolving the inconsistancies in characteristic physical properties of synthetic titanomaghemites. Phys Earth Planet Inter 46:173-183

Mouritsen OG (1984) Computer studies of phase transitions and critical phenomena. Springer-Verlag, Berlin, Heidelberg, New York, Tokyo

Muxworthy AR, Williams W (1999) Micromagnetic models of pseudo-single domain grains of magnetite near the Verwey transition. J Geophys Res 104:29203-29217

Muxworthy AR, McClelland E (2000a) Review of the low-temperature magnetic properties of magnetite from a rock magnetic perspective. Geophys J Inter 140:101-114

Muxworthy AR, McClelland E (2000b) The causes of low-temperature demagnetization of remanence in multidomain magnetite. Geophys J Inter 140:115-131

Nagata T, Uyeda S, Akimoto S (1952) Self-reversal of thermo-remanent magnetization in igneous rocks. J Geomag Geoelectr 4:22

Néel L (1948) Propriétés magnetiques des ferrites; ferrimagnétisme et antiferromagnétisme. Ann Physique 3:137-198

Néel L (1949) Théorie du traînage magnétique des ferromagnétiques en grains fins avec applications aux terres cuites. Ann Géophysique 5:99-136

Néel L (1951) L'inversion de l'aimantation permanente des roches. Annales de Géophysique 7:90-102

Néel L (1955) Some theoretical aspects of rock magnetism. Advances Phys 4:191-243

Nell J, den Hoed P (1997) Separation of chromium oxides from ilmenite by roasting and increasing the magnetic susceptibility of Fe_2O_3-$FeTiO_3$ (ilmenite) solid solutions. *In* Heavy Minerals, South African Institute of Mining and Metallurgy, Johannesburg, p 75-78

Nell J, Wood BJ (1989) Thermodynamic properties in a multi component solid solution involving cation disorder: Fe_3O_4–$MgFe_2O_4$–$FeAl_2O_4$–$MgAl_2O_4$ spinels. Am Mineral 74:1000-1015

Newell AJ, Dunlop DJ, Williams W (1993) A two-dimensional micromagnetic model of magnetizations and fields in magnetite. J Geophys Res 98:9533-9549

Nord GL, Lawson CA (1989) Order-disorder transition-induced twin domains and magnetic properties in ilmenite-hematite. Am Mineral 74:160

Nord GL, Lawson CA (1992) Magnetic properties of ilmenite70-hematite30: effect of transformation-induced twin boundaries. J Geophys Res 97B:10897

O'Donovan JB, O'Reilly W (1980) The temperature dependent cation distribution in titanomagnetites. Phys Chem Minerals 5:235-243

O'Neill H, Navrotsky A (1983) Simple spinels: crystallographic parameters, cation radii, lattice energies, and cation distribution. Am Mineral 68:181-194

O'Neill HSC, Navrotsky A (1984) Cation distributions and thermodynamic properties of binary spinel solid solutions. Am Mineral 69:733-753

O'Neill HSC, Annersten H, Virgo D (1992) The temperature dependence of the cation distribution in magnesioferrite ($MgFe_2O_4$) from powder XRD structural refinements and Mössbauer spectroscopy. Am Mineral 77:725-740

O'Reilly W (1984) Rock and Mineral Magnetism. Blackie, Glasgow, London

O'Reilly W, Banerjee SK (1965) Cation distribution in titanomagnetites. Phys Lett 17:237-238

Otsuka N, Sato H (1986) Observation of the Verwey transition in Fe_3O_4 by high-resolution electron microscopy. J Solid State Chem 61:212-222

Özdemir Ö, Dunlop DJ (1988) Crystallization remanent magnetization during the transformation of maghemite to hematite. J Geophys Res B: Solid Earth 93:6530–6544

Özdemir Ö, Dunlop DJ (1993) Chemical remanent magnetization during γ-FeOOH phase transformations. J Geophys Res B: Solid Earth 98:4191–4198

Özdemir Ö, Dunlop DJ (1998) Single-domain-like behaviour in a 3-mm natural single crystal of magnetite. J Geophys Res 103:2549-2562

Özdemir Ö, Dunlop DJ (1999) Low-temperature properties of a single crystal of magnetite oriented along principal magnetic axes. Earth Planet Science Lett 165:229-39

Razee SSA, Staunton JB, Pinski FJ, Ginatempo B, Bruno E (1998) Effect of atomic short-range order on magnetic anisotropy. Philos Mag B 78:611-615

Readman PW, O'Reilly W (1972) Magnetic properties of oxidised (cation-deficient) titanomagnetites $(Fe,Ti,[\ ])_3O_4$. J Geomag Geoelectr 24:69-90

Rhodes P, Rowlands G (1954) Demagnetizing energies of uniformly magnetized rectangular blocks. Proc Leeds Philos Literary Soc Scientific Section 6:191-210

Riste T, Tenzer L (1961) A neutron diffraction study of the temperature variation of the spontaneous sublattice magnetization of ferrites and the Néel theory of ferrimagnetism. J Phys Chem Solids 19:117-123

Roth WL (1964) Magnetic properties of normal spinels with only A-A interactions. J Physique 25:507-515

Sack RO, Ghiorso MS (1991) An internally consistant model for the thermodynamic properties of Fe-Mg-titanomagnetite-aluminate spinels. Contrib Mineral Petrol 106:474-505

Salje EKH (1990) Phase Transitions in Ferroelastic and Co-elastic Crystals. Cambridge University Press, Cambridge, UK

Schult A (1968) Self-reversal of magnetization and chemical composition of titanomagnetite in basalts. Earth Planet Science Lett 4:57–63

Schult A (1976) Self-reversal above room temperature due to N-type ferrimagnetism in basalt. J Geophys 42:81-84

Shirane G, Cox DE, Takei WJ, Ruby SL (1962) A study of the magnetic properties of the $FeTiO_3$-Fe_2O_3 system by neutron diffraction and the Mössbauer effect. J Phys Soc Japan 17:1598-1611

Stephenson A (1969) The temperature dependent cation distribution in titanomagnetites. Geophys J Royal Astro Soc 18:199-210

Stephenson A (1972a) Spontaneous magnetization curves and curie points of spinels containing two types of magnetic ion. Philos Mag 25:1213-1232

Stephenson A (1972b) Spontaneous magnetization curves and curie points of cation deficient titanomagnetites. Geophys J Royal Astro Soc 29:91-107

Tolédano JC, Tolédano P (1987) The Landau theory of phase transitions. World Scientific, Teaneck, NJ

Trestman-Matts A, Dorris SE, Kumarakrishnan S, Mason TO (1983) Thermoelectric determination of cation distributions in Fe_3O_4-Fe_2TiO_4. J Am Ceramic Soc 66:829-834

Uyeda S (1955) Magnetic interaction between ferromagnetic materials contained in rocks. J Geomag Geoelectrtr 7:9-36

Uyeda S (1957) Thermo-remanent magnetism and coercive force of the ilmenite-hematite series. J Geomag Geoelectrtr 9:61-78

Uyeda S (1958) Thermo-remanent magnetism as a medium of palaeomagnetism, with special reference to reverse thermo-remanent magnetism. Japan J Geophys 2:1–123

van der Woude F, Sawatzky GA, Morrish AH (1968) Relation between hyperfine magnetic fields and sublattice magnetizations in Fe_3O_4. Phys Rev 167:533-535

Varea C, Robledo A (1987) Critical magnetization at antiphase boundaries of magnetic binary alloys. Phys Rev B 36:5561-5566

Veitch RJ (1980) Magnetostatic interaction between a single magnetised particle and a surrounding shell of ulvospinel. Phys Earth Planet Inter 23:215-221

Verhoogen J (1956) Ionic ordering and self-reversal of magnetization in impure magnetites. J Geophys Res 61:201-209

Verwey EJ (1939) Electronic conduction of magnetite (Fe_3O_4) and its transition point at low temperature. Nature 144:327-328

Waychunas GA (1991) Crystal chemistry of oxides and oxyhydroxides. Rev Mineral 25:11-68

Weiss P (1907) L'Hypothèse du champ moleculaire et la propriété ferromagnétique. J Physique 6:661-690

Williams W, Wright TM (1998) High-resolution micromagnetic models of fine grains of magnetite. J Geophys Res 103:30537-30550

Wißmann S, Wurmb Vv, Litterst FJ, Dieckmann R, Becker KD (1998) The temperature-dependent cation distribution in magnetite. J Phys Chem Solids 59:321-330

Wright TM, Williams W, Dunlop DJ (1997) An improved algorithm for micromagnetics. J Geophys Res 102:12085-12094

Wu CC, Mason TO (1981) Thermopower measurement of cation distribution in magnetite. J Am Ceramic Soc 64:520-522

Xu S, Merrill RT (1989) Microstress and microcoercivity in multidomain grains. J Geophys Res 94:10627-10636

Xu S, Dunlop DJ, Newell AJ (1994) Micromagnetic modeling of two-dimensional domain structures in magnetite. J Geophys Res B: Solid Earth 99:9035–9044

Zhang Z, Satpathy S (1991) Electron states magnetism, and the Verwey transition in magnetite. Phys Rev B 44:13319-13331

8 NMR Spectroscopy of Phase Transitions in Minerals

Brian L. Phillips

Department of Chemical Engineering and Materials Science
University of California
Davis, California 95616

INTRODUCTION

Nuclear magnetic resonance (NMR) spectroscopy has been used extensively since the 1960s to study phase transformations. When Lippmaa et al. (1980) presented the first significant high-resolution solid-state NMR spectroscopic study of minerals, the study of structural phase transitions by NMR was a mature field, appearing primarily in the physics literature (e.g. Rigamonti 1984, Blinc 1981). The sensitivity of NMR spectroscopy to short-range structure (first- and second-coordination spheres) and low-frequency dynamics (i.e. frequencies much lower than the thermal vibrations of atoms) make it useful for determining changes in the structure and dynamics of solids that occur near phase transitions. This combination of characteristic time- and distance-scales is not easily accessible by other techniques. Widely studied phase transformations include order-disorder and displacive transitions in compounds with perovskite, antifluorite, β-K_2SO_4 (A_2BX_4), KH_2PO_4 ("KDP"), and other structure types, including some mineral phases (e.g. colemanite; Theveneau and Papon 1976). These studies used primarily single-crystal techniques and were mostly limited to phases that exhibit high symmetry and possess one crystallographic site for the nucleus studied. In some cases, the NMR data can help distinguish order-disorder from purely displacive transition mechanisms. Also, careful measurement of NMR relaxation rates provides information on the spectral density at low frequencies, which reflects softening lattice vibrations or freezing-in of rotational disorder modes.

Many phase transitions interesting to mineralogists became accessible to NMR spectroscopy in the late 1980s with the commercial availability of magic-angle-spinning (MAS) NMR probe assemblies capable of operation at extended temperatures (up to about 600°C) and the development of high-temperature NMR capability at Stanford (Stebbins 1995b). These technical advances enabled many of the structural transitions that occur in minerals to be studied in situ by NMR, including those in phases for which large single crystals are not available. MAS-NMR allows measurement of chemical shifts (which can be related to structural parameters), resolution of multiple crystallographic sites, and detection of changes in the number of inequivalent sites that accompany symmetry changes.

Several of the well-known polymorphic structural phase transitions that occur in minerals at low to moderate temperatures, particularly in framework aluminosilicates, have since been studied by NMR techniques; a brief review of these studies is provided here. Also included is a review of NMR studies of quenchable cation ordering reactions, which further illustrate the types of information on mineral transformations available from NMR experiments. NMR spectroscopy has also been used extensively to study glass transitions and glass-to-liquid transitions, which have been recently reviewed by Stebbins (1995b). We begin with a brief description of NMR spectroscopy, focusing on those aspects that are most useful for studying transitions in minerals.

NMR SPECTROSCOPY

A full account of solid-state NMR spectroscopy is well beyond the scope of this

1529-6466/00/0039-0008$05.00

chapter; for more detailed introductions to NMR spectroscopy and its mineralogical applications, the reader may wish to consult earlier volumes in this series (Kirkpatrick 1988, Stebbins 1988, Stebbins 1995b). Unfortunately, it is difficult to recommend a choice among the available the introductory texts. Most introductory texts intended for the chemistry student focus on aspects relevant to the fluid phase and do not adequately cover the interactions that dominate NMR of solids. Harris (1986) remains a good source of general information, although it is out-of-date in some respects. Akitt and Mann (2000) contains a readable description of the basic NMR experiment and includes some information on solid-state techniques. The recent volume by Fitzgerald (1999) provides a review of applications of NMR to inorganic materials (including minerals) and includes a comprehensive introduction and reviews of recent technical advances in high-resolution and 2-dimensional techniques. Engelhardt and Michel (1987) and Engelhardt and Kohler (1994) review solid-state NMR data for many minerals. Previous reviews of NMR studies of structural phase transitions (e.g. Rigamonti 1984, Blinc et al. 1980, Armstrong and van Driel 1975, Armstrong 1989) are highly technical and difficult to read without some knowledge of both phase transitions and magnetic resonance techniques. From them, however, one can gain an appreciation for the maturity of the field and the types of information available from NMR.

This section presents a phenomenological description of those aspects of solid-state NMR spectroscopy that are most useful for obtaining structural and dynamical information on phase transitions in minerals and the nature of disordered phases. The chemical shift and nuclear quadrupole interactions and their anisotropies receive particular emphasis, because they provide sensitive probes of the short-range structure, symmetry, and dynamics at the atomic position. Frequency shifts arising from these interactions can also serve as physical properties from which order parameters can be obtained for use in Landau-type treatments of the evolution toward a phase transition.

Basic concepts of NMR spectroscopy

The NMR phenomenon arises from the interaction between the magnetic moment of the atomic nucleus (μ) and an external magnetic field ($\mathbf{B_0}$), the energy for which is given by the scalar product: $E = -\mu \cdot \mathbf{B_0}$. The nuclear magnetic moment can take only a few orientations with respect to $\mathbf{B_0}$, so that the energy can be written as $E = -m\gamma h B_0/2\pi$, where γ is the magnetogyric ratio, h is Planck's constant, and m is one of the values $[I, I\text{-}1, \ldots \text{-}I]$, where I is the spin quantum number. The quantum number m simply describes the orientation of the nuclear magnetic moment, as the projection of μ on $\mathbf{B_0}$. The values of I and γ are fundamental nuclear properties and are characteristic for each isotope; e.g. $I = 1/2$ for ^{29}Si (0.048% natural abundance), $I = 0$ for ^{28}Si (NMR inactive). Tables of nuclear properties for isotopes of mineralogical interest can be found in Kirkpatrick (1988), Stebbins (1995a), Fitzgerald (1999), and Harris (1986).

When a sample is placed in a magnetic field of magnitude B_0, each of the magnetic nuclei (those with non-zero I) takes one of the $2I$+1 possible orientations with respect to $\mathbf{B_0}$, which are separated in energy by $\Delta E = \gamma h B_0/2\pi$ (Fig. 1). In response to this energy difference, the nuclei establish a Boltzmann population distribution and transitions between orientations (or, energy levels) can occur by application of radiation with frequency $\nu_0 = \gamma B_0/2\pi$, called the "Larmor frequency," or NMR frequency. The observed NMR signal, however, arises from the net magnetic moment (sum of the moments of the individual nuclei) due to the population differences between the levels. For NMR, ΔE is small compared to thermal energy at ambient conditions, giving population differences that are very small - of the order 10^{-5}. For example, under typical experimental conditions (25°C in a 9.4 T magnet) one million ^{29}Si nuclei yield a net magnetic moment equal to

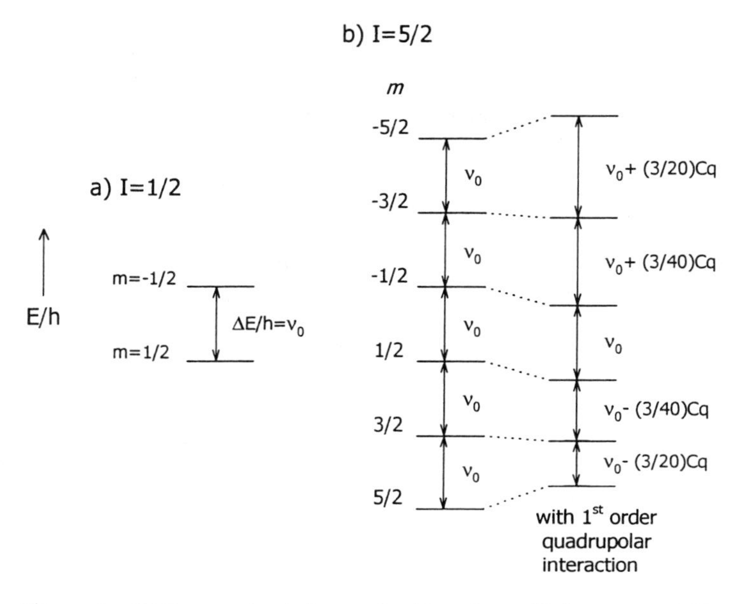

Figure 1. NMR energy level diagram for (a) nucleus with I = 1/2, and (b) nucleus with I = 5/2 with (right) and without (left) shifts due to the first-order quadrupolar interaction. Quadrupolar shifts are exaggerated and shown for $\eta = 0$ and $\theta = 0$ for simplicity (cf. Eqn. 3).

about six nuclei (500003 in the $m = +1/2$ orientation, 499997 in the $m = -1/2$ orientation). This effect contributes to the distinction of NMR as among the least sensitive of spectroscopic techniques.

For most NMR spectrometers currently in use, the value of B_0 is fixed by the current in a large superconducting solenoid, with typical values of from 7 to 14 T, which gives values of ν_0 that range from about 10 to 600 MHz (radio frequencies), depending on the isotope. The value of γ varies widely among the NMR-active isotopes, such that for any spectrometer (constant B_0) the differences in ν_0 between isotopes is much greater than the range of frequencies that can be observed in one NMR experiment. As a result, an NMR spectrum usually contains signal from only one isotope. NMR spectrometers are often identified by ν_0 for protons (^{1}H) at the spectrometer's magnetic field. For example, a "500 MHz" NMR spectrometer has $B_0 = 11.7$ T, from the ^{1}H $\gamma = 2.68 \cdot 10^8$ T^{-1}s^{-1}.

Chemical shifts

Magnetic and electrical interactions of the nucleus with its environment (e.g. other magnetic nuclei or electrons) produce small shifts in the NMR frequency away from ν_0 : $\nu_{obs} = \nu_0 + \Delta\nu$. These interactions, measured by their frequency shifts, convey structural and chemical information. The most useful of these interactions is the so-called "chemical shift", which results from magnetic shielding of the nucleus by nearby electrons, which circulate due to the presence of $\mathbf{B_0}$. The chemical shift (δ) is very sensitive to the electronic structure, and hence the local structural and chemical environment of the atom. The chemical shift is measured by the frequency difference from a convenient reference material, rather than from ν_0 directly: $\delta = 10^6 (\nu_{sample} - \nu_{ref})/\nu_{ref}$. For example, tetra-methylsilane ($Si(CH_3)_4$, abbreviated TMS) serves as the standard frequency reference for

^{29}Si, ^{13}C, and ^{1}H NMR. Values of δ are typically of the order of several parts-per-million (ppm) and reported in relative units, so that δ will be the same on any spectrometer regardless the size of B_0.

Calculation of δ from a known structure requires high-level quantum chemical calculations, because the shielding depends strongly on variations in the occupancy of low-lying excited states. Treatment of these states requires comprehensive basis sets. Structural information is usually obtained from empirical correlations of δ for known structures with short-range structural parameters such as coordination number, number of bridging bonds, bond lengths, bond angles, and ionic potential of counter ions. The *ab initio* calculations of δ are often used to support such correlations and peak assignments. Correlations of δ with structural parameters can be particularly effective when a subset of similar structures is considered, for example correlations for ^{29}Si of δ with Si-O-Si bond angle in framework silicates.

In solids, the chemical shift is a directional property and varies with the crystal's orientation in the magnetic field. This orientation dependence, the chemical shift anisotropy (CSA), can be described by a symmetric second-rank tensor that can be diagonalized and reduced to three principal values: δ_{11}, $\delta_{22,}$ δ_{33}, where the isotropic chemical shift, δ_i , is the average of these values; $\delta_i = 1/3 \cdot (\delta_{11} + \delta_{22} + \delta_{33})$. For a single crystal, or any particular nucleus in a powdered sample, the frequency varies according to:

$$\nu(\theta) = \nu_0 \cdot \left[1 + \delta_i + \frac{1}{3} \sum_{k=1}^{3} (3\cos^2\theta_k - 1)\delta_{kk} \right] \quad (1)$$

where θ_k is the angle between the external magnetic field ($\mathbf{B_0}$) and the principal axis of the chemical shift tensor characterized by value δ_{kk} (Fig. 2).

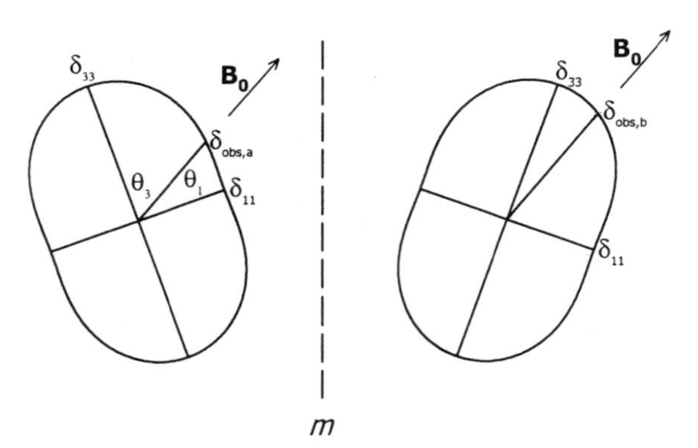

Figure 2. Illustration of magnetic inequivalence for crystallographically equivalent sites of low point symmetry, using the chemical shift anisotropy (CSA). Oval represents a cross-section of the surface for the CSA (Eqn. 1), with axes for the principal axis components corresponding to δ_{11} and δ_{33} (δ_{22} is normal to the page). The observed shift ($\delta_{obs,a}$) corresponds to the distance from the center to the edge, parallel to $\mathbf{B_0}$. Application of a symmetry element (mirror plane) having a general orientation to the principal axes gives a different chemical shift ($\delta_{obs,b}$).

As for all second-rank tensor properties, the orientational variation in chemical shift can be visualized as an ellipsoidal surface for which the symmetry and alignment of the principal axes must conform to the point symmetry at the atom position. For example, axial symmetry (3-fold or higher rotation axis) requires that two of the principal values be equal (e.g. $\delta_{11} = \delta_{22}$) and that the principal axis corresponding to the other value (δ_{33}) be parallel to the symmetry axis. An important point in the context of phase transitions is that the CSA is not invariant under some symmetry operations. For example, as illustrated in Figure 2, for a general position with no symmetry constraints on the CSA, crystallographic sites related by symmetry operations other than a translation (such as the mirror plane in Fig. 2) give separate peaks for most orientations of $\mathbf{B_0}$. The exceptions are those orientations with a special relationship to the symmetry element, for example parallel or perpendicular to the mirror plane in the case shown in Figure 2.

The CSA tensor and its crystallographic orientation can be determined by taking a series of NMR spectra for a single crystal, varying its orientation with respect to $\mathbf{B_0}$ (Fig. 3). These properties are important for understanding NMR of single-crystals, such as for quartz discussed in the following section.

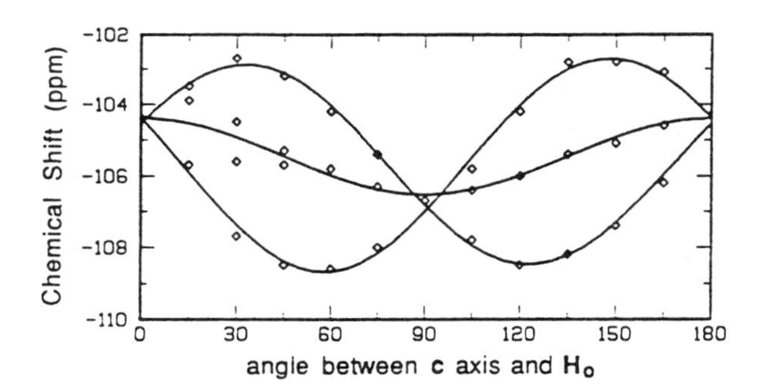

Figure 3. Variation in ^{29}Si NMR peak positions with orientation in the magnetic field for a single crystal of α-quartz, showing the effect of Equation (1). The change in crystal orientation corresponds to rotation about $\mathbf{a}^*$, which is oriented normal to $\mathbf{B_0}$. The three peaks observed for most orientations arise from three magnetically inequivalent orientations of the single crystallographic Si-position, related by a three-fold screw axis parallel to **c**. [Modified from Spearing and Stebbins (1989), Fig. 1, p. 957.]

For polycrystalline samples, the angular dependence of the chemical shift in Equation (1) results in a broad "powder pattern," which can be calculated by assuming a random distribution of θ_k (e.g. a constant increment of θ_k with intensities proportional to $\sin(\theta_k)$). The powder pattern provides information on the size and symmetry of the chemical shift tensor from which useful structural and dynamical information can be obtained. However, it is usually very difficult to resolve separate powder patterns for phases that contain more than one type of crystallographic site, because the full width of the powder pattern (δ_{11} - δ_{33}) typically exceeds the range in δ_i.

Application of "magic-angle-spinning" (MAS), a physical rotation of the sample about a fixed axis, removes the frequency distribution of Equation (1) from the NMR spectrum by averaging the frequency for each nucleus over the rotation path. The angular dependent term in Equation (1) can then be replaced by its value averaged over the rotation:

$$\left(3\cos^2\theta_k - 1\right) = \frac{1}{2}(3\cos^2\beta - 1)(3\cos^2\psi_k - 1) \tag{2}$$

where β is the angle between the rotation axis and $\mathbf{B_0}$, and ψ_k is the angle between the rotation axis and the kk principal axis of the CSA tensor. Setting $\beta = a\cos(1/\sqrt{3}) = 54.7°$, the NMR spectrum contains a single peak for each crystallographic and/or chemically distinct site at its isotropic chemical shift, δ_i, plus a series of "spinning sidebands," spaced at the spinning frequency (ν_{rot}), that approximately span the range of the CSA (from δ_{11} to δ_{33}). The intensity of the spinning sidebands must be included in any quantitative analysis of the NMR spectrum.

In a typical experimental configuration for MAS-NMR the sample is contained in a ceramic cylinder (ZrO_2 and Si_3N_4 are common), 3-7 mm diameter with sample volumes of 0.01 to 0.5 cm^3, which spins while floating on a bearing of compressed gas. This configuration is less than ideal for in situ studies of phase transitions. Large temperature gradients and uncertain temperature calibrations can result, because of frictional heating due to the high velocity of the sample cylinder wall and the distance between the temperature sensor and sample. Recently developed solid-state NMR chemical shift thermometers can be helpful in this respect (van Gorkum et al. 1995; Kohler and Xie, 1997).

Nuclear quadrupole effects

NMR spectra of isotopes for which $I > 1/2$ (quadrupolar nuclei) display additional orientation-dependent frequency shifts and peak broadening through an interaction with the electric field gradient (EFG) at the nucleus. Quadrupolar nuclei include the abundant isotope of many cations of mineralogical interest (e.g., Al and the alkali metals) and the only NMR-active isotope of oxygen (^{17}O, 0.038% natural abundance). Depending on one's perspective, the quadrupolar interaction can be a nuisance to obtaining quantitative, high-resolution NMR spectra and measuring δ_i, or a sensitive probe of the local symmetry and structure. In the context of phase transformations we take the latter view and will not review recently-developed methods for removing or reducing broadening caused by strong quadrupolar interactions. Recent advances in high-resolution NMR of quadrupolar nuclei are reviewed in Fitzgerald (1999) and Smith and van Eck (1999), and more complete descriptions of quadrupole effects in solid-state NMR are presented by Taulelle (1990) and Freude and Haase (1993).

NMR of quadrupolar nuclei is complicated by the presence of several ($2I$) possible transitions that can display distinct frequency shifts from ν_0 (Fig. 1). Most quadrupolar nuclei of mineralogical interest have half-integer I for which it is useful to distinguish two main types of transitions: the "central transition" between the $m = 1/2$ and $m = -1/2$ orientations (denoted (1/2,-1/2)), and the satellite transitions (all others between levels $(m,m\text{-}1)$; only transitions between levels m and $m\pm1$ are usually observed). The frequency shifts depend on the product of the quadrupolar moment of the nucleus, eq (another fundamental nuclear property - the departure from a spherical charge distribution) and the electric field gradient (EFG) at the nucleus, V_{ik}, a traceless and symmetric second rank tensor. In its principal axis system the EFG can be characterized by the values $eQ = V_{zz}$, and $\eta = (V_{xx} - V_{yy})/V_{zz}$, because $V_{xx} + V_{yy} = -V_{zz}$. For describing NMR spectra, the parameters commonly used are the quadrupolar coupling constant $Cq = eq{\cdot}eQ/h$ (in frequency units, where h is Planck's constant) and the asymmetry parameter η, which represents the departure of the EFG from axial symmetry; $\eta = 0$ for sites of point symmetry **3** or higher. Typical values of Cq are from 1 to 10 MHz, although much lower and higher values are not uncommon. As discussed above for the CSA, the orientation and symmetry of the EFG is constrained by the point symmetry at the atom position. For example, cubic point symmetry (intersecting rotation axes 3-fold or higher) requires that $Cq = 0$.

There are several consequences of the quadrupolar interaction on NMR spectra that can be used to probe local structure and that must be considered when interpreting results.

First-order shifts and broadening of the satellite transitions. To first order, the quadrupolar interaction results in very large frequency shifts for the satellite transitions:

$$\nu = \nu_0 \cdot [1+\delta] + Cq\frac{3}{4I(2I-1)}(3\cos^2\theta - 1 + \eta\sin^2\theta\cos 2\phi)(m-\frac{1}{2}) \quad (3)$$

for the transition $(m,m\text{-}1)$, where θ is the angle between $\mathbf{B_0}$ and the principal axis of the EFG (that having value V_{zz}) and ϕ is the angle between the EFG y-axis and the projection of $\mathbf{B_0}$ onto the EFG x-y plane. Note that for the central transition (m = 1/2), the last term in Equation (3) disappears; it is not affected to first-order. Most solid-state NMR spectra of quadrupolar nuclei contain signal only from the central transition.

For a single crystal, the NMR spectrum contains $2I$ equally spaced peaks corresponding to the $(m,m\text{-}1)$ transitions (Fig. 4a), the positions of which depend on the orientation of the crystal in the magnetic field, as described by Equation (3). The EFG and its crystallographic orientation can be determined by measuring $\nu(\theta,\phi)$ for various orientations of the crystal in the magnetic field.

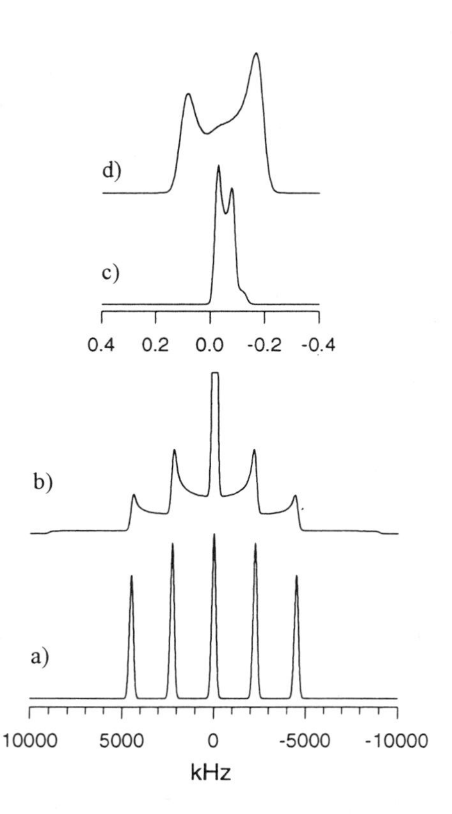

Figure 4. Simulated NMR spectra for a I = 5/2 quadrupolar nucleus with moderate quadrupolar interaction (ν_0 = 100 MHz, Cq = 3 MHz, η = 0), illustrating the effects of the quadrupolar interaction on the satellite and central transitions. (a) Single-crystal oriented at θ = 90° (Eqn. 3). The five peaks correspond to the five possible transitions between the levels (m,m-1) (Fig. 1). (b) Randomly oriented powder, calculated from Equation (3) for all possible values of θ by weighting the intensities proportional to sinθ. 3× vertical exaggeration to show detail in satellite transitions. (c) MAS-NMR and (d) static (without MAS) powder spectra of the central (1/2,-1/2) transition showing the peak shift and broadening due to second-order effects of the quadrupolar interaction. The full width of the powder pattern in (c) can be calculated from Equation 4. Frequency origin corresponds to the chemical shift (δ_i).

In a polycrystalline powder, all combinations of θ and ϕ are present. For each transition, the angular dependent term varies from 2 (θ = 0) to -1 ($\theta = \pi/2$), which spreads the intensity of the satellite transitions over a very wide frequency range, of the order of Cq (Fig. 4b). As a practical matter, the satellite transitions (and therefore much of the total signal intensity) are lost in the baseline and most NMR focuses on the central transition (Fig. 4c-d). In principal, the values of Cq and η can be determined from the width and shape of the full powder spectrum. But in most cases, commercial NMR spectrometers cannot measure the entire spectrum. However, a few exceptions are reviewed below.

In MAS-NMR spectra, the satellite transitions are usually not observed. Although Equation (3) indicates that MAS averages the orientation dependence, available spinning rates (up to about 20 kHz) are much smaller than the typical frequency spread of the individual satellite transitions (100s of kHz). As a result, the satellite transition intensity is distributed among sets of many spinning sidebands that approximately map the full frequency distribution of the powder pattern, which yields intensities for individual spinning sidebands that are very low compared to the central transition. In some cases (e.g. the ±(3/2,1/2) transitions for $I = 5/2$ nuclei, such as ^{27}Al and ^{17}O) these spinning sidebands are more narrow than the central transition and can provide additional information.

Frequency shifts of the central transition. Most solid-state NMR of quadrupolar nuclei concerns the central transition, which is not affected to first-order by the quadrupolar interaction (Eqn. 3, Fig. 1). However, for values of Cq typical for useful nuclei in minerals, second-order effects significantly broaden and shift the central transition peak from the chemical shift, making resolution of peaks from distinct sites difficult and complicating the measurement of δ_i. The second-order quadrupolar frequency shift also depends on orientation; it is more complicated than that of Equation (3) and given in a convenient form by Freude and Haase (1993). For a polycrystalline powder the quadrupolar interaction gives a "powder pattern" for the central transition (Fig. 4d), the width, shape, and position of which depend on Cq, η, and δ_i. MAS does not average the orientation dependence of the quadrupolar broadening from the central transition, but narrows the powder pattern by a factor of about 1/3 (Fig. 4c), giving a full width of:

$$\Delta\nu_{q,MAS} = \frac{9\,Cq^2}{56\,\nu_0}\frac{I(I+1)-\frac{3}{4}}{I^2(2I-1)^2}(1+\frac{\eta}{6})^2 \qquad (4)$$

in frequency units (multiply Eqn. 4 by $10^6/\nu_0$ for relative units, ppm). The dependence of $\Delta\nu_{q,MAS}$ on $1/\nu_0$ results in a decrease in peak widths with increasing B_0 and is a principal source of desire for larger B_0 among solid-state NMR spectroscopists. The MAS-NMR spectrum of the central transition also gives distinct peak shapes that can be fit with calculated lineshapes to obtain Cq, η, and δ_i (see Kirkpatrick 1988). In many cases it is possible to resolve distinct sites by MAS-NMR. New "multiple-quantum" techniques have recently become available (e.g. Baltisberger et al. 1996) that can fully remove the second-order quadrupolar broadening, although the experiments are non-trivial and require significant amounts of spectrometer time.

Dipole-dipole interactions

The direct dipole-dipole interactions between magnetic nuclei also introduce orientation-dependent frequency shifts, because the magnetic field at the nucleus depends slightly on the orientation of the magnetic moments of its neighbors. A nucleus "a" exerts a magnetic field on nucleus "b" with a component parallel to $\mathbf{B_0}$ of approximately:

$$B_{local} = m\frac{\gamma_a \hbar}{r^3}\frac{\mu_0}{4\pi}(3\cos^2\theta - 1) \qquad (5)$$

where γ_a is the magnetogyric ratio for nucleus "a" and m is its orientation (i.e. $m = \pm 1/2$ for $I = 1/2$), r is the internuclear distance, μ_0 is the vacuum permeability, and θ is the angle between the internuclear vector and $\mathbf{B_0}$. The observed NMR frequency for nucleus "b" is then shifted by an amount proportional to B_{local}: $\nu = \nu_0(1+\delta) + \gamma_b B_{local}/2\pi$. For most of the materials discussed below, the frequency shifts due to Equation (5) are small compared to those due to the CSA and quadrupolar interactions and have not been used to study phase transitions in minerals. The primary effect, especially for dilute nuclei with moderate to low values of γ, is a peak broadening that is averaged by MAS (cf. Eqn. 2). However, currently available MAS rates do not completely average strong dipolar coupling between like nuclei (homonuclear coupling). High-resolution NMR spectroscopy of phases with high concentrations of nuclei with large γ, such as ^1H in organic solids and ^{19}F in fluorides, requires considerable effort to remove the dipolar peak broadening.

The dipolar coupling can provide short-range structural information - one of the earliest applications was a determination of the H-H distance in gypsum, making use of the explicit dependence of Equation (5) on r^{-3} (Pake 1948). Dipolar coupling between nuclei can also be manipulated to transfer the large population differences for nuclei with large γ (e.g. ^1H) to those with smaller γ (e.g. ^{29}Si) to improve signal-to-noise ratio (the cross-polarization MAS, CP-MAS, experiment), or to qualitatively measure spatial proximity.

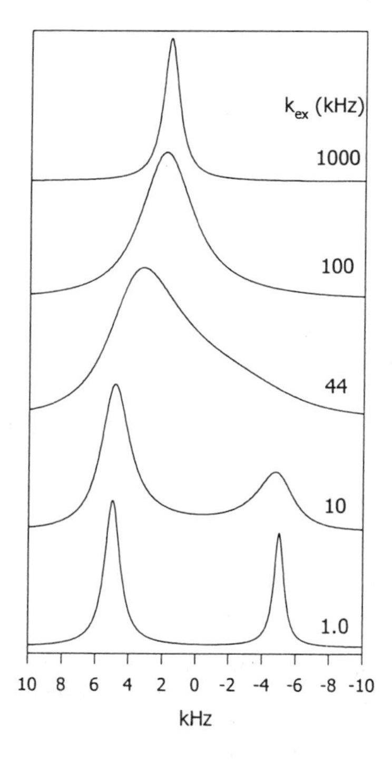

Figure 5. Simulated NMR spectra for an atom undergoing chemical exchange between two sites having relative populations 2:1 and frequency difference $\Delta\nu = 10$ kHz, as a function of the exchange rate $k_{ex} = 1/\tau$, where τ is the average residence time at a particular site, in seconds. No significant effect on the spectrum is observed for $k_{ex} << \Delta\nu$, whereas only a single peak at the weighted average position occurs for $k_{ex} >> \Delta\nu$.

Dynamical effects

NMR spectroscopy can provide information on the dynamics near phase transitions and can constrain the rates of motions responsible for dynamically disordered phases. In this respect, NMR complements most other spectroscopic techniques because it is sensitive to rates of processes that are much slower than thermal vibrations in crystals. Dynamical processes affect NMR spectra by changing the frequency of the nucleus during acquisition of the NMR spectrum. A frequency change can arise from changes in δ_i, for example due to movement of the atom from one site to another, or from a change in orientation that re-orients the CSA or EFG tensors with respect to $\mathbf{B_0}$ (i.e. change of θ in Eqns. 1 and 3). Such processes become apparent in NMR spectra when the rate of frequency change approaches the range of frequencies that the nucleus experiences (Fig. 5).

For example, consider exchange of an atom between two different sites (A and B) having chemical shifts $\delta_{i,A}$ and $\delta_{i,B}$, and therefore the frequency difference $\Delta\nu = \nu_0(\delta_{i,A} - \delta_{i,B})\cdot 10^{-6}$ (Hz), at a rate $k_{ex} = 1/\tau$, where τ is the average lifetime for the atom at either of the sites (Fig. 5). When $k_{ex} \approx \Delta\nu\cdot\pi\sqrt{2}$ (coalescence), the peaks are no longer resolved and the intensity is spread between the chemical shifts for sites A and B. In the rapid-exchange limit, $k_{ex} >> \Delta\nu\cdot\pi$, a single peak occurs at the weighted average frequency, $\langle\nu\rangle = \nu_0(f_A\delta_{i,A} + f_B\delta_{i,B})\cdot 10^{-6}$, where f_A is the fraction of time spent at site A.

The orientation dependence of the CSA and quadrupolar interactions are similarly affected by dynamical processes. For example, in fluids the orientation-dependence of Equations (1) and (3) is fully averaged by isotropic tumbling, because the tumbling rate (10^{10}-10^{12} s^{-1}; Boeré and Kidd, 1982), is much greater than the frequency spread due to CSA (~10^3-10^4 Hz) or quadrupolar coupling (~10^5-10^7 Hz). NMR peak shapes and their variation with the frequency of specific types of motion can be calculated and compared to observed spectra to evaluate the model and/or determine rates of motion. This type of study requires knowledge of the CSA or EFG in the limit of no motion and a specific model for how the motion affects the corresponding tensor orientations, i.e. the angle(s) between rotation axes and the principal axes of the tensor. Several examples are discussed below in which NMR results constrain the lifetimes of any ordered domains in dynamically disordered phases.

Relaxation rates

A property of NMR that has been used extensively to study the details of phase transition dynamics is the time required for the nuclei to establish an equilibrium population distribution among the energy levels, called "spin-lattice" relaxation and denoted by the characteristic time constant T_1. This relaxation time is also important to solid-state NMR in a practical sense, because once a spectrum is acquired one must wait until the nuclei have at least partially re-equilibrated before the spectrum can be acquired again, or fully re-equilibrated to obtain quantitatively correct intensity ratios. Most solid-state NMR spectra represent 100s to 1000s of co-added acquisitions to improve the signal-to-noise ratio.

Relaxation rates in minerals are often very slow, requiring seconds to hours for equilibration. These slow rates arise because transition from one energy level to another requires electrical or magnetic fields that vary at a rate of the order of ν_0 - 10^7 to 10^9 Hz. In most minerals, the activity at these frequencies is very low; thermal atomic vibrations are of the order of 10^{12}-10^{14} Hz. Under normal conditions, typical relaxation mechanisms in minerals are indirect (see Abragam 1961). Coupling to paramagnetic impurities can be an important relaxation mechanism for $I = 1/2$ nuclei, the unpaired electrons providing the needed fluctuating magnetic fields. For quadrupolar nuclei, time-varying EFGs due to thermal vibrations of atoms provide a relaxation mechanism, but the necessary frequency component is usually supplied by frequency differences between interacting vibrations (Raman processes) because of the lack vibrational modes with frequency near ν_0.

Near structural phase transitions, softening of lattice vibrations (displacive transitions) or slowing of rotational disorder modes (order-disorder transitions) can greatly increase the probability of fluctuations with frequency near ν_0 and dramatically increase the relaxation rate. NMR relaxometry has been applied extensively in the physics literature to study the critical dynamics of phase transitions in inorganic materials (see Rigamonti 1984), because it provides a probe of fluctuations at these low frequencies. In favorable cases, specific models for the transition mechanism can be tested by comparing model predictions against NMR relaxation rates measured as functions of temperature and ν_0. Such studies usually require a) pure crystals for which one relaxation mechanism dominates and can be identified; b) precise temperature control and data at temperatures very close to T_c. Although several studies of phase transitions in minerals note increased relaxation rates upon transition to a dynamically disordered phase (e.g. Spearing et al. 1992, Phillips et al. 1993), the data are insufficient to unambiguously distinguish between models for the types of motion. Measurements of NMR relaxation rates have great potential to add to our understanding of structural phase transitions in minerals.

Summary

Some of the specific properties of solid-state NMR spectroscopy for studying mineral transformations include:

Elemental specificity. For any given B_0, ν_0 varies substantially between different isotopes such that signal from only one isotope is present in an NMR spectrum.

Quantitative. NMR spectra can be obtained such that the signal intensities (peak areas) are proportional to the number of nuclei in those respective environments. Such an experiment requires some care to ensure uniform excitation of the spectrum and that the time between acquisitions is long enough (usually, several T_1's) to re-establish the equilibrium Boltzmann population distribution among the energy levels.

Low sensitivity. Low sensitivity arises primarily from the small value of ΔE compared with thermal energy near room-temperature, giving a small population difference between high- and low-energy levels. As a general rule, approximately 1 millimole of nuclei of the isotope of interest are needed in the sample volume (0.1 to 0.5 cm^3). Sensitivity is also reduced by the long times required to re-establish the population difference (relaxation time) after each signal acquisition, before the experiment can be repeated to improve the signal-to-noise ratio. This relaxation time can be long, requiring anywhere from tenths to hundreds of seconds. The total acquisition time for an NMR spectrum varies from a few minutes for abundant isotopes with moderate to high γ (e.g. ^{27}Al) to several hours or days for isotopes with I = 1/2 and low abundance (e.g, ^{29}Si). Also, measurement of broad powder patterns requires much more time than high-resolution MAS spectra.

Short-range structure. The electrical and magnetic interactions among nuclei and between nuclei and electrons introduce small shifts in frequency from ν_0 (a few parts per million, ppm) that can be related to structural parameters. These interactions are effective over distances of up to about 5 Å, making NMR useful as a probe of short-range structure.

Low-frequency dynamics. An "NMR timescale" is defined by the frequency shifts and linewidths which span the range 10^1 to 10^6 Hz, depending on the primary broadening interaction. The effects of dynamical processes on NMR spectra depends on the ratio of the rate of the process to the frequency shift caused by the motion. In absolute terms, these frequency shifts and linewidths are much smaller than other commonly used spectroscopies (IR, Raman, Mössbauer, X-ray), atomic vibrations, and even relatively low-frequency vibrations such as rigid-unit modes.

STRUCTURAL PHASE TRANSITIONS

This section reviews NMR studies of phase transitions in minerals with the goal of illustrating the types of information that can be obtained from NMR spectroscopic data. Most of the studies discussed below do not bear directly on the transition mechanism, but, rather, focus on the temperature dependence of the structure above and below the transition and on the nature of the disorder (static or dynamical) in apparently disordered phases. Information on the transition mechanism requires data very near the transition temperature, whereas precise temperature control is difficult with typical commercial MAS-NMR equipment. This section concentrates on interpretation and structural/dynamical implications of NMR data. For more detailed descriptions of the transitions and the relationship of the NMR results to other experimental techniques, the reader should consult the original papers, or other chapters in this volume.

α-β transition in cristobalite

Cristobalite, the high-temperature polymorph of SiO_2, undergoes a reversible first-order transition $\alpha \leftrightarrow \beta$ near 500 K (see Heaney 1994). The average structure of the high-temperature phase (β) is cubic, but it has been recognized for some time that this phase is probably disordered, because an ordered cubic structure would contain 180° Si-O-Si angles and short Si-O distances (oxygen position "e" in Fig. 6). Structure refinements based on X-ray data (e.g. Wright and Leadbetter, 1975) are improved by distributing the oxygen atom along an annulus (continuously, or in discrete positions) normal to the Si-Si vector (Fig. 6). Debate has centered on whether this disorder is static, comprising very small twin domains of the α-phase, or dynamical, corresponding to motion of the oxygen about the Si-Si vector. Furthermore, if the disorder is dynamical, can it be described as re-orientations of nanoscopic twin domains of α-like symmetry, corresponding to correlated motions of groups of oxygens (Hatch and Ghose 1991), or do the oxygens move more-or-less independently as suggested by lattice dynamical models (Swainson and Dove 1993)? Similar questions arise for a number of minerals and the studies of cristobalite offer a good starting point for illustrating the potential of NMR spectroscopy.

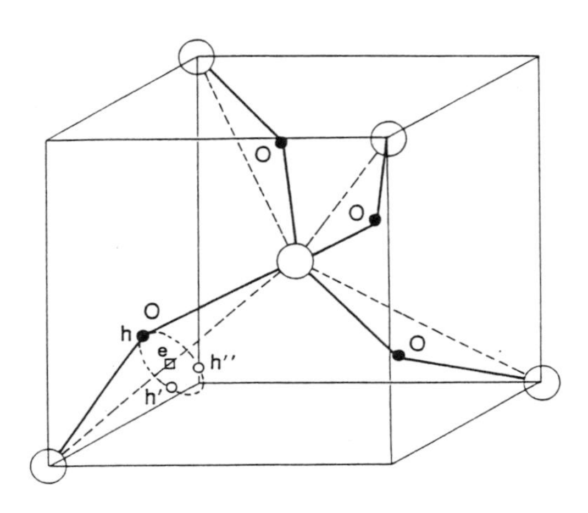

Figure 6. Local structure and disorder in the oxygen positions of β-cristobalite. Large circles correspond to the cation positions, Si for SiO_2, Al at the center and P at the corners for $AlPO_4$ cristobalite. Small circles correspond to oxygen positions for one of the α-like orientational variants. Positions marked "h" are the refined split-atom oxygen positions (occupancy 1/3 each) for $AlPO_4$ cristobalite; those for SiO_2 are similarly distributed, but there are 6 sub-positions. Occupancy of the position marked "e" would correspond to the ordered cubic structure. [Modified after Wright and Leadbetter (1975), Fig. 1, p. 1396.]

From its sensitivity to short-range structure and low-frequency dynamics, NMR spectroscopy provides some constraints on the nature of the β-cristobalite structure. For cristobalite, only powder NMR techniques can be applied, because large single crystals do not survive the α-β transition intact. It is helpful to consider also $AlPO_4$-cristobalite, because it is similar in essential respects to SiO_2 and the cubic point symmetry of the Al-position in the average structure of the β-phase provides some additional constraints.

T-O-T angles from chemical shifts. High-resolution MAS-NMR techniques were applied to the cristobalite phase of both SiO_2 (^{29}Si; Spearing et al. 1992) and $AlPO_4$ (^{27}Al and ^{31}P; Phillips et al. 1993), providing isotropic chemical shifts (δ_i) as a function of temperature through the α-β transition in both materials. Data for all three nuclides show a gradual decrease of δ_i with increasing temperature for the α-phase, and a discontinuous decrease of about 3 ppm at the α-β transition (Fig. 7). The chemical shifts do not change significantly with temperature in the β-phase. Over a span of 15-20°, spectra contain peaks for both phases due to a spread of transition temperatures in the powders, consistent with observations by other techniques.

These spectra can be interpreted in terms of correlations between δ_i and T-O-T bond angles (T = tetrahedral cation) that have been reported for these nuclides in framework structures. Over the range 120-150°, the bridging bond angle correlates approximately

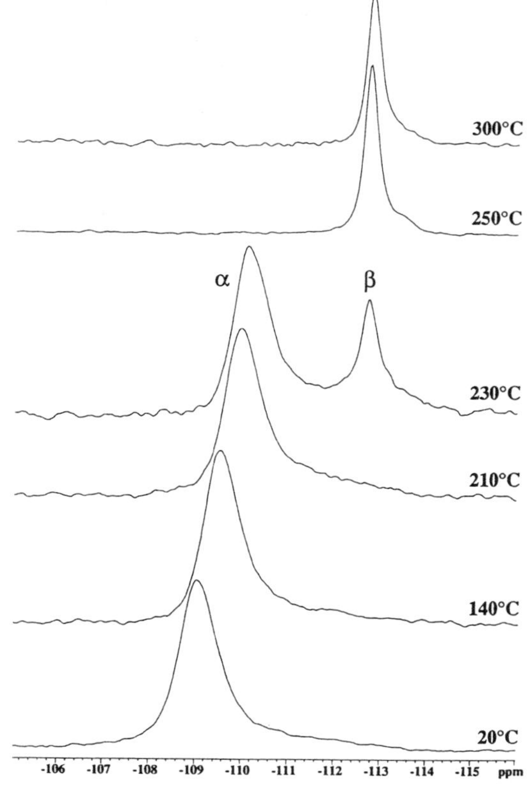

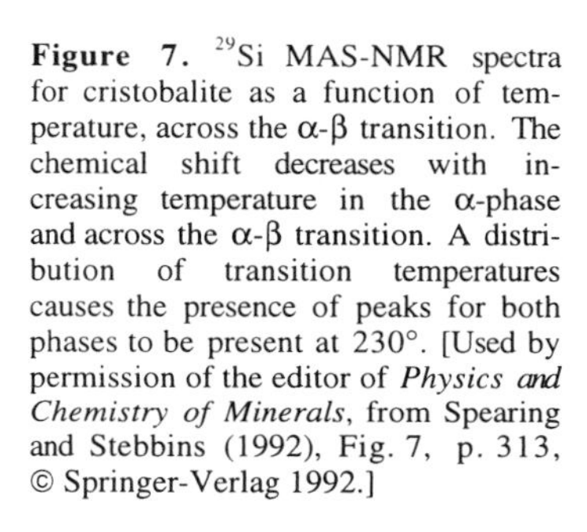

Figure 7. ^{29}Si MAS-NMR spectra for cristobalite as a function of temperature, across the α-β transition. The chemical shift decreases with increasing temperature in the α-phase and across the α-β transition. A distribution of transition temperatures causes the presence of peaks for both phases to be present at 230°. [Used by permission of the editor of *Physics and Chemistry of Minerals*, from Spearing and Stebbins (1992), Fig. 7, p. 313, © Springer-Verlag 1992.]

linearly with δ_i, giving a slope of about -0.5 ppm/degree (Fig. 8). At larger angles, however, the more complicated dependence is expected, as shown in Figure 8 for ^{29}Si in SiO_2 polymorphs. Essentially similar correlations are found for ^{27}Al and ^{31}P in framework aluminophosphates (Müller et al. 1989).

The chemical shift data for the cristobalite forms of both SiO_2 and $AlPO_4$ indicate that the average T-O-T angle increases gradually with temperature through the α-phase and that a further discontinuous increase of about 5° occurs upon transition to the β-phase. These NMR results suggest an average T-O-T angle for

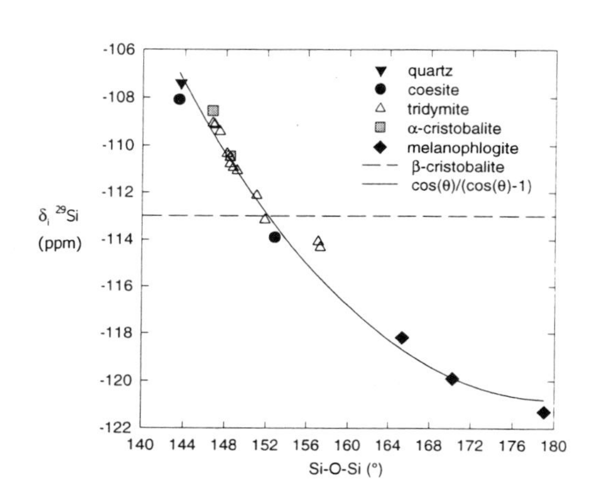

Figure 8. Variation of ^{29}Si NMR chemical shifts for SiO_2 polymorphs with the average Si-O-Si angle. The solid line corresponds to angular dependence predicted by quantum chemical calculations. Dotted line corresponds to δ_i measured for β-cristobalite (Fig. 7), from which an average Si-O-Si angle of 152° can be inferred.

β-cristobalite of about 152° (Fig. 8), much smaller than the 180° required for an ordered cubic structure, and provide direct evidence for a disordered β-cristobalite structure. Note that these NMR data refer to the T-O-T angle averaged over time, not the angle between mean atom positions, which can differ in the presence of large correlated motions of the oxygen. The NMR chemical shift data indicate also that a significant structural change occurs at the cristobalite α↔β transition, corresponding to an increase in the T-O-T angle; the transition to the β-phase requires more than a disordering of α-like domains. Finally, the results obtained for ^{29}Si in SiO_2 (Spearing et al. 1992) and ^{27}Al and ^{31}P in $AlPO_4$ (Phillips et al. 1993) are identical with respect to implications for changes in T-O-T angle. Apparently, at a local level the transition is not affected strongly by the lower symmetry of $AlPO_4$ that arises from ordering of Al and P on the tetrahedral sites. This observation suggests that the additional constraints provided by changes in the ^{27}Al Cq, reviewed below, might also apply to SiO_2-cristobalite.

Dynamical constraints from ^{17}O NMR. Spearing et al. (1992) also obtained ^{17}O NMR data for SiO_2-cristobalite across the α↔β transition. To obtain useful NMR signal levels requires synthesis of a sample isotopically enriched in ^{17}O, a quadrupolar nucleus (I = 5/2) with low natural abundance (0.037%). Because there is only one crystallographic position for oxygen in both the α and β phases, structural information could be obtained from low-resolution powder techniques, without MAS. The ^{17}O NMR spectra (Fig. 9) contain a broad peak corresponding to the powder pattern for the central transition

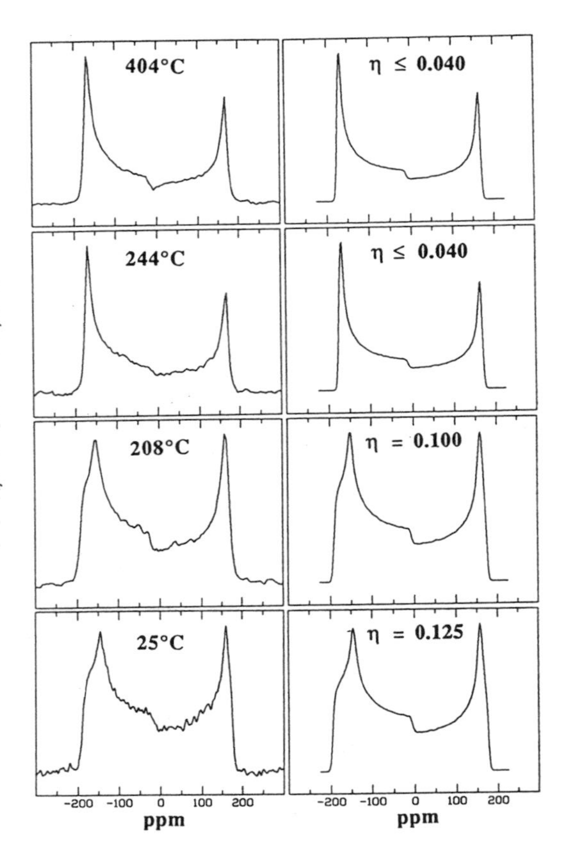

Figure 9. NMR spectra (left) and simulations (right, with corresponding value of the EFG asymmetry, η) of the ^{17}O central transition for SiO_2-cristobalite at the temperatures indicated (without MAS; ν_0 = 54.2 MHz). Spectra at temperatures above the α-β transition are consistent with η = 0, indicating average axial symmetry. [Used by permission of the editor of *Physics and Chemistry of Minerals*, from Spearing and Stebbins (1992), Fig. 11, p. 315, © Springer-Verlag.]

(1/2,-1/2) with a shape characteristic of a strong quadrupolar interaction: Cq = 5.3 MHz and asymmetry parameter η = 0.125 near 298 K (cf. Fig. 4d). With increasing temperature, slight changes occur in the shape of the spectrum that indicate a decrease in η through the α-phase. In the β-phase, the spectrum indicates η = 0, although an upper limit of η = 0.05 was given due to experimental uncertainty.

These changes in the EFG at the ^{17}O nucleus can be interpreted structurally on the basis of theoretical and experimental studies that show correlations of η with the Si-O-Si angle (θ) in SiO_2 polymorphs that vary approximately as $\eta = 1 + \cos(\theta)$ (Grandinetti et al. 1995; Tossell and Lazzeretti 1988). In the α-phase, decreasing η with increasing temperature is consistent with an increasing Si-O-Si angle, as inferred also from the ^{29}Si δ_i data. However, the apparent axial symmetry of the ^{17}O EFG in the β-phase appears at odds with the average Si-O-Si angle of about 152° (Fig. 8), which would correspond to $\eta \approx 0.1$. This difference highlights one of the essential features of the relatively long timescale of NMR spectroscopy. The η measured for ^{17}O represents the EFG averaged over a time on the order of the reciprocal linewidth, which in this case is approximately 1/(21.5 kHz) = 4.7×10^{-5} s (full width 400 ppm, ν_0 = 54.2 MHz for this study). The structural interpretations of the ^{29}Si δ_i and ^{17}O EFG data can be reconciled by axially symmetric motion of the oxygen that retains a time-averaged Si-O-Si angle near 152°, such as along the circle or between the positions marked "h" shown in Figure 6.

Although these ^{17}O NMR data clearly support a dynamical model for disorder in β-cristobalite, they are not sensitive to whether the motions of adjacent oxygens are correlated (as required for a model of re-orienting twin domains), or, whether the motion is continuous or a hopping between discrete positions; they indicate only that the path of each oxygen traces a pattern with 3-fold or higher symmetry over times of the order $4.7 \cdot 10^{-5}$ s. Thus, these results cannot discriminate between models based on RUM's or dynamical twin domains, and place only a lower limit on the timescale of the motions. A tighter restriction on the motions can be obtained from ^{27}Al NMR data for $AlPO_4$ cristobalite.

Dynamical constraints from ^{27}Al NMR. The ^{27}Al nucleus is also quadrupolar (I = 5/2) but is 100% abundant and thus gives a very strong NMR signal. $AlPO_4$-cristobalite contains only one crystallographic position for Al, which is tetrahedrally coordinated and shares oxygens with only $P[O]_4$ tetrahedra (Al at the center and P at the corners in Fig. 6). In the average structure of the β-phase, the cubic point symmetry of the Al position ($\bar{4}3m$)constrains the EFG to Cq = 0, whereas the point symmetry in the α-phase (*2*) does not constrain the magnitude of the EFG. The full width of the ^{27}Al NMR spectrum is much greater than for the ^{17}O central transition discussed above, and places a correspondingly higher constraint on re-orientation frequencies for dynamical models of β-cristobalite.

Low-resolution powder techniques were used to measure the EFG at the ^{27}Al nucleus as a function of temperature through the $\alpha \rightarrow \beta$ transition (Phillips et al. 1993). In the α-phase, partial powder patterns for the ±(3/2,1/2) satellite transitions were obtained (similar to Fig. 4b) that showed a slight decrease of Cq, from 1.2 to 0.94 MHz, with increasing temperature. Although no simple structural correlation exists for the EFG, the relative change is similar in magnitude to those for the chemical shifts, the ^{17}O EFG, and other structural parameters as reflected in the order parameter for SiO_2 cristobalite (Schmahl et al. 1992). Changes in the ^{27}Al NMR signal across the $\alpha \rightarrow \beta$ transition (Fig. 10) are consistent with cubic symmetry in β-cristobalite: the satellite transitions collapse and the intensity of the centerband increases by a factor of approximately four, consistent with the presence of the ±(3/2,1/2) and ±(5/2,3/2) satellite transitions in the centerband of the β-phase, indicating $Cq \approx 0$.

A small complication arises from the presence of crystal defects, which at normal concentrations result in a small residual EFG such that the average Cq is not identically 0. Quadrupolar nuclei in all nominally cubic materials experience small EFGs due to charged defects (e.g. Abragam, p 237-241) and that observed for $AlPO_4$ β-cristobalite is smaller than typically observed for crystals such as alkali halides (^{23}Na in NaCl, ^{79}Br in KBr, ^{129}I in KI) and cubic metals (e.g. ^{63}Cu in elemental Cu). There is some confusion about this aspect (Hatch et al. 1994), but the data for $AlPO_4$ β-cristobalite are fully consistent with cubic site symmetry with defects fixed in space with respect to the ^{27}Al nuclei. Further careful measurements of the nature of the residual broadening indicate that it results from a static distribution of frequencies, as expected for defects, rather than an incomplete dynamical averaging (Phillips et al. 1993).

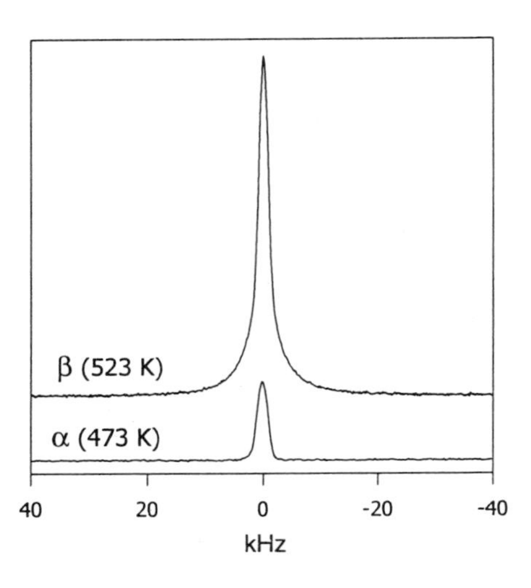

Figure 10. ^{27}Al NMR spectra (without MAS) of $AlPO_4$ cristobalite taken just below (473 K) and above (523 K) the α-β transition, with y-axis scaling proportional to absolute intensity. Spectrum of the α-phase contains only the central transition, whereas that for the β-phase contains also the satellite transitions, indicating Cq = 0 (cf. Fig. 4) and consistent with cubic point symmetry. For the α-phase, the central transition comprises 9/35 of the total intensity; the remainder occurs in the satellite transition powder spectrum that spans a wide frequency range according to Equation (3), with Cq = 0.9 MHz. Origin of the frequency scale is arbitrary. [Redrawn from data of Phillips et al. (1993).]

The implications for these ^{27}Al NMR results are similar to those discussed above for ^{17}O, but further reduce the timescale of the motions. Axial motion of the oxygens about the Al-P vector, as described above for SiO_2, can give an average Al-O-P angle <180°, but apparent cubic symmetry at the Al site, if the motions are fast enough. The timescale for the motions was tested with a model comprising re-orienting twin domains of α-like symmetry (Fig. 11), but the results apply equally to uncorrelated motion of the oxygens. A powder spectrum of the ±(3/2,5/2) satellite transitions (those with the broadest frequency distribution, Eqn. 3) was calculated assuming that the crystals re-orient, with respect to a fixed B_0, between the twelve possible twin- and anti-phase domains of the α-phase at a rate $k = 1/\tau$, where τ is the average time spent in any one orientation (Fig. 11). A limiting, instantaneous value of $Cq = 0.6$ MHz was estimated by extrapolating a correlation of δ_i with Cq for the α-phase across the transition. The simulations show that reduction of the linewidth to that observed for the β-phase (cf. Figs. 10 and 11) requires re-orientation frequencies (k) greater than 1 MHz (i.e. a lifetime for any ordered domain in one orientation less than 10^{-6} s). These ^{27}Al NMR data are fully consistent also with a RUM model (Dove, this volume), in which rotations of the tetrahedra are relatively low in frequency for lattice modes, but at 10^{12} Hz are much greater than needed to fully average the EFG at the Al site (Swainson and Dove 1993).

α-β quartz

Quartz (SiO_2) undergoes a series of reversible structural phase transitions near 570°C from the α-phase at low temperatures to the β-phase, via an intervening incommensurate phase (INC) stable over about 1.8° (e.g. Heaney 1994). The transition from the α- to the

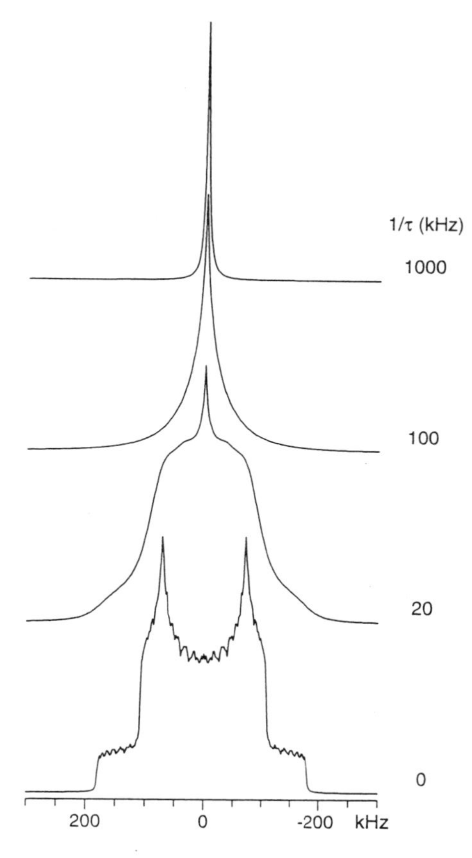

Figure 11. Simulated spectra for the outer [±(3/2,5/2)] satellite transitions for ^{27}Al in $AlPO_4$ β-cristobalite for a model of re-orienting twin- and anti-phase domains of α-like symmetry. Spectral width for the ordered domains (bottom, corresponding to Cq = 0.6 MHz) was obtained by extrapolation from its temperature dependence in the α-phase. Rapid fluctuation of domain orientations can produce average cubic symmetry at the Al-site (giving averaged Cq = 0), corresponding locally to a distribution of the oxygens over the sub-positions "h" in Figure 6, but only if the re-orientation frequency $1/\tau > 1$ MHz, where τ is the average lifetime of a particular configuration (in seconds).

β-phase corresponds to addition of a 2-fold symmetry axis parallel to *c* (e.g. space groups $P3_221$ (α) to $P6_222$ (β), and their enantiomorphs) that also relates the Dauphiné twin orientations of the low temperature phase. To the SiO_4 tetrahedra, the transition corresponds to a rotation about the ⟨100⟩ axes such that a line connecting the Si atom and the bisector of the O-O edge, which corresponds to a 2-fold axis in the β-phase, rotates away from *c* by an angle θ (Fig. 12). The Dauphiné twin orientations correspond to rotations of the opposite sense (i.e. θ and -θ). The nature of β-quartz—whether an ordered phase with relatively small displacement of the oxygens from their position in the average structure, or a disordered structure that can be represented by a space and/or time average of the two Dauphiné twin domains—remains a matter of debate. The observation by TEM of increased density of Dauphiné twin domain boundaries and their spontaneous movement near the transition temperature suggests re-orienting domains of α-like symmetry (e.g. van Tendeloo et al. 1976, Heaney and Veblen 1991). However, Raman spectroscopic data indicates the absence of certain vibrational modes expected for α-like clusters (Salje et al. 1992). More recent work (Dove, this volume) suggests the dynamical properties of β-quartz correspond to RUMs, which are relatively low frequency rotations of the SiO_4 tetrahedra.

Spearing et al. (1993) present ^{29}Si NMR spectra for quartz from 25 to 693°C, across the α-β transition, using single-crystal techniques (Fig. 13). Based on an earlier determination of the ^{29}Si CSA (Spearing and Stebbins 1989), they chose an orientation of the crystal that gave three NMR peaks separated by about 2.5 ppm near 25°C, corresponding to an angle of 150° between c and B_0 in Figure 3. In α-quartz, the crystallographically equivalent Si sites related by three-fold screw axes differ in their CSA orientation with respect to B_0, because the point symmetry of the Si site (*1*) does not constrain the orientation of the CSA with respect to the symmetry axis (analogous to Fig. 2). For most crystal orientations these different CSA orientations give slightly different peak positions according to Equation (1) and the parameters given by Spearing and Stebbins (1989). From a local

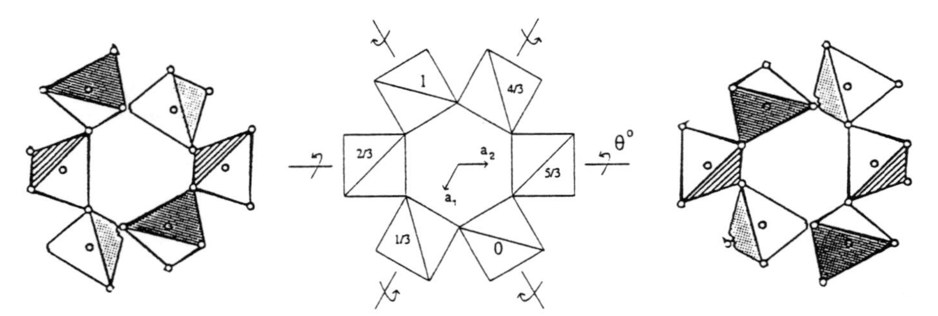

Figure 12. Polyhedral representation of a fragment of the quartz structure showing the local relationship of the two Dauphiné twin orientations of the α-phase (left and right) to the average structure of β-quartz (center). The orientation of the $Si[O]_4$ tetrahedra in α-quartz results from a rotation about the **a**-axes from the β-quartz structure ($\theta = 16.5°$ at 25°C). Dauphiné twin domains are related by rotations of opposite sense. [Modified after Heaney (1994), Fig. 3, p. 8 and Fig. 5, p. 11.]

perspective, exchange of Dauphiné twin domain orientations corresponds to a change in the angle between c and B_0 in Figure 3 from 30° to 150°, hence a change in the observed ^{29}Si chemical shift for those Si giving a peak at -102.7 ppm to -107.7 ppm, and vice versa.

With increasing temperature, the two outermost peaks gradually converge with the middle peak. At 693°C only a single peak is observed, consistent with the average symmetry of the β-phase. In β-quartz, the Si position has point symmetry *2*, which requires the CSA orientations of the equivalent Si sites (related by 3-fold screw

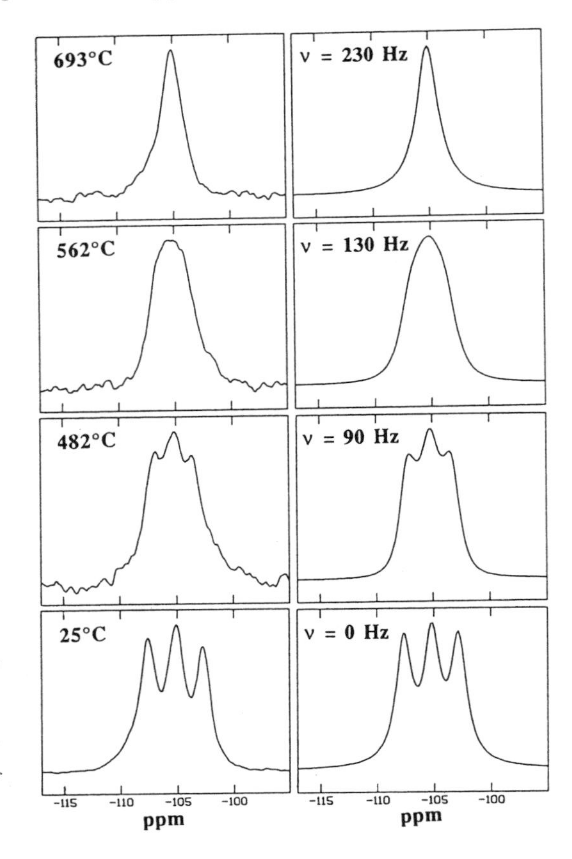

Figure 13. Single-crystal ^{29}Si NMR spectra (taken at $\nu_0 = 79.6$ MHz) for quartz at the temperatures indicated, above and below the α-β transition (573°C). For α-quartz, the Si positions related by the 3-fold screw axis (see Fig. 12) are magnetically inequivalent in this orientation, which corresponds to an angle 150° in Figure 3, giving three peaks. Presence of a 2-fold axis at the Si position, parallel to the 3-fold screw axes, in β-quartz makes all Si-positions magnetically equivalent in this orientation. Spectra on right are simulations assuming convergence of the outer peaks is due solely to dynamical alternation of Dauphiné twin domain orientations at the indicated frequency (i.e., change of angle between 30° and 150° in Fig. 3B). [Used by permission; Spearing and Stebbins (1992), *Physics and Chemistry of Minerals*, Fig. 16, p. 313, © Springer-Verlag 1992.]

axes that parallel the point symmetry axis) to be the same.

The changes in the ^{29}Si NMR spectra are fit equally well with two physical models. Because exchange of Dauphiné twin orientations causes Si nuclei giving rise to one of the outer peaks to change to the frequency corresponding to the other outer peak, Spearing et al. (1992) fit these data to a model of re-orienting Dauphiné twin domains. To fit the data solely with a dynamical model, the average re-orientation frequency increases from 90 Hz at 482°C to 230 Hz at 693°C, assuming that the static peak positions are the same as at 25°C. The spectra can be fit equally well with three peaks of equal intensity, the separation of which decreases with temperature, corresponding to a gradual movement of the two outer curves in Figure 3 towards the middle curve. Physically, the peak convergence should correlate with the change in the absolute rotation angle ($|\theta|$) of the SiO_4 tetrahedra in the α-phase (Fig. 12), which decreases from 16.3° near room temperature to 8.5° just below the transition temperature, and is 0° in the β-phase.

The spectra for β-quartz, containing only one peak, are consistent with both ordered and disordered models for its structure. Because the peak separations for ordered domains in the α-phase are small ($\Delta\nu \approx 400$ Hz; 5 ppm, $\nu_0 = 79.46$ MHz), the frequency of domain re-orientation required to average the differences in CSA orientation are correspondingly quite low. The data of Spearing et al. (1992) place a lower limit on the lifetimes of any α-like domains in the β-phase of 2.5 milliseconds (1/400 Hz), assuming these domains have the structure of the α-phase at 25°C. Static structural changes with increasing temperature would lengthen this lower limit even more. Further information might be obtained from single-crystal ^{17}O NMR, because the absolute frequency differences between the Dauphiné twin domains due to the quadrupolar interaction is likely to be much larger than for the ^{29}Si CSA. Such a study, however, would require a crystal significantly enriched in ^{17}O.

Cryolite (Na_3AlF_6)

The mineral cryolite (Na_3AlF_6), a mixed fluoride perovskite, undergoes a reversible structural transition near 550°C between a pseudo-tetragonal orthorhombic phase (β) stable at high temperatures to a monoclinic structure (α) (Yang et al. 1993). Many perovskite-type compounds exhibit a sequence of phase transitions cubic ↔ tetragonal ↔ orthorhombic that can be related to rotations of the polyhedra (Fig. 14) and does not involve a change in the number of crystallographic sites. Although both Na and Al occupy octahedral sites in cryolite (corresponding to a perovskite structural formula $Na_2[NaAl]F_6$), they alternate such that the topology is compatible with cubic symmetry. Phase transitions in perovskites have been extensively studied by NMR spectroscopy (e.g. Rigamonti 1984), because most have simple structural chemistry, cubic symmetry in the high-temperature phase, moderate transition temperatures, and transition mechanisms that appear to span purely displacive to dynamical order-disorder (e.g. Armstrong 1989). Cryolite is interesting also because it provides an example of a "lattice melting" transition in which the F (and Na) atoms become mobile at relatively low temperatures, well below the α-β structural transition. All three components of cryolite have sensitive NMR-active nuclei (^{27}Al, ^{23}Na, ^{19}F), which Spearing et al. (1994) studied as a function of temperature through both the lattice melting and structural transitions.

Near 25°C, the EFG at the Al site is small enough ($Cq = 0.6$ MHz) that the entire powder pattern for the ^{27}Al satellite transitions can be observed (Fig. 15; cf. Fig. 4b). With increasing temperature, the full width of the ^{27}Al NMR spectrum decreases, corresponding to a decrease of Cq that is approximately linear with temperature, to $Cq \approx 0.2$ MHz at 531°C, just below T_c. At 629°C, above the monoclinic-to-

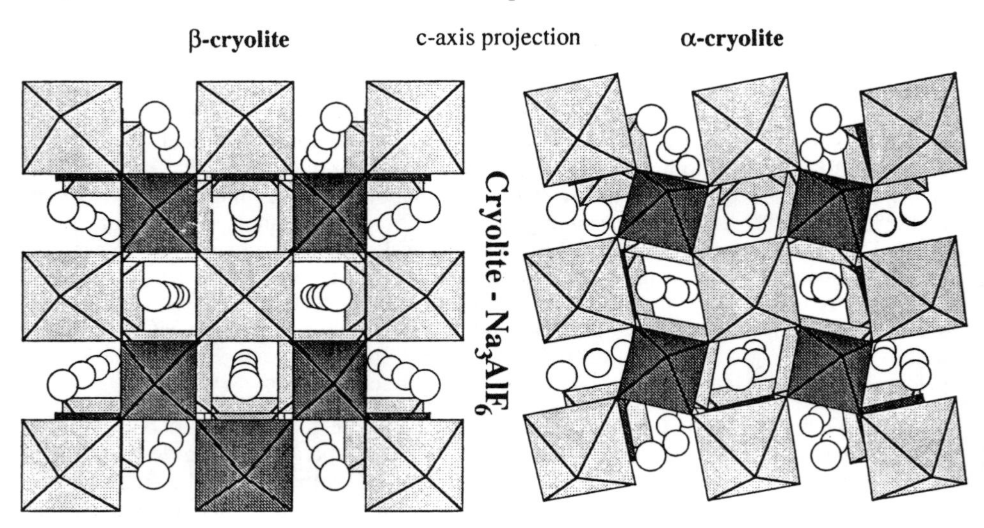

Figure 14. Polyhedral representation of the cryolite structure ($Na_2(NaAl)F_6$ perovskite). Lightly shaded octahedra are $Al[F]_6$, dark octahedra are $Na[F]_6$, circles are 8-coordinated Na. The pseudo-tetragonal β-phase is orthorhombic (*Immm*). [Used by permission of the editor of *Physics and Chemistry of Minerals*, from Spearing et al. (1994), Fig. 1, p. 374, © Springer-Verlag 1994.]

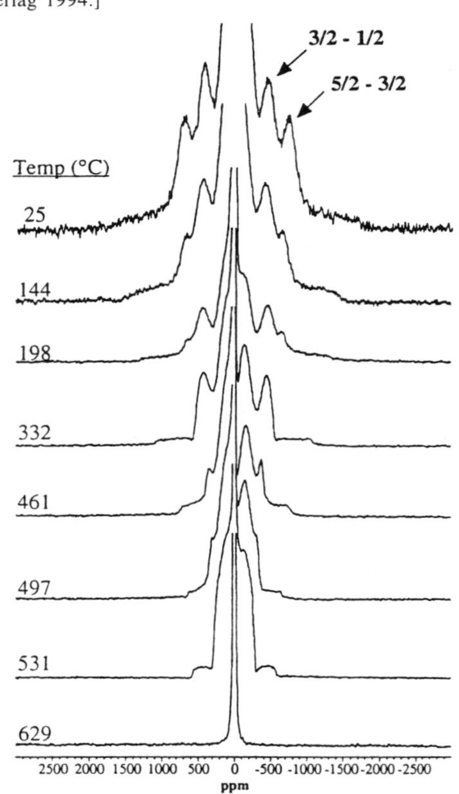

Figure 15. ^{27}Al NMR spectra (taken at ν_0 = 104.2 MHz, without MAS) of cryolite polycrystalline powder, with vertical exaggeration to emphasize the ±(3/2,1/2) and ±(5/2,3/2) satellite transitions. Decrease of spectral width with increasing temperature corresponds to decrease of Cq (Eqn. 3). In the β-phase all the transitions occur in the narrow center band, indicating $Cq = 0$ and average cubic symmetry at the Al-position. [Used by permission of the editor of *Physics and Chemistry of Minerals*, from Spearing et al. (1994), Fig. 6, p. 378, © Springer-Verlag 1994.]

orthorhombic transition, the static ^{27}Al NMR spectrum contains only a single, narrow peak that contains all of the transitions ($Cq = 0$), which would be consistent with cubic point symmetry. These data were not fit to a specific model for the structural changes, but are consistent with a gradual rotation of the AlF_6 octahedron with increasing temperature in the α-phase toward its orientation in an idealized cubic phase. Dynamical rotations of the AlF_6 octahedra at temperatures below the α-β transitions are more likely to cause sudden collapse of the satellite powder pattern upon reaching a frequency of the order of the full linewidth (in Hz), as discussed above for $AlPO_4$ cristobalite (cf. Fig. 11). The apparent cubic point symmetry of the Al-site in the β-phase, combined with the orthorhombic (pseudo-tetragonal) X-ray structure, strongly suggests that the orthorhombic domains re-orient dynamically (e.g. corresponding to exchange of crystallographic axes) at a rate that is rapid on the NMR timescale. An estimate for the maximum lifetime of ordered orthorhombic domains can be obtained as the inverse of the full width of the ^{27}Al satellite transitions just below the transition temperature: 8 μs = 1/(125000 Hz) (from 1200 ppm full width, ν_0 = 104.2 MHz).

Interpretation of these ^{27}Al spectra might be affected by a partial averaging of the EFG due to Na exchange between the 6- and 8-coordinated sites and motion of the F ions at temperatures well below the α-β structural transition. Near 25°C, the ^{23}Na MAS-NMR peaks for the 6- and 8-coordinated sites are well-resolved (Fig. 16), separated by about 12 ppm, or $\Delta\nu \approx 1300$ Hz (ν_0 = 105.8 MHz). With increasing temperature, the peaks

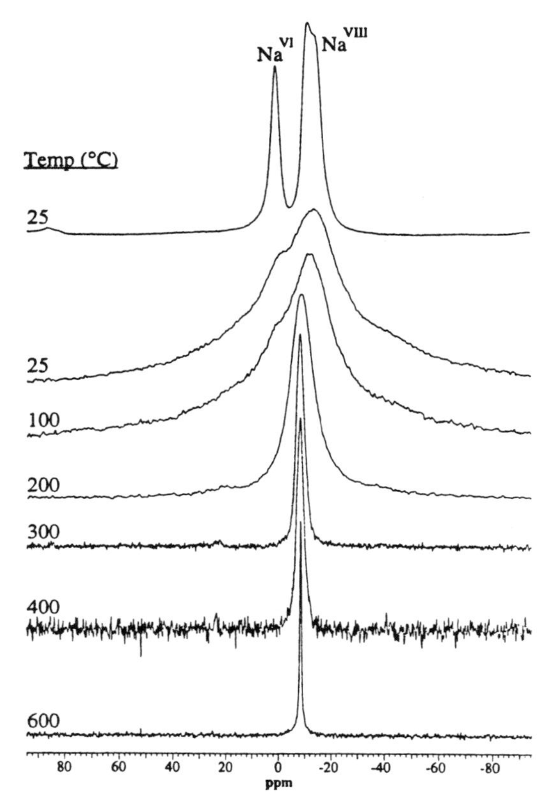

Figure 16. ^{23}Na MAS-NMR spectra (taken at ν_0 = 105.8 MHz) of cryolite. Upper spectrum was taken at a spinning rate of 10.2 kHz, whereas the spinning rate of the lower spectra (3 kHz) was insufficient to remove dipolar coupling to ^{19}F, giving poorer resolution. Spectra taken above 200° show rapid exchange of Na between 6- and 8-coordinated sites (cf. Fig. 5). [Used by permission of the editor of *Physics and Chemistry of Minerals*, from Spearing et al. (1994), Fig. 3, p. 376, © Springer-Verlag 1994.]

coalesce and by 300°C only a single, narrow peak is observed, indicating chemical exchange of Na between 6- and 8-coordinated sites at rate $k > 10 \cdot \Delta\nu$, or 13000 s^{-1} (cf. Fig. 5). A motional narrowing is also observed for static ^{19}F NMR powder spectra below 150°C, but could not be fully analyzed because the source of the peak broadening is uncertain.

Order parameters: The $P\bar{1}$-$I\bar{1}$ transition in anorthite ($CaAl_2Si_2O_8$)

Of the several structural phase transitions that occur in feldspars (Carpenter 1994), the $P\bar{1} - I\bar{1}$ transition of anorthite ($CaAl_2Si_2O_8$) occurs at a relatively low temperature ($T_c \approx 510$ K) and is easily accessible by MAS-NMR techniques. This transition appears to be mostly displacive (Redfern and Salje 1992); loss of the body-centering translation with decreasing temperature splits each crystallographic position of the $I\bar{1}$ phase (four each for Si and Al) into two positions related by translation of ~(1/2,1/2,1/2). Anti-phase domain boundaries (c-type) separate regions with opposite sense of distortion, which are offset by the (1/2,1/2,1/2) translation. Several models for the transition mechanism and the nature of the high-temperature phase of anorthite have been proposed. For example, TEM observations of mobile c-type domain boundaries near T_c (van Tendeloo et al. 1989) have been interpreted as evidence for a disordered $I\bar{1}$ structure, comprising a space- and time-average of the $P\bar{1}$ anti-phase domain orientations. However, the feldspar framework can also exhibit rigid-unit-modes (see Dove, Chapter 1 in this volume).

In an early study, Staehli and Brinkmann (1974) used single-crystal NMR techniques to follow the ^{27}Al central transitions as a function of temperature. In favorable crystal orientations, peaks for each of the eight crystallographically distinct Al sites in the $P\bar{1}$ phase can be resolved at $T < T_c$. The central transitions could be resolved because of a wide variation in the magnitude and orientation of the EFGs, and the small values of B_0 used. For single crystals, resolution is better at low B_0 because the second-order quadrupolar shifts of the central transition (in Hz) are inversely proportional to B_0. At temperatures above T_c, only four peaks were observed, consistent with the average $I\bar{1}$ structure. These data provided strong evidence against static disorder models for the structure of $I\bar{1}$ anorthite. With increasing temperature through the $P\bar{1}$ phase, the peak positions change systematically such that pairs of peaks appear to converge. These data aided peak assignments, because the converging pairs of peaks likely correspond to sites that become equivalent in the $I\bar{1}$ phase. Just below T_c Staehli and Brinkmann report separations between converging pairs of peaks as large as 17 kHz. This observation constrains the lifetime for any domains of $P\bar{1}$ symmetry above T_c to be less than about $\tau < 1/(17{,}000$ Hz), or ~60 μs.

Essentially similar results were obtained for the Si sites using ^{29}Si MAS-NMR techniques (Phillips and Kirkpatrick 1995). In this study the separation between converging pairs of peaks in the $P\bar{1}$ phase could be related to an order parameter (Q) for a Landau-type analysis. Near 298 K, ^{29}Si MAS-NMR spectra of Si,Al ordered anorthite contain six peaks for the eight inequivalent crystallographic Si sites, because some of the crystallographic sites give nearly the same chemical shift (Fig. 17c,d). Correlation of chemical shift with average Si-O-Al bond angle, similar to that in Figure 8, suggests peak assignments to crystallographic sites. These assignments are consistent with the pair-wise convergence of peaks observed with increasing temperature, shown schematically in Figure 17. Spectra taken at temperatures above T_c contain only four peaks of approximately equal intensity, consistent with a decrease in the number of inequivalent Si positions from eight to four with increasing temperature across the $P\bar{1} \rightarrow I\bar{1}$ transition. The peak separations just

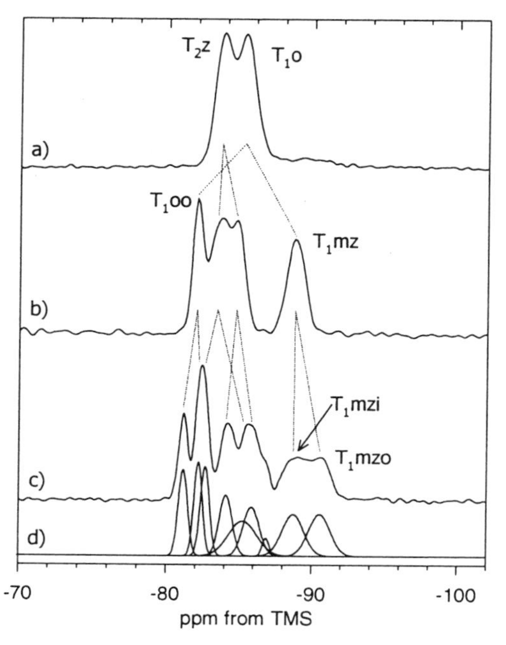

Figure 17. ^{29}Si MAS-NMR spectra for feldspars of various symmetry. (a) synthetic $SrAl_2Si_2O_8$ feldspar, *I*2/c, containing two inequivalent Si positions. Broad peak near -90 ppm arises from Si(3Al)-type local configurations due to a small amount of Si,Al disorder. (b) $I\bar{1}$-phase of a well-ordered anorthite ($CaAl_2Si_2O_8$; four inequivalent Si positions), taken at 400°C. (c) $P\bar{1}$ phase of the same sample as in (b), taken at 25°C. (d) Fit of the spectrum in (c) with eight peaks of equal intensity, corresponding to the eight inequivalent Si-positions of anorthite, plus a small peak near -87 ppm assigned to Si(3Al) environments. Dotted lines relate peaks assigned to sites that become equivalent with an increase in symmetry. [Redrawn from data of Phillips and Kirkpatrick (1995) and Phillips et al. (1997).]

below T_c (less than 100 Hz) are much smaller than for the ^{27}Al results of Staehli and Brinkmann (1974) and do not further constrain the lifetime of any ordered domains present above T_c .

For the best resolved peaks corresponding to a pseudo-symmetric pair of sites (those at -88.7 and -90.5 ppm, for T_1mzi and T_1mzo, respectively) the difference in chemical shift could be related to the order parameter (Q) for the transition. The difference in chemical shift ($\Delta\delta$) between a site in the $P\bar{1}$ phase (e.g., δ_{T1mzo} for the T_1mzo site) and its equivalent in the $I\bar{1}$ phase just above T_c (δ_{T1mz} for T1mz; $\Delta\delta_{T1mzo} = \delta_{T1mz} - \delta_{T1mzo}$) can be expressed in terms of Q by a power series expansion:

$$\Delta\delta = a_0 + a_1 Q + a_2 Q^2 + a_3 Q^3 + \ldots \tag{6}$$

where the coefficients a_i are independent of temperature and can vary among the crystallographic sites. The temperature dependence of the chemical shifts in the $P\bar{1}$ phase can be attributed to that of Q, which is assumed to vary such that $Q = 1$ for the $P\bar{1}$ phase at 0 K and $Q = 0$ in the $I\bar{1}$ phase:

$$Q(T) = \left(\frac{T_c - T}{T_c}\right)^{\beta} \tag{7}$$

where T_c is the observed transition temperature (in K) and β is the critical exponent, which describes the thermodynamic character of the transition. Use of the difference in chemical shift between the pseudo-symmetric pair of sites, $\Delta_{T1mz} = \Delta\delta_{T1mzo} - \Delta\delta_{T1mzi}$, simplifies the analysis by removing the chemical shift for the equivalent site in the $I\bar{1}$ phase. Furthermore, the expression for Δ_{T1mz} should contain only odd terms:

$$\Delta = a_1' Q + a_3' Q^3 + \ldots \tag{8}$$

because for any position in the crystal a change of antiphase orientation, represented by replacing Q by $-Q$, should change the chemical shift to that of the other pseudosymmetric site (e.g. $\Delta_{T1mzo}(Q) = -\Delta_{T1mzi}(-Q)$).

A fit of the data for anorthite to Equations (7) and (8) describes the change in chemical shift with temperature in the $P\bar{1}$ phase (Fig. 18) and yields a value for the critical exponent, $\beta = 0.27(\pm0.04)$, that is consistent with measurements using techniques sensitive to much longer length scales, such as X-ray diffraction (Redfern et al. 1987), that indicate the $P\bar{1}$-$I\bar{1}$ transition in Si,Al ordered anorthite is tricritical.

A similar analysis applies to the triclinic-monoclinic ($I\bar{1}$-$I2/c$) transition that occurs at 298 K across the compositional join $CaAl_2Si_2O_8$-$SrAl_2Si_2O_8$ near 85 mol % Sr, except that the order parameter varies with composition with a form similar to Equation (7) (Phillips et al. 1997). With increasing Sr-content, the ^{29}Si MAS-NMR spectra (Fig. 17) clearly show a decrease in the number of peaks that corresponds to a change in the number of crystallographically distinct Si sites from four ($I\bar{1}$) to two ($I2/c$). The order-parameter could be related to the difference in chemical shift between the T_1o site of the $I2/c$ phase (-85.4 ppm) and the peak for the T_1mz site of the $I\bar{1}$ samples, which is well-resolved and moves from -89.5 to -86.7 ppm with increasing Sr-content. These results yielded a critical exponent $\beta = 0.49\pm0.2$, consistent with the second-order character of the transition.

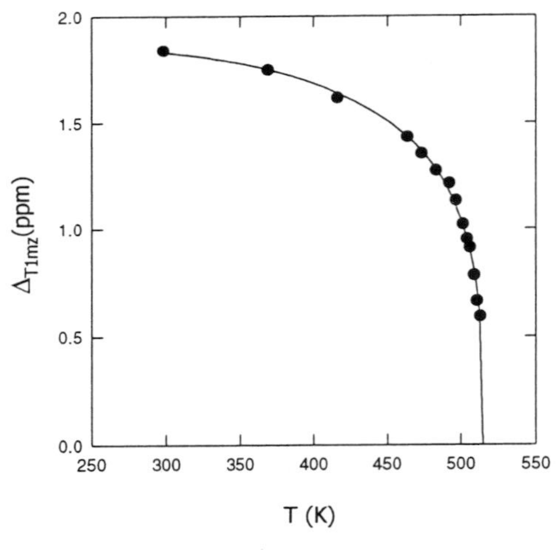

Figure 18. Temperature variation of the difference in chemical shift between peaks assigned to T_1mzi and T_1mzo for $P\bar{1}$ anorthite ($\Delta_{T1mz} = \delta_{T1mzo} - \delta_{T1mzi}$; Fig. 17), which is related to the order parameter for the $P\bar{1}$-$I\bar{1}$ transition, Q, by Equation 8. Line is a least-squares fit to Equations 7 and 8. [Redrawn from data of Phillips and Kirkpatrick (1995).]

Melanophlogite

The ^{29}Si MAS-NMR study by Liu et al. (1997) of the silica clathrate melanophlogite ($SiO_2\cdot(CH_4, CO_2, N_2)$; the volatiles were removed for this study by heating) illustrates how the quantitative nature of NMR peak intensities can help constrain the space-group classification of phases related by displacive transitions. Low-temperature structural studies of melanophlogite are difficult because the crystals are finely twinned at room temperature, due to a series of displacive structural transitions that occur upon cooling from the temperature of formation. Above about 160°C melanophlogite is cubic, space group *Pm*3*n*, the structure of which contains three inequivalent Si positions with multiplicities 24, 16, and 6. The ^{29}Si MAS-NMR spectrum of this phase at 200°C (Fig. 19) shows three peaks with relative intensity ratios 12:8:3, consistent with the structure. Below about 140°C, the two most intense peaks (Si(1) and Si(2)) each appear to split into two peaks of equal intensity and the splitting increases with decreasing temperature. These observations suggested the presence of a previously unrecognized cubic-cubic transition to a *Pm*3 phase and narrowed the possibilities for the space-group of the room-temperature phase. A second transition occurs near 60°C that splits each of the two peaks from Si(1) into two additional peaks with 1:2 intensity ratios. These NMR data indicate that the evolution of the space-group for melanophlogite should give multiplicities for the *Pm3n* Si(1) site of 24→12:12→4:8:4:8, suggesting an orthorhombic (*Pmmm*) structure at room-temperature.

INCOMMENSURATE PHASES

The NMR chemical shift, being sensitive to short-range structure, provides a unique, local probe of transitions to incommensurate (INC) phases and the nature of their structural modulations. The wave-like structural modulations that give rise to satellite peaks in diffraction patterns produce distinct NMR peak shapes, corresponding to a range of NMR parameters (e.g. chemical shifts) that quantitatively reflect the spatial distribution of local structural environments. The intensities and frequency shifts obtained from the NMR spectra can be compared to calculations based on specific models for the structural modulation. NMR techniques have long been applied to INC phases, but until recently they have focused mostly on single-crystal data for quadrupolar nuclei (Blinc 1981). With the availability of high-resolution solid-state techniques, several INC phases of mineralogical interest have been studied that illustrate the power of combining techniques of differing length scales (diffraction vs. NMR) to characterize these types of transitions.

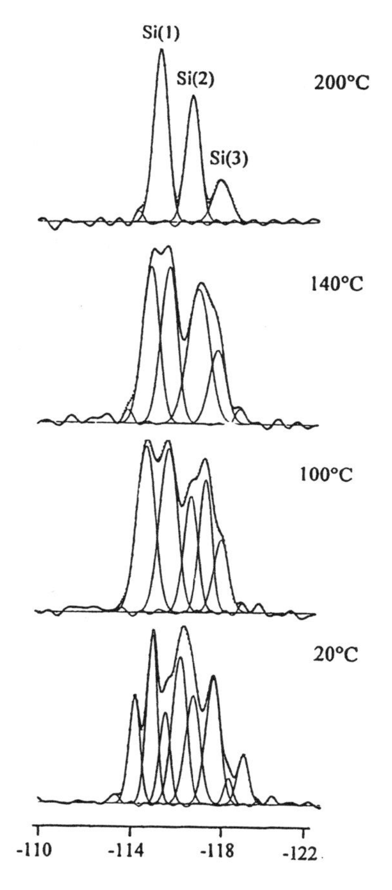

Figure 19. ^{29}Si MAS-NMR spectra of the silica clathrate melanophlogite, after heating to remove volatile guest molecules. Intensity ratios at 200°C (12:8:3) are consistent with the *Pm3n* cubic phase. Splitting of the Si(1) and Si(2) sites (140 and 100°C) indicates transformation to cubic *Pm*3 (intensity ratios 6:6:4:4:3). Further 1:2 splitting of each of the Si(1) sites at 20°C is consistent with transformation to a phase with orthorhombic symmetry (*Pmmm*). Frequency scale is ppm from tetramethylsilane. [Modified from Liu et al. (1997), Fig. 1, p. 2812.)

Sr_2SiO_4

Sr-orthosilicate (Sr_2SiO_4) provides a particularly striking example of the effect of incommensurate structural modulations on NMR spectra. It undergoes a transition from a monoclinic form (β) stable near room temperature to an INC phase (α'_L) near 70°C that is easily accessible for MAS-NMR (Phillips et al. 1991). The structure of Sr_2SiO_4 (Fig. 20) is closely related to that of larnite, Ca_2SiO_4, (as well as a class of materials of the "β-K_2SO_4" structure type) which undergoes a similar transition near 670°C upon cooling, although the β-phase is metastable and eventually transforms to γ, which is stable at ambient conditions. For Sr-orthosilicate, the INC phase is stable over a very large temperature-range, about 400°, and transforms at higher temperatures to the orthorhombic α-form depicted in Figure 20. The presence of only one crystallographic site for Si in both the α- and β-phases simplifies interpretation of the ^{29}Si MAS-NMR spectra.

Figure 20. Polyhedral representation of the structure of β-Sr_2SiO_4 (left, T < 70°C) and α-Sr_2SiO_4 (right, T > 500°C). Tetrahedra are $Si[O]_4$ and large circles are Sr atoms. An intervening incommensurate phase (α'_L) is stable over about 400°.

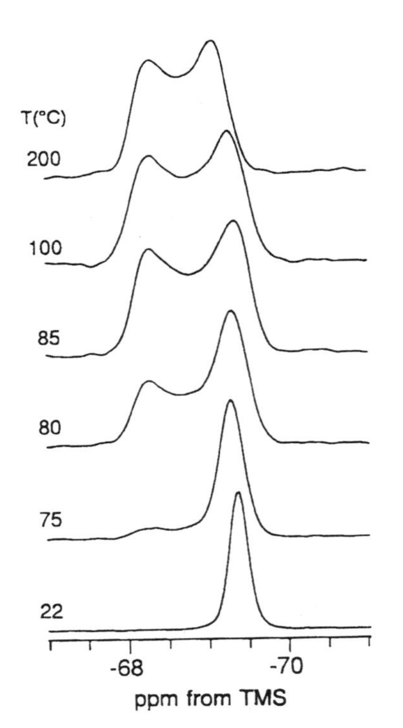

The effect of the structural modulation present in the INC phase on the NMR spectra can be visualized as a spatial variation in the rotation of the $Si[O]_4$-tetrahedron away from its orientation in the α–phase (Fig. 20). In the orthorhombic (α) phase, the Si-position has point symmetry *m* and one edge of the Si-tetrahedron is perpendicular to **c**, whereas this edge is rotated by about 10° in the monoclinic (β) phase. One view of the INC phase involves a variation of this rotation angle along the crystallographic **b** direction giving a modulation wavelength of approximately 3b. This rotation of the tetrahedron away from its position in the α–phase is reflected in the ^{29}Si chemical shift.

The ^{29}Si MAS-NMR spectra of the β-phase contain a single peak near -69.4 ppm (Fig. 21). Upon transformation to the INC phase near 75°C, a shoulder develops near -68.3 ppm, the intensity of which increases with temperature. The chemical shift of the peak at -68.3 ppm does not change with temperature, because it corresponds to Si atoms with the local structure of the α-phase (Fig. 20b). The peak near –69.4 ppm moves to higher chemical shifts with temperature. This observation suggests that the amplitude of the modulation decreases with increasing temperature. Physically, the amplitude

Figure 21. ^{29}Si MAS-NMR spectra of Sr_2SiO_4 taken at temperatures below (22°C) and above the β-INC phase transition. Spectral profiles in the INC phase are typical of those expected for a single crystallographic site with a non-linear modulation wave (outside the plane-wave limit). [Redrawn from data of Phillips et al. (1991).]

can be interpreted as the maximum rotation angle of the $Si[O]_4$ tetrahedra along the modulation.

At temperatures where the INC phase is stable, the NMR spectrum shows a distinct peak shape with two sharp edges of unequal intensity that cannot be fit with a sum of two symmetrical peaks. Because the left- and right- rotations of the Si-tetrahedron give the same chemical shifts (corresponding to twin domains of the β-phase), a plane wave modulation can only give a symmetrical peak shape. The unequal intensity for the two peaks in this case implies a non-linear variation of phase. A model for the calculation of the ^{29}Si NMR peak shape is shown in Figure 22, based on a "soliton" model containing domain walls of finite thickness. The domain walls, shown near phase values $\varphi = 2\pi(n + 1/2)$, roughly correspond to regions where the phase angle (φ) varies rapidly, between regions with local structure of the β-phase. The plane wave limit, corresponding to a soliton density of unity, would show a linear spatial variation of the phase (straight line in Fig. 22) and give a symmetrical peak shape, with two "horns" of equal intensity. Calculation of the ^{29}Si NMR lineshape (Fig. 22) assumes that the Si atoms are distributed uniformly along the spatial dimension (x; normalized by the wavelength of the modulation, λ) and that δ can be expressed as a power series expansion of the phase angle. Using this model, simulation of the spectra in Figure 21 yields the variation with temperature of the soliton density (which is the order parameter for the INC→β transition), and the modulation amplitude.

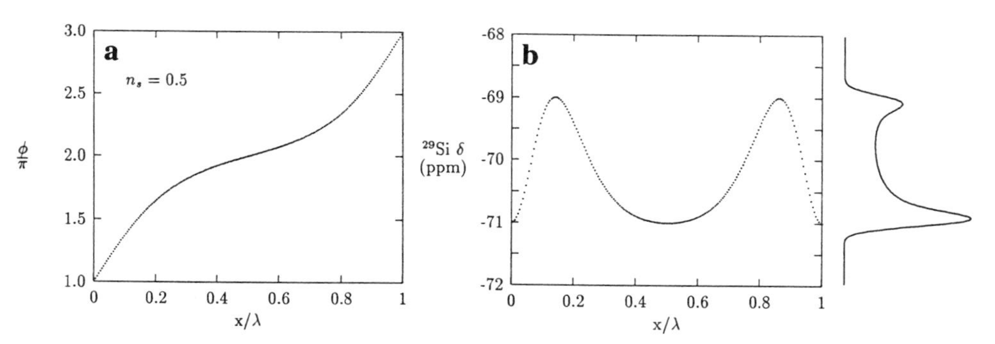

Figure 22. Model calculation of the NMR spectrum for a non-linear modulation wave, corresponding to the INC phase of Sr_2SiO_4 with a soliton density $n_s = 0.5$. The phase of the modulation (φ) can be related to the rotation of the Si-tetrahedron from its orientation in the α-phase ($\phi = n\pi + 1/2$) to that similar to the β-phase ($\phi = n\pi$; Fig. 20). NMR spectral intensities shown in (b) are proportional to the number of Si atoms (dots), which are assumed to be distributed uniformly along the modulation wave (x/λ, where λ is the wavelength of modulation). A plane wave, in which ϕ varies linearly with x, gives a symmetrical spectral profile. Compare to spectrum taken near 80°C in Figure 21.

Åkermanite

Somewhat similar ^{29}Si MAS-NMR peak shapes were obtained by Merwin et al. (1989) for the INC phase of åkermanite, $Ca_2MgSi_2O_7$, although these data were not analyzed using incommensurate lineshape models. End-member akermanite transforms to an INC phase upon cooling through 85°C. The INC phase is stable at ambient conditions and its modulation wavelength varies with temperature (Seifert et al. 1987). The Si in akermanite occupy Q^1-type sites (having one bridging oxygen). The significant changes in the ^{29}Si NMR spectrum suggests that the structural modulation involves variation in the Si-O-Si angle, which has a large effect on ^{29}Si chemical shifts. More detailed spectral interpretations might be complicated by the presence of two crystallographically distinct Si positions in the commensurate phase. However, the ^{29}Si NMR spectrum at temperatures above the stability range of the INC phase contain only one, relatively narrow peak, suggesting that the two sites give very similar chemical shifts.

Tridymite

The familiar, idealized hexagonal structure of tridymite exists only above about 400°C, probably as the average of a dynamically disordered phase. At lower temperatures, tridymite exhibits many different structural modifications, including several incommensurate phases, related by reversible structural transitions. The sequence of phases varies for tridymite samples differing in origin, extent of order, stacking faults, and thermal history (see Heaney 1994). At least three different forms have been reported at 25°C: monoclinic MC-1, orthorhombic PO-n, and incommensurate MX-1, which exhibits a monoclinic subcell. Each of these forms appears to undergo a distinct series of structural transitions with increasing temperature. Several studies have explored these phase transitions by ^{29}Si MAS-NMR spectroscopy with the hope of further characterizing the crystallographic relationships and the incommensurate phases of tridymite.

Xiao et al. (1995) obtained ^{29}Si MAS-NMR data for the INC phase of MX-1 type tridymite, in addition to following its evolution across several displacive transitions between 25 and 540°C. The MX-1 phase can be produced from MC-type tridymite by mechanical grinding or rapid cooling from elevated temperature, although a small amount of the MC phase usually remains. Diffraction data indicate that the INC phase of MX-1 tridymite, which exists below about 65°C, contains a two dimensional structural modulation with principal components along a^* and c^*. This two-dimensional modulation gives ^{29}Si MAS-NMR spectra that are more complicated than those of Sr_2SiO_4 described above (Fig. 23). Xiao et al. (1995) fit these spectra with a two-dimensional plane wave model that returns estimates for the amplitudes of the principal components of the modulation in terms of the average Si-Si distance. The shape of the calculated spectrum is also very sensitive to the relative phases of the two modulation components, but in such a way that the amplitudes and relative phase could not be determined independently. Two sets of compatible values both give amplitudes in terms of average Si-Si distance of the order 0.02 Å.

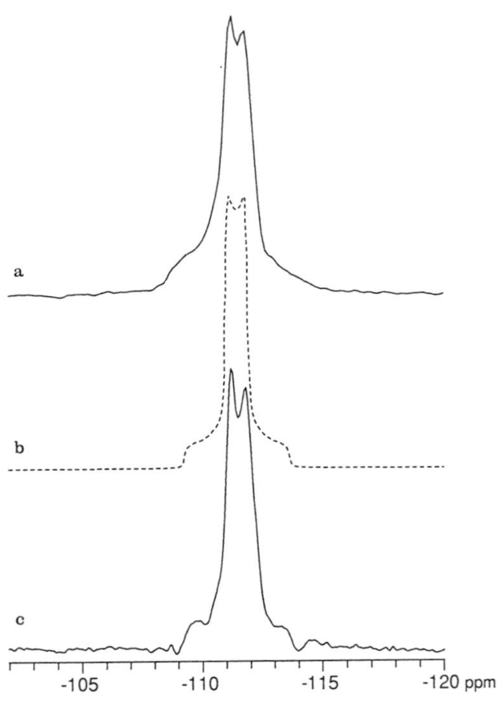

Figure 23. ^{29}Si MAS-NMR spectra of MX-1 type tridymite near 298 K (top and bottom) and a simulated lineshape for a two-dimensional plane wave. Fit of the spectra yields the amplitudes of the two modulation waves, in terms of Si-Si distance, and their relative phases. [Used by permission of the editor of *Physics and Chemistry of Minerals*, from Xiao et al. (1995), Fig. 8, p. 36, © Springer-Verlag 1995.]

Kitchin et al. (1996) examined a remarkably well-ordered MC-type tridymite with temperature through the complete set of structural transitions to the high-temperature hexagonal phase by ^{29}Si MAS-NMR, up to 450°C (Fig. 24). Near room-temperature the spectrum of the monoclinic phase contains a series of narrow peaks between -108 and -115 ppm that can be fit with a sum of twelve peaks of equal intensity, consistent with the number of distinct sites indicated by crystal structure refinements. The correlation of δ_i with average Si-O-Si bond angle suggests an assignment scheme for these twelve sites and is shown in Figure 8.

The transition to an orthorhombic phase (OP), near 108°C, corresponds to a large change in the ^{29}Si NMR spectrum. Most structure refinements of OP tridymite give six crystallographic Si positions, but the NMR spectral profile cannot be fit with a sum of six peaks of equal intensity. More recent structural models suggest that the number of inequivalent positions might be twelve or 36. A reasonable fit with twelve equally intense peaks could be obtained only if the peak widths were allowed to vary, which would imply the presence static disorder (e.g., in the mean Si-O-Si angle) that varies among the sites. A fit of the spectrum to 36 curves would be under-constrained. These authors could obtain a reasonable fit with six equally intense peaks by including a one-dimensional plane wave incommensurate modulation, which fits the fine-structure at more negative chemical shifts, although there is no other evidence for an incommensurate structure.

The ^{29}Si MAS-NMR spectrum narrows considerably upon transformation to the incommensurate OS phase (near 160°C), which has a single crystallographic site. The spectrum of this phase is characteristic of a non-linear modulation wave with low soliton density (cf. Fig. 24c with Figs. 21 and 22). Both OP (210 to 320°C) and hexagonal (LHP; above

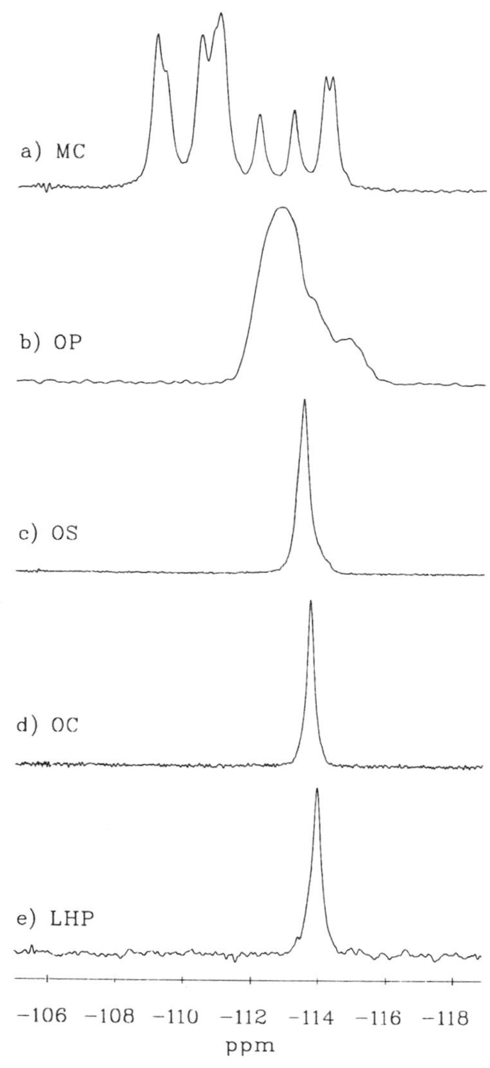

Figure 24. ^{29}Si MAS-NMR spectra for the sequence of stable phases produced upon heating a well-ordered MC-1 type tridymite. (a) monoclinic MC phase near 25°C, which contains 12 inequivalent Si positions. (b) metrically orthorhombic OP-phase, taken at 142°C. (c) Incommensurate OS-phase at 202°C, consistent with a single crystallographic site and non-linear structural modulation characterized by low soliton density. (d) Orthorhombic OC phase at 249°C, which contains one crystallographic Si position. (e) Hexagonal LHP phase at 401°C. The -114 ppm chemical shift suggests an average Si-O-Si angle of about 152°, consistent with dynamical disorder of the oxygen position. [From Kitchin et al. (1996), Fig. 1, p. 552.]

320°C) tridymite phases give a single, narrow ^{29}Si NMR peak, consistent with one inequivalent Si position. The chemical shift of the hexagonal phase (-114 ppm) suggests a mean Si-O-Si bond angle near 153°, compared to 180° for the angle between mean atom positions. This result is consistent with dynamical disorder in the oxygen positions of high tridymite.

ORDERING/DISORDERING TRANSITIONS

Cation order/disorder reactions can have a large effect on the thermochemistry and relative stability of minerals. In addition, the state of order can affect the nature of structural transitions through coupling of their strains. The sensitivity of NMR chemical shifts to short-range structure and the quantitative nature of the peak areas in NMR spectra can help quantify the state of order and/or changes in cation distribution that accompany an ordering reaction. Information from NMR spectroscopy is particularly helpful for cases in which cations that are difficult to distinguish by X-ray diffraction (Mg and Al in spinel; Si and Al in aluminosilicates) are disordered over a crystallographic site. NMR spectroscopy can also quantify short-range order and help determine whether the distribution of local configurations differs from that expected for a statistical distribution of average site occupancies.

Due to the limited temperature range accessible for MAS-NMR and the sluggishness of cation diffusion in many minerals, samples typically are prepared by annealing under the desired conditions and the reaction arrested by quenching to ambient conditions to collect the NMR spectra. This experimental method limits the temperature range to that over which cation diffusion is slow compared to the quenching rate.

For cases in which the NMR peaks corresponding to different structural/chemical environments can be resolved, the distribution of local configurations and/or site occupancies can be determined as a function of annealing time or temperature, because the peak areas are proportional to the number of nuclei in those environments. In this sense NMR spectroscopy provides a continuous picture of the ordering reaction, complementary to X-ray diffraction, which ordinarily detects changes in symmetry or site occupancies averaged over large portions of the crystal. Interpretation of the NMR spectra can be very difficult if the ordering reaction involves multiple crystallographic sites, although in some instances careful data analysis has yielded useful results. Furthermore, the presence of paramagnetic impurities in natural specimens reduces spectral resolution and usually limits NMR studies to synthetic material.

Si,Al ordering in framework aluminosilicates

The state of Al,Si order in aluminosilicates contributes significantly to the energetics and relative stability of these phases. For example, the net enthalpy change for the reaction $NaAlO_2 + SiO_2 = NaAlSiO_4$ (albite), -50 kJ/mole, is only about twice that for complete Al,Si ordering in albite (2Si-O-Al + Al-O-Al + Si-O-Si = 4Si-O-Al), about -26 kJ/mole using recent estimates (Phillips et al. 2000). MAS-NMR spectroscopy provides information that complements the average site occupancies obtained from structure refinements of diffraction data. For phases that yield resolved peaks for different crystallographic sites, the peak areas give the relative populations on those sites. In addition, NMR spectra, especially of ^{29}Si, can provide the distribution of local configurations.

For aluminosilicates, ^{29}Si NMR has proven to be extremely useful, because it gives naturally narrow peaks and exchange of an Si for Al in an adjacent tetrahedral site results in a large change in the ^{29}Si chemical shift, about -5 ppm, that is linearly additive (Fig. 25). Thus, for framework structures with Al and Si disordered on a crystallographic site, the

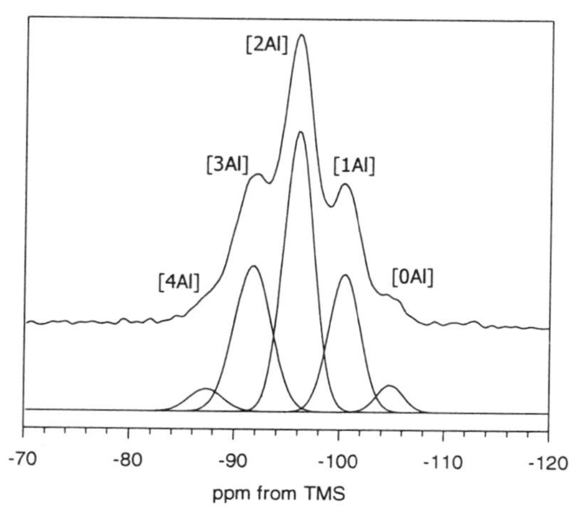

Figure 25. ^{29}Si MAS-NMR spectrum of the cubic phase of a Cs-substituted leucite ($CsAlSi_2O_6$), obtained at 150°C. In the cubic phase Si and Al are disordered on one crystallographic position. The ^{29}Si chemical shift depends on the number of Al in the adjacent framework sites (from zero to four), giving distinct peaks for the same crystallographic site, but with different local configurations. [Redrawn from data of Phillips and Kirkpatrick (1994).]

^{29}Si NMR spectrum contains five peaks corresponding to Si having from 0 to 4 Al in the adjacent sites [denoted in the "Q" notation as Q^4(0Al), Q^4(1Al), ... Q^4(4Al), where "Q" represents quadrifunctional (four-coordination) and the superscript denotes the number of bridging bonds to other four-coordinated sites; i.e. 4 for a framework, 3 for a sheet-structure, etc.]. High-resolution ^{17}O NMR techniques now becoming available might prove very useful (e.g., Stebbins et al. 1999), because the ^{17}O *Cq* and δ_i in framework structures depends primarily on the local configuration, allowing resolution of Si-O-Si, Si-O-Al, Al-O-Al environments. Particularly exciting is the possibility of directly detecting and counting Al-O-Al linkages, the presence of which has been inferred from the observation of Q^4(3Al) environments in phases with composition Si/Al = 1 (see below), but otherwise difficult to detect directly.

Using ^{29}Si MAS-NMR, Si,Al order has been measured for many framework and sheet-structure aluminosilicate minerals (See Engelhardt 1987; Engelhardt and Koller 1994). For some aluminosilicates however, the combination of Al,Si disorder and multiple crystallographic sites severely complicates the spectra, because each crystallographic site can give a series of five peaks, making it difficult or impossible to obtain quantitative information (e.g., Yang et al. 1986). Specific Al,Si ordering reactions have been studied for cordierite (Putnis et al. 1987), anorthite (Phillips et al. 1992), and β-eucryptite ($LiAlSiO_4$, quartz structure; Phillips et al. 2000), each of which illustrates a different solution to the resolution problem.

β-eucryptite β-eucryptite ($LiAlSiO_4$, quartz structure) contains four crystallographically distinct framework sites, two each for Si and Al. An order/disorder transition appears to occur along the $LiAlSiO_4$-SiO_2 compositional join near 30 mol % SiO_2. In well-ordered β-eucryptite, the two Si sites appear to give nearly the same ^{29}Si chemical shift. The ^{29}Si MAS-NMR spectra of Si,Al disordered samples, prepared by crystallization from a glass of the same composition, contain a series of evenly spaced peaks due to the local Q^4(nAl) ($0 \leq n \leq 4$) configurations (Fig. 26). However, the long- range order (i.e., distribution of Si and Al over the four crystallographic sites) cannot be determined. Because the compositional ratio Si/Al = 1, the number of Al-O-Al linkages could be determined from these spectra as a function of annealing time. Correlation of changes in the number of Al-O-Al linkages with solution calorimetric data obtained for the same samples gives an estimate of -26 kJ/mole for the enthalpy of the reaction:

$$\text{Al-O-Al} + \text{Si-O-Si} \rightarrow 2(\text{Al-O-Si}) \quad (9)$$

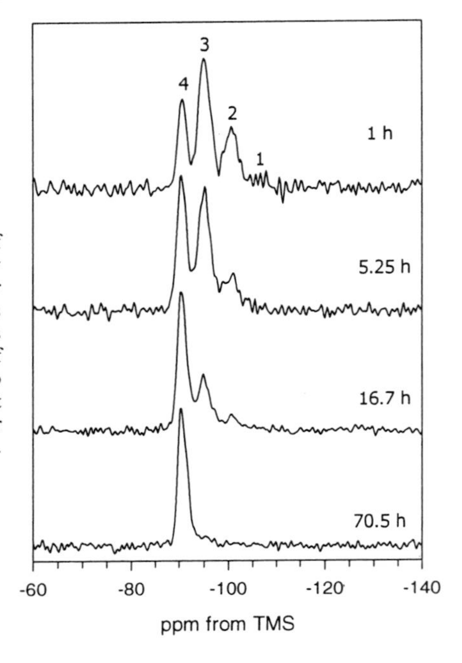

Figure 26. The ^{29}Si MAS-NMR spectra of β-eucryptite ($LiAlSiO_4$) for samples prepared by crystallizing from glass and annealing at 900°C for the times indicated. Presence of Si,Al disorder gives peaks for Si with fewer than four Al neighbors (as indicated for the 1 h sample), the intensities of which decrease with increasing Si,Al order. The two crystallographic Si positions of β-eucryptite cannot be resolved because they exhibit very similar chemical shifts. [From Phillips et al. (2000), Fig. 2, p. 183.]

Anorthite. Anorthite ($CaAl_2Si_2O_8$) also has a compositional ratio Si/Al = 1, but its crystal structure contains 16 inequivalent framework cation sites. Even for the most ordered samples, all the crystallographic sites cannot be resolved in ^{29}Si NMR spectra (Fig. 17). In this case, the extent of short-range disorder was estimated from changes in the weighted average chemical shift of the spectrum relative to that for a well-ordered natural sample. In a perfectly ordered anorthite all the tetrahedral sites adjacent to Si contain Al (Q^4(4Al) only), so it was assumed that each substitution of an Si for an Al in the sites adjacent to Si results in a -5 ppm change in chemical shift. For example, if each Si has one Si neighbor (all Q^4(3Al)), the average chemical shift would differ by -5 ppm from that for a perfectly ordered sample. In this way, quantitative estimates for the number of Al-O-Al linkages (which must equal the number of Si-O-Si configurations) could be obtained. These values were also correlated with calorimetric data to obtain an enthalpy of -39 kJ/mole for the reaction in Equation (9). Also, it was found that even the first-formed crystals contain a large amount of short-range Si,Al order, with the average Si atom having only 0.2 Si neighbors, compared to 2.0 for complete short-range Si,Al disorder.

Cordierite Unlike the previous two examples, a symmetry change accompanies Si,Al ordering in cordierite ($Mg_2Al_4Si_5O_{18}$). The transition from the fully Si,Al disordered hexagonal phase, which is stable above about 1450°C, to the ordered orthorhombic phase occurs upon isothermal annealing at lower temperatures. The tetrahedral framework of cordierite contains two topologically distinct sites, hexagonal rings of tetrahedra (T_2-sites) that are cross-linked by T_1-sites, with each T_1-site connecting four different rings. In the fully ordered orthorhombic phase, Si occupies two of the three distinct T_2-type sites, each with multiplicity of two and adjacent to three Al-sites (Si(3Al)), and one of the two distinct T_1-type sites, Si(4Al) and multiplicity of one. The ^{29}Si NMR spectrum of fully ordered cordierite contains two peaks, one at -79 ppm for the T_1 site and another at -100 ppm for the T_2-type sites with four times the intensity, consistent with the crystal structure (Putnis et al. 1987). The crystallographically distinct T_2 sites could not be resolved.

Putnis et al. (1987) obtained ^{29}Si MAS-NMR spectra of samples as a function of annealing time to quantitatively determine the extent of Si,Al order. Because of the large chemical shift difference between the T_1 and T_2 sites, a series of peaks could be resolved for each type of site corresponding to Si with different numbers of Al neighbors. Thus both the long-range order (distribution of Si and Al over the T1 and T2 sites) and the short range order (e.g., number of Al-O-Al linkages) could be obtained as a function of annealing time and correlated with X-ray diffraction data obtained for the same samples. The decrease in the number of Al-O-Al linkages with annealing time could also be correlated with enthalpies of solution from an earlier calorimetric study to obtain a net enthalpy of -34 kJ/mole for Equation (9).

Cation ordering in spinels.

Oxide spinels, with general stoichiometry AB_2O_4, can accommodate a large variety of different cations with different charges. In "normal" spinels the A cation occupies four-coordinated sites and the octahedral sites are filled by the B cation, whereas "inverse" spinels have B cations on the tetrahedral sites and [AB] distributed over the octahedral sites. Most spinels display some disorder, which is represented by the "inversion parameter," x, for the structural formula $({}^{[4]}A_{1-x}{}^{[4]}B_x)({}^{[6]}A_x{}^{[6]}B_{2-x})O_4$. Some inverse spinels undergo order/disorder transitions corresponding to ordering of A and B over the octahedral sites. Spinels are of considerable petrologic interest as geothermometers and geobarometers, and the energetics of the cation disordering can have a large effect on spinel stability and therefore mineral assemblage, especially at high temperatures.

Ordering of $MgAl_2O_4$-composition spinels has been studied by ^{27}Al NMR (Gobbi et al. 1985; Wood et al. 1986; Millard et al. 1992; Maekawa et al. 1997), because of the distinct chemical shift ranges for $^{[4]}Al$ and $^{[6]}Al$ in oxides. In spinel, $^{[4]}Al$ gives a narrow peak near +70 ppm, whereas the $^{[6]}Al$ gives a broader peak near +10 ppm (Fig. 27). The tetrahedral site in spinel has cubic point symmetry which requires Cq = 0, but crystal defects and the [Mg,Al] disorder results in a small distribution of EFG's at the Al site, so that the peak at 70 ppm contains only the central transition. The earlier studies suffered from low MAS rates, which yield spinning sidebands that overlap the centerbands, and unequal excitation of the two ^{27}Al NMR signals. Later studies quantified the $^{[4]}Al/^{[6]}Al$ ratio to determine the inversion parameter as a function of annealing temperature.

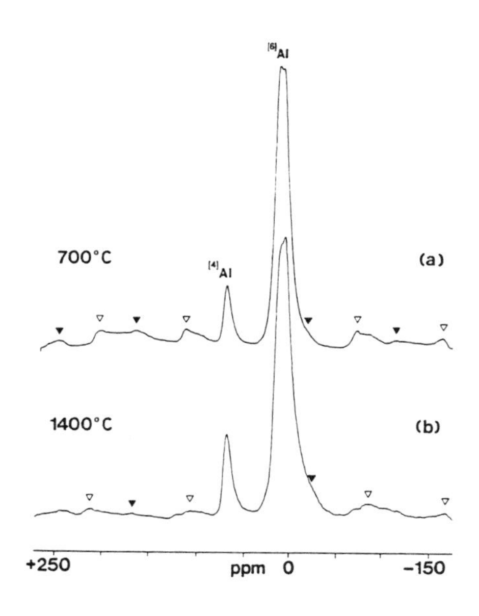

Figure 27. ^{27}Al MAS-NMR spectra (central transition only) of $MgAl_2O_4$ spinel samples quenched from different temperatures. Peak centered near 70 ppm arises from Al on the tetrahedral sites ($^{[4]}Al$) and that near 0 ppm from octahedral Al ($^{[6]}Al$). The peak areas are proportional to the number of atoms on those sites. Perfectly normal spinel would contain only $^{[6]}Al$. Spinning sidebands are denoted by triangles. [From Millard et al. (1992), Fig. 1, p. 46.]

In principal, the distribution of Al

over the tetrahedral and octahedral sites can be accurately determined from the MAS-NMR spectra. The main problem with this technique is that cation exchange between the sites prevents the equilibrium state of order from being quenched from temperatures above about 1100°C. Maekawa et al. (1997) attempted to overcome this problem by making in situ measurements at high temperatures. Below about 700°C, the ^{27}Al static powder spectra show two poorly resolved peaks (central transition only) for $^{[4]}Al$ and $^{[6]}Al$. With increasing temperature these peaks merge into one peak that appears to narrow further with temperature, apparently due to exchange of the Al between tetrahedral and octahedral sites. At high temperatures, the inversion parameter was estimated from the weighted average frequency by extrapolating a correlation of the weighted average frequency with inversion parameter for the spectra taken below 1100°C. Unfortunately, the change in average frequency with temperature is small compared to the width of the peak, which yields large uncertainties in the inversion parameter.

The ^{17}O MAS-NMR spectra of $MgAl_2O_4$ spinels show two, relatively narrow spectral features. But, it could not be ascertained whether they result from two chemically distinct O environments or a sum of quadrupolar MAS powder patterns (Millard et al. 1992).

Millard et al. (1995) studied order/disorder transitions in the inverse titanate spinels Mg_2TiO_4 and Zn_2TiO_4 by ^{17}O MAS-NMR techniques. For both compositions, the quenched cubic phase gave a relatively broad MAS-NMR peak, whereas the peaks for the ordered (tetragonal) phases were narrow, indicating small ^{17}O *Cq* values (Fig. 28). The broad peak of the cubic phase probably results from a distribution of chemical shifts and/or *Cq*'s that result from the different local configurations. In a two-phase region, the spectra showed a sum of broad and narrow peaks that could be fit to obtain relative proportions of the cubic and tetragonal phases in the sample.

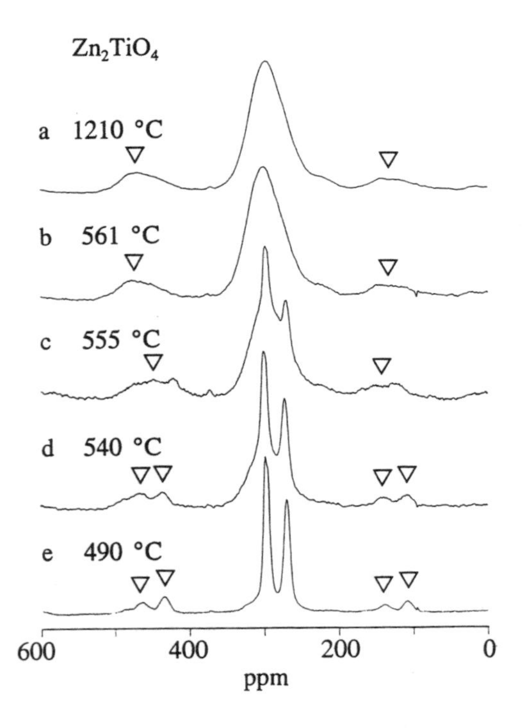

Figure 28. ^{17}O MAS-NMR spectra (central transition only) of Zn_2TiO_4 spinel for samples quenched from the temperatures indicated. A cubic-tetragonal ordering transition occurs near 550°C. Spectra of the cubic phase (1210 and 561°C) give a broad peak due to a distribution of EFG's resulting from disorder of [Zn,Ti] over the octahedral sites. The tetragonal phase (490°C) gives two peaks for two crystallographic sites, which exhibit small Cq values. Triangles denote spinning sidebands. [From Millard et al. (1995), Fig. 3, p. 891.]

CONCLUSIONS

The purpose of this review is to describe the many ways in which NMR spectroscopy can be applied to the study of mineral transformations. The characteristics of NMR spectroscopy make it a versatile tool for studying many types of the transformation processes that occur in minerals, from reversible structural transitions to ordering reactions that involve diffusion of cations. However, the results of NMR experiments are most useful when combined with those of other theoretical and experimental studies, such as those described in the remainder of this volume. A complete picture usually requires structural and dynamical information over a range of times and distances that cannot be supplied by any one experimental technique. Information available from NMR complements that from other techniques available to the mineralogist, because it presents a view of the short-range structure (first and second coordination sphere) averaged over a period that is long compared to that of the thermal vibrations of atoms. Furthermore, NMR can be used to measure or constrain the rates dynamical processes such as chemical exchange and molecular-scale reorientation. Although NMR spectroscopy is limited to certain NMR-active isotopes and (usually) to diamagnetic materials, many systems of mineralogical interest are accessible. Other limitations include a relatively poor sensitivity and difficulty in performing NMR experiments at high temperatures and pressures. NMR spectroscopy requires a large, homogeneous magnetic field over the sample and access to radio-frequency fields, which restrict the materials that can be used to hold the sample. Despite these limitations, further applications of NMR spectroscopy to the study of mineral transformation processes await a resourceful experimentalist.

ACKNOWLEDGMENTS

I thank Bill Casey, Jean Tangeman, and Jim Kirkpatrick for their reviews of the manuscript and suggestions for improvement. I am grateful to Jonathan Stebbins for providing original figures and the permission to reprint them here. The general description of NMR spectroscopy has evolved through interaction with the diverse body students who have endured EMS 251, Applications of NMR Spectroscopy, at the University of California, Davis. This work benefited from support provided by the US National Science Foundation and Department of Energy, and from the facilities of the W.M. Keck Solid-State NMR Laboratory at UCD.

REFERENCES

Abragam A (1961) Principles of Nuclear Magnetism. Clarendon Press, Oxford, UK

Akitt JW, Mann BE (2000) NMR and Chemistry: an Introduction to Modern NMR Spectroscopy, Stanley Thornes Pub Ltd

Armstrong RL (1989) Displacive order-disorder crossover in perovskite and antifluorite crystals undergoing rotational phase transitions. Prog NMR Spectros 21:151-173

Armstrong RL, van Driel HM (1975) Structural phase transitions in RMX_3 (perovskite) and R_2MX_6 (antifluorite) compounds. Adv Nuc Quad Reson 2:179-253

Baltisberger JH, Xu Z, Stebbins JF, Wang SH, Pines A (1996) Triple-quantum two-dimensional ^{27}Al magic-angle spinning nuclear magnetic resonance spectroscopic study of aluminosilicate and aluminate crystals and glasses. J Am Chem Soc 118:7209-7214

Blinc R (1981) Magnetic resonance and relaxation in structurally incommensurate systems. Phys Rep 79:331-398

Boeræ RT, Kidd G (1982) Rotational correlation times in nuclear magnetic relaxation. *In* GA Webb (ed.) Ann. Rep NMR Spectros 13:319-385

Carpenter MA (1994) Subsolidus phase relations of the plagioclase feldspar solid solution. *In* I Parson (ed.) Feldspars and Their Reactions. NATO ASI Series C 421:221-269

Engelhardt G, Michel D (1987) High-resolution Solid State NMR of Silicates and Zeolites. Wiley, New York, 485 p

Engelhardt G, Koller H (1994) ^{29}Si NMR of inorganic solids. *In* B Blümich (ed) Solid-State NMR II: Inorganic Matter. Springer-Verlag, Berlin, p 1-30

Fitzgerald JJ, DePaul SM (1999) Solid-state NMR spectroscopy of inorganic materials: an overview. *In* Solid-state NMR spectroscopy of inorganic materials, JJ Fitzgerald (ed) Am Chem Soc Symp Series 717:2-133

Freude D, Haase J (1993) Quadrupole effects in solid-state nuclear magnetic resonance. NMR Basic Principles Progress 29:1-90

Gobbi GC, Christofferesen R, Otten MT, Miner B, Buseck PR, Kennedy GJ, Fyfe CA (1985) Direct determination of cation disorder in $MgAl_2O_4$ spinel by high-resolution ^{27}Al magic-angle-spinning NMR spectroscopy. Chem Lett, p 771-774

Grandinetti PJ, Baltisberger JH, Farnan I, Stebbins JF, Werner U, Pines A (1995) Solid-state ^{17}O magic-angle and dynamic-angle spinning NMR study of the SiO_2 polymorph coesite. J Phys Chem 99:12341-12348

Harris RK (1986) Nuclear Magnetic Resonance Spectroscopy: a Physicochemical View. Longman Scientific, Essex, UK, 260 p

Hatch DM, Ghose S (1991) The α-β phase transition in cristobalite, SiO_2 cristobalite: symmetry analysis, domain structure and transition dynamics. Phys Chem Minerals 21:67-77

Hatch DM, Ghose S, Bjorkstam JL (1994) The α-β phase transition in $AlPO_4$. Phys Chem Minerals 17:554-562

Heaney PJ (1994) Structure and chemistry of the low-pressure silica polymorphs. *In* PJ Heaney, CT Prewitt, GV Gibbs (eds) Silica: Physical Behavior, Geochemistry, and Materials Applications. Rev Mineral 29:1-4

Heaney PJ, Veblen DR (1991) Observations of the α-β transition in quartz: A review of imaging and diffraction studies and some new results. Am Mineral 76:1018-1032

Hua GL, Welberry TR, Withers RL, Thompson JG (1988) An electron diffraction and lattice-dynamical study of the diffuse scattering in β-cristobalite, SiO_2. J Appl Crystallogr 21:458-465

Kirkpatrick, RJ (1988) MAS NMR spectroscopy of minerals and glasses. *In* Spectroscopic Methods in Mineralogy and Geology. FC Hawthorne (ed.) Rev Mineral 18:341-403

Kitchin SJ, Kohn SC, Dupree R, Henderson CMB, Kihara K (1996) I ^{29}Si MAS NMR studies of structural phase transitions of tridymite. Am Mineral 81:550-560

Köhler FH, Xie X (1997) Vanadocene as a temperature standard for ^{13}C and ^{1}H MAS NMR and for solution-state NMR spectroscopy. Mag Reson Chem 35:487-492

Lippmaa E, Magi M, Samoson A, Engelhardt G, Grimmer A-R (1980) Structural studies of silicates by solid-state high-resolution ^{29}Si NMR. J Am Chem Soc 102:4889-4893

Liu SX, Welch, MD, Klinowski J, Maresch, WV (1996) A MAS NMR study of a monoclinic/triclinic phase transition in an amphibole with excess OH - $Na_3Mg_5Si_8O_{21}(OH)_3$. Eur J Mineral 8:223-229

Liu, SX; Welch, MD; Klinowski, J (1997) NMR study of phase transitions in guest-free silica clathrate melanophlogite. J Phys Chem B 101:2811-2814

Maekawa H, Kato S, Kawamura K, Yokikawa T (1997) Cation mixing in natural $MgAl_2O_4$ spinel: a high-temperature ^{27}Al NMR study. Am Mineral 82:1125-1132

Millard RL, Peterson RC, Hunter BK (1992) Temperature dependence of cation disorder in $MgAl_2O_4$ spinel using ^{27}Al and ^{17}O magic-angle spinning NMR. Am Mineral 77:44-52

Millard RL, Peterson RC, Hunter BK (1995) Study of the cubic to tetragonal transition in Mg_2TiO_4 and Zn_2TiO_4 spinels by ^{17}O MAS NMR and Rietveld refinement of X-ray diffraction data. Am Mineral 80:885-896

Merwin LH, Sebald A, and Seifert F (1989) The incommensurate-commensurate phase transition in åkermanite, $Ca_2MgSi_2O_7$, observed by *in situ* ^{29}Si MAS NMR spectroscopy. Phys Chem Minerals 16: 752-756

Müller D, Jahn E, Ludwig G, Haubenreisser U (1984) High-resolution solid-state ^{27}Al and ^{31}P NMR: Correlation between chemical shift and mean Al–O–P angle in $AlPO_4$ polymorphs. Chem Phys Lett 109:332-336

Pake GE (1948) J Chem Phys 16:327

Phillips BL, Kirkpatrick RJ (1994) Short-range Si-Al order in leucite and analcime: Determination of the configurational entropy from ^{27}Al and variable temperature ^{29}Si NMR spectroscopy of leucite, its Rb- and Cs-exchanged derivatives, and analcime. Am Mineral 79:1025-1031

Phillips BL, Kirkpatrick RJ (1995) High-temperature ^{29}Si MAS NMR spectroscopy anorthite ($CaAl_2Si_2O_8$) and its $P\bar{1}-I\bar{1}$ structural phase transition. Phys Chem Minerals 22:268-276

Phillips BL, Thompson JG, Kirkpatrick RJ (1991) ^{29}Si magic-angle-spinning NMR spectroscopy of the ferroelastic-to-incommensurate transition in Sr_2SiO_4. Phys Rev B 43:1500

Phillips BL, Kirkpatrick RJ, Carpenter MA (1992) Investigation of short-range Al,Si order in synthetic anorthite by ^{29}Si MAS NMR spectroscopy. Am Mineral 77:490-500

Phillips BL, Thompson JG, Xiao Y, Kirkpatrick RJ (1993) Constraints on the structure and dynamics of the cristobalite polymorphs of $AlPO_4$ and SiO_2 from ^{31}P, ^{27}Al, and ^{29}Si NMR spectroscopy to 770 K. Phys Chem Minerals 20:341-352

Phillips BL, McGuinn, MD, Redfern, SAT (1997) Si/Al order and the $I\bar{1}$-$I2/c$ structural phase transition in synthetic $CaAl_2Si_2O_8$-$SrAl_2Si_2O_8$ feldspar: A ^{29}Si MAS-NMR spectroscopic study. Am Mineral 82:1-7

Putnis A (1987) Solid state NMR spectroscopy of phase transitions in minerals. *In* EKH Salje (ed) Physical Properties and Thermodynamic Behaviour of Minerals. NATO ASI Series C 225:325-358

Putnis A, Salje E, Redfern SAT, Fyfe CA, Strobl H (1987) Structural states of Mg-cordierite I: Order parameters from synchrotron X-ray and NMR data. Phys Chem Minerals 14:446-454

Redfern SAT, Salje E (1987) Thermodynamics of plagioclase II: temperature evolution of the spontaneous strain at the $P\bar{1}-I\bar{1}$ phase transition in anorthite. Phys Chem Minerals 14:189-195

Redfern SAT, Salje E (1992) Microscopic dynamic and macroscopic thermodynamic characte of the $P\bar{1}-I\bar{1}$ phase transition in anorthite. Phys Chem Minerals 18:526-533

Rigamonti A (1984) NMR-NQR studies of structural phase transitions. Adv Phys 33:115-191

Salje EKH, Ridgwell A, Göttler B, Wruck B, Dove MT, Dolino G (1992) On the displacive character of the phase transition in quartz: A hard-mode spectroscopic study. J Phys Cond Matter 4:571-577

Schmahl WW, Swainson IP, Dove MT, Graeme-Barber A (1992) Landau free energy and order parameter behaviour of the α/β phase transition in cristobalite. Z Kristallogr 210:125-145

Seifert F, Czank M, Simons B, Schmahl W (1987) A commensurate-incommensurate phase transition in iron-bearing akermanites. Phys Chem Minerals 14:26-35

Smith ME, van Eck ERH (1999) Recent advances in experimental solid state NMR methodology for half-integer spin quadrupolar nuclei. Progr NMR Spectros 34:159-201

Spearing DR, Stebbins JF (1989) The Si-29 NMR shielding tensor in low quartz. Am Mineral 74:956-959

Spearing DR, Farnan I, Stebbins JF (1992) Dynamics of the alpha-beta phase transitions in quartz and cristobalite as observed by *in situ* high temperature Si-29 NMR and O-17 NMR. Phys Chem Minerals 19:307-321

Spearing DR, Stebbins JF, Farnan I (1994) Diffusion and the dynamics of displacive phase transitions in cryolite (Na_3AlF_6) and chiolite ($Na_5Al_3F_{14}$)—multi-nuclear NMR studies. Phys Chem Minerals 21:373-386

Staehli JL, Brinkmann, D (1974) A nuclear magnetic resonance study of the phase transition in anorthite, $CaAl_2Si_2O_8$. Z Kristallogr 140:360-373

Stebbins JF (1988) NMR spectroscopy and dynamic processes in mineralogy and geochemistry. *In* FC Hawthorne (ed.) Spectroscopic methods in mineralogy and geology, Rev Mineral 18:405-429

Stebbins JF (1995a) Nuclear magnetic resonance spectroscopy of silicates and oxides in geochemistry and geophysics. *In* Mineral Physics and Crystallography a Handbook of Physical Constants. TJ Ahrens (ed) American Geophysical Union, Washington DC, p 303-331

Stebbins JF (1995b) Dynamics and structure of silicate and oxide melts: Nuclear magnetic resonance studies. *In* JF Stebbins, PF McMillan, DB Dingwell (eds) Structure, Dynamics and Properties of Silicate Melts. Rev Mineral 32:191-246

Stebbins JF, Lee SK, Oglesby JV (1999) Al-O-Al oxygen sites in crystalline aluminates and aluminosilicate glasses: High-resolution oxygen-17 NMR results. Am Mineral 84:983-986

Swainson IP, Dove MT (1993) Low-frequency floppy modes in β-cristobalite. Phys Rev Lett 71:193-196

Taulelle F (1990) NMR of quadrupolar nuclei in the solid state. *In* P Granger, RK.Harris (eds) Multinuclear magnetic resonance in liquids and solids, chemical applications. NATO ASI Series C, 322:393-413

Tossell JA, Lazzeretti P (1988) Calculation of NMR parameters for bridging oxygens in H_3T-O-$T'H_3$ linkages (T,T′ = Al, Si, P), for oxygen in SiH_3O^-, SiH_3OH and SiH_3OMg^+ and for bridging fluorine in $H_3SiFSiH_3^+$. Phys Chem Minerals 15:564-569

van Gorkom LCM, Hook JM, Logan MB, Hanna JV, Wasylishen RE (1995) Solid-state lead-207 NMR of lead(II) nitrate: localized heating effects at high magic angle spinning speeds. Mag Reson Chem 33:791-795

van Tendeloo G, Ghose S, Amelinckx S (1989) A dynamical model for the $P\bar{1}-I\bar{1}$ phase transition in anorthite, I. Evidence from electron microscopy. Phys Chem Minerals 16:311-319

van Tendeloo G, van Landuyt J, Amelinckx S (1976) The $\alpha \rightarrow \beta$ phase transition in quartz and $AlPO_4$ as studied by electron microscopy and diffraction. Phys Stat Sol 33:723-735

Wright AF, Leadbetter AJ (1975) The structures of the β-cristobalite phase of SiO_2 and $AlPO_4$. Philos Mag 31:1391-1401

Wood BJ, Kirkpatrick RJ, Montez B (1986) Order-disorder phenomena in $MgAl_2O_4$ spinel. Am Mineral 71:999-1006

Xiao YH, Kirkpatrick RJ, Kim YJ (1993) Structural phase transitions of tridymite—a ^{29}Si MAS NMR investigation. Am Mineral 78:241-244

Xiao YH, Kirkpatrick RJ, Kim YJ (1995) Investigations of MX-1 tridymite by ^{29}Si MAS NMR —modulated structures and structural phase transitions. Phys Chem Minerals 22:30-40

Yang H, Ghose S, Hatch DM (1993) Ferroelastic phase transition in cryolite, Na_3AlF_6, a mixed fluoride perovskite: high temperature single crystal X-ray diffraction study and symmetry analysis of the transition mechanism. Phys Chem Minerals 19:528-544

Yang W-H, Kirkpatrick RJ, Henderson, DM (1986) High-resolution ^{29}Si ^{27}Al, and ^{23}Na NMR spectroscopic study of Al-Si disordering in annealed albite and oligoclase. Am Mineral 71:712-726

9

Insights into Phase Transformations from Mössbauer Spectroscopy

Catherine A. McCammon

Bayerisches Geoinstitut
Universität Bayreuth
95440 Bayreuth, Germany

INTRODUCTION

The Mössbauer effect is the recoilless absorption and emission of γ-rays by specific nuclei in a solid (Mössbauer 1958a,b), and provides a means of studying the local atomic environment around the nuclei. It is a short-range probe, and is sensitive to (at most) the first two coordination shells, but has an extremely high energy resolution that enables the detection of small changes in the atomic environment. Mössbauer spec-troscopy therefore provides information on phase transformations at the microscopic level.

Mössbauer spectra of materials can be recorded under a large range of conditions, including temperatures from near absolute zero to at least 1200°C, and pressures to at least 100 GPa. Spectra can also be collected under different strengths of external magnetic field, currently to at least 15 T. This enables *in situ* observations of changes to the atomic environment before, during and after phase transformations under varying conditions. For phase transformations that are quenchable, it is possible to characterise changes between polymorphs at conditions where spectral resolution is optimal. Over 100 different Mössbauer transitions have been observed, although unfavourable nuclear properties limit the number of commonly used nuclei. The 14.4 keV transition in ^{57}Fe is by far the most studied, and will be the focus of this chapter since iron is the most relevant nucleus for mineralogical applications.

The aim of this chapter is to provide a brief background to Mössbauer spectroscopy within the context of phase transformations. The relevant parameters are summarised and the effect of temperature and pressure are discussed, particularly with reference to identifying phase transformations and characterising the electronic and structural environment of the Mössbauer nuclei. Instrumentation is summarised, particularly as it relates to *in situ* measurements of phase transformations, and a brief survey of applications is given. The appendix includes a worked example that illustrates the methodology of investigating a phase transformation using *in situ* Mössbauer spectroscopy. Numerous textbooks and review chapters have been written on Mössbauer spectroscopy, and a selection of the most relevant ones as well as some useful resources are listed in Table 1.

MÖSSBAUER PARAMETERS

The interactions between the nucleus and the atomic electrons depend strongly on the electronic, chemical and magnetic state of the atom. Information from these hyperfine interactions is provided by the hyperfine parameters, which can be determined experimentally from the line positions in a Mössbauer spectrum. The following section gives only a brief description of the parameters themselves, since this information is widely available in the references listed in Table 1. The section focuses on the structural and electronic information available from the parameters, and on the influence of

1529-6466/00/0039-0009$05.00

Table 1. List of resources for Mössbauer spectroscopy.

Type	*Reference*
Book	Bancroft, G.M. Mössbauer Spectroscopy. An Introduction for Inorganic Chemists and Geochemists. McGraw Hill, New York, 1973.
	Berry, F.J. and Vaughan, D.J. (eds.) Chemical Bonding and Spectroscopy in Mineral Chemistry, Chapman and Hall, London, 1986. See Chapter on Mössbauer spectroscopy in mineral chemistry, A.G. Maddock, p. 141-208.
	Cranshaw, T.E., Dale, B.W., Longworth, G.O. and Johnson, C.E. Mössbauer Spectroscopy and its Applications, Cambridge University Press, Cambridge, 1986.
	Dickson, D.P. and Berry, F.J. (eds.) Mössbauer Spectroscopy, Cambridge University Press, Cambridge, 1986.
	Gibb, T.C. Principles of Mössbauer Spectroscopy, Chapman and Hall, London, 1977.
	Gonser, U. (ed.) Mössbauer Spectroscopy, Topics in Applied Physics, Vol. 5, Springer-Verlag, Berlin, 1975.
	Greenwood, N.N. and Gibb, T.D. Mössbauer Spectroscopy, Chapman and Hall, London, 1971.
	Gütlich, P., Link, R. and Trautwein, A., Mössbauer Spectroscopy and Transition Metal Chemistry, Springer-Verlag, Berlin, 1978.
	Hawthorne, F.C. (ed.) Spectroscopic Methods in Mineralogy and Geology, Rev. Mineral. Vol. 18, Mineralogical Society of America, 1988. See Chapter on Mössbauer Spectroscopy, F.C. Hawthorne, p. 255-340.
	Long, G.L. and Grandjean, F. (eds.) Mössbauer Spectroscopy Applied to Inorganic Chemistry, Vols. 1-3; Mössbauer Spectroscopy Applied to Magnetism and Materials Science, Vols. 1-2, Plenum Press, New York and London, 1984, 1987, 1989, 1993, 1996.
	Robinson, J.W. (ed.) Handbook of Spectroscopy, Vol. 3, CRC Press, Inc., Boca Raton, USA, 1981. See Chapter on Mössbauer Spectroscopy, J.G. Stevens (ed.), p. 403-528.
Journal	Analytical Chemistry (American Chemical Society, Washington DC) contains biennial reviews (starting in 1966) of Mössbauer spectroscopy, see for example Vol. 62, p. 125R-139R, 1990.
	Hyperfine Interactions (J.C. Baltzer AG, Basel) publishes proceedings from various Mössbauer conferences, see for example Vol. 68-71, 1992.
	Mössbauer Effect Reference and Data Journal (Mössbauer Effect Data Center, Asheville, NC) contains references and Mössbauer data for nearly all Mössbauer papers published.
Data Resource	Mössbauer Effect Data Center Mössbauer Information System (maintained by the Mössbauer Effect Data Center, Asheville, NC) contains extensive bibliographic and Mössbauer data entries compiled from the Mössbauer literature. Searches of the database are possible; contact the Mössbauer Effect Data Center for details (http://www.unca.edu/medc).
	Stevens, J.G., Pollack, H., Zhe, L., Stevens, V.E., White, R.M. and Gibson, J.L. (eds.) Mineral: Data and Mineral: References, Mössbauer Handbook Series, Mössbauer Effect Data Center, University of North Carolina, Asheville, North Carolina, USA, 1983.
	Mössbauer Micro Databases (Mössbauer Effect Data Center, Asheville, NC) cover many topics including Minerals. Databases are set up to run on IBM-compatible microcomputers and can be searched using various options.
	Mössbauer Information Exchange (MIX) is a project of the KFKI Research Institute for Particle and Nuclear Physics, Budapest, Hungary. It includes an e-mail list, archives of programs and other documents, databases, current information and relevant links, all accessible through their world wide web site (http://www.kfki.hu/~mixhp/).

temperature and pressure. Often phase transformations involve only small changes in the atomic environment, so in these cases it is only through observation of hyperfine parameter trends with temperature or pressure that a transformation can be identified using Mössbauer spectroscopy.

Isomer shift

The isomer shift arises through an electric monopole interaction between the positive nuclear charge and the electric field of the surrounding electrons. This interaction causes a shift of the nuclear energy levels compared to an unperturbed nucleus, where the magnitude of the shift is a function of the difference in *s*-electron densities between the source and absorber nuclei (Fig. 1). In addition to the intrinsic (or chemical) isomer shift, δ_0, there is also a small shift due to atomic vibrations, called the second-order Doppler shift, δ_{SOD}. The second-order Doppler shift is a strong function of temperature and decreases to zero at 0 K. Experimentally the total shift of the Mössbauer spectrum from the reference zero point is simply $\delta_0 + \delta_{SOD}$. To avoid confusion, the experimentally observed shift is generally referred to as the *centre shift*, and is generally given the unsubscripted symbol δ. The contribution from δ_{SOD} is similar for most standard materials, so for purposes of comparison the isomer shift is often taken to be equal to the centre shift. All energy shifts are measured relative to a reference zero point, which is conventionally taken to be the centre shift of α-Fe at room temperature and pressure.

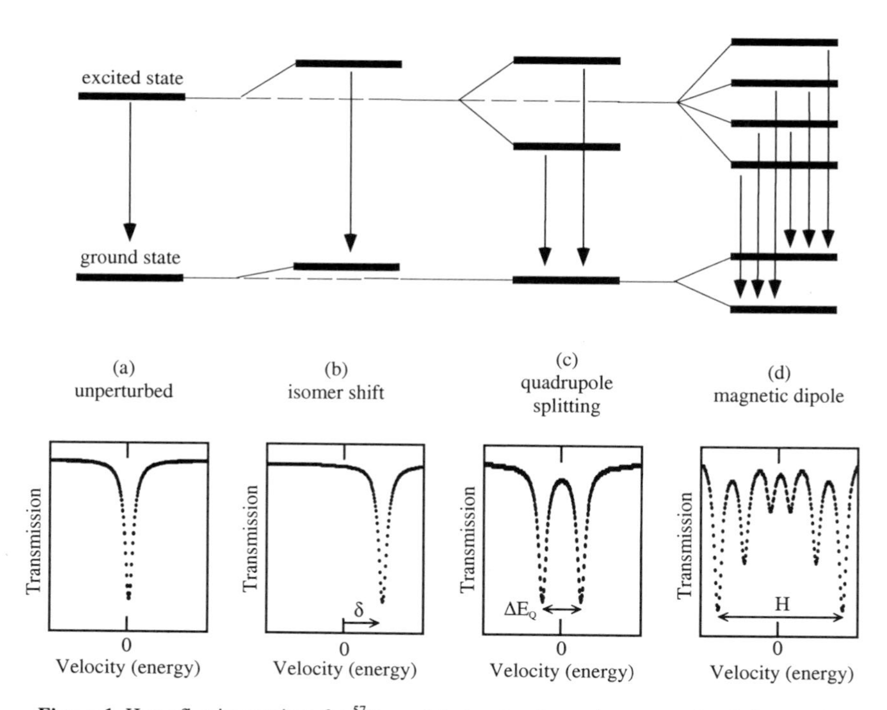

Figure 1. Hyperfine interactions for ^{57}Fe nuclei, showing the nuclear energy level diagram for (a) an unperturbed nucleus; (b) electric monopole interaction (isomer shift); (c) electric quadrupole interaction (quadrupole splitting); and (d) magnetic dipole interaction (hyperfine magnetic splitting). Each interaction is shown individually, accompanied by the resulting Mössbauer spectrum.

The isomer shift is related to the *s*-electron density, which is in turn affected by *p*- and *d*-electrons through shielding effects. The observed centre shift can therefore be used to discriminate between different valence states, spin states and coordinations of the absorber atoms. Ideally we would like to be able to calculate the centre shift expected for a specific structural and electronic configuration, but advances in computational methods have not yet progressed to the point where this is possible for most minerals. Empirically it has been noted that centre shifts for different configurations of Fe fall into specific ranges (Fig. 2), and although there is some overlap, they generally serve to identify the spin state, valence state and (usually) coordination number corresponding to a given iron site.

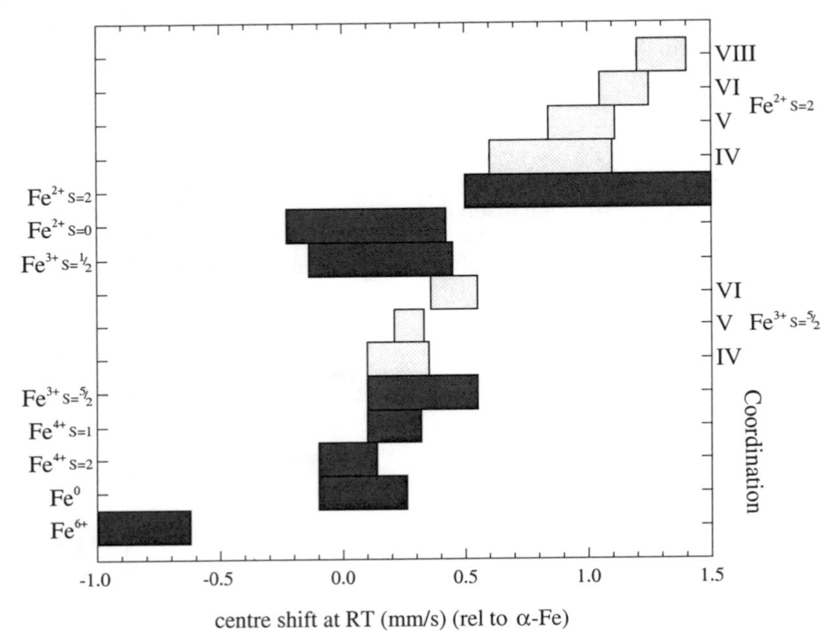

Figure 2. Approximate range of room temperature centre shifts (relative to α-Fe) observed for iron compounds. Black bars refer to data compiled from Greenwood and Gibb (1971), Maddock (1985) and Hawthorne (1988). Grey bars represent ranges for high-spin Fe^{2+} and Fe^{3+} in minerals reported by Seifert (1990), where additional data from Burns and Solberg (1990) have been added for pentacoordinated Fe^{3+}.

The temperature dependence of the isomer shift can be written

$$\delta(T) = \delta_0(T) + \delta_{SOD}(T). \quad (1)$$

To a first approximation, the variation of δ with temperature reflects changes in δ_{SOD} only, which can be expressed using the Debye model as

$$\delta_{SOD}(T) = -\frac{9}{16}\frac{k_B\Theta_M}{mc} - \frac{9}{2}\frac{k_B T}{mc}\left(\frac{T}{\Theta_M}\right)^3 \int_0^{\Theta_M/T} \frac{x^3\,dx}{e^x - 1}, \qquad T < \Theta_M \quad (2)$$

where m is the mass of the Mössbauer isotope, k_B is the Boltzmann constant, c is the speed of light and Θ_M is the characteristic Mössbauer temperature (Pound and Rebka 1960). Θ_M should not be confused with the Debye temperature, Θ_D, determined from specific heat measurements, which is based on a different weight of the phonon frequency distribution (e.g. Kolk 1984). The isomer shift, δ_0, also has a temperature dependence through the volume effect and the change in population of the electron orbitals (e.g. Perkins and Hazony 1972), but generally this is significantly smaller than the temperature dependence of δ_{SOD}. A phase transformation may be recognised from a plot of centre shift versus temperature either (1) as a discontinuity if there is significant structural rearrangement and hence a change in the value of $\delta_0(T)$; or (2) as an anomalous value of Θ_M fitted to the data if there is a more subtle or a continuous change to the atomic environment. Typical values for Θ_M are given by De Grave and Van Alboom (1991).

Both the isomer shift, δ_0, and the second-order Doppler shift, δ_{SOD}, are pressure dependent through the volume dependence of the electronic charge density at the nucleus. The pressure dependence of the centre shift at constant temperature can be approximated by

$$\partial\delta/ P = \partial\delta_0 /\partial P + \partial\delta_{SOD}/\partial P. \tag{3}$$

The volume dependence of δ_{SOD} can be estimated following the method of Williamson (1978):

$$\frac{\partial\,\delta_{\mathbf{SOD}}}{\partial \ln V} = \gamma\frac{9\mathbf{k_B}\Theta_\mathbf{D}}{16m\mathbf{c}}F(T/\Theta_\mathbf{D}) \tag{4}$$

where γ is the lattice Grüneisen parameter, Θ_D is the lattice Debye temperature and the other symbols are as for Equation (1). The function F is given by:

$$F(T/\Theta_D) = 1 + \frac{8}{e^{\Theta_D/T}-1} - 24(T/\Theta_D)^4 \int_0^{\Theta_D/T} \frac{x^3}{e^x - 1}dx. \tag{5}$$

The volume dependence of δ_{SOD} is positive, implying that δ_{SOD} decreases with increasing pressure, but increases with increasing temperature. The effect of pressure on the isomer shift is more difficult to determine, and for ^{57}Fe compounds can be affected by (1) changes in orbital occupation; and (2) distortion of the wavefunctions—either compression of the *s* electrons or spreading out of the 3*d* electrons (increase in covalency) (Drickamer et al. 1969). The second factor, an increase in covalency, causes an increase in *s* electron density and hence a decrease in δ_0. The first factor includes transfer of electrons between the iron 3*d* and 4*s* levels, between different iron 3*d* levels, and between iron orbitals and orbitals of the coordinating anions. It can either increase or decrease δ_0 depending on the nature and direction of transfer. Generally the pressure effect on δ_0 is significantly greater than for δ_{SOD}, so the latter is often taken to be zero (Amthauer 1982), although there are exceptions (e.g. McCammon and Tennant 1996). A phase transformation may be recognised from a plot of centre shift versus pressure either (1) as a discontinuity if there is a significant change in bonding and/or crystal structure; or (2) as a change in slope if the change to the atomic environment is more subtle.

Quadrupole splitting

Quadrupole splitting arises through a quadrupole interaction between the nuclear quadrupolar moment and the electric field gradient (EFG) at the nucleus due to the surrounding electrons, causing a splitting of the nuclear energy states. In the case of ^{57}Fe the excited state is split into two levels, giving rise to a doublet with equal component linewidths and areas in the ideal random absorber case (Fig. 1). For the case of an oriented absorber, the areas vary according to the angle between the principal axis of the EFG (V_{zz}) and the propagation direction of the γ-ray. The quadrupole splitting, ΔE_Q, measures the difference in energy between the excited states, and can be expressed as

$$\Delta E_Q = \frac{1}{2} eQV_{zz}\sqrt{1+\frac{\eta^2}{3}} \tag{6}$$

where Q is the nuclear quadrupole moment, e is the electron charge, V_{zz} is the principal component of the EFG tensor and η is the asymmetry parameter, where $\eta = (V_{xx} - V_{yy})/V_{zz}$ and $0 \le \eta \le 1$. According to the static crystal-field model of Ingalls (1964), the EFG components can be related to crystal and electronic structural factors through the following:

$$V_{zz} = (1-R)eq_{val} + (1-\gamma_\infty)eq_{lat}$$

$$V_{xx} - V_{yy} = (1-R)e\eta_{val}q_{val} + (1-\gamma_\infty)e\eta_{lat}q_{lat}, \tag{7}$$

where the subscript "val" refers to the asymmetric charge distribution of the valence electrons, and the subscript "lat" refers to the deviation from cubic symmetry of the neighbouring atoms in the crystalline lattice. R and γ_∞ are the Sternheimer shielding and anti-shielding factors, respectively, and are added to correct for polarisation effects. The valence and lattice contributions can be cast in terms of a reduction function, F:

$$\Delta E_Q = \Delta E_{Q,val}(0)F(\Delta_1, \Delta_2, \lambda_0, \alpha^2, T) + \Delta E_{Q,lat} \tag{8}$$

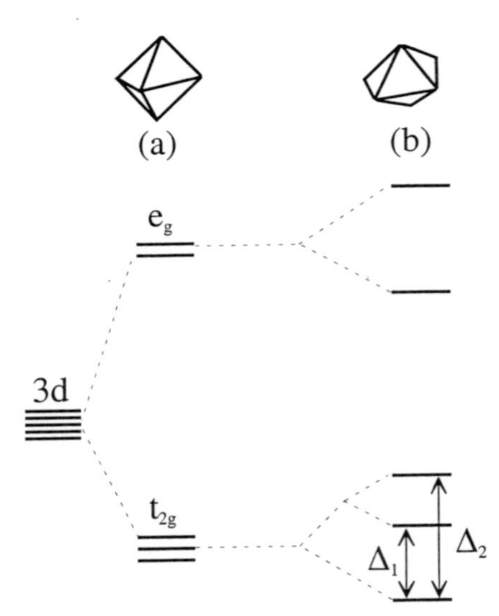

Figure 3. Energy levels of 3d electrons in (a) regular octahedron; (b) highly distorted octahedron. The splitting of energy levels in the ground state are given by Δ_1 and Δ_2 (adapted from Burns 1993).

where $\Delta E_{Q,val}(0)$ is the valence contribution at 0 K, α^2 is a covalency factor, λ_0 is the spin-orbit coupling constant for the free ion, T is the temperature, and Δ_1 and Δ_2 are the ground state splittings of the crystal field levels (Fig. 3). The valence and lattice contributions to the quadrupole splitting for ^{57}Fe are always of opposite sign. Figure 4 illustrates the schematic variation of ΔE_Q with increasing values of the ground state splitting, which in the simplest case can be linked to the distortion of the site from cubic symmetry. At low distortions the lattice contribution is small, so ΔE_Q increases with increasing distortion, but at high degrees of distortion the lattice term dominates, and ΔE_Q remains constant or even decreases with increasing distortion.

The quadrupole splitting depends on the valence and spin state of the absorber atoms, as well as the coordination and degree of distortion of the crystallographic site. Values of ΔE_Q extend from zero to a maximum of approximately 3.7 mm/s for ^{57}Fe compounds, but are less diagnostic of valence and coordination compared to the centre shift. The quadrupole splitting is more sensitive to small changes in the atomic environment, however, and there are many instances where phase transformations cause a change in the quadrupole splitting, but none in the centre shift.

The quadrupole splitting is sensitive to temperature primarily through the valence contribution, which reflects the temperature dependence of electrons between different

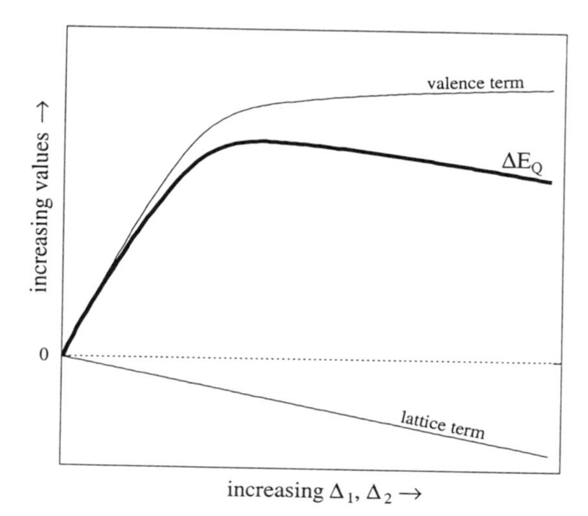

Figure 4. Schematic variation of the valence and lattice term contributions to the quadrupole splitting with increasing values of the $3d$ ground state splittings, Δ_1 and Δ_2. The quadrupole splitting, ΔE_Q, is the sum of the valence and lattice terms (adapted from Ingalls 1964).

energy levels. The lattice term is sensitive to temperature only through changes in the relative position of neighbouring atoms, and such changes are generally negligible unless a phase transformation takes place. The crystal field model of Ingalls (1964) can be used to fit experimental quadrupole splitting data as a function of temperature in order to extract parameters such as the ground-state splittings, and these can be useful for comparison on both sides of a phase transformation to provide an insight into changes to electronic structure through the transformation.

The quadrupole splitting is sensitive to pressure through both the valence and lattice terms as well as the covalency effect, and may increase or decrease with pressure, depending on the relative magnitude of these contributions. Similar to the temperature variation, the pressure variation of the quadrupole splitting is more likely than the centre shift to show either a discontinuity or a change in slope when a phase transformation occurs.

Hyperfine magnetic splitting

Magnetic splitting arises through a dipole interaction between the nuclear magnetic dipole moment and a hyperfine magnetic field at the nucleus, causing a splitting of the nuclear energy states. In the case of ^{57}Fe the excited state is split into four levels and the ground state into two levels, giving eight possible transitions. In the absence of a quadrupole interaction, the selection rules for magnetic dipole interactions allow only six transitions to occur, which in the ideal random absorber case give a sextet with equal component linewidths and relative line areas in the ratio 3:2:1:1:2:3 (Fig. 1). For the case of an oriented absorber, the areas vary according to the angle between the hyperfine field direction and the propagation direction of the γ-ray. In the absence of a strong quadrupole interaction, the energy difference between the outer two lines is proportional to the magnitude of the hyperfine magnetic field.

The hyperfine magnetic field seen at the nucleus can be written as the vector sum of three components:

$$\mathbf{H} = \mathbf{H}_c + \mathbf{H}_{dip} + \mathbf{H}_{orb}. \tag{9}$$

The first term, $\mathbf{H}_c$, generally has the largest magnitude, and is related to the spin density at the nucleus (Fermi contact term). $\mathbf{H}_{dip}$ and $\mathbf{H}_{orb}$ arise from the dipolar interaction of the magnetic moment of the atom with the nucleus (dipolar term), and the orbital angular momentum of the atom (orbital term), respectively. The relative size of each term determines the sign and magnitude of the internal field, where the orbital and dipolar terms have opposite signs to the Fermi contact term. In the case of ^{57}Fe, magnetically-ordered compounds containing high-spin Fe^{3+} tend to have large negative hyperfine magnetic fields due to the large value of the Fermi contact term, while magnetically-ordered high-spin Fe^{2+} compounds tend to have smaller fields which may be positive or negative, depending on the values of the competing orbital and dipolar fields.

Hyperfine magnetic splitting is seen in a Mössbauer spectrum when the hyperfine magnetic field is present at the nucleus over a period longer than the Larmor precession time. In the case of paramagnetic absorbers, spin-spin and spin-lattice relaxation times are generally much faster than the Larmor precession time, hence the average field seen at the nucleus is zero and no magnetic splitting is observed. Below the Curie or Néel temperature, however, magnetic exchange interactions establish a hyperfine magnetic field on a significantly longer timescale, hence magnetic splitting is observed. If the timescale of magnetic fluctuations is comparable to the Larmor precession time, relaxation effects cause broadening and changes to the lineshape of the resulting Mössbauer spectrum.

The variation of the magnetic hyperfine field with temperature provides information on the bulk magnetic properties. For absorbers where $\mathbf{H}$ is dominated by the Fermi contact term (e.g. high-spin Fe^{3+} compounds), the hyperfine magnetic field measured by Mössbauer spectroscopy is proportional to the sublattice magnetisation, and the simplest model that describes the variation of magnetisation with temperature is the Weiss molecular field theory (originally described by Weiss 1906). Using this model the temperature dependence of the hyperfine magnetic field is given by the equation

$$H(T) = H_0 B_J\left(\frac{3J}{(J+1)}\frac{H(T)}{H_0}\frac{T_m}{T}\right) \tag{10}$$

(see Morrish 1965 for a derivation) where H_0 is the saturation field at $T = 0$ K, T_m is the magnetic transition temperature, J is the total angular momentum, and the Brillouin function $B_J(x)$ is defined as

$$B_J(x) = \frac{2J+1}{2J}coth\left[\frac{2J+1}{2J}x\right] - \frac{1}{2J}coth\left(\frac{x}{2J}\right). \tag{11}$$

T_m represents the temperature at which the thermal energy becomes equal to the magnetic exchange energy, and is related to the strength of exchange interactions between atomic dipoles. In the case where there are substantial contributions to the hyperfine magnetic field from $\mathbf{H}_{orb}$ and $\mathbf{H}_{dip}$ (e.g. high-spin Fe^{2+} compounds), the hyperfine magnetic field may not be strictly proportional to the magnetisation, and hence T_m in Equation (10) might not reflect the bulk magnetic transition temperature. In this case, however, it is possible to determine the latter quantity from relative line intensities of spectra collected under applied magnetic fields (e.g. Johnson 1989). The value of $|\mathbf{H}|$ is likely to change at phase transformations that involve structural rearrangement, but may also change at magnetic transitions such as a spin reorientation. In the case where changes to |**H**| are more subtle, a plot of hyperfine field versus temperature fitted to Equation (10) may indicate the transition temperature.

The hyperfine magnetic field is sensitive to pressure mainly through the value of the Fermi contact term, and can be affected by the competing effects of enhanced spin density due to reduced interatomic distances, or decreased spin density due to changes in interatomic angles that reduce magnetic exchange interactions. A phase transformation may be recognised from a plot of hyperfine field versus pressure either as a discontinuity or a change in slope.

Table 2. Appearance of magnetic hyperfine Mössbauer spectra with ideal sextet intensity ratios for various magnetic structures as a function of applied magnetic field.

External field	*Ferromagnet*	*Antiferromagnet*	*Ferrimagnet*	*Canted antiferromagnet*
	↑↑↑↑	↑↓↑↓	↑↓↑↓	erer&&
$\mathbf{H}_{ext} = 0$	one subspectrum	one subspectrum	two subspectra	one subspectrum
$\mathbf{H}_{ext}$ // easy axis	one subspectrum $\mathbf{H}_{obs} = \mathbf{H}_{hf} + \mathbf{H}_{ext}$ 3:0:1:1:0:3	two subspectra $\mathbf{H}_{obs} = \mathbf{H}_{hf} \pm \mathbf{H}_{ext}$ 3:0:1:1:0:3	two subspectra $\mathbf{H}_{obs} = \mathbf{H}_{hf} \pm \mathbf{H}_{ext}$ 3:0:1:1:0:3	one subspectrum $\mathbf{H}_{obs} = \mathbf{H}_{hf} + \mathbf{H}_{ext}$ 3:x:1:1:x:3

In addition to variations in temperature and pressure, application of an external magnetic field can also induce phase transformations. These involve changes to the magnetic structure of an ordered array of atomic spins that is described by the relative alignment of spins to each other, and the orientation of the spins to the crystallographic axes. Mössbauer spectroscopy can be used to deduce the arrangement of atomic spins through one or more of the following: (1) spectral fitting to determine the number of magnetic subspectra compared to the number of iron sites in the crystal lattice; (2) application of an external magnetic field to determine the effect on the resulting hyperfine field; and (3) application of an external magnetic field to determine the change in the relative areas of component lines in the magnetic subspectra (Table 2). Similar methods can be used to deduce changes to the magnetic structure under the influence of an externally applied magnetic field.

Relative area

The area of the absorption line is related to the number of iron atoms per unit area of the absorber. The expression is complex in most cases, however, so Mössbauer spectroscopy is generally not a practical method for determining absolute iron concentrations. The situation is more favourable, however, for determination of relative abundances based on relative spectral areas. For this discussion it is useful to define the dimensionless absorber thickness

$$t_a = \sigma_0 f_a n_a \tag{12}$$

where σ_0 is the cross-section at resonance for the Mössbauer transition (= 2.56×10^{-18} cm^2 for ^{57}Fe), f_a is the recoil-free fraction of the absorber, and n_a is the number of ^{57}Fe atoms per cm^2. A corresponding dimensionless thickness can be defined for the source. In the thin source and absorber approximation (t_a, $t_s << 1$), the total spectral area can be written as the sum of the individual subspectral contributions, where the area of each subspectrum is proportional to the total number of ^{57}Fe atoms occupying the corresponding site. The thin source approximation is usually satisfied for conventional ^{57}Co sources (but not point sources!), while the thin absorber approximation as given above usually is not. The extent to which the condition $t_a << 1$ can be relaxed in the case of intrinsically broadened linewidths (arising from differences in next-nearest neighbour configurations, for example) is discussed by Ping and Rancourt (1992).

In the thin source and absorber approximation, the area of the subspectrum corresponding to the ith site can be written

$$A_i \propto f_s t_{a,i} \Gamma_i = \sigma_0 f_s \Gamma_i f_{a,i} n_{a,i} \tag{13}$$

where Γ_i is the linewidth, which according to Margulies and Ehrmann (1961) can be represented for absorber thicknesses up to $t_a \approx 4$ by

$$\Gamma_i = \Gamma_0 (2 + 0.27 t_a), \tag{14}$$

where Γ_0 is the natural linewidth (= 0.097 mm/s for ^{57}Fe). In the case of non-overlapping lines, the relative areas can be used through the above equations to determine the relative abundance of the Mössbauer nuclei in the sites corresponding to the different subspectra. Equation (13) is not valid when the thin absorber approximation is not fulfilled, however. Calculations by Rancourt (1989) illustrate the extent to which relative area fractions are overestimated with increasing absorber thickness, and a subsequent paper by Rancourt et al. (1993) provides guidelines for calculating absorber thicknesses where the thin absorber approximation is still approximately valid.

The relative area of the Mössbauer spectrum depends on temperature primarily through changes in the recoil-free fraction of the absorber (Eqn. 13). The variation of f_a with temperature can be determined from the Debye model

$$f(T) = exp\left\{ -\frac{3}{2} \frac{E_R}{k_B \Theta_M} \left[1 + 4 \left(\frac{T}{\Theta_M} \right)^2 \int_0^{\Theta_M/T} \frac{x dx}{e^x - 1} \right] \right\} \tag{15}$$

(Mössbauer and Wiedemann 1960) where E_R is the recoil energy (= 3.13425×10^{-22} J for the 14.4 keV transition of ^{57}Fe), k_B is the Boltzmann constant, and Θ_M is the characteristic Mössbauer temperature, where Θ_M can be determined from temperature dependant measurements of the centre shift as described by Equation (2). The recoil-free fraction of the absorber can also be determined through temperature dependant measurements of the relative area according to Equation (13). In either case, a phase transformation can be recognised as either a discontinuity or a change in slope of the recoil-free fraction plotted as a function of temperature.

The relative area of the Mössbauer spectrum varies with pressure also primarily through changes in the recoil-free fraction of the absorber (Eqn. 13). The volume dependence of the recoil-free fraction is similar to that of δ_{SOD}, since both are related to the value of Θ_M (Eqns. 2 and 15). The recoil free fraction therefore increases with pressure, and similar to the temperature variation discussed above, a phase transformation could cause either a discontinuity or a change in slope in the variation of recoil-free fraction with pressure.

INSTRUMENTATION

A Mössbauer apparatus is relatively simple and can be divided into three parts – the source, the absorber and the detector. The modular nature of the spectrometer enables Mössbauer spectroscopy to be performed under varying conditions in a number of different configurations. Many of these can be interchanged and are listed in Figure 5.

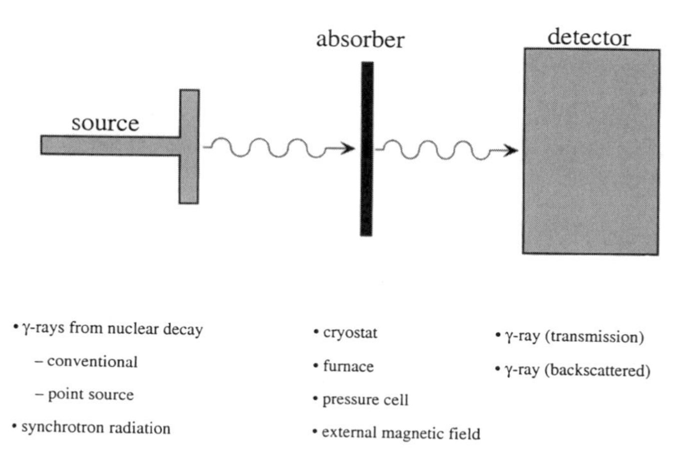

Figure 5. Schematic view of a Mössbauer spectrometer showing various possibilities for the source, absorber and detector. Nearly all the individual configurations are independent, enabling numerous combinations to be made.

Variables that can be applied to the absorber include temperature, pressure and magnetic field. The absorber can be placed inside a cryostat (e.g. Williams 1975) or a furnace (e.g. Kolk et al. 1985), or it can be placed inside a pressure device such as a diamond anvil cell (DAC) (e.g. Pasternak and Taylor 1996). Simultaneous P,T measurements can be made through combination of the above, such as a DAC + cryostat (Hearne et al. 1994) and a DAC + furnace (Hålenius et al. 1996). External magnetic fields can be applied to the absorber (e.g. Craig 1965) and combined with studies under high pressure and variable temperature.

The source is generally a radioactive parent of the Mössbauer isotope, and γ-rays are produced through nuclear decay. For radioactive parents that are magnetically ordered, such as ^{57}Co, a non-magnetic matrix is used (e.g. Rh) to dilute the atoms sufficiently such that the emitted radiation is a single energy with no detectable broadening by magnetic interactions and a low probability of self-absorption (i.e. absorption by another nucleus in the source). The latter ensures that the source thickness (which affects the linewidth and recoil-free fraction) does not increase significantly with time. The source diameter is then determined by the number of parent nuclei required to achieve a sufficient count rate, and for conventional ^{57}Co sources is of the order 1 cm. This diameter can be significantly reduced if the density of parent nuclei is increased to the level where broadening due to magnetic interactions just starts to occur. For ^{57}Co nuclei, this enables the source to be reduced to a diameter of ca. 500 μm without significant loss in count rate. The penalty, however, is a rapid increase in source thickness, which limits the useable lifetime of the source. Such concentrated sources, so-called "point sources", are essential for high-pressure experiments, but can also be used to record Mössbauer spectra of microscopic samples ($d \geq 50$ μm) at ambient conditions (McCammon et al. 1991, McCammon 1994).

Source radiation can also be generated by a particle accelerator, such as a synchrotron facility. There are many advantages to such an experiment, including the possibility for a small beam size with high signal/noise ratio that does not suffer from the limitations described above. Some recent reviews are given by Arthur (1996) and Ruffer and Chumakov (1996).

The geometry of the Mössbauer spectrometer can be varied depending on the type of experiment. Transmission studies require a linear arrangement where the detector sits behind the source and absorber, such as illustrated in Figure 5, and provide information on all ^{57}Fe nuclei in the absorber. The detector can also be placed in a backscattering geometry between the source and absorber, enabling the study of only the surface nuclei in the absorber. Recent reviews of the latter technique, called Conversion Electron Mössbauer Spectroscopy (CEMS) are given by Nomura et al. (1996) and Gancedo et al. (1997).

Many of the above possibilities for source, absorber and detector can be combined, enabling a wide range of instrumental possibilities to use Mössbauer spectroscopy in the study of phase transformations. Additional possibilities are provided through absorber and source preparation for Mössbauer isotopes such as ^{57}Fe that have a high absorption cross section. For example, ^{57}Fe has a natural abundance of only 2.14%, yet adequate signal/noise ratios can be obtained for absorbers containing as little as 5 wt % FeO. This enables enriched ^{57}Fe to be used for preparation of synthetic samples with barely detectable amounts of Fe; hence phase transformations can be studied in nearly iron-free systems. ^{57}Fe can be added homogeneously to study bulk properties, or it can be added inhomogeneously (e.g. deposited in a thin layer) to study surface properties. The same enrichment possibilities exist for ^{57}Co as a source (in this case a single-line absorber would be used), although experimentally it is more difficult due to the need to work with radioactive materials.

Collection of ^{57}Fe Mössbauer spectra with adequate signal/noise ratios requires a minimum of several hours under ideal conditions, extending to several days or weeks for experiments where conditions are less than optimal. This is most relevant for *in situ* experiments, because it limits the number of temperatures and/or pressures at which spectra can be collected. One technique that avoids this limitation is thermal scanning (e.g. Nolle et al. 1983), although the information provided is generally limited to only the phase transformation temperature.

APPLICATIONS

Since its discovery in 1958, the Mössbauer effect has been used to characterise the nature of numerous phase transformations. As of 1999, the Mössbauer Effect Data Center has reported nearly 5000 papers that were published with keywords relating to phase transformations. The sensitivity of the Mössbauer effect enables its use not only for detecting phase transformations, but through the hyperfine parameters and their variation with temperature, pressure and magnetic field, allows a detailed characterisation of the nature of the phase transformation. The following provides a brief survey of applications of Mössbauer spectroscopy for studying different types of phase transformations. It is not intended to be a comprehensive review, but rather provides a sense of the current state of the art. A number of reviews have appeared on the application of Mössbauer

(Shenoy 1973, Wertheim 1973, Preston 1978, Johnson 1989, Seifert 1990).

Structural transformations

Structure modification. Phase transformations that involve a change in coordination number of the site containing the Mössbauer nucleus generally involve large changes to the hyperfine parameters, while those that involve modification only in the second coordination shell may show smaller changes. The quadrupole splitting and hyperfine magnetic field (if present) are generally the most sensitive to such changes. This is illustrated by pressure-induced phase transformations in the compound $Fe_xMg_{1-x}SiO_3$, which for values of $x \approx 0.1$ include the following structures:

orthopyroxene → clinopyroxene → garnet → perovskite

(e.g. Akaogi et al. 1992, Woodland and Angel 1997). The transformation from orthopyroxene to clinopyroxene involves changes primarily in the second coordination shell, and cannot be quenched to ambient conditions. McCammon and Tennant (1996) studied the transition in $FeSiO_3$ *in situ* as a function of pressure at room temperature, and were able to identify the phase transformation from a discontinuity in the quadrupole splitting variation with pressure, while the centre shift data showed (at most) a break in slope. From the relative changes to the hyperfine parameters, they were able to extract information regarding the electronic structure and site distortion that could be correlated with high-pressure data from optical spectroscopy and X-ray diffraction. The phase transformation of $Fe_xMg_{1-x}SiO_3$ from the pyroxene to the garnet and perovskite structures involves a change to the coordination number of the iron sites, and both can be quenched to ambient conditions. Table 3 lists the hyperfine parameters of all polymorphs at room temperature, and illustrates (1) the minimal change to hyperfine parameters between the orthopyroxene and clinopyroxene structures, the largest being the quadrupole splitting of the M1 site which was used by McCammon and Tennant (1996) to identify the phase transformation at high pressure; (2) the slightly larger change to the centre shift of Fe^{2+} between six-fold and eight-fold coordination compared to the negligible change between six-fold and the 8-12 coordinated site of the perovskite structure, the latter due to factors including site distortion and reduced Fe^{2+}-O distance; and (3) the large difference in quadrupole splitting of Fe^{2+} for sites with the same coordination number, due primarily to differences in distortion between the sites.

Table 3. Room temperature hyperfine parameters for $Mg_{1-x}Fe_xSiO_3$ polymorphs.

Structure	x_{Fe}	*Site*	δ *(mm/s)*	ΔE_Q *(mm/s)*	*Reference*
opx	0.1	$^{VI}Fe^{2+}$ M1	1.171(5)	2.49(1)	[1]
		$^{VI}Fe^{2+}$ M2	1.142(5)	2.11(1)	
cpx	0.11	$^{VI}Fe^{2+}$ M1	1.175(5)	2.58(1)	[1]
		$^{VI}Fe^{2+}$ M2	1.139(5)	2.11(1)	
garnet	0.1	$^{VI}Fe^{2+}$	1.146(5)	1.26(1)	[1]
		$^{VIII}Fe^{2+}$	1.259(5)	3.60(1)	
perovskite	0.09	$^{VIII-XII}Fe^{2+}$	1.13(1)	1.77(2)	[2]

References: [1] McCammon, unpublished data; [2] Lauterbach et al. (2000)

Order-disorder. Phase transformations that involve the long-range ordering of cations over two or more lattice sites can generally be studied using Mössbauer spectroscopy if one of the atoms has a Mössbauer transition, and the resulting spectrum is sufficiently resolved that robust relative areas can be determined. One classic example is the pyroxene solid solution $MgSiO_3$-$FeSiO_3$, which has been studied in both the ortho- (e.g. Virgo and Hafner 1969, Skogby et al. 1992, Domeneghetti and Stefan 1992) and clino- (Woodland et al. 1997) structure modifications. Careful analysis of the relative areas enabled site occupancies to be extracted that were consistent with structure refinements, and allowed a long-range order parameter to be calculated. Short-range ordering of isovalent cations involving the Mössbauer atom is more difficult to study, since one is looking for differences in next-nearest neighbour configurations. These generally cause slight variations in hyperfine parameters that give rise to broadened, asymmetrical spectral lines. Spectra can be analysed and compared with results expected from possible arrangements, and in some cases it is possible to deduce whether atoms are randomly distributed or arranged in clusters, such as in the system $FeCr_2S_4$-$FeRh_2S_4$ thiospinel (Riedel and Karl 1980) and $MgSiO_3$-$FeSiO_3$ clinopyroxene (Angel et al. 1998). A similar situation exists for ordering studies that involve non-Mössbauer isovalent cations, since effects also occur only in the next-nearest neighbour shell. In these cases, however, it is generally not even possible to distinguish between short- and long-range ordering from the Mössbauer data, since the hyperfine parameters are only affected by, at most, the second coordination shell.

Ordering of anions or lattice defects (which for the purposes of this discussion include heterovalent cation and anion substitutions) generally involve greater changes to the environment of the Mössbauer nucleus, and hence to the hyperfine parameters. A classic example is wüstite, Fe_xO, where broad quadrupole splitting distributions in the spectra can be interpreted in terms of the most likely defect cluster arrangements (e.g. Greenwood and Howe 1972, McCammon and Price 1985). Broadening of Mössbauer spectra has been interpreted to indicate short-range ordering of heterovalent cations in aluminous orthopyroxene (Seifert 1983). A different approach can be taken if additional information is available, such as from structure or composition. McCammon et al. (2000) were able to deduce short-range ordering of oxygen vacancies in the system $CaTiO_3$-$CaFeO_{2.5}$ from comparison of octahedral, pentacoordinated and tetrahedral Fe^{3+} site occupancies with composition data.

Glass transitions. Phase transformations that involve a transition between a glass and a supercooled liquid state generally modify the recoil-free fraction at the transition, and may be accompanied by changes in quadrupole splitting and linewidth. An early review by Ruby (1973) discusses the nature of glass transitions and their observation by Mössbauer spectroscopy. One example is a high temperature study by Bharati et al. (1983) of alkali borate, alkali borosilicate and vanadate glasses containing ^{57}Fe, where distinct softening of the lattice at the glass transition temperature is observed. Changes in the degree of polymerisation and coordination of Fe in glass can also be studied using Mössbauer spectroscopy, where *in situ* measurements are preferable to studies of quenched melts. One can combine the latter with *in situ* measurements using complementary techniques, however, such as in experiments by Wang et al. (1993) who used high temperature Raman spectroscopy to study changes to polymerisation with temperature, and then used Mössbauer spectra from quenched melts to obtain information on coordination.

Pressure-induced amorphisation studies using Mössbauer spectroscopy have been briefly reviewed by Pasternak and Taylor (1996). One example is the amorphisation of Fe_2SiO_4 above 40 GPa at room temperature where antiferromagnetic ordering appears to be preserved in the glassy state (Kruger et al. 1992). This behaviour is quite different to magnetic glasses produced by rapid quenching from the melt, where magnetic order is either absent or significantly inhibited due to the loss of structural order.

Electronic transitions

Transitions that involve a change in electronic structure of the Mössbauer atom can generally be detected using Mössbauer spectroscopy, regardless of whether changes in atomic positions are involved or not. The isomer shift is particularly sensitive to the spin state of the iron atom (Fig. 2) and is one of the primary diagnostic tools for distinguishing spin transitions. Recent examples include $CaFeO_3$, which has been inferred to undergo a first-order transition from the high-spin (Fe^{4+}: $S = 2$) to the low-spin (Fe^{4+}: $S = 1$) state near 30 GPa from a combined *in situ* Mössbauer and X-ray diffraction study (Takano et al. 1991). The change in isomer shift at the transition point was greater than 0.2 mm/s.

Changes in isomer shift can also indicate variations in valence state, including time-dependent fluctuations. In these cases there is generally no observed change in crystal structure. For example, a fraction of Fe^{2+} and Fe^{3+} atoms in $(Mg,Fe)SiO_3$ perovskite are involved in electron transfer that occurs faster than the mean lifetime of the excited state ($\approx 10^{-7}$ s for ^{57}Fe), giving rise to a subspectrum with isomer shift intermediate between Fe^{2+} and Fe^{3+}. The process is thermally activated, as shown by temperature dependant measurements of Fei et al. (1994), who showed that the relative area of the quadrupole doublet assigned to the average valence state increases dramatically with increasing temperature. Isomer shift data combined with the observed collapse of two subspectra (corresponding to Fe^{3+} and Fe^{5+}) into a single spectrum (corresponding to Fe^{4+}) have been used to infer a second-order transition in $CaFeO_3$ with increasing temperature involving the delocalisation of one of the *d* electrons (Takano et al. 1977, Takano et al. 1991).

Mössbauer spectroscopy generally cannot provide independent evidence for insulator-metal transitions, but combined with other measurements such as electrical conductivity, Mössbauer data can provide important data to elucidate aspects such as transition mechanisms. Recent electrical resistance measurements of α-Fe_2O_3 show a large drop in resistivity above 50 GPa, and combined with *in situ* high-pressure Mössbauer measurements that show a loss of magnetic order at the transition, was interpreted by Pasternak et al. (1999) to indicate a Mott transition resulting from the breakdown of *d-d* electron correlations. In contrast, Rozenberg et al. (1999) found an insulator-metal transition in the perovskite $Sr_3Fe_2O_7$ above 20 GPa using electrical measurements where magnetic moments, as measured by Mössbauer spectroscopy, were preserved through the phase transition. This behaviour was interpreted to indicate band closure due to *p-p* gap closure that does not involve the *d* electrons.

Magnetic transitions

Transitions that involve a change to the magnetic structure of an ordered array of atomic spins can sometimes be detected by Mössbauer spectroscopy, depending on the nature of the transition. The largest effect is seen in transitions between the magnetically-ordered and non-ordered state (Curie or Néel point), providing that the magnetically-ordered state exists over a sufficient timescale (see section above). In these cases the change between a singlet or quadrupole doublet and a magnetically-split spectrum are usually unambiguous, enabling such studies to be conducted under marginal signal:noise conditions such as at very high pressures. Fe_xO is paramagnetic at ambient conditions, but transforms to the antiferromagnetic phase above 5 GPa according to *in situ* Mössbauer measurements of Zou et al. (1980). However at significantly higher pressures (> 90 GPa), it transforms to a non-magnetic phase, inferred by Pasternak et al. (1997) to be diamagnetic.

Transitions involving only a change in the relative alignment of spins can be detected using Mössbauer spectroscopy, although oriented measurements are usually necessary to determine the spin direction. A classic example is the Morin transition in hematite, α-Fe_2O_3, which involves a transition from a collinear antiferromagnet with spin direction along the *c* axis, to a canted structure (weak ferromagnet) with spins oriented in the basal plane. There is a change in quadrupole splitting as well as a discontinuity in the variation of hyperfine magnetic field with temperature and pressure at the Morin transition, which allowed Bruzzone and Ingalls (1983) to extract information regarding the location of atoms within the unit cell that compared well with results from high-pressure X-ray diffraction. Mössbauer measurements of oriented single crystals of the rare earth orthoferrite $TbFeO_3$ were made by Nikolov et al. (1996) at 4.2 K as a function of externally applied magnetic field. From the change in relative area of the magnetic sextet components, they were able to characterise the direction and nature of spin-reorientation transitions.

CONCLUDING REMARKS

This chapter has presented a review of the parameters involved in Mössbauer spectroscopy within the context of phase transformations. Although the Mössbauer effect can be considered a "mature" technique, now more than forty years old, technical developments continue to expand the experimental possibilities. Spatial resolution has improved within the last decade. The development of the Mössbauer milliprobe, for example, has enabled spatial resolution to be increased by more than two orders of magnitude (McCammon et al. 1991, McCammon 1994). Further improvement of spatial resolution may be anticipated with advances in nuclear forward scattering. Other possibilities on the horizon include development of a Mössbauer electron microscope which would focus conversion electrons using conventional electron optics (Rancourt and Klingelhöfer 1994).

One aspect of Mössbauer spectroscopy that has not been widely exploited in phase transformation studies is time resolution. Studies with conventional techniques are possible over a wide range of time scales, starting from the intrinsic time scale of the ^{57}Fe Mössbauer effect ($t \approx 10^{-8}$ s) to investigate processes such as electron transfer, to time scales of $t \approx 10^{-6}$ s to measure diffusion, to longer timescales of $t \geq 10^{3}$ s to study phase transitions, oxidation and other chemical reactions.

The range of *P*,*T* conditions over which Mössbauer spectroscopy can be performed continues to expand. Existing equipment already allows experiments to be performed to 120 GPa, but there are no obvious theoretical limitations that restrict advances to higher pressures. Experiments at high temperatures are ultimately limited by lattice vibrations, but since Debye temperatures generally increase with pressure, technical advances could increase temperature limits beyond 1200°C at high pressure, raising the intriguing possibility of *in situ* Mössbauer measurements at *P*,*T* conditions approaching those of the Earth's mantle.

REFERENCES

Akaogi M, Kusaba K, Susaki J, Yagi T, Matsui M, Kikegawa T, Yusa H, Ito E (1992) High-pressure high-temperature stability of αPbO_2-type TiO_2 and $MgSiO_3$ majorite: Calorimetric and *in situ* X-ray diffraction studies. *In* Y Syono, MH Manghnani (eds) High Pressure Research: Application to Earth and Planetary Sciences. Terra Scientific Publ Co/Am Geophys Union, Tokyo/Washington DC, p 447-455

Amthauer G (1982) High pressure ^{57}Fe Mössbauer studies on minerals. *In* W Schreyer (ed) High-Pressure Researches in Geoscience. E. Schweizerbart'sche Verlagsbuchhandlung, Stuttgart, p 269-292

Angel RJ, McCammon CA, Woodland AB (1998) Structure, ordering and cation interactions in Ca-free $P2_1/c$ clinopyroxenes. Phys Chem Minerals 25:249-258
Arthur J (1996) How to do resonant nuclear experiments with synchrotron radiation. Il Nuovo Cimento Soc Ital Fis 18D:213-220
Bharati S, Pathasarathy R, Rao KJ, Rao CNR (1983) Mössbauer studies of inorganic glasses through their glass transition temperatures. Sol State Comm 46:457-460
Bruzzone CL, Ingalls RL (1983) Mössbauer-effect study of the Morin transition and atomic positions in hematite under pressure. Phys Rev B 28:2430-2440
Burns RG (1993) Mineralogical Applications of Crystal Field Theory, 2nd ed. Cambridge University Press, Cambridge, UK
Burns RG, Solberg TC (1990) ^{57}Fe-bearing oxide, silicate, and aluminosilicate minerals. *In* LM Coyne, SWS McKeever, DF Blake (eds) Spectroscopic Characterization of Minerals and their Surfaces, Vol 415. American Chemical Society, Washington DC, p 262-283
Craig JR (1965) Superconducting magnets: Applications to the Mössbauer effect. *In* IJ Gruverman (ed) Mössbauer Effect Methodology, Vol. 1. Plenum Press, New York, p 135-145
De Grave E, Van Alboom A (1991) Evaluation of ferrous and ferric Mössbauer fractions. Phys Chem Min 18:337-342
Domeneghetti MC, Steffen G (1992) M1, M2 site populations and distortion parameters in synthetic Mg-Fe orthopyroxenes from Mössbauer spectra and X-Ray structure refinements. Phys Chem Min 19:298-306
Drickamer HG, Vaughn RW, Champion AR (1969) High-pressure Mössbauer resonance studies with iron-57. Accounts Chem Res 2:40-47
Fei Y, Virgo D, Mysen BO, Wang Y, Mao HK (1994) Temperature dependent electron delocalization in $(Mg,Fe)SiO_3$ perovskite. Am Mineral 79:826-837
Gancedo JR, Davalos JZ, Gracia M, Marco-Sanz JF (1997) The use of Mössbauer spectroscopy in surface studies. A methodological survey. Hyper Inter 110:41-50
Greenwood NN, Gibb TD (1971) Mössbauer Spectroscopy. Chapman and Hall, London
Greenwood NN, Howe AT (1972) Mössbauer studies of Fe(1-x)O. Part I. The defect structure of quenched samples. J Chem Soc Dalton Trans:110-116
Hålenius E, Annersten H, Jönsson S (1996) *In situ* ^{57}Fe Mössbauer spectroscopy of iron and olivine at high pressure and temperature. *In* MD Dyar, CA McCammon, M Schaeffer (eds) Mineral Spectroscopy: A Tribute to Roger G. Burns, Vol 5. Geochemical Society, USA, p 255-260
Hawthorne FC (1988) Mössbauer spectroscopy. *In* FC Hawthorne (ed) Spectroscopic Methods in Mineralogy and Geology. Rev Mineral 18:255-340
Hearne GR, Pasternak MP, Taylor RD (1994) ^{57}Fe Mössbauer spectroscopy in a diamond-anvil cell at variable high pressures and cryogenic temperatures. Rev Sci Instrum 65:3787-3792
Ingalls R (1964) Electric-field gradient tensor in ferrous compounds. Phys Rev 133:A787-A795
Johnson CE (1989) The Mössbauer effect and magnetic phase transitions. Hyper Inter 49:19-42
Kolk B (1984) Studies of dynamical properties of solids with the Mössbauer effect. *In* GK Horton, AA Maradudin (eds) Dynamical Properties of Solids, Vol 5. North Holland, Amsterdam, p 3-328
Kolk B, Bleloch AL, Hall DB, Zheng Y, Patton-Hall KE (1985) High-temperature Mössbauer-effect measurements with a precision furnace. Rev Sci Instrum 56:1597-1603
Kruger MB, Jeanloz R, Pasternak MP, Taylor RD, Snyder BS, Stacy AM, Bohlen SR (1992) Antiferromagnetism in Pressure-Amorphized Fe_2SiO_4. Science 255:703-705
Lauterbach S, McCammon CA, van Aken P, Langenhorst F, Seifert F (2000) Mössbauer and ELNES spectroscopy of $(Mg,Fe)(Si,Al)O_3$ perovskite: A highly oxidised component of the lower mantle. Contrib Mineral Petrol 138:17-26
Maddock AG (1985) Mössbauer spectroscopy in mineral chemistry. *In* FJ Berry, DJ Vaughan (eds) Chemical Bonding and Spectroscopy in Mineral Chemistry. Chapman and Hall, London, p 141-208
Margulies S, Ehrman JR (1961) Transmission and line broadening of resonance radiation incident on a resonance absorber. Nucl Instr Meth 12:131-137
McCammon CA (1994) A Mössbauer milliprobe: Practical considerations. Hyper Inter 92:1235-1239
McCammon CA, Price DC (1985) Mössbauer spectra of Fe_xO ($x > 0.95$). Phys Chem Min 11:250-254
McCammon CA, Tennant C (1996) High-pressure Mössbauer study of synthetic clinoferrosilite, $FeSiO_3$. *In* MD Dyar, CA McCammon, M Schaeffer (eds) Mineral Spectroscopy: A Tribute to Roger G. Burns, Vol 5. Geochemical Society, USA, p 281-288
McCammon CA, Chaskar V, Richards GG (1991) A technique for spatially resolved Mössbauer spectroscopy applied to quenched metallurgical slags. Meas Sci Technol 2:657-662
McCammon CA, Becerro AI, Langenhorst F, Angel R, Marion S, Seifert F (2000) Short-range ordering of oxygen vacancies in $CaFe_xTi_{1-x}O_{3-x/2}$ perovskites ($0 \leq x \leq 0.4$). J Phys: Cond Matt 12:2969-2984
Morrish AH (1965) The physical properties of magnetism. John Wiley and Sons, Inc., New York, p 262-264

Mössbauer RL (1958a) Kernresonanzfluoresent von Gammastrahlung in Ir^{191}. Z Phys 151:124-143
Mössbauer RL (1958b) Kernresonanzfluoresent von Gammastrahlung in Ir^{191}. Naturwiss 45:538-539
Mössbauer RL, Wiedemann WH (1960) Kernresonanzabsorption nicht Doppler-verbreiterter Gammastrahlung in Re^{187}. Z Phys 159:33-48
Nikolov O, Hall I, Barilo SN, Mukhin AA (1996) Field-induced reorientations in $TbFeO_3$ at 4.2 K. J Magn Magn Mater 152:75-85
Nolle G, Ullrich H, Muller JB, Hesse J (1983) A microprocessor controlled spectrometer for thermal scan Mössbauer spectroscopy. Nucl Instr Meth Phys Res 207:459-463
Nomura K, Ujihira Y, Vertes A (1996) Applications of conversion electron Mössbauer spectrometry (CEMS). J Radioanal Nucl Chem 202:103-199
Pasternak MP, Taylor RD (1996) High pressure Mössbauer spectroscopy: The second generation. *In* GJ Long, F Grandjean (eds) Mössbauer Spectroscopy Applied to Magnetism and Materials Science, Vol. 2. Plenum Press, New York and London, p 167-205
Pasternak MP, Taylor RD, Jeanloz R, Li X, Nguyen JH, McCammon CA (1997) High pressure collapse of magnetism in $Fe_{0.94}O$; Mössbauer spectroscopy beyond 100 GPa. Phys Rev Lett 79:5046-5049
Pasternak MP, Rozenberg GR, Machavariani GY, Naaman O, Taylor RD, Jeanloz R (1999) Breakdown of the Mott-Hubbard state in Fe_2O_3: A first order insulator-metal transition with collapse of magnetism at 50 GPa. Phys Rev Lett 82:4663-4666
Perkins HK, Hazony Y (1972) Temperature-dependent crystal field and charge density: Mössbauer studies of FeF_2, $KFeF_3$, $FeCl_2$, and FeF_3. Phys Rev B 5:7-18
Ping JY, Rancourt DG (1992) Thickness effects with intrinsically broad absorption lines. Hyper Inter 71:1433-1436
Pound RV, Rebka Jr. JA (1960) Variation with temperature of the energy of recoil-free gamma rays from solids. Phys Rev Lett 4:274-277
Preston RS (1978) Isomer shift at phase transitions. *In* GK Shenoy, FE Wagner (eds) Mössbauer Isomer Shifts. North-Holland Publishing Co., Amsterdam, p 281-316
Rancourt DG (1989) Accurate site populations from Mössbauer spectroscopy. Nucl Instr Meth Phys Res B44:199-210
Rancourt DG, Klingelhöfer G (1994) Possibility of a Mössbauer resonant-electron microscope. Fourth Seeheim Workshop on Mössbauer Spectroscopy, p 129
Rancourt DG, Mcdonald AM, Lalonde AE, Ping JY (1993) Mössbauer absorber thicknesses for accurate site populations in Fe-bearing minerals. Am Mineral 78:1-7
Riedel E, Karl R (1980) Mössbauer studies of thiospinels. I. The system $FeCr_2S_4$-$FeRh_2S_4$. J Sol State Chem 35:77-82
Rozenberg GR, Machavariani GY, Pasternak MP, Milner AP, Hearne GR, Taylor RD, Adler P (1999) Pressure-induced metallization of the perovskite $Sr_3Fe_2O_7$. Phys Stat Sol B 211:351-357
Ruby SL (1973) Mössbauer studies of acqueous liquids and glasses. *In* SG Cohen, M Pasternak (eds) Perspectives in Mössbauer Spectroscopy. Plenum Press, New York-London, p 181-194
Ruffer R, Chumakov AI (1996) Nuclear Resonance Beamline at ESRF. Hyper Inter 97/98:589-604
Seifert F (1983) Mössbauer line broadening in aluminous orthopyroxenes: Evidence for next nearest neighbors interactions and short-range order. Neues Jahrbuch Miner Abh 148:141-162
Seifert F (1990) Phase transformation in minerals studied by ^{57}Fe Mössbauer spectroscopy. *In* A Mottana, F Burragato (eds) Absorption Spectroscopy in Mineralogy. Elsevier, Amsterdam, p 145-170
Shenoy GK (1973) Mössbauer effect studies of phase transitions. *In* SG Cohen, M Pasternak (eds) Perspectives in Mössbauer Spectroscopy. Plenum Press, New York-London, p 141-169
Skogby H, Annersten H, Domeneghetti MC, Molin GM, Tazzoli V (1992) Iron distribution in orthopyroxene: A comparison of Mössbauer spectroscopy and x-ray refinement results. Eur J Mineral 4:441-452
Takano M, Nakanishi N, Takeda Y, Naka S, Takada T (1977) Charge disproportionation in $CaFeO_3$ studied with the Mössbauer effect. Mater Res Bull 12:923-928
Takano M, Nasu S, Abe T, Yamamoto K, Endo S, Takeda Y, Goodenough JB (1991) Pressure-induced high-spin to low-spin transition in $CaFeO_3$. Phys Rev Lett 67:3267-3270
Virgo D, Hafner SS (1969) Fe^{2+},Mg order-disorder in heated orthopyroxenes. Mineral Soc Am Spec Paper 2:67-81
Wang Z, Cooney TF, Sharma SK (1993) High temperature structural investigation of $Na_2O \cdot 0.5Fe_2O_3 \cdot 3SiO_2$ and $Na_2O \cdot FeO \cdot 3SiO_2$ melts and glasses. Contrib Mineral Petrol 115:112-122
Weiss P (1906) La variation du ferromagnétisme avec la température. Comptes Rendus des Séances de L'Académie des Sciences 143:1136-1139
Wertheim GK (1973) Phase transitions in Mössbauer spectroscopy. *In* HK Henisch, R Roy (eds) Phase Transition-1973. Pergamon Press, New York, p 235-242

Williams JM (1975) The Mössbauer effect and its applications at very low temperatures. Cryogenics:307-322

Williamson DL (1978) Influence of pressure on isomer shifts. *In* GK Shenoy, FE Wagner (eds) Mössbauer Isomer Shifts. North-Holland Publishing Co., Amsterdam, p 317-360

Woodland AB, Angel RJ (1997) Reversal of the orthoferrosilite-High-P clinoferrosilite transition, a phase diagram for $FeSiO_3$ and implications for the mineralogy of the Earth's upper mantle. Eur J Mineral 9:245-254

Woodland AB, McCammon CA, Angel RJ (1997) Intersite partitioning of Mg and Fe in Ca-free high-pressure C2/c clinopyroxene. Am Mineral 82:923-930

Zou GT, Mao HK, Bell PM, Virgo D (1980) High pressure experiments on the iron oxide wüstite. Carnegie Inst Washington Yrbk 79:374-376

APPENDIX

Worked example: Incommensurate-normal phase transformation in Fe-doped åkermanite

Introduction. Fe-åkermanite, $Ca_2(Fe,Mg)Si_2O_7$, crystallises in the melilite structure, which consists of Si_2O_7 dimers connected via tetrahedrally coordinated Mg^{2+}/Fe^{2+} sites to form a sheet-like arrangement. At high temperature it exhibits a normal (N) structure with space group $P\bar{4}2_1m$, which transforms reversibly to an incommensurate (IC) structure with decreasing temperature. Theoretical analysis indicates that there are two component structures within the modulation with symmetries $P\bar{4}$ and $P2_12_12$ (McConnell 1999), which correlates with earlier work by Seifert et al. (1987) who showed that Mössbauer spectra of the IC structure showed two quadrupole doublets. Since Mössbauer spectroscopy is one of the few techniques able to distinguish between the component structures in the IC structure, McConnell et al. (2000) undertook a study using *in situ* Mössbauer spectroscopy at variable temperature and at high pressure to further characterise the IC structure and the IC-N phase transformation in Fe-åkermanite.

Sample preparation and data collection. In order to study the phase transformation in a sample close to the $Ca_2MgSi_2O_7$ endmember, the minimum amount of enriched ^{57}Fe required to obtain an adequate signal must be calculated. First, the optimum thickness for 90% ^{57}Fe enrichment was determined using Long et al. (1983) for a series of Fe-doped åkermanites, $Ca_2Fe_xMg_{1-x}Si_2O_7$. The results are listed in Table A1 in terms of the dimensionless thickness (Eqn. 12, this Chapter), and show that addition of 0.005 ^{57}Fe atoms p.f.u. would be sufficient to obtain an adequate signal for conventional transmission measurements (a dimensionless thickness of 2.6 corresponds to an

Table A1. Optimum Fe concentration for $Ca_2Fe_xMg_{1-x}Si_2O_7$ Fe-åkermanite based on Long et al. (1983) and ^{57}Fe enrichment of 90%

x	*Optimum Fe concentration (mg $^{57}Fe/cm^2$)*	*Dimensionless thickness*
0.005	0.15	2.6
0.01	0.30	5.1
0.02	0.54	10.1
0.03	0.88	15.0
0.04	1.16	19.8
0.05	1.44	24.5

unenriched Fe concentration of 7 mg Fe/cm^2, which is sufficient to give an adequate signal:noise ratio). The optimum thickness calculation assumes that the physical thickness of the absorber is unrestricted, i.e. that a sufficient amount of sample can be added to the absorber in order to achieve the desired iron concentration. For high pressure experiments, however, this is generally not possible, since the physical thickness of the absorber is constrained by the gasket to be no more than approximately 100 μm. The calculation of optimum thickness was then repeated using this constraint, showing that a concentration corresponding to x = 0.03 would be needed to give an unenriched Fe concentration near 7 mg Fe/cm^2 for a 100 μm thick absorber.

Mössbauer spectra were collected as a function of temperature at room pressure, and as a function of pressure at room temperature. This was accomplished using a continuous flow cryostat for temperatures from 4.2 to 293 K, a vacuum furnace for temperatures from 293 to 425 K, and a diamond anvil cell for pressures from 0 to 5.3 GPa. Further details are given in McConnell et al. (2000).

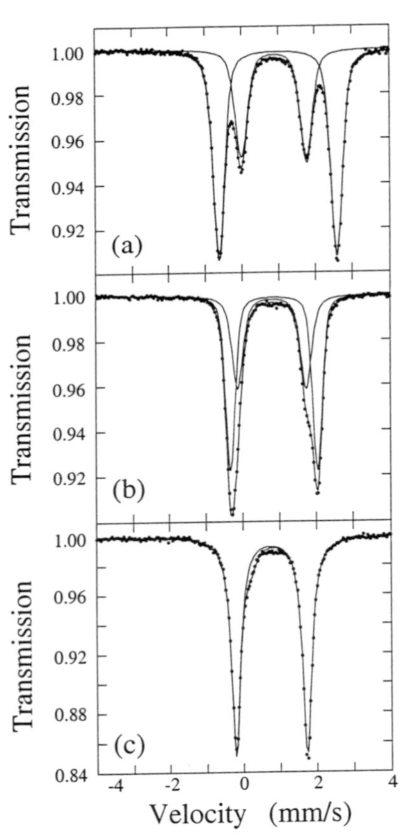

Figure A1. Mössbauer spectra of $Ca_2Fe_xMg_{1-x}Si_2O_7$ Fe-åkermanite at atmospheric pressure and (a) 4.2 K; (b) 293 K; and (c) 400 K. The outer two lines refer to $P\bar{4}$ symmetry while the inner two lines refer to $P2_12_12$ symmetry. The small amount of additional absorption in the spectrum recorded at 400 K is due to impurities in the furnace windows (data taken from McConnell et al. 2000).

Fitting models. Visually the spectra consist of two quadrupole doublets, where resolution improves with decreasing temperature. Above the IC-N transformation there is only one quadrupole doublet (Fig. A1). A general approach to fitting is to select the simplest model that is physically realistic, yet provides a good statistical fit to the experimental data. The conventional constraint of equal component areas of the quadrupole doublets was applied, since effects such as preferred orientation or anisotropic recoil-free fraction are not expected to be present. Component linewidths were also assumed to be equal, despite evidence at 4.2 K for a slight asymmetry in the single N-structure doublet. Doublet separation is not sufficient at higher temperatures to enable a unique determination of the asymmetry, but fortunately neglect of the asymmetry has a negligible effect on the line positions. Nine variables were fit to each spectrum: the baseline and four variables for each doublet (linewidth, area, centre shift and quadrupole splitting). For spectra recorded at 4.2 and 80 K, fits using Lorentzian lineshapes gave large residuals, so Voigt lineshapes were used instead. Voigt lines represent a Gaussian distribution of Lorentzian lines, and are expected when there are variations in the iron environment for a particular site, such as for solid solutions. These added an additional variable per doublet, the Gaussian standard deviation, and an F test was used (e.g. Bevington 1969) to confirm that addition of these variables was statistically significant to better than the 99% level. This was not the case for spectra recorded at higher temperatures, and for all spectra recorded at high pressure, so those were fit to Lorentzian lineshapes only.

Transition temperature and pressure. The IC-N transformation is recognised from the Mössbauer spectra as the point at which the two subspectra collapse to a single quadrupole doublet. This point is difficult to recognise from the two-doublet fits, however, due to the large line overlap just below the transition. A different approach was therefore used to determine the IC-N phase transformation point. Following the method of Seifert et al. (1987), two singlets were fit to the Mössbauer spectra, meaning that the high-velocity components of both subspectra were fit to a single line, and likewise for the low-velocity components. While the singlets do not provide a good statistical fit to the data in the IC structure region, their linewidths provide a measure of the splitting between the two subspectra. The linewidths are large at low temperature where the two subspectra are well resolved, and smaller at high temperature where subspectra overlap more (Fig. A1). The difference between the two singlet linewidths was used to monitor the approach to the IC-N phase transformation, where the transition point is seen as a discontinuity in the slope of linewidth difference versus temperature, and a break in the slope of linewidth difference versus pressure (Fig. A2). The transition temperature at room pressure is consistent with the linear trend shown by Seifert et al. (1987) as a function of iron composition, and the transition pressure at room temperature is consistent with the results from high-pressure X-ray diffraction reported by McConnell et al. (2000).

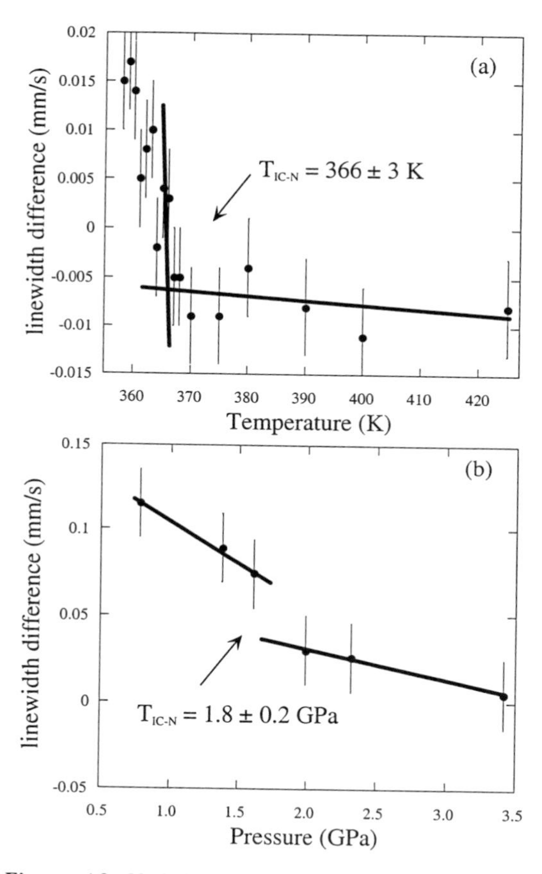

Figure A2. Variation of the difference in linewidth between high- and low-velocity components as a function of (a) temperature; and (b) pressure. The linewidths were determined based on a fit of two singlets to the Mössbauer spectra. The IC-N transition is recognised as either a discontinuity or a break in the slope in the variation near the transition.

Application of the Debye model. The centre shift data collected at room pressure and variable temperature were fit to the Debye model (Eqn. 2, this Chapter). The integral can be evaluated using a series approximation, where only a few terms of the series are needed to achieve an accuracy that exceeds the experimental one (Heberle 1971). At low temperatures ($z = \Theta_M/T$ 2.5) the following expression approximates the integral:

$$D_3(z) = \int_0^z \frac{x^3}{e^x - 1} dx$$

$$\approx \frac{\pi^4}{15} + z^3 \, ln(1 - e^{-z}) - 3e^{-z}(z^2 + 2z + 2) - \frac{3}{16} e^{-2z}(4z^2 + 4z + 2), \quad \text{(A1)}$$

while at high temperatures ($z = \Theta_M/T$ 2.5) the following expression is used:

$$D_3(z) = \int_0^z \frac{x^3}{e^x - 1} dx \approx \frac{z^3}{3}\left(1 - \frac{3}{8}z + \frac{z^2}{20} - \frac{z^4}{1680} + \frac{z^6}{90720}\right). \quad \text{(A2)}$$

The equation describing the variation of centre shift with temperature (Eqn. 1, this Chapter) contains two adjustable parameters, Θ_M and δ_0. Using a non-linear least squares method, the values of these parameters which best fit the data were determined (Table A2). The centre shift data together with the Debye model calculations are plotted in Figure A3.

Table A2. Best-fit Debye model parameters for $Ca_2Fe_{0.03}Mg_{0.97}Si_2O_7$ Fe-åkermanite derived from Mössbauer centre shift data

	δ_0 (mm/s)	Θ_M (K)
IC symmetry #1 $\rightarrow P\bar{4}$	1.101(3)	340(18)
IC symmetry #2 $\rightarrow P2_12_12$	0.995(3)	774(31)
N structure $P\bar{4}2_1m$	1.004(3)	791(23)

Relative area. The relative areas of the quadrupole doublets are related to the relative abundance of each component, which can be estimated according to the thin absorber approximation (Eqn. 13, this Chapter) in cases where the approximation is valid. Rancourt et al. (1993) provide a means of estimating how closely the thin absorber approximation is satisfied for a given absorber based on the attenuation of spectral areas for individual lines (see Fig. 3 in their paper). McConnell et al. (2000) used an iron density in the high temperature experiments corresponding to an unenriched concentration of 5 mg Fe/cm^2, which is equivalent to a dimensionless effective thickness of 2.0. The attenuation of spectral

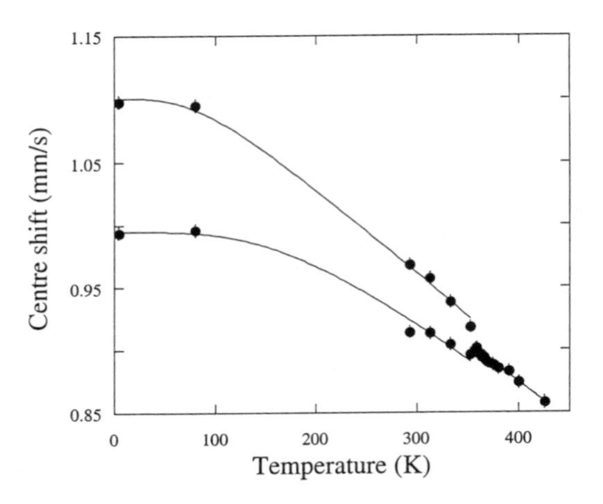

Figure A3. Variation of centre shift with temperature based on spectral fits incorporating two quadrupole doublets. The solid line represents the Debye model fit (Eqns. 1 and 2, this Chapter) to the centre shifts (data taken from McConnell et al. 2000).

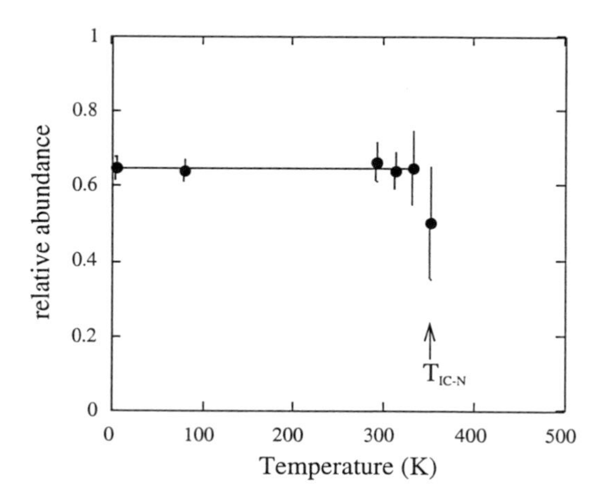

Figure A4. Relative abundance of the IC structure component with high quadrupole splitting (doublet #1) plotted as a function of temperature. The relative abundance was calculated from the relative areas as described in the text, and the solid line represents a constant y-axis value of 0.64.

areas increases as the relative abundance of the individual components becomes more different, so the maximum thickness effect can be estimated assuming the greatest possible difference in component abundances. The maximum abundance ratio is estimated to be 2:1, which implies effective thicknesses of 1.34 and 0.66 for the two components. The effective thickness is halved for each line of a quadrupole doublet, hence the relevant numbers for the x-axis in Figure 3 of Rancourt et al. (1993) are 0.67 and 0.33. The y-axis value for Figure 3 is calculated from $\Gamma_{obs} - 2\Gamma_0$, where the observed linewidth is 0.36 mm/s, which implies a y-axis value of 0.17 mm/s. This gives area attenuations of approximately 0.93 and 0.96 for the two components, which can be considered to lie close to the thin absorber approximation (Rancourt et al. 1993).

Another consideration for calculating relative abundances from the relative areas is the difference in recoil-free fractions corresponding to the specific sites. The Debye model allows recoil-free fractions to be estimated using Equation (15) (this Chapter), and combined with the results listed in Table A2, gives recoil-free fractions of the two components at 4.2 K of 0.91 and 0.96. This results in, at most, a 2% correction to the relative areas in order to obtain the relative abundances. The situation is different at room temperature, however, where recoil-free fractions of the two components are calculated to be 0.70 and 0.92, which results in a correction of more than 10%.

The relative abundance (determined as described above) of the component corresponding to doublet #1 is plotted in Figure A4. The relative abundance remains relatively constant within the IC structure stability region until just before the transition, where it apparently decreases to approximately 50%. This is consistent with theory that requires the abundance of the two components to be equal at the phase transformation (McConnell 1999).

Interpretation. The Mössbauer spectra obtained for the IC structure in Fe-åkermanite comprise two and only two subspectra over the temperature range down to 4.2 K. In terms of the theoretical model (McConnell 1999), they may be identified with the two symmetries $P\bar{4}$ and $P2_12_12$. The former involves a simple rotation of the tetrahedral site with respect to the Si_2O_7 dimers and hence no change to the coordination environment, while the latter involves an orthorhombic site distortion. The quadrupole doublets observed in the IC structure spectrum may be characterised as follows:

Doublet #1 – high ΔE_Q, large decrease in ΔE_Q with increasing temperature, normal Θ_M
Doublet #2 – low ΔE_Q, small increase in ΔE_Q with increasing temperature, anomalous Θ_M

Doublet #1 may be assigned unambiguously to $P\bar{4}$ symmetry, since the hyperfine parameters follow the expected behaviour for Fe^{2+} in a relatively undistorted tetrahedral site. The high value of quadrupole splitting indicates a high valence contribution, and displays the expected temperature variation for tetrahedral environments (Gibb 1968). The characteristic Mössbauer temperature Θ_M is in the range expected for Fe^{2+} (De Grave and Van Alboom 1991).

Doublet #2 may be assigned unambiguously to $P2_12_12$ symmetry. The small quadrupole splitting indicates a high lattice contribution, which is consistent with the small temperature dependence of the quadrupole splitting (Fig. 4, this Chapter). The anomalously high value of Θ_M is likely related to constraints on lattice vibrations imposed by symmetry. Similar constraints evidently exist also in the normal structure, since the value of Θ_M is comparable.

Summary. Mössbauer spectroscopy confirms results from the theoretical analysis of incommensurate structures using full space group theory that the IC structure of melilite comprises two (and only two) component structures (McConnell 1999). Mössbauer spectra of Fe-åkermanite contain two quadrupole doublets at low temperature and pressure (IC structure) and one quadrupole doublet at high temperature and pressure (N structure). Transition temperatures and pressures, as well as parameters relating to the structures of the IC and N structures, were determined from *in situ* Mössbauer measurements. The characteristic Mössbauer temperatures of the N structure and the $P2_12_12$ component of the IC structure are anomalous, which is likely related to symmetry-imposed constraints on lattice vibrations. Quadrupole splitting values and their temperature dependence are consistent with the symmetry-imposed distortions of the two components of the IC structure. The relative abundance of the two components appears to be equal at the transition temperature, as required by theory, while at low temperatures the relative abundance of the $P\bar{4}$ symmetry is larger, indicating that it is the favoured configuration.

APPENDIX REFERENCES

Bevington PR (1969) Data Reduction and Error Analysis for the Physical Sciences. McGraw-Hill, New York

De Grave E, Van Alboom A (1991) Evaluation of ferrous and ferric Mössbauer fractions. Phys Chem Minerals 18:337-342

Gibb TC (1968) Estimation of ligand-field parameters from Mössbauer spectra. J Chem Soc A:1439-1444

Heberle J (1971) The Debye integrals, the thermal shift and the Mössbauer fraction. *In* IJ Gruverman (ed) Mössbauer Effect Methodology 7:299-308. Plenum Press, New York

Long GL, Cranshaw TE, Longworth G (1983) The ideal Mössbauer effect absorber thickness. Möss Effect Ref Data J 6:42-49

McConnell JDC (1999) The analysis of incommensurate structures in terms of full space group theory, and the application of the method to melilite. Z Kristallor 214:457-464

McConnell JDC, McCammon CA, Angel RJ, Seifert F (2000) The nature of the incommensurate structure in åkermanite, $Ca_2MgSi_2O_7$, and the character of its transformation from the normal structure. Z Kristallogr (in press)

Rancourt DG, Mcdonald AM, Lalonde AE, Ping JY (1993) Mössbauer absorber thicknesses for accurate site populations in Fe-bearing minerals. Am Mineral 78:1-7

Seifert F, Czank M, Simons B, Schmahl W (1987) A commensurate-incommensurate phase transition in iron-bearing Åkermanites. Phys Chem Minerals 14:26-35

10 Hard Mode Spectroscopy of Phase Transitions

U. Bismayer

Mineralogisch-Petrographisches Institut
Universität Hamburg
Grindelallee 48
20146 Hamburg, Germany

INTRODUCTION

Phonon spectra respond, in principle, to any change of the crystal structure or lattice imperfections of a material. Traditionally optical soft modes, which can show relatively large variations, have been widely used for the analysis of displacive phase transitions. A soft mode is a particular mode of vibration of the high symmetry phase, whose frequency tends towards zero as the critical point is approached. In other cases the variations can be small; typical examples are the weak phonon renormalization of high temperature superconductors at the superconducting phase transition or isosymmetric phase transitions. Typical relative changes $\Delta\omega/\omega$ of phonon frequencies which do not directly correspond to characteristic response modes (such as soft modes) are between 0.001 and 0.02. Similar changes are observed for the intensities of phonon signals and, rather often, for the spectral line widths.

Although such relative changes are small, their absolute values (e.g. 5 cm^{-1} for a phonon signal at 500 cm^{-1}) can be well resolved experimentally by most Raman and infrared spectrometers (~0.1 cm^{-1}). Over the last two decades, several phase transitions were either first discovered or analysed in detail using the renormalization of optical phonon signals as the primary analytical tool which was named 'hard mode spectroscopy' (Petzelt and Dvorak 1976, Bismayer 1988, Güttler et al. 1989, Bismayer 1990, Salje and Yagil 1996, Zhang et al. 1996, 1999; Salje and Bismayer 1997).

Following the phenomenological theory of structural phase transitions by Landau and Lifshitz (1979) the order parameter Q is non-zero below the transition point and increases on cooling (in the case of temperature as state variable) according to $Q(T) \sim |T-T_c|^{\beta}$ with a critical exponent ß. In terms of thermodynamics, the soft mode theory relates directly to Landau theory in the quasi-harmonic approximation in which the order parameter is associated with the mean value of the eigenvector of the soft mode (Blinc and Zeks 1974). Using the hard mode concept in order to describe structural instabilities we consider the phonon self energy which changes by a small relative fraction. The anharmonic Hamiltonian of the phase transition with order parameter Q is coupled with the harmonic phonon Hamiltonian, e.g. $\frac{1}{2}\Sigma\chi_i^{-1}Q_i^2$, where χ_i^{-1} is the susceptibility of the i-th phonon via a coupling term which can be expanded in powers of the phonon coordinates Q_i. The symmetry change determines the order of the coupling such as $\lambda_i^{(m)}Q_iQ_jQ^m$, where m is the faintness exponent of the order parameter. Diagonalizing the total Hamiltonian leads to a change of the phonon energy or the self energy (Bismayer 1990, Salje and Bismayer 1997) and a Q dependence according to

$$\Delta(\omega^2) \propto \Delta\chi^{-1} \propto AQ + BQ^2 \quad (1)$$

or

1529-6466/00/0039-0010$05.00

$$\Delta(\omega^2) \propto A_1 Q^2 + B_1 Q^4 \quad (2)$$

where only the two lowest order terms are considered. If the product representation $\Gamma(Q_i) \otimes \Gamma(Q_j) \otimes \Gamma(Q_l)$ contains the identity representation, linear coupling is symmetry allowed. It applies only to a small number of cases (e.g. Güttler et al. 1989). Other types of coupling ($\propto Q^2$ and Q^4) are compatible with all symmetry constraints and the quadratic Q dependence of $\Delta(\omega^2)$ accounts most commonly. The same applies for the integrated intensity ΔI and the line width Γ (Salje and Bismayer 1997); the intrinsic value $\Delta\Gamma$ is often found to be increased by the influence of defects. Further changes of the line profiles may stem from coupling with other excitations such as domain boundary movements or other low-energy phonon branches.

A major advantage of hard mode spectroscopy has been described by Salje (1992) which stems from the short correlation length of high frequency phonons. The dispersion of the hard mode is expressed by

$$\omega^2 = \omega_0^2 + \alpha^2 q^2 \quad (3)$$

and the phonon amplitude follows from the dissipation-fluctuation theorem

$$\left(Q_1^2\right) \propto \frac{kT}{\omega_0^2 + \alpha^2 q^2} \quad \text{for} \quad kT \rangle\rangle h\omega. \quad (4)$$

In real space the correlation (Q_iQ_j) of phonon amplitudes at the sites i and j is described by the Ornstein-Zernike function

$$(Q_iQ_j) \propto \frac{1}{r_i - r_j} exp\left(\frac{-\omega_o(r_i - r_j)}{\alpha}\right) \quad (5)$$

with the characteristic length $\xi = \alpha/\omega_0$. For typical values of α and ω_0 one finds $\xi \le d$, where d is the lattice repetition unit. The estimate shows that high frequency phonons with modest dispersion act almost as Einstein oscillators in real space. Hence, their renormalization probes the changes of the local states and no further corrections for correlation effects are necessary under most circumstances. In the following the concept of hard mode spectroscopy is discussed on the basis of some selected experimental applications.

THE ANALYSIS OF PHONON SPECTRA

IR powder spectra

The optical reflectance of bulk, single crystal and thick film samples followed by Kramers-Kronig analysis yields the phonon spectra as well as free and bound charge carrier contributions. Single crystal studies can be rather complex and sample preparation is often easier in the case of powder absorbance measurements. On the other hand a quantitative analysis of powder data can also be quite difficult. Several methods for the numerical analysis of powder spectra and their advantages and disadvantages were described in detail by Salje and Bismayer (1997). A small amount of sample powder (typically less than 1% volume) is mixed with an infrared window (such as KB or CsI) and compressed into a pellet. The absorbance spectra is defined as $A = -\log(T_S/T_R)$ where T_S, T_R are the optical transmittances of the sample and the reference pellet, respectively. In the dilute limit both scattering and specular reflectance of the reference and the sample pellet are almost identical and therefore cancel out. Hence, the absorbance is directly related to the losses in the sample. It was shown (Yagil et al. 1995) that the powder absorbance technique provides useful information even in the NIR regime when the grain sizes are comparable with the optical wavelength. The weak renormalization of the phonon signals due to structural variations is not affected by the embedding material so that relative changes $\Delta\omega/\omega$, $\Delta T/T$ and $\Delta I/I$ need not be corrected in most cases (Salje 1994).

Phonon signals which do not overlap too much with other bands in the spectrum can be analysed using Least Squares routines. For most applications it is sufficient to use an overall line width parameter Γ for TO and LO phonons and to approximate the spectral function by an oscillator profile which is a simple Lorentzian for $\Gamma << \omega_0$ according to

$$\varepsilon'' \approx \frac{2\omega\Gamma}{(\omega_0 - \omega)^2} \tag{6}$$

where ω_0 is the peak position. Additional structural fluctuations and instrumental line broadening are taken into account by extending this line profile to a Voigt-profile with a Gaussian and a Lorentzian contribution. Most commercial instruments use such Voigt-profiles for their data analysis which can be applied directly in hard mode spectroscopy. The line profiles of atomic ordering processes are often Gaussian and reflect the various relevant ordering configurations in a disordered or partially ordered system. The line profile of the fully ordered material is again Lorentzian.

In order to analyse spectra with several overlapping phonon bands, the integral line width is used as relevant parameter. In case of overlapping bands, spectral regions are selected and a baseline is defined where the intensity between these regions is weak. Typical spectral regions are, for example, related to molecular vibrations, tetrahedral stretching modes, structural bending modes, etc. If no breaks between the spectral regions allow for the determination of the baseline, the total spectrum has to be taken as such a region. The auto-correlation function for a spectrum $S^i(\omega)$ in a region i is (Salje and Bismayer 1997)

$$A^i(\omega) = \int_0^\infty S^i(\omega)\; S^i(\omega + \omega')d\omega' \tag{7}$$

Outside the spectral region $S^i(\omega) = 0$. $A^i(\omega)$ reflects the average peak profiles in the spectral region and for small values of ω the spectral function of the auto-correlation function can be approximated as

$$A^i(\omega) = A^i(0) - \frac{1}{2}\gamma \cdot \omega^2 + \frac{1}{4}\delta \cdot \omega^4 + \ldots . \tag{8}$$

The generalized line width $\overline{\Gamma}$ is described by the parameter γ

$$\overline{\Gamma} = 1/\sqrt{\gamma}\,. \tag{9}$$

After calibration with the spectrum of a fully ordered and a fully disordered sample the order parameter can be determined directly from $\overline{\Gamma}$ as the scaling relation $\overline{\Gamma}(Q)$ is the same as for the individual line widths.

Raman spectra

The Raman technique is often applied for single crystals studies. The Raman effect occurs when a beam of monochromatic light with frequency ω_i passes through a crystal leading to inelastic scattering by the phonons of the crystal with frequency shifts ω_{ph} of ~10 cm^{-1} to 3000 cm^{-1} according to $\omega_s = \omega_i \pm \omega_{ph}$ where ω_s is the scattered beam frequency. The signs - and + represent Stokes and anti-Stokes processes, respectively.

frequency. The signs - and + represent Stokes and anti-Stokes processes, respectively. For an incident beam (λ = 514 nm) the wave vector of the incident light $k_i = 2\pi/\lambda$ is 1.2×10^{-2} nm^{-1} while the reciprocal lattice vector $q = 2\pi/a$ is 6.2 nm^{-1} for a lattice constant a = 1 nm. Accordingly, Raman excitations stem from sections of the Brilllouin zone close to its centre (Γ point). The Raman scattering cross section is characterised by the correlation tensor which determines the scattering into a polarization ($\beta\delta$) from an incident beam polarized along ($\alpha\beta$) (Cowley 1964)

$$I_{\alpha\beta\gamma\delta}(\omega) = \frac{1}{2\pi N}\int_{-\infty}^{\infty}\left\langle P^*_{\alpha\beta}(0)\, P_{\gamma\delta}(t)\right\rangle \exp i\omega t\, dt. \tag{10}$$

$P_{\gamma\delta}(t)$ describes the deviation from the equilibrium polarisability of a crystal of N unit cells. In the harmonic approximation, the nuclear motion is discussed on the basis of normal coordinates. Assuming that the displacements of the atoms from their equilibrium positions are small, the polarisability can be expanded in a Taylor series with respect to the displacements (Bruce and Cowley 1981)

$$\begin{aligned} P_{\alpha\beta}(k) &= P^0_{\alpha\beta}(k)\Delta(\mathrm{k}) + (\frac{1}{N})^{\frac{1}{2}} \sum_{qj} P_{\alpha\beta}\begin{pmatrix} q \\ j \end{pmatrix} Q(qj)\,\Delta(k+q) \\ &+ \frac{1}{N}\sum_{q_1q_2j_1j_2} P_{\alpha\beta}\begin{pmatrix} q_1 & q_2 \\ j_1 & j_2 \end{pmatrix} Q(q_1j_1)\,Q(q_2j_2)\,\Delta(k+q_1+q_2) + \ldots \end{aligned} \tag{11}$$

k is the light-wave vector transfer and $Q(qj)$ describes the mode coordinates of the wave vector q and phonon branch j. In a displacive phase transformation the structural distortions may be described by an n-fold degenerate mode labelled by the wave vector q and branch indices $j_S = 1, \ldots n$ where j_h and j_s denote the hard- and soft-mode branches, respectively. The static amplitude of the distortion corresponds to the order parameter. Raman active hard modes which are symmetry equivalent to the soft mode with $q = q_S$, but branch indices $j_h \neq j_s$, contribute to the first-order Raman scattering by the following term

$$I_{\alpha\beta\gamma\delta}(\Omega) = 4\sum_{j_hj_s} P_{\alpha\beta}\begin{pmatrix} q_s & -q_s \\ j_h & j_s \end{pmatrix} P_{\gamma\delta}\begin{pmatrix} q_s & -q_s \\ j_h & j_s \end{pmatrix} Q^2_{j_s}\, G^{ph}(q_s j_h \Omega) \tag{12}$$

where G is the spectral function. It is seen that the integrated intensity of the Raman scattering from the hard modes (qj_h) varies with the square of the macroscopic order parameter when the critical point is approached. However, this correlation function represents only that part of scattering which corresponds to long-range order. In the vicinity of the critical point, fluctuations of the coordinates Q_i about their equilibrium value, which are comparable to the order-parameter value itself, dominate and provide the (short-range) precursor order contribution to the scattering (Bismayer 1990).

The spectral function leading to a quasi-first-order Raman spectrum close to the critical point of a displacive system contains basically three contributions from 1) the soft mode, 2) the central peak and 3) higher order Raman scattering arising from processes involving hard and soft phonons

$$G^{ph}(qj_s, \omega) = G^{lr} + G^{cp} + G^{ph'}. \tag{13}$$

G^{lr} is the spectral function of a long range phonon sideband with a quasiharmonic frequency that softens on approaching the transition point but which saturates at a finite frequency value at the critical point. G^{cp} describes a central component leading to quasi-first-order Raman scattering from the hard modes. The latter can be induced by a slow finite range component related to phonon density of states fluctuations or to cluster dynamics. A measure of the local distortion (quasi static on phonon time scales) within the dynamic clusters is the precursor order parameter. The clusters may be considered as pre-ordered regions. Accordingly, short-range precursor order induces, in the high symmetry phase, similar effects to those which static long-range order induces in the low-symmetry phase. Such pretransitional effects due to the occurrence of short range order at temperatures above the transition point are sometimes encountered in systems with 'intermediate phases' (Bismayer et al. 1982, Salje et al. 1993). $G^{ph'}$ represents higher order Raman contributions to the spectral density involving hard and soft quasi-harmonic phonons which leads, close to the transition point, to quasi-first-order features characterised by overdamped soft phonons.

In this work we focus on the short range order in ferroelastic lead phosphate, $Pb_3(PO_4)_2$, in Sr-doped lead phosphate crystals, in antiferroelectric titanite, $CaTiSiO_5$, and isosymmetric effects in malayaite, $CaSnSiO_5$.

EXAMPLES OF SHORT-RANGE ORDER IN STRUCTURAL PHASE TRANSITIONS

Precursor in $Pb_3(PO_4)_2$

Lead phosphate, $Pb_3(PO_4)_2$, lead phosphate-arsenate, $Pb_3(P_{0.77}As_{0.23}O_4)_2$, and lead strontium phosphate, $(Pb_{1-x}Sr_x)_3(PO_4)_2$, exhibit a ferroelastic phase transition between a rhombohedral para phase ($R\bar{3}m$) and a monoclinic ferro phase ($C2/c$) (Brixner et al. 1973, Bismayer and Salje 1981, Joffrin et al. 1979, Bismayer et al. 1982, Salje and Devarajan 1981, Salje et al. 1993, Bismayer et al. 1995). The existence of an intermediate pseudo-phase is indicated by X-ray and neutron diffraction (Joffrin et al. 1979), Raman and IR spectroscopy (Salje and Bismayer 1981, Salje et al. 1983) and electron microscopy (Roucau et al. 1979). Bismayer et al. (1982) reinvestigated the intermediate pseudo-phase using high-resolution neutron scattering, Raman spectroscopy and diffuse X-ray scattering. It was concluded that whereas the deformations in the intermediate phase of $Pb_3(PO_4)_2$ are purely dynamic in character, the deformations in the mixed crystals $Pb_3(P_{0.77}As_{0.23}O_4)_2$ and $(Pb_{1-x}Sr_x)_3(PO_4)_2$ are static just above the ferroelastic phase transition temperature and are superposed by reorientational motions at higher temperatures. Hence, the 'pseudo-phase' is a 'phase' in the crystallographic sense for the mixed crystals, and an anharmonic regime with short range order exists in pure lead phosphate (Salje and Wruck 1983). Originally, Torres (1975) suggested that the phase transition takes place due to instabilities at the L points of the rhombohedral Brillouin zone and he expressed the Gibbs free energy in terms of three order parameters (η_1, η_2, η_3) stemming from the three arms of the L points. Salje and Devarajan (1981) explained the phase transition in terms of a three-states Potts model containing the order parameter components Q_1, Q_2 and Q_3 which accounted well for the intermediate phase. The authors suggested, that in the case of $Pb_3(PO_4)_2$, there is a first phase transition around 560 K involving the order parameter Q_3. This is equivalent to the condensing of the three zone boundary L modes, causing static shifts of the lead positions from the threefold inversion axis in the three possible and symmetrically equivalent directions of the binary axis of the monoclinic symmetry. At 451 K Q_1 and Q_2 become critical. They describe the orientation of the binary axis in the ferrophase $C2/c$. The 'critical excitation' in the intermediate phase corresponds to the spontaneous thermal switching between the three binary axes. In order to study the short range order in the lead phosphate-type material hard mode

A strong and appropriate phonon signal is the ramanactive A_g-mode near 80 cm^{-1} (ferrophase *C2/c*) which transforms into an E_g-mode in the high symmetry phase $R\bar{3}m$. The temperature evolution of the spectra is shown in Figure 1. The integrated intensity of the phonon band near 80 cm^{-1} is displayed together with the temperature behaviour of the morphic optical birefringence along the rhombohedral triad in Figure 2. Both curves show the same temperature evolution ($\propto Q^2$) at temperatures below the transition point. At temperatures above the ferroelastic transition point (451 K), the optical birefringence disappears in a defect free crystal, which agrees with a macroscopic rhombohedral symmetry for the sample. The hard mode intensity, however, does not converge to the extrapolated value of the E_g-phonon. The persisting phonon intensity indicates that locally the crystal structure is not rhombohedral but preserves structural distortions of the same type as in the monoclinic phase (i.e. displaced lead atoms). The same precursor effect of structural distortions under heating was observed by the increased specific heat in the intermediate pseudo phase (Salje and Wruck 1983) and for the excess molar volume of lead phosphate (Salje et al. 1993).

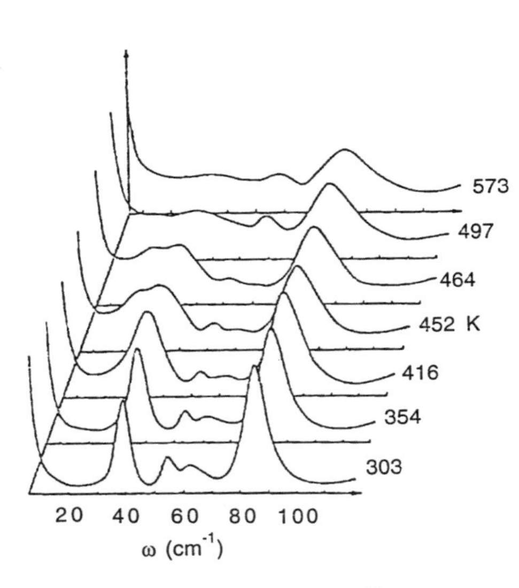

Figure 1. Raman spectra (A_g^{bb}) of $Pb_3(PO_4)_2$. A ferroelastic transition takes place at 451 K; it is monitored by the change of the scattering profile of the 80 cm^{-1} phonon.

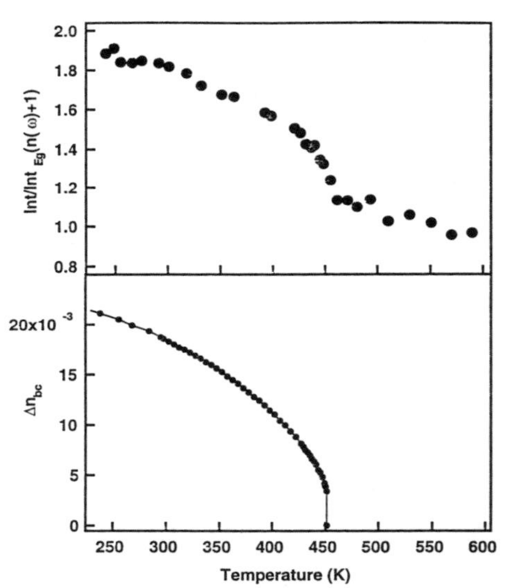

Figure 2 . Temperature-evolution of the intensity of the A_g^{bb} phonon at 80 cm^{-1} relative to the E_g-phonon at $T \gg T_c$ (top) and morphic optical birefringence Δn_{bc} along (100) in $Pb_3(PO_4)_2$. The ferroelastic phase transition is weakly first order and appears as a step in Δn_{bc} at 451 K. No such step occurs for the phonon intensity indicating short range order decreasing gradually with increasing temperature.

In lead phosphate (short-range) precursor order contributes to the scattering. Hence, the scattering at frequencies close to the hard modes arises from terms that involve two coordinates belonging to a hard branch, and two belonging to a soft branch. The relevant terms are (Bruce et al. 1980)

$$I_{\alpha\beta\gamma\delta}(\Omega) \approx 4 \sum_{j_h j_s} P_{\alpha\beta}\begin{pmatrix} q & -q \\ j_h & j_s \end{pmatrix} P_{\gamma\delta}\begin{pmatrix} q & -q \\ j_h & j_s \end{pmatrix} Q_{j_s}^2 \, G^{ph}(q_s j_h \Omega)$$
$$\times \sum_q \frac{1}{2\pi N} \langle Q(qj_s,0)Q(qj_s,t)\rangle \times \langle Q(qj_h,0)Q(-qj_h,t)\rangle \exp(i\Omega t)dt \qquad (14)$$

$Q(qj_h)$ represent the hard-mode coordinates and $Q(qj_s)$ the soft-mode coordinates. The correlation function of the hard mode coordinates follows

$$\langle Q(qj_h,0)Q(-qj_h,t)\rangle = \int_{-\infty}^{\infty} G^{ph}(qj_h,\Omega)\, exp(-i\Omega t)d\Omega. \tag{15}$$

This leads to the spectral function G^{ph} which is approximated by a damped quasi-harmonic oscillator (Salje and Bismayer 1997)

$$G^{ph}(qj,\Omega) = \frac{n+1}{\pi}\frac{\Gamma(qj)}{\left(\omega_{\infty}^{2}(qj_h)-\Omega^2\right)^2-\Gamma^2(qj_h)\Omega^2}, \tag{16}$$

where n is the Bose-Einstein factor, q is the wavevector; j_h and j_s denote the hard- and soft-mode branches, respectively, $\Gamma\,(qj_h)$ represents the damping constant and $\omega_{\infty}(qj_h)$ is the hard-mode harmonic frequency. The pair correlation corresponding to the soft coordinates can be separated in static and dynamic contributions

$$\langle Q(qj_s,0)Q(-qj_s,t)\rangle = Q^2{}_{js}(q)+\langle Q^f(qj_s,0)Q^f(-qj_s,t)\rangle \tag{17}$$

where $Q_{js}(q)$ is the displacive order parameter of the ferrophase, associated with the order parameter component Q_3 in the Potts model (Salje and Devarajan 1981). $Q^f(qj_s)$ represents fluctuations in the soft coordinates whose correlation function is given by

$$\langle Q^f(qj_s,0)Q^f(-qj_s,t)\rangle = \int_{-\infty}^{\infty} G(qj_s,\Omega)\, exp(-i\Omega t)d\Omega. \tag{18}$$

Contributions to the scattering cross-section arising from processes involving hard and soft phonons (see also Eqn. 13) lead to the following expression

$$\begin{aligned} I_{\alpha\beta\gamma\delta}(\Omega) \approx\ & CG^{ph}\left(\Gamma(0j_h)\omega_{\infty}(0j_h)\Omega\right) \\ & + DQ_{j_s}^2\ G^{ph}\left(\Gamma(0j_h)\omega_{\infty}(0j_h)\Omega\right) \\ & + EQ_{j_{qs}}^2\ G^{ph}\left(\Gamma(0j_h)\omega_{\infty}(0j_h)\Omega\right) \\ & + F\frac{1}{N}\sum_q G^{flip}\left(qjQ_{jq_3}\right) G^{ph}\left(\overline{\Gamma}(qj_h)\omega_{\infty}(qj_h)\Omega\right). \end{aligned} \tag{19}$$

The parameters C, D, E and F involve components of polarizability derivative tensors. Q_{js} is the displacive order parameter in the low-temperature phase and Q_{jq^3} corresponds to Q_3 which is present in the intermediate pseudo-phase. The damping constant $\Gamma\,(qj_h)$ includes the effect of coupling between the hard mode and the reorientational relaxation of displaced lead atoms.

The deviation of the Pb(II) position from the threefold inversion axis can be taken as a useful measure for the order parameter. At room temperature (in the monoclinic phase) this atom is displaced by 0.335 Å from this axis (Guimaraes 1979), whereas the rhombohedral symmetry of $R\bar{3}m$ requires the atom to be located strictly on the threefold inversion axis. At which distance from the triad do we expect Pb(II) in the regime of short range order near to the transition point? From the intensity of the phonon signal one determines a distance of 0.182Å which is in good agreement with the scaled value of the optical birefringence at temperatures just below T_c. We can now scale the order parameter Q at room temperature with Q (300 K) = 0.3. The resulting maximum order parameter in the regime of short range order is then ~0.16 which decreases further with increasing temperature above 451 K.

Renormalization phenomena are induced in lead phosphate by fields conjugate to the ferroelastic order paramter. The energy contributions of impurities can generally be described by field terms H_d (Levanyuk et al. 1979, Salje 1990)

$$\Delta G \propto \int dr \left\{ \frac{A_d}{2}(r)\langle Q\rangle^2 - H_d(r)\langle Q\rangle - g_d \nabla\langle Q\rangle \right\}. \tag{20}$$

Fields conjugate to the order parameter lead to the renormalization of the quadratic terms in the Gibbs free energy and, therefore, to a change of the critical temperature. Renormalization phenomena depend on the type of impurities (Bismayer et al. 1986). The temperature dependence of the specific heat of the Sr-diluted crystals follows predictions by Levanyuk et al. (1979) and is compatible with experimental observations in isostructural lead phosphate-arsenate (Salje and Wruck 1983). Figure 3 shows that higher defect concentrations lead to a more λ-shaped singularity. The maximum excess specific heat capacity at the ferroelastic transition point appears at reduced temperatures with increasing x-values. In the diluted samples the critical point is renormalized to lower temperatures and the transition appears to be smeared. Raman spectroscopy shows an extended stability range of the intermediate pseudo-phase where the local symmetry above the macroscopic ferroelastic transition point is still monoclinic and thermal reorientations between the three binary axes are thermally activated. The integrated intensity of the A_g-phonon near 85 cm^{-1} becomes virtually temperature independent where this mode merges into the E_g mode of the rhombohedral phase (Fig. 4).

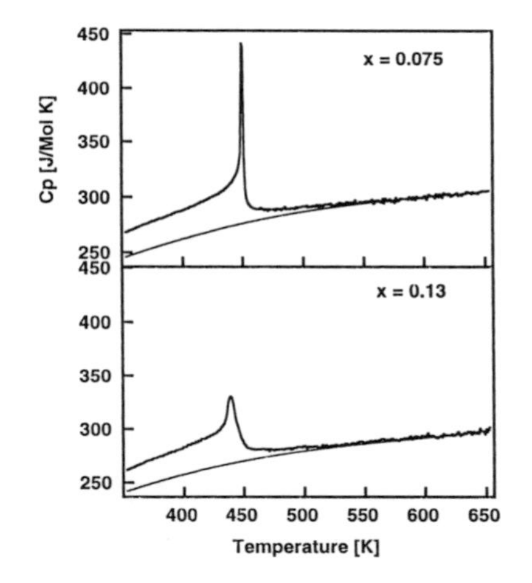

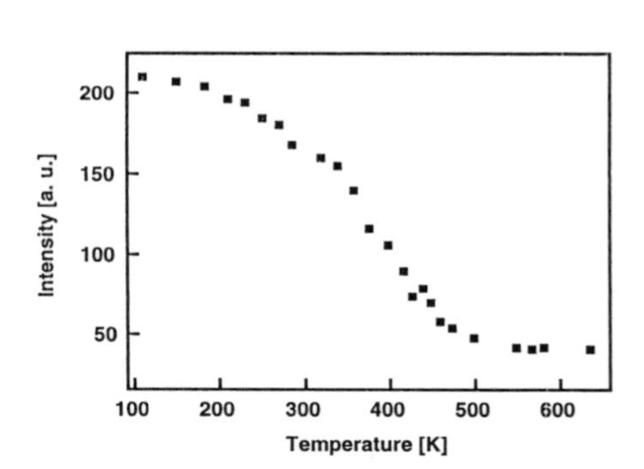

Figure 3 (left). Specific heat of $(Pb_{1-x}Sr_x)_3(PO_4)_2$ showing a peak at the ferroelastic transition point and excess specific heat at higher temperatures.

Figure 4 (right). Evolution of the scattering intensity of the A_g^{bb} mode near 85 cm^{-1} in $(Pb_{1-x}Sr_x)_3(PO_4)_2$ (x = 0.075) with temperature.

Phase transitions in synthetic titanite, natural titanite and malayaite

Synthetic titanite, $CaTiSiO_5$, undergoes structural phase transitions near 500 K and ~825 K (Salje et al. 1993, Zhang et al. 1996, Kek et al. 1997, Malcherek et al. 1999a). The phases were described by Chrosch et al. (1997) as α for the room-temperature structure. In a continuous transformation the phase β forms under heating at 496 K (Kek et al. 1997). The β phase, in turn transforms into the γ phase near 825 K. The symmetry properties of the phases α and β were discussed by Speer and Gibbs (1976) and Taylor and Brown (1976). Most experimental work has focussed on the α–β transition (Ghose et al. 1991, Bismayer et al. 1992, Kek et al. 1997), with unusual pseudo-spin characteristics (Bismayer et al. 1992) and the non-classical effective order parameter exponent (Bismayer et al. 1992, Salje et al. 1993). Further studies concerned its antiferroelectric distortion pattern (Taylor and Brown 1976, Zhang et al. 1995), the pseudo-symmetry of natural samples with chemical substitution of Ti by Fe and Al (Higgins and Ribbe 1976, Hollabaugh and Foit 1984, Meyer et al. 1996), and metamictization behaviour (Hawthorne et al. 1991).

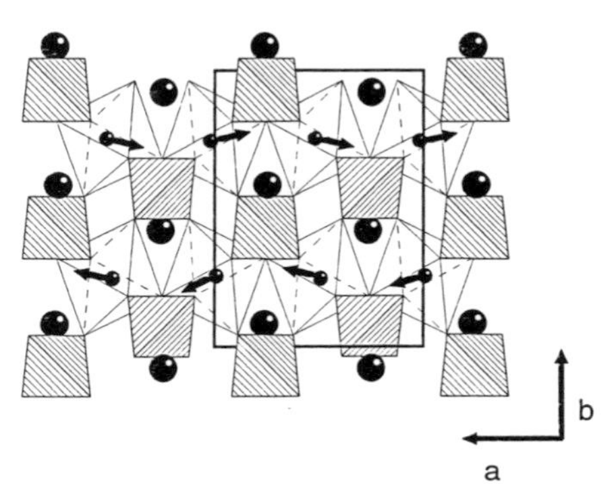

Figure 5. Crystal structure of titanite projected onto the *ab*-plane in the $P2_1/a$ phase. The arrows indicate the antiparallel Ti-displacements in neighbouring TiO_6 chains. Made with CrystalMaker, a crystal structures program for Macintosh computers.
CrystalMaker Software, P.O. Box 183
Bicester, Oxfordshire, OX6 7BS, UK
(http://www.crystalmaker.co.uk).

The crystal structure of titanite is shown in Figure 5. The structure is characterized by chains of corner-sharing TiO_6 polyhedra aligned parallel to the crystallographic *a* axis. The octahedra are cross-linked to 7-coordinated Ca atoms via SiO_4-tetrahedra. Atomic rearrangements at the transition from the high symmetry phase to the low-temperature phase involve displacements of Ti atoms to off-centre positions within the TiO_6 octahedra and additional shifts of Ca atoms from their high-temperature positions (Kek et al. 1997). The displacement pattern of $P2_1/a$ leads to antiparallel sublattices of TiO_6 octahedral chains with larger Ti-shifts along *a* and smaller shifts of Ti along the *b* and *c* axes. The smallest component along *b* can be mapped onto a spin-coordinate and the order parameter behaviour of the system has been described on the basis of Ising-type interactions (Stanley 1971, Bismayer et al. 1992).

The phase transition of synthetic titanite near 496 K is characterized by the effective critical exponent $ß \approx 1/8$, which was obtained by fitting the temperature evolution of the order parameter (Bismayer et al. 1992, Salje et al. 1993). A λ-anomaly of the specific heat (Zhang et al. 1995) and persisting diffuse X-ray intensities at temperatures $T > 496$ K have been observed experimentally (Van Heurck et al. 1991, Ghose et al. 1991, Bismayer et al. 1992, Kek et al. 1997). Raman-spectroscopic studies (Salje et al. 1993), and IR-spectroscopy (Zhang et al. 1996) showed that local distortions from the global $A2/a$ symmetry exist at temperatures above 496 K. The precursor ordering and diffuse X-ray diffraction signals observed at 530 K (Kek et al. 1997) also indicate that the two phase transitions are related to different transformation mechanisms. Group theory allows for symmetry breaking from $A2/a$ ($C2/c$) to $P2_1/a$ ($P2_1/n$) via an intermediate disordered phase. In this context Kek et al. (1997) discussed the existence of atomic split positions above 496 K. During the breaking of translational symmetry within the same point group, the critical Z-point of the high-temperature Brillouin zone transforms to the origin (Γ-point) of the low-temperature Brillouin zone. Raman spectroscopy and infrared spectroscopy show that small structural changes in titanite correspond to strong changes of the phonon spectra so that hard mode techniques are an ideal tool to detect subtle atomic rearrangements on a local length scale.

Samples and experiments. The titanite single crystal used for the scattering experiments was synthesized by Tanaka et al. (1988) using the floating zone technique. The colourless, optically clear crystal was polished to a parallelepiped with dimensions 3 × 3 × 4 mm. In addition, natural samples from the Rauris locality in Austria (Fe 1.8%, Al 3.8%), and another sample from the Smithsonian Institute in Washington (B20323 with Fe 2.7% and Al 0.4%) were studied (Meyer et al. 1996). The malayaite sample came from a skarn north of Ash Mountain, in northern British Columbia, Canada, with a composition corresponding to 0.98 Si, 1.01 Ca, 0.01 Fe, and 1.01 Sn atoms per formula unit based on five oxygen positions (Meyer et al. 1998). Powder samples for IR experiments were obtained by ball-milling the crystals in a Spex micro-mill for ~30 minutes. The even-sized fine-grained powder was kept in an oven at 390 K to avoid absorption of water. KBr and CsI were used as matrix materials. Raman spectroscopic investigations were carried out using a 4 W Argon laser and a Jarrell Ash double-spectrometer. The line fitting procedures were performed as described earlier (Bismayer 1990, Salje 1992) using an appropiate oscillator function (Eqn. 16). The absorption spectra were recorded under vacuum using a Bruker 113-v FT-IR spectrometer (Zhang et al. 1996). Experiments were performed between 200 cm^{-1} and 5000 cm^{-1} at room-temperature and in the spectral region 500 cm^{-1} and 5000 cm^{-1} on heating.

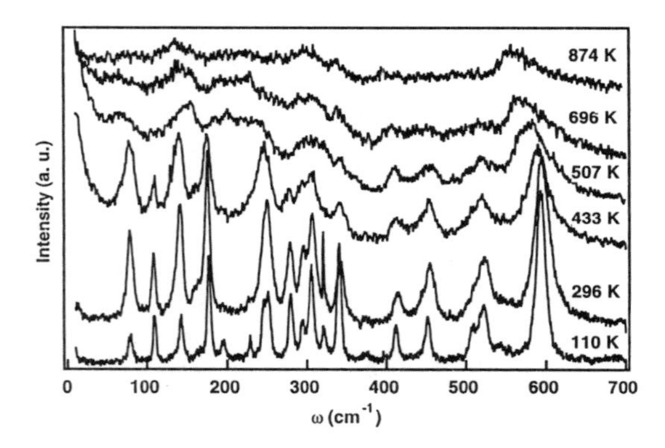

Figure 6. Unpolarized first order Raman spectrum of synthetic titanite at different temperatures.

Synthetic titanite. The first order Raman spectrum of titanite at different temperatures is shown in Figure 6. The line widths of all phonons are approximately 10 cm^{-1} at room temperature, which is a typical value for weak thermal anharmonicity. On heating, the peak intensities decrease rapidly while the apparent line widths increase. Above 825 K several Raman lines disappear. The evolution of the peak intensities on heating is shown in Figure 7 for the mode near 455 cm^{-1}. With increasing temperature the scattering intensitiy decreases strongly and disappears at an extrapolated value of ~496 K. This agrees well with the extrapolated T_c of linear optical birefringence studies (Bismayer et al. 1992) and specific heat measurements shown in Figure 8 (Zhang et al. 1995). At temperatures between ~500 K < T < 825 K, the excess intensity decreases gradually with increasing temperature. The 455 cm^{-1} A_g mode disappears close to 825 K and its intensity variation is directly related to local structural modifications of the system.

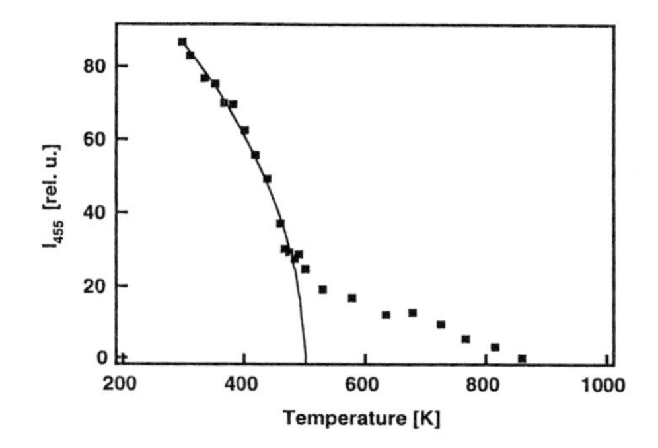

Figure 7. Thermal evolution of the peak intensity of the 455 cm^{-1} A_g-mode.

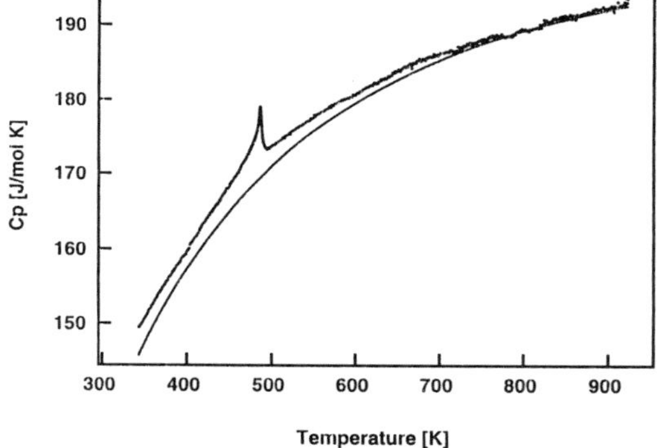

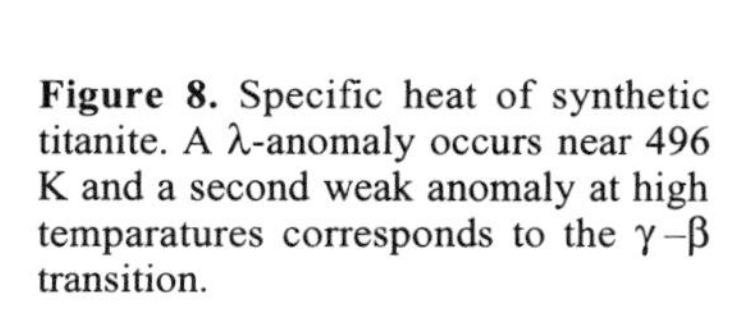

Figure 8. Specific heat of synthetic titanite. A λ-anomaly occurs near 496 K and a second weak anomaly at high temparatures corresponds to the γ–β transition.

The $3n$ vibrational modes of the first order phonon spectra of titanite in the phases $A2/a$ and $P2_1/a$ are described by the optical active representations

$$\Gamma^{A2/a}_{\text{optic}} = 9\,A_g + 12\,B_g + 11\,A_u + 13\,B_u$$
$$\Gamma^{A2/a}_{\text{acoustic}} = 1\,A_u + 2\,B_u$$
$$\Gamma^{P2_1/a}_{\text{optic}} = 24\,A_g + 24\,B_g + 23\,A_u + 22\,B_u$$
$$\Gamma^{P2_1/a}_{\text{acoustic}} = 1\,A_u + 2\,B_u$$

Twenty-one ramanactive modes are expected in $A2/a$ and 48 ramanactive modes in $P2_1/a$. The mode at 455 cm^{-1} is raman inactive in the high symmetry phase. From the thermal evolution of its scattering intensity (Fig. 7) we find that the low-temperature values extrapolate to zero at 496 K. At higher temperatures, strong scattering intensity persists and decreases linearly with increasing temperature. The extrapolated scattering intensity vanishes at the γ–β transition point near 825 K. At $T < 496$ K the quantitative relationship between the intensity and the thermodynamic order paramter Q for the β–α transition is described by

$$\Delta I \propto AQ^2 + BQ^4 \tag{21}$$

where A and B are coefficients independent of temperature. For modes which are symmetry forbidden in the high-temperature modification and do not transform as the order parameter, the coefficient of the quadratic term A vanishes ($A = 0$). An effective exponent β can then be estimated from

$$\Delta I \propto Q^4 \propto (T_c - T)^{4\beta}. \tag{22}$$

The experimental data are described by an effective exponent of ß = 1/8. As shown in Figure 9 the quantity $(\Delta I)^2 \propto (T_c - T)^{8\beta}$ follows a linear temperature dependence below 496 K. The analysis of the intensity data with T_c = 496 K leads to ß = 0.12 ± 0.03. The local distortions above T_c can be described by an order paramter Q' which scales the intensity of the Raman signals in the intermediate regime 496 K ≤ T ≤ 825 K. The experimental results show that the intensity is a linear function of T in this regime. In agreement with the Raman scattering data the volume expansion of titanite (Fig. 10) shows a discontinuity near 825 K which corresponds to the transition γ–β (Malcherek et al. 1999a).

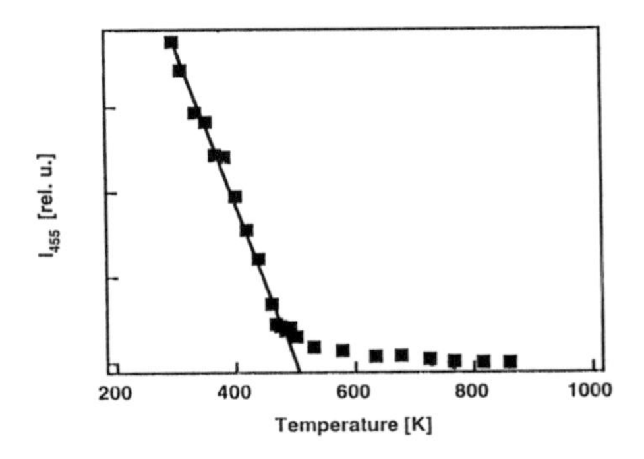

Figure 9. Squared peak intensity of the 455 cm^{-1} A_g mode as a function of temperature.

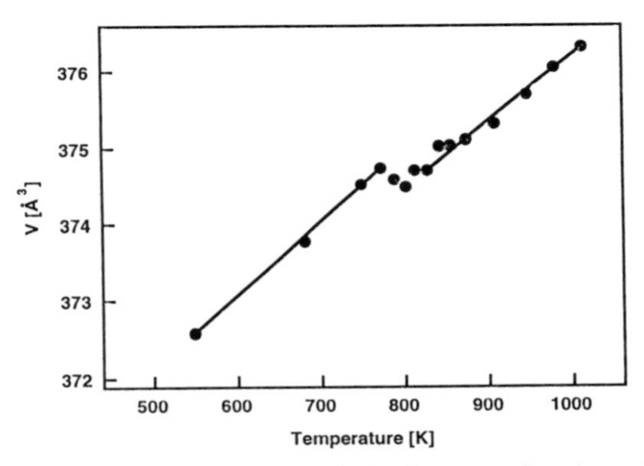

Figure 10. Cell volume of titanite as a function of temperature. The two lines are guides to the eye, indicating a discontinuity at the γ–β transition.

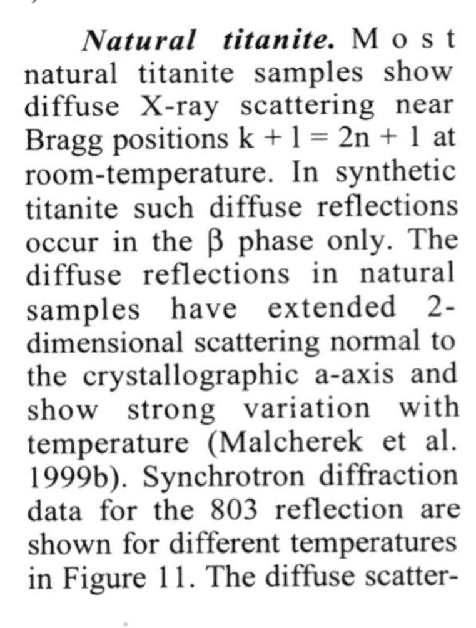

Natural titanite. Most natural titanite samples show diffuse X-ray scattering near Bragg positions k + l = 2n + 1 at room-temperature. In synthetic titanite such diffuse reflections occur in the β phase only. The diffuse reflections in natural samples have extended 2-dimensional scattering normal to the crystallographic a-axis and show strong variation with temperature (Malcherek et al. 1999b). Synchrotron diffraction data for the 803 reflection are shown for different temperatures in Figure 11. The diffuse scattering becomes weaker on heating and disappears at ~830 ± 20 K. A distinct anomaly, as observed for synthetic titanite near 500 K, does not occur in the natural material. Similarly, a smeared transition was found in the Raman spectra of a natural sample from the Rauris locality at 95 K, 300 K and 680 K (Fig. 12). The temperature evolution of the phonon spectra between 375 cm^{-1} and 725 cm^{-1} in Figure 13 indicate that the Raman signals disappear at temperatures well above 500 K, which is in agreement with earlier observations in synthetic titanite

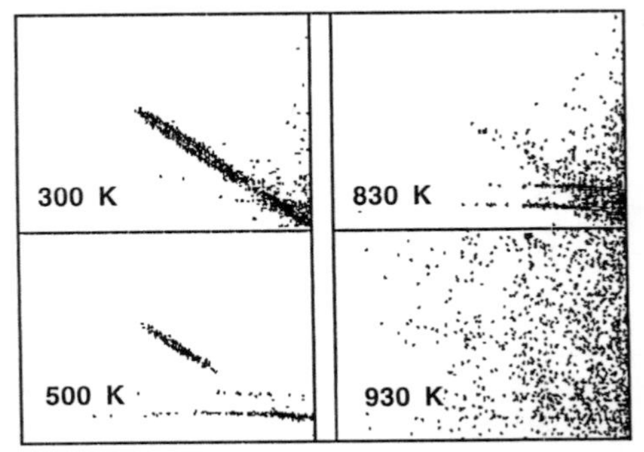

Figure 11. Image-plate Laue exposures of the 803 reflection of a natural titanite crystal from Rauris.

(Salje et al. 1993). The integrated intensity of the 470 cm^{-1} mode is plotted in Figure 14. Two temperature regimes can be identified. A linear decrease of the integrated intensity with increasing temperature occurs between 95 K and ~500 K. At temperatures above 500 K a similar linear decrease was found, however, with a smaller slope. The Raman intensities of the measured bands disappear at temperatures above ~830 K. Hence, it can be concluded that an intermediate phase might also exist in natural titanite at temperatures between ~496 K < T < 825 K. The effective exponent of the order parameter near 500 K is strongly renormalized by impurities, whereas the thermal behaviour at high temperatures seems hardly influenced by structural imperfections.

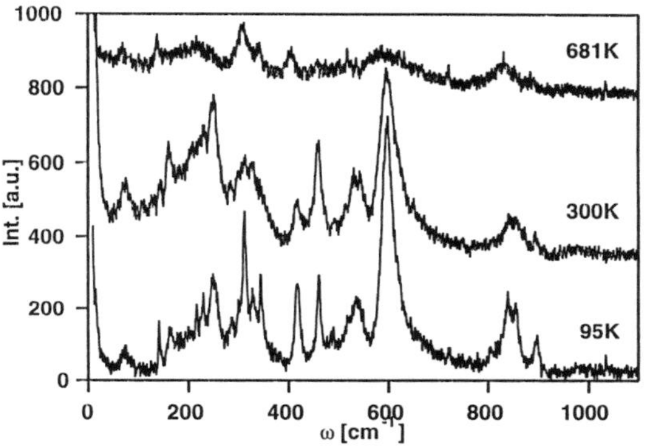

Figure 12. Unpolarized Raman spectra of a Rauris crystal at different temperatures.

Infrared spectroscopy of synthetic and natural titanite - compared with the isosymmetric instability in malayaite. The thermal anomalies and short range order observed in the Raman spectra of titanite is also seen using IR spectroscopy. The direct experimental observation of the structural anomalies near 496 K and 825 K in synthetic and natural titanites is discussed and compared with results found in malayaite, $CaSnSiO_5$. The room-temperature IR spectra of Rauris, B20323, and the synthetic titanite are plotted in Figure 15. The spectra are dominated at high frequencies by the IR band near 900 cm^{-1} which is attributed mainly to SiO_4 stretching modes. The spectra of the natural samples (Rauris and B20323) are very similar to that of the synthetic sample. However, the various heights of Ti-O phonon signals near 685 cm^{-1} result from the coupled substitution of Al and Fe for Ti and F for O in the natural samples (Meyer et al. 1996). The temperature evolution of the Mid IR spectrum of synthetic titanite is shown in Figure 16. The effect of temperature is most clearly seen as an increase in band widths, a decrease in band intensity and a softening of phonon bands. With increasing temperature we find that the Si-O bending at 563 cm^{-1} and the Si-O stretching band near 900 cm^{-1} exhibit softening in frequency below 825 K. On further heating they harden again at temperatures above 825 K. Figure 17 shows the thermal evolution of the mode near 563 cm^{-1} for all titanite samples where the structural phase transition at 825 K is clearly seen

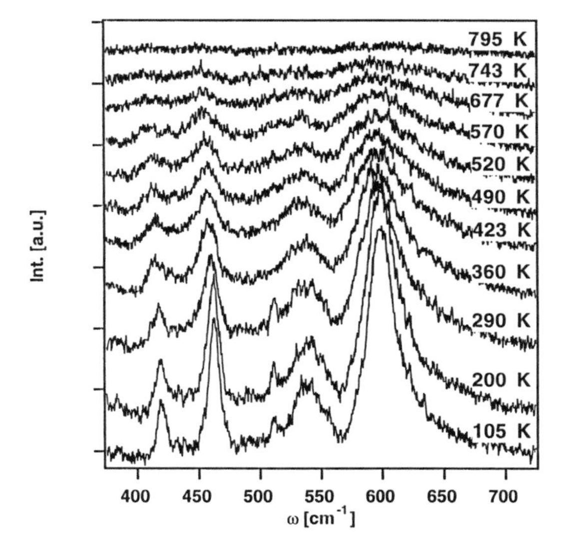

Figure 13. Thermal evolution of the Raman bands near 470 cm^{-1} (Rauris).

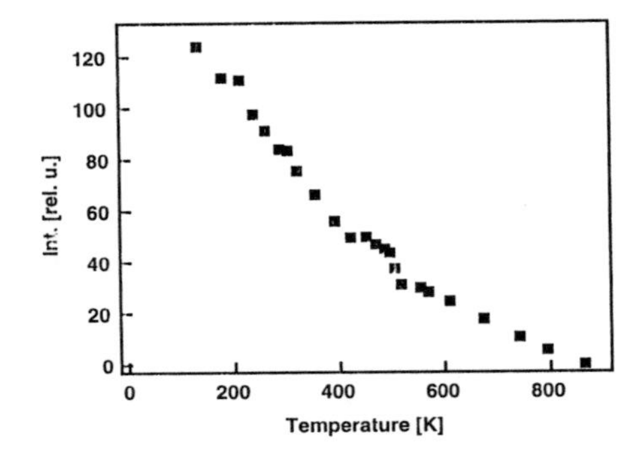

Figure 14. Integrated intensity of the Raman signal of the Rauris crystal at 470 cm^{-1}.

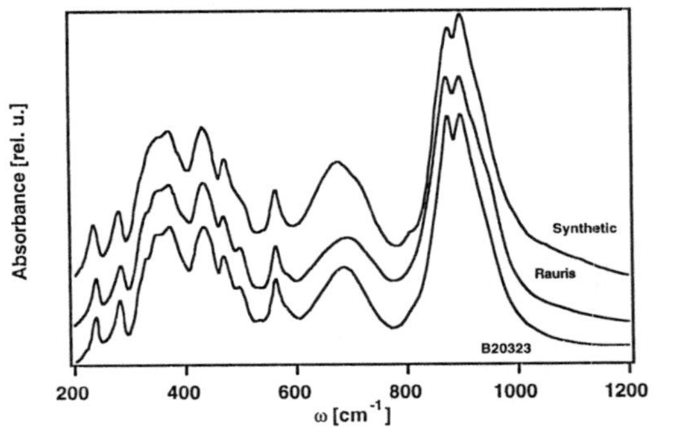

Figure 15. Comparison between the absorption spectra of the natural titanite (Rauris sample, B20323) and that of synthetic titanite.

as a break of the temperature dependence of the phonon frequencies. In the natural titanite the transition at 496 K is smeared. Using the correlation

$$\Delta\omega_i^2(T) = (\omega_i^0 - \omega_i)\,(\omega_i^0 + \omega_i) \sim \delta_i Q^2(T) \tag{23}$$

where Q is the order parameter corresponding to the phonons. With $(\omega_i^0 + \omega_i) \approx$ constant, the frequency shift becomes

$$\Delta\omega_i^2(T) = (\omega_i^0 - \omega_i) \sim Q^2(T). \tag{24}$$

As shown in Figure 17 for the IR band near 563 cm^{-1}, there is a strong change of $\delta\omega/\delta T$ at $T_c \approx 825$ K which is associated with a structural phase transition. Scaling the data at $T > 500$ K, the temperature dependencies of $\delta\omega(T)$ and $Q^2(T)$ can be written as

$$\Delta\omega_i^2(T) \propto Q^2 \propto (T_c - T)^{2\beta},\ T < T_c. \tag{25}$$

The experimental data are then fitted with the exponent, $\beta = 1/2$. This corresponds to a classical second-order behaviour which has been predicted earlier (Salje et al. 1993).

The antiferroelectric displacement of Ti in the TiO_6-octahedra seems to be the essential structural process of the α - β transition. Kek et al. (1997) reported that in the ß phase the Ca atoms occupy 8(f) split positions (Ca-Ca distance 0.14 Å at 530 K) whereas in the low-temperature phase $P2_1/a$ Ca is located in 4(d) Wyckoff positions. The

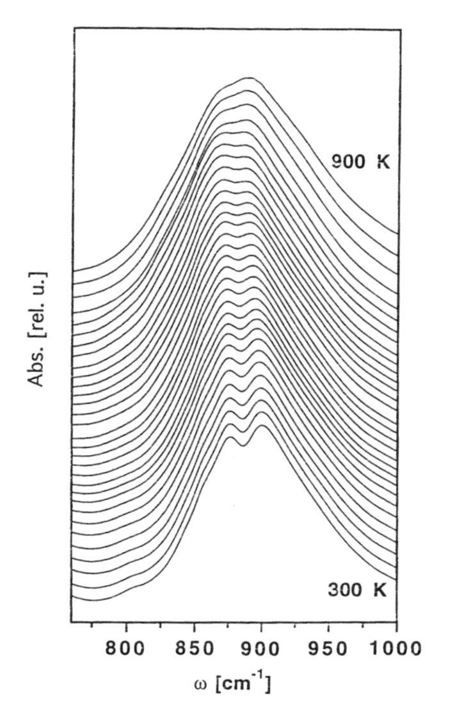

Figure 16. Temperature evolution of IR spectra of synthetic titanite between 720 cm^{-1} and 1000 cm^{-1} showing details of the spectral changes during the structural phase transition. The temperature intervals are 20K.

"overall" space group in the β state is then $A2/a$ and the diffuse scattering at positions $hkl{:}k+l = 2n + 1$ has been attributed to disordered Ca atoms with occupancy close to 0.5 at the 8(*f*) positions. However, in their high-temperature X-ray diffraction study Malcherek et al. (1999a) showed that local displacements of Ti may also play a role in the stability field between 496 and 825 K.

Let us now ask, what are the effects of the impurities in natural titanites on the phase transitions? With substitution of Al and Fe for Ti, the low-temperature phase transition around 496 K is smeared in natural samples and the transition point can not be precisely determined. On the other hand, the impurities in the natural titanites exhibit much less effect on the phase transition near 825 K. The natural samples show a similar thermal behaviour, within experimental resolution, as the pure synthetic sample (Fig. 17). Both, the Rauris and B20323 samples show nearly the same transition temperature as that of the pure titanite. The different dependencies of the two transition mechanisms on doping the Ti-positions allows us to characterize the driving forces of the phase transitions even further. The phase transition at 496 K is antiferro-electric on a macroscopic scale. The transition depends strongly on the chemical occupancy of the position which is filled by Ti in the pure material. This result is correlated with the observation that the antiferroelectric displacement of Ti in the TiO_6 octahedral is the essential driving mechanism of the α–β phase transition. The high temperature phase transition shows different features. The thermal evolution of the phonons depends only weakly on the atoms replacing Ti. These observations indicate that the TiO_6 octahedron plays a minor role in the mechanism of the high temperature phase transition. If we assume that the changes of the phonons related

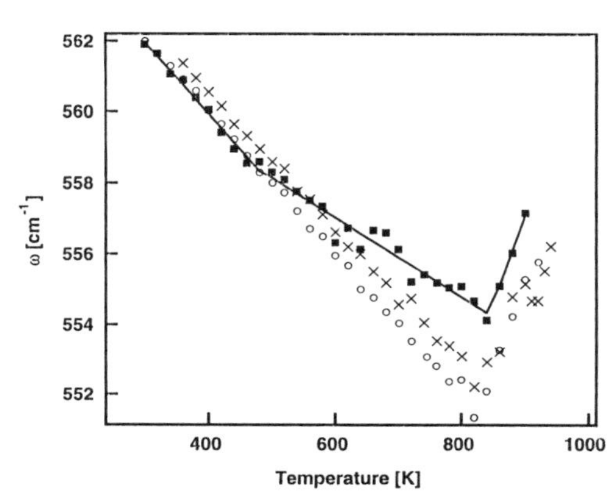

Figure 17. The temperature dependence of the Si-O bending band near 560 cm^{-1}. The data points correspond to synthetic titanite (dark squares), the Rauris sample (crosses), and the smple B20323 (open circles). Two discontinuities are seen in synthetic titanite whereas the slope change near 825 K indicates a second-order phase transition in all samples. The average errors for the peak positions are 0.25 cm^{-1} near room temperature and 1.0 cm^{-1} for high temperatures.

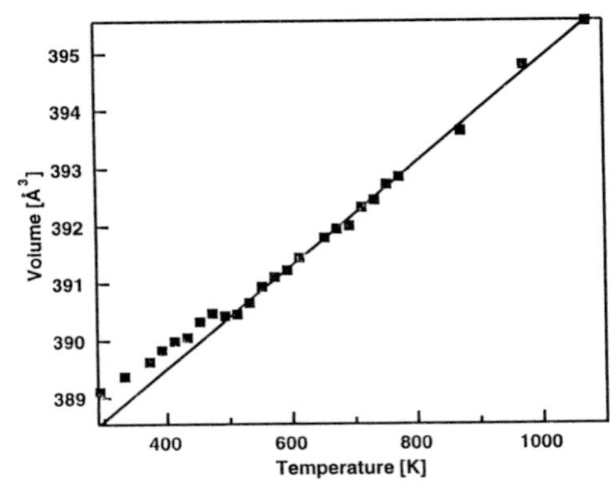

Figure 18. Temperature dependence of the cell volume of malayaite. The high-temperature values are linearly extrapolated to low temperatures.

to the SiO_4 tetrahedra can be explained by depolarisation effects by Ca (and very small geometrical changes of the tetrahedra), we may conclude that the Ca-polyhedra play the essential role in the structural phase transition near 825 K (Malcherek et al. 1999a).

Titanite is isostructural with malayaite, $CaSnSiO_5$, a rare mineral found in skarn deposits (Higgins and Ribbe 1977, Groat et al. 1996). Previous studies on malayaite revealed a thermal discontinuity near 500 K. No further anomaly and no indication of deviations from *A2/a* symmetry between 100 and 870 K could be detected (Groat et al. 1996, Meyer et al. 1998). The thermal evolution of the volume of malayaite as shown in Figure 18 is most sensitive to structural modifications. The cell volume shows a distinct break in slope near 493 K and the extrapolation of the high-temperature data gives a difference of the measured and the extrapolated volume at 300 K of ~0.18%. The experimentally observed volume at room-temperature is larger than the extrapolated one. A structure refinement at different temperatures revealed that the atomic parameters show only a slight adjustment of the average structure with increasing temperature (Zhang et al. 1999). In order to investigate the phonon spectra of malayaite, Bismayer et al. (1999) studied its IR absorption as a function of temperature. The room-temperature IR spectrum of malayaite is shown in Figure 19. The temperature evolution of the infrared spectrum of malayaite between 400 and 600 cm^{-1} is shown in Figure 20. Five phonon bands occur in this frequency range. The infrared signals become broader and the band frequencies shift with increasing temperature. The band near 499 cm^{-1} has been analysed in more detail and its shift with temperature is plotted in Figure 21. Two regions of different linear slope can clearly be distinguished. A change occurs near 500 K and above this temperature the slope is reduced.

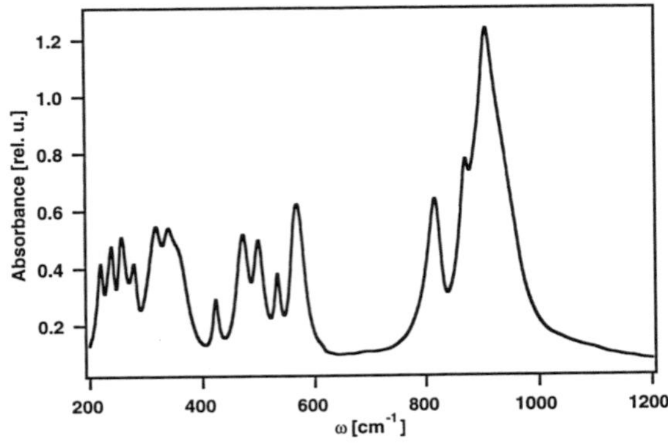

Figure 19. Infrared vibrational spectrum of malayaite at room temperature in the region of 200 to 1200 cm^{-1}.

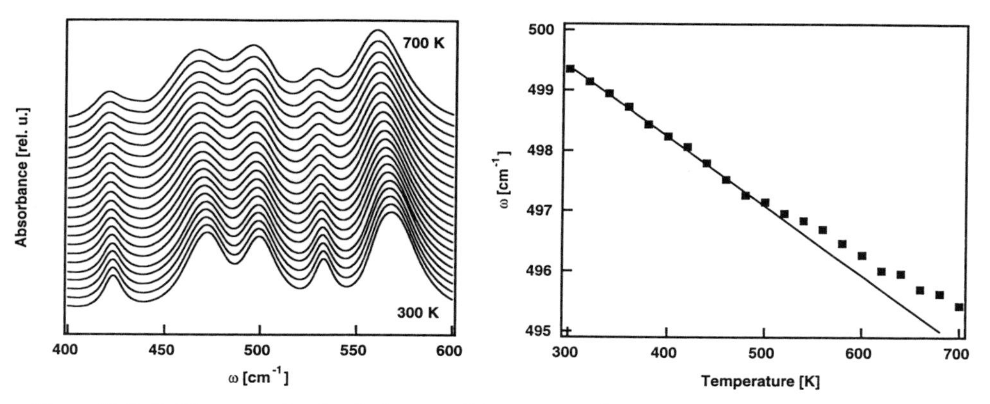

Figure 20 (left). IR spectra of malayaite between 400 and 600 cm^{-1} at temperatures between 300 and 700 K in steps of 20 K.

Figure 21 (right). The temperature-evolution of the peak position of the 499 cm^{-1} IR band of malayaite. The line is fit to the data below 500 K. Note the change in slope near 500 K.

Let us now compare the thermal behaviour of titanite and malayite. The discontinuities observed in the IR spectra and the specific heat of synthetic titanite correspond to the α–β transition at 496 K and to the β–γ transition near 825 K. The diffuse scattering of the intermediate regime at positions $k + l$ = odd disappears during the β–γ transition. In natural titanite the transition at 496 K is smeared or fully suppressed for higher impurity content, but the order parameter behaviour of the high-temperature transition near 825 K remains unchanged. Although a phonon anomaly near 500 K can also be observed in malayaite, the general features of the structural anomaly are different from those observed at the α–β transition in titanite, which is governed by the antiferroelectric displacement of Ti in the TiO_6 octahedra. No Sn displacement inside the SnO_6 polyhedra could be detected and the structural change in malayaite corresponds to an isosymmetric transformation. We may therefore relate the room temperature structure of malayaite to the paraelectric β phase of titanite with an overall symmetry *A*2/*a*. The structure refinement of malayaite and the displacement of the atomic positions at different temperatures by Zhang et al. (1999) shows that the isosymmetric transition of malayaite is characterised by a tumbling motion of the SnO_6 octahedra which also alters the oxygen sphere around the Ca positions. We may speculate that a similar process takes place during the β–γ transformation in titanite.

ACKNOWLEDGMENTS

The author is grateful to the DFG, ARC and the BMBF (05SM8 GUA9).

REFERENCES

Bismayer U (1988) New developments in Raman spectroscopy on structural phase transitions. *In* E Salje (ed) Physical Properties and Thermodynamic Behaviour in Minerals, NATO ASI C225:143-183 Reidel, Dordrecht

Bismayer U (1990) Hard mode Raman spectroscopy and its application to ferroelastic and ferroelectric phase transitions. Phase Trans 27:211-267

Bismayer U, Salje EKH (1981) Ferroelastic phases in $Pb_3(PO_4)_2$-$Pb_3(AsO_4)_2$ X-ray and optical experiments. Acta Crystallogr A37:145-153

Bismayer U, Salje EKH, Joffrin C (1982) Reinvestigation of the stepwise character of the ferroelastic phase transition in lead phosphate-arsenate, $Pb_3(PO_4)_2$-$Pb_3(AsO_4)_2$. J Physique 43:119-128

Bismayer U, Salje EKH, Glazer AM, Cosier J (1986) Effect of strain-induced order-parameter coupling on the ferroelastic behaviour of lead phosphate-arsenate. Phase Trans 6:129-151

Bismayer U, Schmahl W, Schmidt C, Groat LA (1992) Linear birefringence and X-ray diffraction studies on the structural phase transition in titanite, $CaTiSiO_5$. Phys Chem Minerals 18:260-266

Bismayer U, Hensler J, Salje EKH, Güttler B (1994) Renormalization Phenomena in Ba-diluted Ferroelastic lead Phosphate, $(Pb_{1-x}Ba_x)_3(PO_4)_2$. Phase Trans 48:149-168

Bismayer U, Röwer RW, Wruck B (1995) Ferroelastic phase transition and renormalization effect in diluted lead phosphate, $(Pb_{1-x}Sr_x)_3(PO_4)_2$ and $(Pb_{1-x}Ba_x)_3(PO_4)_2$. Phase Trans 55:169-179

Bismayer U, Zhang M, Groat LA, Salje EKH, Meyer H-W (1999) The β–γ phase transition in titanite and the isosymmetric analogue in malayaite. Phase Trans 68:545-556

Blinc R, Zeks B (1974) Soft modes in ferroelectrics and antiferroelectrics. *In* EP Wohlfarth (ed) Selected Topics in Solid State Physics. North Holland, Amsterdam, p 10-18

Brixner LH, Bierstedt PE, Jaep WF, Barkley J R (1973) α-$Pb_3(PO_4)_2$—a pure ferroelastic. Material Res Bull 8:497-504

Bruce AD, Taylor W, Murray AF (1980) Precursor order and Raman scattering near displacive phase transitions. J Phys C: Solid State Phys 13:483-504

Bruce A, Cowley RA (1981) Structural Phase Transitions. Taylor and Francis, London, 326 p

Chrosch J, Bismayer U, Salje EKH (1997) Anti-phase boundaries and phase transitions in titanite: an X-ray diffraction study. Am Mineral 82:677-681

Cowley RA (1964) The theory of Raman scattering from crystals. Proc Phys Soc 84:281-296

Ghose S, Ito Y, Hatch DM (1991) Paraelectric-antiferroelectric phase transition in titanite, $CaTiSiO_5$: I. A high temperature X-ray diffraction study of the order parameter and transition mechansim. Phys Chem Minerals 17:591-603

Groat LA, Kek S, Bismayer U, Schmidt C, Krane HG, Meyer H, Nistor L, Van Tendeloo G (1996) A synchrotron radiation, HRTEM, X-ray powder diffraction, and Raman spectroscopic study of malayaite, $CaSnSiO_5$. Am Mineral 81:595-602

Guimaraes DMC.(1979) Ferroelastic transformations in lead orthophosphate and its structure as a function of temperature. Acta Crystallogr A 35:108-114

Güttler B, Salje EKH, Putnis A (1989) Structural states of Mg-cordierite III: infrared spectroscopy and the nature of the hexagonal-modulated transition. Phys Chem Minerals 16:365-373

Hawthorne FC, Groat LA, Raudsepp M, Ball NA, Kimata M, Spike FD, Gaba R, Halden NM, Lumpkin GR, Ewing RC, Greegor RB, Lytle FW, Ercit TS, Rossman GR, Wicks FJ, Ramik RA, Sheriff BL, Fleet ME, McCammon C (1991) Alpha-decay damage in titanite. Am Mineral 76:370-396

Higgins JB, Ribbe PH (1976) The crystal chemistry and space groups of natural and synthetic titanites. Am Mineral 61:878-888

Higgins JB, Ribbe, PH (1977) The structure of malayaite, $CaSnSiO_4$, a tin analog of titanite. Am Mineral 62:801-806

Hollabaugh CL, Foit FF (1984) The crystal structure of Al-rich titanite from Grisons, Switzerland. Am Mineral 69:725-732

Joffrin C, Benoit JP, Currat R, Lambert M (1979) Transition de phase ferroelastique du phosphat de plomb. Etude par diffusion inelastique des neutrons. J Physique 40:1185-1194

Kek S, Aroyo M, Bismayer U, Schmidt C, Eichhorn K, Krane HG (1997) Synchrotron radiation study of the crystal structure of titanite ($CaTiSiO_5$) at 100K, 295K and 530K: Model for a two-step structural transition. Z Kristallogr 212:9-19

Landau LD, Lifshitz EM (1979) Lehrbuch der Theoretischen Physik V, Statistische Physik I. Akademie Verlag, Berlin, p 423-490

Levanyuk AP, Osipov VV, Sigov AS, Sobyanin AA (1979) Change of defect structure and the resultant anomalies in the properties of substances near phase-transition points. Soviet Phys JEPT 49:176-188

Malcherek T, Domeneghetti CM, Tazzoli,V, Salje EKH, Bismayer U (1999a) A high temperature diffraction study of synthetic titanite $CaTiOSiO_4$.Phase Trans 69:119-131

Malcherek T, Paulmann C, Bismayer U (1999b) Diffuse scattering in synthetic titanite, $CaTiOSiO_4$. HASYLAB Ann Report I:517-518

Meyer H-W, Zhang M, Bismayer U, Salje EKH, Schmidt C, Kek S, Morgenroth W, Bleser T (1996) Phase transformation of natural titanite: An infrared, Raman spectroscopic, optical birefringence and X-ray diffraction study. Phase Trans 59:39-60

Meyer H-W, Bismayer U, Adiwidjaja G, Zhang M, Nistor L, Van Tendeloo G (1998) Natural titanite and malayaite: Structural investigations and the 500 K anomaly. Phase Trans 67:27-49

Petzelt J, Dvorak V (1976) Changes of infrared and Raman spectra induced by structural phase transitions: 1. General considerations. J Phys C: Solid State Phys 9:1571-1586

Roucau C, Tanaka M, Torres J, Ayroles R (1979). Etude en microscopie electronique de la structure liee aux proprietes ferroelastique du phosphat de plomb, $Pb_3(PO_4)_2$. J Micros Spectros Electronique 4:603-612

Salje EKH (1990) Phase Transitions in Ferroelastic and Co-elastic Crystals. Cambridge University Press, Cambridge, UK, 366 p

Salje EKH (1992) Hard mode spectroscopy: experimental studies of structural phase transitions. Phase Trans 37:83-110

Salje EKH (1994) Phase transitions and vibrational spectroscopy in feldspars. *In* I Parson (ed) Feldspars and Their Reactions. Kluwer, Dordrecht, The Netherlands, p 103-160

Salje EKH, Bismayer U (1981) Critical behaviour of optical modes in ferroelastic $Pb_3(PO_4)_2$-$Pb_3(AsO_4)_2$, Phase Trans 2:15-30

Salje EKH, Bismayer U (1997) Hard mode spectroscopy: the concept and applications. Phase Trans 63: 1-75

Salje EKH, Devarajan V (1981) Potts model and phase transitions in lead phosphate, $Pb_3(PO_4)_2$. J Phys C: Solid State Phys 14:L1029-L1035

Salje EKH, and Wruck B (1983) Specific-heat measurements and critical exponents of the ferroelastic phase transition in $Pb_3(PO_4)_2$ and $Pb_3(P_{1-x}As_xO_4)_2$. Phys Rev B28:6510-6518

Salje EKH, and Yagil Y (1996) Hard mode spectroscopy for the investigation of structural and superconducting phase transitions. J Phys Chem Solids 57:1413-1424

Salje EKH, Devarajan V, Bismayer U, Guimaraes DMC (1983) Phase transitions in $Pb_3(P_{1-x}As_xO_4)_2$: influence of the central peak and flip mode on the Raman scattering of hard modes. J Phys C: Solid State Phys 16:5233-5243

Salje EKH, Devarajan V (1986) Phase transitions in systems with strain-induced coupling between two order parameters. Phase Trans 6:235-248

Salje EKH, Graeme-Barber A, Carpenter, MA, Bismayer U (1993) Lattice parameters, spontaneous strain and phase transitions in $Pb_3(PO_4)_2$. Acta Crystallogr B49:387-392

Salje EKH, Schmidt C, Bismayer U (1993) Structural phase transition in titanite, $CaTiSiO_5$: A ramanspectroscopic study. Phys Chem Minerals 19:502-506

Speer JA, Gibbs GV (1976) Crystal structure of synthetic titanite $CaTiSiO_5$, and the domain textures of natural titanites. Am Mineral 61:238-247

Stanley HE (1971). Introduction to phase transitions and critical phenomena. Oxford University Press, Oxford, UK

Tanaka I, Obushi T, Kojima, H (1988) Growth and characterization of titanite $CaTiSiO_5$ single crystals by the floating zone method. J Cryst Growth 87:169-174

Taylor M, Brown GE (1976) High-temperature structural study of the $P2_1/a$ <=> A2/a phase transition in synthetic titanite, $CaTiSiO_5$. Am Mineral 61:435-447

Torres J (1975) Symmetrie du parametere d'ordre de la transition de phase ferroelastique du phosphat de plomb. Phys Status Solidi 71:141-150

Van Heurck C, Van Tendeloo G, Ghose S, Amelinckx S (1991) Paraelectric-antiferroelectric phase transition in titanite, $CaTiSiO_5$: II. Electron diffraction and electron microscopic studies of transition dynamics. Phys Chem Minerals 17:604-610

Yagil Y, Baundenbacher F, Zhang M, Birch JR, Kinder H, Salje EKH (1995) Optical properties of $Y_1Ba_2Cu_3O_{7-\delta}$ thin films. Phys Rev B 52:15582-15591

Zhang M, Salje EKH, Bismayer U, Unruh HG, Wruck B, Schmidt C (1995) Phase transition(s) in titanite $CaTiSiO_5$: an infrared spectrosopic, dielectric response and heat capacity study. Phys Chem Minerals 22:41-49

Zhang M, Salje EKH, Bismayer U (1996) Structural phase transition near 825 K in titanite: evidence from infrared spectroscopic observation. Am Mineral 82:30-35

Zhang M, Meyer H-W, Groat LA, Bismayer U, Salje EKH, Adiwidjaja G (1999) An infrared spectroscopic and single-crystal X-ray study of malayaite, $CaSnSiO_5$. Phys Chem Minerals 26:546-553

11 Synchrotron Studies of Phase Transformations

John B. Parise

Center for High Pressure Research
Department of Geosciences and Department of Chemistry
State University of New York
Stony Brook, New York 11794

INTRODUCTION AND OVERVIEW

Material properties are dependent upon atomic arrangement. The structure provides a basis for calculation and for the interpretation of experiment. In the Earth context, mineral structure is composition, pressure and temperature sensitive, and changes in the phases present cause profound changes in properties. Examples include the relationship between sound velocities (elasticity) and structure as a function of depth in the Earth (Chen et al. 1999, Li et al. 1998, Liebermann and Li 1998, Parise et al. 1998). The selective sorption of ions and isotopes on mineral surfaces, and on particular faces on one growing surface, are excellent examples of how sorption is controlled by atomic arrangement (Reeder 1991, 1996). It is desirable therefore that material properties and structure be studied together, and if possible under the conditions of operation for that material. For the geosciences, this means studies of crystal structure and physical properties at elevated temperatures and pressures. For the solid state physics, chemistry and planetary science communities, it might also include high pressure/low temperature investigations of magnetic structures. Access to synchrotron radiation has made these types of studies routine. To address topical concepts and issues, beamlines have become more versatile and more easily reconfigured. This has decreased the cycle time between blue-sky concept and reality, and it will radically alter the mineral chemistry and mineral physics cultures, as national facilities are now firmly established at the cutting edge of solid state research (Hemley 1999). The science enabled by these facilities is impossible to reproduce in the home laboratory, as is the culture of interdisciplinary activity, interdependence between groups of collaborators, and speedy application of basic X-ray and neutron physics to applied problems. This mode of science is strange to many. The change however, is inevitable and will continue to gather momentum.

The advent of readily accessible bright synchrotron radiation has revolutionized our abilities to determine structure as a function of P and T. Bright VUV and X-ray sources allow us to carry out laboratory-style measurements on smaller samples, to higher resolution and with greater precision and accuracy. The determination of structure under the conditions of pressure, temperature and chemistry of the Earth, are mainstays of modern mineral physics. Synchrotron radiation, because of its inherent brilliance, provides opportunities to include time in these studies. For example, as advanced instruments, with high data acquisition rates, become operational, state-of-the-art scattering will involve the construction of an "image" in some multidimensional space rather than the collection and analysis of a single diffraction pattern. Possible dimensions of this "image space" include chemical composition, temperature, pressure, and the evolution in time of structure and physical properties such as elasticity, rheology, dielectric constant, etc. The "instrument" will consist of environment apparatus, diffractometer and its computer control hardware, as it does in the laboratory setting. Software, the data acquisition system, and techniques for data visualization will allow the "image" to be viewed on the time scale of the experiment in progress. Some steps toward this situation are already in hand, and the concept is further illustrated toward the end of

1529-6466/00/0039-0011$05.00

the chapter. My primary aims are to introduce experiments made possible by synchrotron X-radiation, and to describe source characteristics, and access to the beams, which facilitate unique studies of phase transitions.

OVERVIEW: DIFFRACTION AND SPECTROSCOPIC TECHNIQUES FOR STUDYING TRANSITIONS

The most versatile, testable and reproducible methods, providing detailed knowledge of the atomic arrangement for crystalline and amorphous materials, are based upon X-ray and neutron scattering techniques. These include X-ray diffraction (XRD). I will concentrate on "Bragg" XRD during this review. Spectroscopic techniques are excellent supplements to XRD and provide unique insights into local structure for disordered materials. The National Synchrotron Light Source (NSLS) is unique in this regard (Table 1), since it provides two storage rings in one building; one dedicated to hard X-ray work and the other to infrared and VUV spectroscopy. The unique properties of these sources are covered in a number of recent reviews and papers (Hanfland et al. 1994, 1992; Hemley et al. 1996, 1998a; Nagara and Nagao 1998, Nanba 1998, Reffner et al. 1994) and Volume 37 in the *Reviews in Mineralogy* series (Hemley et al. 1998b, Mao and Hemley 1998). A brief overview will suffice here.

X-ray absorption fine structure

XAFS is a powerful means of obtaining information on local structure (Crozier 1997, Henderson et al. 1995, Young and Dent 1999) in ferroelastic phases (Henderson et al. 1995, Sicron and Yacoby 1999, Sicron et al. 1997) for example. This method allows studies of glassy materials (Akagi et al. 1999, Shiraishi et al. 1999). Also it is used to study changes in coordination for ions in hydrothermal systems (Mayanovic et al. 1999, Mosselmans et al. 1999, Seward et al. 2000) and site-selective behavior in other systems (Ishii et al. 1999). XAFS includes the analysis of X-ray absorption near the edge

Table 1. Web addresses for some synchrotron radiation facilities

Location	*E(GeV)*	*Address (http://)*
CLS, Saskatoon, Canada*	2.5-2.9	cls.usask.ca
BEPG, Beijing, China	1.5-2.8	solar.rtd.utk.edu/~china/ins/IHEP/bsrf/bsrf.html
Hsinchu, Taiwan	1.3-1.5	www.srrc.gov.tw/
Daresbury, England	2	www.dl.ac.uk/SRS/index.html
Diamond*	3	www.clrc.ac.uk/NewLight/
ESRF, Grenoble, France	6	www.esrf.fr/
LURE, Orsay, France	1.8	www.lure.u-psud.fr/
DORISIII, Hamburg	4.5-5.3	info.desy.de/hasylab/
PETRA II*, Hamburg	7-14	info.desy.de/hasylab/
PF, KEK, Tsukuba, Japan	2.5	pinecone.kek.jp
SPring-8, Hyogo, Japan	8	www.spring8.or.jp/
Middle East	1?	www.weizmann.ac.il/home/sesame/
APS, Argonne, IL	7	epics.aps.anl.gov
ALS, Berkeley, CA	1.5-1.9	www-als.lbl.gov
CHESS, Ithaca, NY	5.5	www.chess.cornell.edu
SSRL, Stanford, CA	3	www-ssrl.slac.stanford.edu
NSLS, Upton, NY	2.8	www.nsls.bnl.gov

*Designed or only partially built at time of publication

(XANES), where changes in oxidation state are monitored, as well as in the extended (EXAFS) region, where analysis of fine struc-ture provides information on bonding geometry for the absorbing atom.

Other spectroscopic techniques

Modern synchrotron sources have allowed weak interactions to be studied routinely. Brighter beams have allowed the construction of high-energy resolution spectrometers to study of the Mössbauer effect (Zhang et al. 1999) and electronic spectra (Badro et al. 1999, Rueff et al. 1999). In the latter case, the use of Be gaskets in the diamond anvil cell (DAC) permit studies of spectra at energies below 10 keV, and consequently of the important high spin-low spin transitions in Fe containing materials (Hemley et al. 1998b, Pasternak et al. 1999). Results of these investigations, using emission spectroscopy (Badro et al. 1999, Hemley et al. 1998b, Rueff et al. 1999) confirm the phase transition in FeS to the FeS-III structure (Nelmes et al. 1999b) accompanied by a HS-LS transition. This result is confirmed by resistivity measurements (Takele and Hearne 1999). This same transition however was *not* observed in FeO (Badro et al. 1999). The results obtained so far have been encouraging, and suggest X-ray-based specroscopies of other strongly correlated electronic systems at pressure will show similarly interesting results (Hemley et al. 1998b).

Synchrotron radiation allows the development of the energy/angular dispersive diffraction setups, which can be used for anomalous diffraction. In diffraction anomalous fine structure (DAFS) studies, this allows the experimentor to collect diffracted intensities around an absorption edge over a continuous energy range. This technique has recently been reviewed by Yacoby (1999). Its application to the study of phase transitions is well illustrated by the study of the mechanism of the structural transformation in Fe_3O_4 during its Verwey transition at 120 K carried out by Frenkel et al. (1999). In their experiments, information obtained using DAFS was used to separately solve the local structures around the octahedral and tetrahedral sites in the spinel structure of magnetite.

In the next decade, application of inelastic X-ray scattering (IXS) will become more routine at synchrotron sources (Caliebe et al. 1996, 1998; Hamalainen et al. 1995, 1996; Hill et al. 1996, Kao et al. 1996, Krisch et al. 1995). Like inelastic optical and neutron scattering, IXS requires the amount of energy added to or subtracted from the incident beam energy, ΔE, be measured with sufficient accuracy. In the neutron case, the optics for carrying out these experiments at cold sources (low energy, long wavelength neutron sources) has been available for sometime. For X-rays, the more demanding optic requirements have necessitated the use of intense X-ray sources. The application to the study of phase transitions is clear, since changes in elementary excitations, such as phonons, play an important role.

Pair distribution analysis

PDA (Toby and Egami 1992) for the study of local structure, particularly when teamed with the synchrotron's tunability to absorption edges to exploit anomalous scattering effects (Hu et al. 1992), is a somewhat under-utilized technique compared to XAFS. This technique is useful for the study of local structure and transitions in amorphous cryptocrystalline and crystalline materials (Dmowski et al. 2000, Imai et al. 1996, 1999; Tamura et al. 1999) mainly those involving order-disorder phenomena. It offers several advantages over XAFS including a more robust data/parameter ratio, though it lacks the sensitivity and is considerably more demanding of time for data collection. When information on the local structure of glasses at high pressure is required, PDA has a several advantages (Hemley et al. 1998b, Mao and Hemley 1998) over XAFS.

Chief amongst these, is the difficulty of performing absorption studies at the Si-K edge (1.8 keV) with XAFS when the sample is in the high-pressure cell. In order to maximize k-resolution, data for PDA are best collected at the high energies required to minimize absorption from the sample and cell assembly.

Scattering studies of surfaces

Changes in fluid-solid interactions are a growing area of interest as well and synchrotrons have distinct advantages in these studies (Bahr 1994, Etgens et al. 1999, Henderson et al. 1995, Moller et al. 1998, Peng et al. 1997, Tomilin 1997, Vollhardt 1999). The techniques used to study surfaces in the laboratory include atomic force microscopy and a plethora of vacuum spectroscopic techniques including low energy electron diffraction (LEED), TEM and electron energy loss spectroscopy (EELS). All these techniques provide important insights. The synchrotron-based scattering techniques include X-ray standing wave (XSW) (Kendelewicz et al. 1995, Qian et al. 1994) and lower dimensional X-ray diffraction (Robinson et al. 1992, 1994). All are in their infancy and have only recently been applied to minerals. In the future the utility of these techniques, some of which can be applied under conditions more realistic for the earth, will continue to grow.

Since the time of the Braggs, bulk mineral structures at ambient conditions and as a function of PTX, have provided the basis for the interpretation of the manner in which minerals and rocks behave. The structures of the more important rock forming minerals on this planet, at ambient pressure and temperature are, for the most part, well known. Details of the structure of glassy materials and the poorly crystalline, or cryptocrystalline, mineral are on-going concerns of the mineralogical community. With greater information available from planetary probes, the details of the extra-terrestrial minerals, such as clathrates and gas alloys will become more important. In order to emphasize some of the opportunities now available with synchrotron sources, I will concentrate on the crystalline materials for the most part, and on powder and single-crystal diffraction techniques at high pressure and/or temperatures predominantly. Many of the resources outlined below will be equally applicable to spectroscopic techniques. Recognizing the synergy and commonality of beamline optics between the various techniques has led to the construction of a number of hybrid beamlines capable of easy reconfiguration. The reader may wish to explore the following web sites as examples of how future multipurpose beamlines, which are discipline rather than individual technique based, are evolving:

http://cars.uchicago.edu and http://www.anl.gov/OPA/whatsnew/hpcatstory.htm

Overview of the diffraction-based science from bulk samples

The remainder of this chapter will concentrate on descriptions of the sources of X-rays available, how to access and prepare to use these sources, and some of the science enabled by them. The increased brightness and energies (Fig. 1) available to experimentalists at synchrotron storage rings facilitates new classes of X-ray scattering experiments with increased energy, angular, spatial and time resolution. For *ex situ* study of phase transitions, data suitable for full structure determination on sub-micron single crystals or single domains quenched from high pressures and temperatures, is now possible (Neder et al. 1996b). Weak scattering phenomena, such as the diffuse scattering arising from incipient order-disorder, can be studied with unprecedented precision with small single crystals (Renaud et al. 1995). When only powdered samples are available, high-resolution powder diffraction data are providing precise crystal structure determinations using a variety of detectors (McMahon et al. 1996). Resonant X-ray scattering studies of single crystals and powders are used to enhance contract between structural features such as atomic site occupancy factors (Cox and Wilkinson 1994).

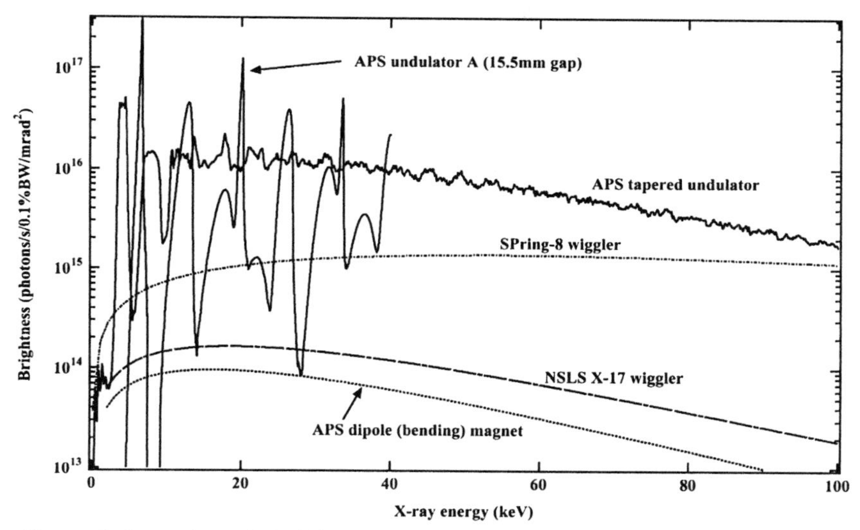

Figure 1. Comparison of the brightness from selected synchrotron sources. A powerful laboratory sources (rotating anode, W target operating at 150 keV and 500 mA, would produce a spectrum peaked at the Kα line with a brightness of about 10^{10}. A free electron laser will produce peak brightness $>10^{22}$.

Time-resolved diffraction provides information for the interpretation of the mechanism of reaction (Chen et al. 1996b, Parise et al. 1998). With appropriate environmental cell, beamline and detector designs, physical and crystallographic properties are being measured simultaneously; suggesting this may become the default mode of study for the more important systems. Some of these points will be illustrated below along with the experimental designs allowing them.

Keep in mind that not *every* experiment requires the latest generation of synchrotron sources (not every experiment requires synchrotron radiation). The user community at synchrotrons worldwide is growing, and there is an increasing realization that the use of these facilities can offer much for the understanding fundamental processes occurring in materials. However, the mode of operation is very different from that which prevails at a home laboratory. Therefore careful planning of experiments before resorting to the use of the synchrotron is required. This is especially true for *in situ* experiments for studying phase transitions. Here the experimentalist needs to evaluate requirements related to the X-ray beam (beam stability, brightness, energy, resolution, etc.), as well as those related to the sample (scattering from furnaces or pressure vessels, sample strain, among others).

SYNCHROTRON RADIATION SOURCES

General characteristics

Synchrotron storage rings have been used as intense X-ray sources for about 30 years. Electromagnetic radiation is emitted when high-energy electrons, or positrons, are accelerated in a magnetic field. This radiation is highly collimated in the vertical direction as the charged particles approach the speed of light. Compared to conventional sealed tube sources, synchrotron radiation[§] is 10^4 to 10^{12} times brighter. Comparable

[§] Several excellent texts on synchrotron radiation and crystallography are also available (Coppens 1992, Finger 1989, Kunz 1979, Young 1995).

counting statistics can be obtained from sample orders of magnitude smaller, or say 10^6 times faster, on similarly sized samples. The latter point implies that experiments requiring a 10-second exposure at the synchrotron require over 100 days exposure for a laboratory sealed-tube source. Practically, this allows the study of smaller samples under more extreme conditions and as a function of time. These are obvious advantages in the study of phase transitions.

There are several Figures-of-Merit (FOM), which can be used to compare the photon output from synchrotron radiation sources (Table 1). The *flux* is the number of photons/second/horizontal angle/bandwidth, integrated over the entire vertical opening angle. This is the appropriate FOM for an experiment, which can use all of the photons in a particular energy interval from a synchrotron port. Examples include X-ray absorption fine structure (XAFS) (Crozier 1997, Young and Dent 1999) and powder diffraction experiments on large samples (Cox 1992, Cox et al. 1988, Finger 1989). The *brightness* is the flux/vertical angle. The brightness is thus the number of photons per solid angle, and is the appropriate FOM for an experiment, which uses a collimator to define a small beam. Examples include X-ray diffraction in a diamond cell and other high spatial resolution spectroscopic studies, such as XAFS of interfaces. The flux and the brightness depend upon the intrinsic opening angle of the radiation and upon the electron or positron beam divergence. They are nearly independent of the size of the electron beam however, because the intrinsic opening angle for all of the sources is large enough that at the end of a typical beamline the fan of radiation is much larger than the electron beam. The *brilliance* is the brightness/source area. The brilliance is the appropriate FOM when a beam line contains focusing optics, because the source dimensions then limit the minimum focused spot size, which can be produced. The limits to sample size, and the trade-offs in designing a particular experiment, will depend on these source characteristics; on the flux, brightness and brilliance. These characteristics will in turn depend on the design of the storage rings and beamlines (Table 1).

Properties of undulators, wigglers and bending magnets

Storage rings are divided into classes, largely along generational lines (Table 1). First-generation X-ray storage rings (CHESS, SSRL) are parasitic, dependent on the accelerator and particle physics communities to produce X-rays as an unwanted by product of their experiments. The second-generation sources (NSLS, Photo Factory, Daresbury, and an upgraded SSRL) were built as facilities dedicated to the production of synchrotron radiation and optimized for beam lifetime and user access. The third-generation sources are optimized to provide bright high-energy beams by maximizing the number of straight sections available for insertion devices (Chavanne et al. 1998, Gluskin 1998, Kitamura 1998). Insertion devices, in particular undulators, have many properties that make them extremely attractive sources of electromagnetic radiation. For example, The Advanced Photon Source (APS) provides significant increases in photon intensity over second-generation sources (Fig. 1). Undulators increase the flux in the 3-40 keV range by several orders of magnitude with relatively narrow energy bandwidths. Wigglers and bending magnets provide much higher flux in the energy range above 40 keV with a smooth energy spectrum.

Bending magnets . The radiation spectrum emitted from a bending magnet (Fig. 1) is a smooth function of energy characterized by a critical or half-power energy given by

$$\varepsilon_c = 0.665 \cdot B \cdot E^2 \tag{1}$$

where B is the magnetic field in Tesla and the E is the storage ring energy in GeV. The brightness of a bending magnet source is given by

$$B_r = 1.33 \cdot 10^{13} \cdot E^2 \cdot I \cdot H_2(\varepsilon/\varepsilon_c) \tag{2}$$

where I is the storage ring current in A, ε is the X-ray energy and $H_2(\varepsilon/\varepsilon_c)$ is a Bessel function with a value near 1 at $\varepsilon/\varepsilon_c = 10$. Hence the critical energy of a bending magnet source varies with the square of the storage ring energy for a given dipole magnet field. The brightness at the critical energy also increases with the square of the ring energy and linearly with the storage ring current. Radiation is emitted from a bending magnet with a very small vertical opening angle, typically less than 0.01° (Fig. 1). The main advantage of third-generation radiation sources is clearly the greater brightness at high energies (Fig. 1). The brightness of the APS bending magnets is 10 times greater than that of the NSLS bending magnets at 15 keV, but because the APS beam lines must be twice as long, the actual gain in throughput is only a factor of 2.5. At 60 keV, however, the actual gain is more than a factor of 1000. There are several experiments that require this high brightness at high energy. High-pressure and/or high temperature X-ray diffraction or absorption requires high energy X-rays both to pass through the environmental cell, and to sample a large volume of reciprocal space when using the energy or angle dispersive techniques. The brightness of the APS at 60 keV is comparable to that of the second-generation NSLS at 15 keV and within a factor of 2 of that of the NSLS superconducting wiggler, X-17.

Wigglers. The on-axis spectrum from a wiggler is, to a good approximation, equal to that from a bending magnet with the same field scaled by the number of poles in the wiggler. The wiggler field can be chosen to be greater or less than the field of the bending magnets (Fig. 1). The radiation from a wiggler is linearly polarized, which can be useful for XAFS experiments, where scattered backgrounds can be minimized by taking advantage of the polarization and where bonding characteristics in single-crystal or textured samples vary with sample orientation (Manceau et al. 2000a,b; Schlegel et al. 1999).

Undulators. By far the greatest attraction of the third-generation sources is the hard X-ray undulator. An undulator is a periodic, low magnetic field device whose output has peaks in its energy spectrum, and is highly collimated in both the vertical *and* horizontal directions (Walker and Diviacco 2000). For the APS undulator "A"', the brightness at 10 keV is a remarkable 14,000 times greater than that of an NSLS bending magnet (Fig. 1). Thus, one could reduce the spot size by a factor of 100, to well below 1 μm, and retain the same experimental sensitivity as demonstrated at NSLS for approximately 10 μm-spots. Alternatively, one can use less efficient detectors with much better resolution and lower backgrounds to dramatically improve sensitivities, and reduce interference for surface-scattering studies or measurements of diffuse scattering. The high undulator source brilliance means that one can use focusing optics to achieve even higher intensity on the sample. It should be quite possible to achieve sub-micrometer beams with $>10^{10}$ photons/sec. This source permits fundamentally new experiments in X-ray diffraction and with unprecedented time resolution. It is possible to "taper" an undulator, e.g. set the gap larger at one end than the other. The tapered undulator produces a spectrum with much flatter tops on the harmonic peaks, e.g. greater bandwidth in each harmonic. Analysis has shown that the integrated flux in each harmonic decreases very little although the peak brightness of course decreases (Fig. 1).

Sources optimized for undulator sources (Spring-8, the APS and ESRF; Table 1) are large because of the length of the undulator and because the device needs to be inserted into a straight section of the storage ring. Recent technological developments at the NSLS (Stefan et al. 1998, Tatchyn 1996) and other sources (Walker and Diviacco 2000), provides strong evidence that small-gap, short-period undulators will play an important role in the future. These devices have many of the advantages of current-generation undulators, but require neither the high energies nor amount of straight section of the third-generation sources. The new storage rings now being considered in Europe (Diamond), or new upgrades to the second-generation sources such as the NSLS, may rely heavily on this new technology.

Next-generation sources. An undulator beam from a third-generation source (Fig. 1) while highly collimated and reasonably monochromatic, does not have the full coherence of an optical laser. Free Electron Lasers (FEL), ultimately the fourth-generation synchrotron sources, have the potential (Lumpkin and Yang 1999, Prazeres et al. 1999) to provide ten orders of magnitude improvement in brilliance, with time structures in the sub-picosecond regime. The science enabled by these devices is in the area of "blue sky" at present. However, experience over the past 20 years suggests leaps in capability have preceded explosions in the new solid state science carried out at synchrotron facilities. The FELs (Nuhn and Rossbach 2000) will undoubtedly continue this trend: build it and they will come.

Access

Beamline construction. Facilities come in a variety of "flavors" though most are governed by variations on two basic philosophies dictated mainly by how the funding for their construction was obtained. The NSLS, and most of the APS beamlines, were constructed by consortia ("Participating Research Teams" or PRTs for the NSLS, "Collaborating Access Teams" or CATs at the APS). These teams are responsible for writing the original proposal, and periodic proposals there after, which fund beamline operations. For this yeoman-work the PRTs are allocated a percentage (say 75% but negotiable with the particular light source) of the available beamtime with the remainder going to "outside" users who apply through a proposal process. The advantages of this system is that the peer review process basically picks the beamlines to be constructed based upon the usual criteria of national need, relevance and scientific excellence. The disadvantage is that multiple groups, which institution- rather than discipline-based, tend to build very similar facilities. The "European and Japanese" models tend to favor a more centralized decision-making process, with the facility funding science-based beamlines dedicated to one type of experiment or technique. This tends to produce well-staffed beamlines that offer a maximum of available time for outside users. The middle road involves beamlines built by consortia that are beholden to outside users because they are set up as user facilities. They are constructed with the help of individuals who write the original and renewal grants and justifies construction for an identified user base (for example geoscientists or protein crystallographers). Once constructed, the beamline is turned over to outside users who gain access through beamline proposals. In some ways this is the worst of all possible worlds for those involved in the early planning stages; the good citizenship of those involved in the construction phase of the beamline is scarcely rewarded before beamtime becomes essentially unavailable because of the crush of outside users. Yet another variant is Britain's "ticket" system, an attempt to make synchrotron beamtime cost "fully recoverable" and to include them in individual research proposals. Individual proposals for beamtime are evaluated along with regular research proposals and then tickets are allocated. These are then turned over to the facilities that allocate time. The user can consult web resources (Table 1) to be sure they know which regime they need to follow.

Unlike the laboratory setting, where decisions are made with a small group of intimate colleagues, user facilities bring together people from a wide variety of disciplines, often with competing agenda. It is important for communities with an interest in maintaining access to these resources, also identify individuals prepared to represent them at early stages of decision-making processes. Unfortunately for disciplines that have long been single investigator based, the prospect of significant funding being diverted to user facilities is wrenching. It is important to promote the availability of these facilities. For investigators involved in work on phase transitions, the interdisciplinary atmosphere at these facilities adds real value, difficult or impossible to reproduce at a single-investigator laboratory.

Gaining access and planning. For first-time or infrequent user, the most pressing problems are associated with the unfamiliar working environment at national facilities, gaining access and knowing what to say in a proposal to gain that access. Preliminary experiments and calculations help in determining the likelihood of success, and to convince reviewers of beamline proposals. Some of the preliminaries might include, identifying the (P, T) conditions needed, calculating the powder diffraction

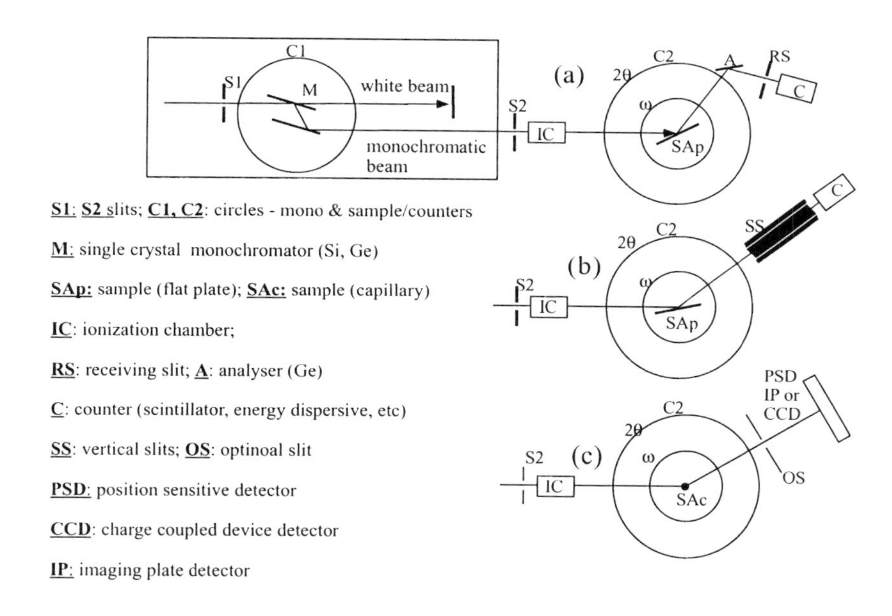

Figure 2. Schematics for synchrotron powder diffractometry in monochromatic mode. A double-crystal monochromator (M), either channel cut or independently translatable and rotatable, is used to select the incident energy. A mirror or bent crystal optics (not shown) is often used to focus the beam on the sample. Because the synchrotron beam decays with time an incident beam monitor (IC) is needed to normalize data to the same relative intensity scale. A movable beam stop can be inserted after the monochromator to select either monochromatic beam or while beam transmitted through the first crystal of the monochromator. Two common sample geometries, flat plate (not shown) and capillary (SAc) are often combined with three detection schemes (shown in boxes): crystal analyzer (**A**) mode (Cox 1992, Cox et al. 1988, Cox and Wilkinson 1994), vertical (Soller) slit (**SS**) assembly (both coupled to point counter, (**C**) and detection with a position sensitive detector (**PSD**) or area detector such as an imaging plate or charge coupled device (**CCD**). In the case of the area detectors it is difficult to use slits and/or an analyzer crystal to better define the scattering from the sample and contamination from the sample assembly (see above) can be a problem. The resolution in this later case is often dominated by the diameter of SAc—or area detectors such as an IP or CCD detector (Hammersley et al. 1996). The symbols have the following meaning: **S1**: S2 slits; C1, **C2**: circles—mono & sample/counter; **M**: single crystal monochromator (Si, Ge); **IC**: ionization chamber; **SAc**: sample (capillary); **OS** and **SS**: optional slit for area or PSD detectors or Soller Slit for scintillation counter; **A**, optional crystal analyzer.

patterns, comparing resolution function and peak-to-background (Fig. 2) and calculation of any systematic errors, such as P- and T-calibration, sample absorption and interference from the sample cell.

All beamlines require productivity. The PRTs and CATs depend upon it for renewed funding, the central facilities need to justify the money spent on resources. Most facilities will allow sufficient beamtime to test a good idea. A program can not long survive without productivity in terms of publication in refereed journals. This makes choosing the correct problem and instrument configuration critical. Fortunately, the amount of information available on the web, has made managing the bewildering number of choices (see above) more manageable.

Web resources

A comprehensive list of the world's synchrotron radiation sources, including many still on the drawing boards, can be found at many of the web-sites maintained by the Light Sources (e.g. http://www.nsls.bnl.gov/AccPhys/synchros.htm). Table 1 provides the web addresses for some synchrotron facilities. These sites provide up to date information on characteristics of the beams provided, key experimental expertise, future construction plans, and resources for users including how to obtain access.

DIFFRACTION STUDIES AT SYNCHROTRON SOURCES

General considerations

The study of phase transitions might be divided into their detection, their characterization, determination of the crystal structure, the evolution of the phase with time, and the measurement of physical properties. Future studies of phase transitions will be increasingly carried out with the aim of measuring a number of characteristics simultaneously. The technology exists now to concurrently measure all of these at a minimum of one beamline (see below) and once technical challenges are overcome, such measurements will become the norm. A review of some of the currently available tools required for the successful experiment at the synchrotron is given below. Several groups maintain web sites, which are intended to provide the user an introduction to the "art" and the interested reader is encouraged to visit these and other sites (Table 1) for detailed descriptions. A good example of such sites are those maintained by the International Union of Crystallography (IUCr) and its Commission for High Pressure and Commission for Powder Diffraction (http://www.iucr.org).

Diffraction from single- and from micro-crystals

Monochromatic studies on microcrystals at ambient conditions. Single-crystal diffraction is the most informative and definitive technique for the study of the structure of materials. Unfortunately, many materials form single crystals a few microns to sub-micron in diameter, and they might tend to twin. Until recently, crystallographic studies of samples with crystallites in this size range could only be carried out using powder diffraction techniques. Indeed, this crystallite size is well matched to the resolution function of modern synchrotron-based powder diffractometers (see below). The single-crystal technique remains the most reliable for *ab initio* structure determination and refinement. The availability of data in three dimensions allows angular separation of reflections with identical or nearly identical d-spacings, particularly in high symmetry space groups. It also allows straightforward visualization of diffraction effects due to incommensurabilities, stacking faults or other sources of diffuse X-ray scattering. Despite developments in the analysis of powder diffraction data (Burger and Prandl 1999, Knorr and Madler 1999, Kubota et al. 1999, Nishibori et al. 1999, Ostbo et al. 1999) single-crystal data provides more reliable electron density distributions. The determination of unit cell, space group and crystal structure are also more routine using this methodology.

When twinning accompanies a phase transition as in the case of ferroelastic materials, de-twinning strategies (Koningsveld et al. 1987) involving application of differential pressure can be used to obtain near single-domain samples. For the zeolite ZSM-5, this strategy paid off handsomely since pseudo-symmetry, along with the orthorhombic-monoclinic transition in this important catalyst, had stymied precise studies of its structure (Koningsveld et al. 1987, 1989a,b). Once obtained, structures from these crystals provided constraints on more convenient, though less precise, powder investigations (Eylem et al. 1996, Parise 1995, Parise et al. 1993). The ability to look at much smaller crystals at the synchrotron will make the task of studying "single" domain specimens more straightforward.

Despite the obvious advantages of bright, tunable X-rays sources for single-crystal diffraction, early progress in this area was limited by the speed with which point counters could be used to collect complete data sets. Problems included beam drift and large spheres of confusion for the diffractometers used. Area detectors and more stable beamline components which have allowed reliable intensities to be collected for structure solution and refinement on micro-crystalline samples (Gasparik et al. 1995, Tan et al. 1995) have overcome many of these difficulties. The data obtained from these beamlines are comparable to those obtained from the laboratory on much larger samples (Gasparik et al. 1995). Weak scattering phenomena, such as diffuse scattering arising from imperfect or developing short-range order in a system undergoing an order-disorder transition (Proffen and Welberry 1997, 1998; Welberry and Butler 1994, Welberry and Mayo 1996, Welberry and Proffen 1998) are also of great interest. These types of studies have also recently been carried out on surfaces (Aspelmeyer et al. 1999, Bahr 1994, Etgens et al. 1999, Idziak et al. 1994, Peng et al. 1997, Tomilin 1997, Vollhardt 1999). While area detectors will increasingly be used for these investigations, better peak-to-background discrimination is still afforded by scintillation counters coupled to tight collimation (Eng et al. 1995, Robinson et al. 1992, 1994). The higher brightness available at the synchrotron (Fig. 1) will facilitate these experiments since point counting can be carried out more quickly at these sources. The development of higher resolution energy-discriminating area detectors (Arfelli et al. 1999, Sarvestani et al. 1999) is now required to keep up with the capabilities of the new X-ray sources.

Studies of micro-single crystals have benefited from developments in software (Hammersley et al. 1996, Otwinowski and Minor 1995) and area detector technology (Borman 1996). The possibilities for this type of work were demonstrated in a set of experiments on kaolinite clay particles down to 0.4 μm^3 by Neder and coworkers (Neder et al. 1996a, Neder et al. 1996b) who described techniques for mounting the sub-micron crystals. They also described ways of decreasing background by carrying out diffraction in a vacuum, thereby minimizing air scattering (Neder et al. 1996c) and increasing the signal to noise discrimination on the charge coupled device (CCD) detector used to record the diffraction pattern. Similar diffraction studies using imaging plate (IP) detectors are also possible (Ohsumi et al. 1991). The structure refined for kaolinite from this data (Neder et al. 1996b, 1999) suggest routine structure refinements for sub-micron crystals in the near future.

Equally interesting was the clear observation (Neder et al. 1996b, 1999) of diffuse scattering from the kaolinite crystals studied at the ESRF. This has implications for clays and materials where diffuse scattering is important (Welberry and Butler 1995).

Conventional analysis of Bragg peaks provides information about individual atomic sites averaged over the number of unit cells contributing to diffraction. This average crystal structure provides information such as atomic coordinates, site occupancies and displacement parameters. Diffuse scattering on the other hand, contains information about how pairs of atoms behave and is therefore a rich source of information about how atoms and molecules interact. This type of diffuse scattering is well known in electron microscopy, and this technique has been used to advantage to characterize a number of disordered materials (Cowley 1992, Dorset 1992). A quantitative treatment (Warren 1969) is complicated for electron diffraction by multiple scattering effects. Electron diffraction also suffers the disadvantage that only sections of reciprocal space passing through the origin are accessible. Third-generation sources provide sufficient brightness to study diffuse scattering from crystals in the 1 μm size-range (Neder et al. 1996a,b,c; 1999). This suggests a more general application to order-disorder transitions (Welberry and Butler 1994, 1995; Welberry and Mayo 1996) perhaps in real time.Comparatively little new conceptual work is required to extend these crystallographic studies to high temperature and/or pressure and to be able to follow these diffraction effect through phase transitions.

Other advantages for diffraction at synchrotron sources include the minimization of systematic errors, which limit the accuracy with which crystallographic models can be refined. Both extinction and absorption are strongly dependent on crystal size and wavelength with the primary extinction characterized by an extinction length:

$$L_e = \frac{V_u}{r_e \lambda \; |F_{hkl}|} \tag{3}$$

where V_u is the unit cell volume, r_e is the electron radius λ is the wavelength and F_{hkl} is the structure factor for the lattice plane (*hkl*). Equation (3) suggests that crystals smaller than L_e are essentially free of extinction. This size is a few microns at $\lambda = 0.5$Å for materials with unit cell volumes of about 1000 Å^3. In those cases where absorption is also a problem, it may be advantageous to also perform the experiment at short wavelengths. However, the efficiency of elastic scattering drops proportional to λ^3, and the intensity of Compton scattering increases, causing additional background. With synchrotron radiation and appropriate beamline optics the experimental setup can be optimized to provide the best possible data for the particular resolution (energy, spatial, angular and time) and for the specific sample under investigation. This flexibility is unique to this source of X-rays and, although certain aspects can be approached with laboratory based systems, (Atou and Badding 1995) it can not be duplicated.

Microcrystals at non-ambient conditions. Apparatus to allow studies in environmental cells are similar to those used in the home laboratory (Angel et al. 1992, LeToullec et al. 1992). The difficulties encountered because of a lack of diffracted beam collimation (Fig. 2), when the usual point counter is replaced by either an imaging plate (IP) or charge coupled device (CCD), are also similar. Parasitic scattering from the cell is difficult to avoid without resorting to elaborate slit designs (Yaoita et al. 1997). Diffraction with a polychromatic beam, where no attempt is made to select specific energy with a monochromator, provides unique information (Somayazulu et al. 1996, Zha et al. 1993a,b) and avoids scattering from the cell. The d-spacings are observed by using a multi-channel analyzer with a fixed diffraction angle. A discussion of ED powder diffraction is given below.

Powder diffraction studies

Powder, textured and "single-crystal" samples. Unfortunately, many phase transitions lead to a degradation in the quality of single-crystal samples through the formation of twins and an increase in mosaicity, or from the texturing of powdered samples. Considerable progress has been made in the routine handling of twinned crystals (Gasparik et al. 1999, Herbst-Irmer and Sheldrick 1998, Kahlenberg 1999) especially with the advent of area detectors, which record all of reciprocal space rather than a subset of it biased by the choice of unit cell. Indeed, the growth of twinned, multiple crystal or highly textured samples (Brenner et al. 1997, Nelmes et al. 1999a) has recently been advantageous for the solution of crystal structures for materials which have defied solution from powder diffraction data. These techniques, developed for growing textured samples in the diamond anvil cell (DAC), may be widely applicable (Nelmes et al. 1999a). In many instances powder diffraction is the only recourse.

There is considerable loss of information when 3-D single-crystal data are condensed into a 1-D powder diffraction pattern. Peak overlap can be alleviated somewhat by using the highest possible angular resolution (Fig. 2), and peak deconvolution or Rietveld refinement techniques (Rietveld 1969, Larson and von Dreele 1986). In high-pressure transitions involving powders carried out at temperatures below the strength limit of the sample (Weidner et al. 1994a), deviatoric stress will *always* be a problem; *even when supposedly "hydrostatic" conditions involving liquid pressure transmitting media are used.* This leads to a loss of information, especially at high angles (low d-spacing) which can severely limit both the ability to solve structures and to refine them successfully. Maximum likelihood (entropy) techniques (Burger et al. 1998, Burger and Prandl 1999, Cox and Papoular 1996, Fujihisa et al. 1996, Ikeda et al. 1998, Kubota et al. 1999) can ameliorate the loss of information, for well-crystallized powdered materials. Recent studies in which electron density distributions were followed through the high-pressure metallization of powdered iodine (Fujihisa et al. 1996) suggests other phase transitions can be similarly treated with powder diffraction data.

Monochromatic powder diffraction in general. Scattering experiments from powders using laboratory sources are almost always carried out in the angle dispersive mode (Klug and Alexander 1974) with X-rays of a fixed wavelength (λ) and variable diffraction angle (θ). The d-spacings are then determined from the observed angles using the Bragg equation,

$$\lambda = 2d_{hkl}\sin\theta \qquad (4)$$

Because of the smooth spectrum of energies available at the synchrotron (Fig. 1) a single-crystal monochromator (Fig. 2) is used to select a small spread of energies (typically with $\Delta E/E \approx 10^{-3}$). In other respects, most of the laboratory hardware required for diffraction experiments at the synchrotron is similar to that used in the laboratory and will not be dealt with in detail here. Several excellent texts and at least one comprehensive volume (Klug and Alexander 1974), detail the requirements for powder diffractometry.

Resolution. The study of phase transitions with powder diffractometry often involves the observation and measurement of small deviations from aristotype symmetry. Excellent examples are the symmetry of the perovskites with changes in P, T and X (Burns and Glazer 1990, Glazer 1972, Park et al. 1998, Woodward 1997a,b,c). Observation of weak superlattice peaks mark the onset of a new phase (Burns and Glazer 1990, Glazer 1972) and may also be accompanied by peak splitting. In order to follow transitions of this type, it is important to design an instrument optimized for both high peak-to-background discrimination and high resolution ($\Delta d/d$). Considerations of the resolution afforded by simple beamline optics at a synchrotron source (such as those illustrated schematically in Fig. 2) are well known (Cox et al. 1988) and have been implemented to some extent on most synchrotron X-ray sources (Table 1). It is advantageous that the resolving capabilities of the beamline be well known before studies involving changes in symmetry are attempted. This can be either calculated or calibrated using a "standard" material. The peak full-width at half-maximum (Γ) as a function of 2θ, neglecting contributions from the Darwin width of the monochromator, is given by the expression (Cox et al. 1988):

$$\Gamma = \{\phi_v^2(2\tan\theta / \tan\theta_M - 1)^2 + \delta^2\}^{1/2} \tag{5}$$

where ϕ_v is the vertical divergence of the incident beam (typically 0.01°), θ_M is the monochromator angle and δ is the divergence of the collimator, defined as the spacing between the foils divided by the length of the collimator (Fig. 2b). This equation also applies when a position-sensitive detector (PSD), area detector or conventional receiving slit (Fig. 2c) are used in the diffracted beam. For this condition (Cox et al. 1988):

$$\delta \approx \left(\omega_S^2 + \omega_R^2\right)^{1/2} / D_{SR} \tag{6}$$

where w_S *and* w_R are the width of the sample (or the incident beam for flat plate geometry) and the width of the receiving slit (or spatial resolution of the PSD) respectively. D_{SR} is the distance between the sample and the receiving slit (or PSD). Thus for a 0.2 mm-diameter sample with a PSD at 1 m- and with 0.2 mm-resolution, $\delta \approx 0.02°$. The peak profile should be approximately Gaussian in shape if effects due to particle size broadening and strain are small. From the Scherrer formula (Warren 1969) $\Delta 2\theta \approx \lambda / L\cos\theta$, this implies a mean particle size L of at least 2 μm. This approaches the limits currently used for single-crystal studies (Neder et al. 1996a).

When a perfect analyzer crystal is used in the diffracted beam (Fig. 2a) Cox et al. (1988) derive the following expression for non-dispersive geometry:

$$\Gamma = \{\phi_v^2(2\tan\theta / \tan\theta_M - \tan\theta_A / \tan\theta_M - 1)^2 + \Gamma_{\min}^2\}^{1/2} \tag{7}$$

This geometry is particularly useful for studies for phase transitions. The analyzer acts as a fine slit, increasing resolution. Further, Equation (7) implies the focusing minimum now occurs at $2\tan\theta = \tan\theta_A + \tan\theta_M$, and this provides a very convenient means of varying where that minimum will occur, by changing analyzer crystal. In practice, the beam energy is chosen to optimize resolution, the energy required to minimize beam attenuation in sample cells and where brightness is maximized in the incident spectrum (Fig. 1). Many of these issues are covered in recent reviews of powder diffractometry at synchrotron sources (Cox 1992, Cox and Wilkinson 1994, Finger 1989, McCusker et al. 1999).

It is preferable to use monochromatic radiation for the determination of crystal structure from powder diffraction data for all but the most simple of close packed structures (Somayazulu et al. 1996). Even in those cases where transitions were thought to involve "simple" close packed structures, more careful examination with monochromatic diffraction, which provides superior peak-to-background discrimination and angular resolution, prompted considerable revision of an entrenched literature (McMahon et al. 1996). The use of imaging plate technology (Sakabe et al. 1989, Shimomura et al. 1989) has revolutionized the study of high-pressure phases from powder data. Developed initially in Japan (Sakabe et al. 1989, Shimomura et al. 1989) this technology was combined with careful Rietveld refinement techniques at the Daresbury Laboratory (Table 1) by the Edinburgh group (Fujihisa et al. 1996, McMahon and Nelmes 1993, McMahon et al. 1996, Nelmes et al. 1999a, Nelmes et al. 1992, Nelmes and McMahon 1994, Nelmes et al. 1999b, Piltz et al. 1992). From there it spread to several installations (Andrault et al. 1997, Dubrovinsky et al. 1999, Fei et al. 1999, Hammersley et al. 1996, Hanfland et al. 1999, Knorr et al. 1999, Loubeyre et al. 1996, Mezouar et al. 1999, Schwarz et al. 1998, 1999; Thoms et al. 1998, Tribaudino et al. 1999, Yoo et al. 1999). When the structure needs to be solved *ab initio* from high-pressure data, systematic error, broadening due to deviatoric stress and other sample effects, can produce severe peak overlap and peak shifts (Meng et al. 1993). This makes *ab initio* structure determination from these data very difficult, except in the simplest cases. Fortunately, the number of structures adopted by solid state materials at high pressure is restricted. Simple isostructural relationships recognized from the similarity in reduced unit cell volumes and cell parameters, often suffice to allow determination of structure (Fei et al. 1999, Kunz et al. 1996). More often, high-pressure structures involve atomic rearrangement of simple sub-lattices. Provided the correct unit cell is chosen, and the space group symmetry correctly identified based on superior monochromatic data (Nelmes et al. 1999b), the derivation of the distorted high-pressure structure is obtainable (McMahon and Nelmes 1993, McMahon et al. 1996, Nelmes et al. 1992, Nelmes and McMahon 1994). It is critical that peak-to-background discrimination be maximized in order to observe the weak peaks (McMahon et al. 1996, Nelmes and McMahon 1994, Nelmes et al. 1999b) which result from subtle distortions of the sub-structure or ordering of ions with low scattering contrast. Monochromatic data will, in general, provide the best chance to observe such phenomena.

The disadvantage of the monochromatic technique, when PSD's (including IP and CCD; Fig. 2) are used, is that it does not take advantage of the full spectral brightness available from bending, wiggler or tapered undulator source (Fig. 1). Further, parasitic scattering from sample cell is difficult to avoid (Yaoita et al. 1997). In cases where the structure is well known and information on the unit cell volume is sufficient, for the calculation of the equation of state for example, energy dispersive diffraction techniques offer a number of advantages.

Energy-dispersive (ED) studies

In ED studies, a polychromatic beam (usually filtered with graphite to remove low energy radiation and decrease heat-load on the sample) irradiates a powder or single-crystal specimen. An energy dispersive detector records the diffracted X-rays. Pulses from this detector are amplified, fed into and accumulated in a multi-channel analyzer, which has been calibrated with a set of known energy standards such as appropriate radioactive sources and fluorescence lines. The polychromatic source not only generates the familiar X-ray diffraction pattern, but also excites fluorescence lines from the samples, and often from sundry items (Pb shielding) in the enclosure protecting experimenters from exposure. It is the presence of these fluorescence lines, variations in absorption, a non uniform response function (Klug and Alexander 1974) and several other difficulties, which make ED data difficult to use for accurate and precise structure determination. Furthermore, the resolution of ED diffractometers is invariably inferior to that of wavelength dispersive diffractometers, due to the relatively low energy-resolution of X-ray detectors. Attempts to use ED data, made prior to the popularization of IP and CCD detectors (Sakabe et al. 1989, Shimomura et al. 1989) did produce results (Yamanaka and Ogata 1991). Energy dispersive diffraction patterns (Fig. 3) are obtained with the detector at a fixed angle and all lines accumulating simultaneously. It is possible to record a recognizable diffraction pattern in a few seconds, and this makes the device particularly useful for dynamic studies (Evans et al. 1994, O'Hare et al. 1998a,b; Parise and Chen 1997).

For deriving *d* values from an ED diffraction pattern, Bragg's equation (Eqn. 4) can be expressed in terms of the photon energy. Photon energy is related to wavelength:

$$E = hc/\lambda = 12.398/\lambda \quad (8)$$

Combining Equations (4) and (8) we obtain:

$$d = 6.199/(E\sin\theta) \quad (9)$$

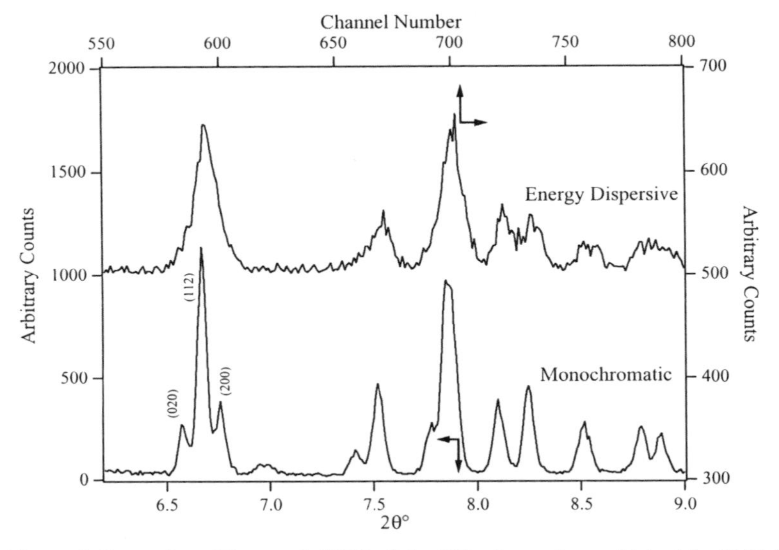

Figure 3. Comparison (Zhao et al. 1994) of the diffraction patterns obtained for $NaMgF_3$ (perovskite phase) in ED mode and in monochromatic [λ=0.3100(7)]. The familiar perovskite "triplet" is indexed for the orthorhombic (***Pbnm***) setting of the unit cell. Although the relative intensities in the two patterns are similar, indicating that in this case problems associated with absorption, detector sensitivity and so on are minimized (Yamanaka and Ogata 1991) the features of the monochromatic pattern are clearly better resolved. Indeed because of the ability to heat the sample, deviatoric stresses built up during pressure loading have disappeared in the monochromatic pattern. The halfwidths (0.06°) are close to those calculated for this beamline configuration (Figs. 2b and 4). On the other hand the ED data, although more poorly resolved, were collected in 30 seconds and produce comparable counting statistics to the monochromatic data, collected in step-scan mode with a point detector in 15 hours. This disadvantage has virtually disappeared with the use of IP and CCD detectors, though at the cost of increased background and interference from the sample cell.

Tools for the collection and analysis of powder diffraction data.

Much of the hardware required for diffraction work at the synchrotron are relatively straightforward modifications of what is found in the laboratory setting. Furnaces and diamond anvil cells (DAC) of well knows design (Angel et al. 1992, Mao and Hemley 1998) are used routinely at synchrotron sources. Exceptions are large volume devices, where the ability to place a high-energy beam (or polychromatic beam) on the sample permits the design of cells with materials that would be regarded as opaque in the laboratory setting. Examples include large volume high-pressure devices (Chen et al. 1998a, Chen et al. 1997, Mezouar et al. 1999, Parise et al. 1998, Shimomura et al. 1985, Weidner 1994, Weidner and Mao 1993) and hydrothermal cells (Clark et al. 1994, Evans et al. 1994, O'Hare et al. 1998a,b; Shaw et al. 2000). Several groups maintain web sites containing detailed descriptions of hardware and facilities for the study of phase transitions. These include:

The IUCr High-Pressure Commission
(http://www.iucr.org/iucr-top/comm/chp/index.html)
The NSF-funded Center for High Pressure Research
(http://www.chipr.sunysb.edu/)

On the other hand, the configuration of the beamline, especially for powder diffraction, is dependent on the experiment being performed, and especially on the resolution requirement (see above). Modern beamlines should be easily configured to optimize one, or a combination of spatial, angular and time resolution. An excellent example of this philosophy is the X-7A beamline (Cox 1992, Cox et al. 1988, Cox and Wilkinson 1994, Finger 1989) at the NSLS and those based upon it (Fitch 1996, Mora et al. 1996).

Software for monochromatic and ED studies. When phases with unknown structures are involved during transitions, software for the determination of unit cell, space group and solution of structure is required. The ready access to integrated software packages (Table 2) has lead to an explosion in the "routine" determination of structures from powders. The following references provide a sampling of recent accomplishments by the community in this area: (Andreev and Bruce 1998, Andreev et al. 1997a, Andreev et al. 1997b, Brenner et al. 1997, Burger et al. 1998, Csoka et al. 1998, Dinnebier et al. 1999, Engel et al. 1999, Falcioni and Deem 1999, Gies et al. 1998, Harris 1999, Harris et al. 1998a,b,c; Kariuki et al. 1998, 1999; Langford and Louer 1996, Lasocha and Schenk 1997, Meden 1998, Putz et al. 1999, Rius et al. 1999, Rudolf 1993, Sawers et al. 1996, Shankland et al. 1997). Along with the web sites list in Table 2, these references contain much of the state of the art. Unfortunately some of these methodologies are not readily available. As they become either commercialized or freely available to the community, they should be posted on sites such as: CCP-14, the Collaborative Computational Project on powder and single-crystal diffraction instituted to "assist universities in developing, maintaining and distributing computer programs and promoting the best computational methods" (http://www.ccp14.ac.uk/) maintained by Lachlan Cranswick); Rietveld Users Mailing list (http://www.unige.ch/crystal/stxnews/riet/welcome.htm), and Rietveld web archive (http://www.mail-archive.com/rietveld_l@ill.fr/).

PHASE TRANSITIONS AND SYNCHROTRON RADIATION: CASE STUDIES

Some examples will now be used to illustrate some of the considerations going into planning, carrying out experiments and analyzing experimental results for samples undergoing phase transitions. Many of these studies have appeared in the literature and so no great emphasis is placed on the ramifications of the structures determined. The reader is directed to the literature for more detailed descriptions of these aspects of the projects.

Time resolved diffraction studies

Experimental equipment for the time resolved experiment. The choice of radiation source, beamline optics, sample cell and software are dependent on the system studied and the goal of the study. Some of the considerations are discussed above as are some of the experimental configurations for monochromatic and energy dispersive powder diffraction at synchrotron sources (Fig. 2). Angular resolution with full-width at half-maximum (Γ) < 0.01° is afforded by the crystal analyzer configuration (Fig. 2). However, the decreased integrated intensity dictates that this mode of operation is reserved for the most crystalline of samples and for the *ab initio* determination of crystal structure from powder diffraction data (Cox 1992). Increased time-resolution, without compromising angular resolution, can be achieved by coupling several detectors and crystal analyzers to the same axis, so the pattern can be recorded in several segments simultaneously (Hodeau et al. 1998, Siddons et al. 1998, 1999; Toraya et al. 1996).

Table 2. Useful web-sites* for computer programs for the analysis of scattering data

Program	*Author*	*Purpose*	*Web site*
ATOMS	Dowty	Structure drawing	www.shapesoftware.com
BGMN	Bergmann	Rietveld Refinement	www.bgmn.de/
CMPR	Toby	Diffraction toolkit	www.ncnr.nist.gov/programs/crystallography/
ConvX	Bowden	data conversion	www.ceramics.irl.cri.nz/Convert.htm
crush	Dove	Modeling	rock.esc.cam.ac.uk/mineral_sciences/crush/
crysfire	Shirley	Powder Indexing	www.ccp14.ac.uk/tutorial/crys/
C'Maker	Palmer	Structure Drawing	www.crystalmaker.co.uk/
crystals	Watkin	Single Crystal	www.xtl.ox.ac.uk/
Diamond	Putz	Structure drawing	www.crystalimpact.com
DIRDIF	Beurskens	Structure solution	www-xtal.sci.kun.nl/xtal/documents/software/dirdif.html
Endeavour	Putz	Structure - powder data	www.crystalimpact.com
ESPOIR	Le Bail	Structure - powder data	sdpd.univ-lemans.fr/
EXPO	Giacovazzo	Structure - powder data	www.ba.cnr.it/IRMEC/SirWare_main.html
FIT2D	Hammersley	Image Plate Data	www.esrf.fr/computing/expg/subgroups/data_analysis/FIT2D/
Fullprof	Rodriguez	Rietveld Refinement	ftp://charybde.saclay.cea.fr/pub/divers/
GSAS	von Dreele	Rietveld/single crystal	ftp://ftp.lanl.gov/public/gsas/
LAPOD	Langford	Unit Cell Refinement	www.ccp14.ac.uk/ccp/web-mirrors/lapod-langford/
LHPM	Hunter	Rietveld Refinement	ftp://ftp.ansto.gov.au/pub/physics/neutron/rietveld/
LMGP	Laugier/Bochu	Powder Diffraction	www.ccp14.ac.uk/tutorial/lmgp/
MAUD	Lutterotti	Rietveld/strain	www.ing.unitn.it/~luttero/
ORTEX	McArdle	Single Crystal Suite	www.nuigalway.ie/cryst/
Overlap	Le Bail	intensities from powder	sdpd.univ-lemans.fr/
PDPL	Cockcroft	Powder suite	ftp://img.cryst.bbk.ac.uk/pdpl/
PLATON	Spek	structure toolkit	www.cryst.chem.uu.nl/platon/
Powder	Dragoe	powder diffract. tools	www.chem.t.u-tokyo.ac.jp/appchem/labs/kitazawa/dragoe
P'Cell	Kraus/Nolze	powder pattern calc.	www.bam.de/a_v/v_1/powder/e_cell.html
PowderX	Dong	Powder pattern analysis	www.ccp14.ac.uk/ccp/web-mirrors/powderx/Powder/
Rietan	Izumi	Rietveld refinement	ftp://ftp.nirim.go.jp/pub/sci/rietan/
RMC	McGreevy	structure modeling	www.studsvik.uu.se/rmc/rmchome.htm
SHELX	Sheldrick	solution/refinement	shelx.uni-ac.gwdg.de/SHELX/
SImPA	Desgreniers	Imaging Plate Analysis	www.physics.uottawa.ca/~lpsd/simpa/
Simref	Ritter	Rietveld refinement	www.uni-tuebingen.de/uni/pki/
TriDraw	Hualde	Phase Diagrams	www.geol.uni-erlangen.de/html/software/
unitcell	Holland/Redfern	Unit cell refinement	www.esc.cam.ac.uk/astaff/holland/UnitCell.html
Valence	Brown	Bond Valence	www.ccp14.ac.uk/ccp/web-mirrors/valence/
Valist	Wills	Bond Valence	ftp://ftp.ill.fr/pub/dif/valist/
WINFIT	Krumm	Powder Diffraction	www.geol.uni-erlangen.de/html/software/
WinGX	Farrugia	Single Crystal	www.chem.gla.ac.uk/~louis/wingx/
XFIT	Cheary/Coelho	Powder Diffraction	www.ccp14.ac.uk/tutorial/xfit-95/xfit.htm
XLAT	Rupp	Unit cell refinement	ftp://www-structure.llnl.gov/
XRDA	Desgreniers	Energy Dispersive	http://www.physics.uottawa.ca/~lpsd/xrda/xrda.htm

* The CCP14 (www.ccp14.ac.uk) and IUCr (www.iucr.org) sites are a useful sources of comprehensive and updated information on available computer programs for crystallography in general. These sites are mirrored world wide: it is faster to use a local-mirror.

An increase in time-resolution usually involves compromising angular resolution. Parallel blade collimators (Fig. 1), especially when coupled with multiple detectors, can give moderate resolution with angular acceptances of 0.03-0.07° and high throughput. Receiving slits are particularly useful when samples are studied under non-ambient pressures and temperatures, and it is desirable to exclude diffraction from sample environment containers such as high-pressure cells, furnaces and cryostats (Yaoita et al. 1997).

Over the past decade, area detectors with a large dynamic range and which are capable of rapidly imaging large sections of reciprocal space have become the recorders of choice for single-crystal diffraction. They are increasingly being used for powder diffractometry (Hammersley et al. 1996) and provide a number of advantages (Thoms et al. 1998). Since they record a large portion of the Debye-Scherrer ring, the effects of preferred-orientation can be clearly observed. By changing the angle of inclination of the sample to the incident beam, intensity variations around the ring can be modeled to define its effects on sample texture. Another advantage of recording a large portion of the ring is the better powder averaging attainable upon integration; the low divergence of the synchrotron X-ray beam places greater restrictions on the minimum particle size acceptable for recording powder data.

There are several disadvantages when area detectors are used without collimation, some of them due to the non-discriminating nature of both the IP and CCD. The synchrotron beam inevitably excites fluorescence within the hutch and care must be taken to shield these devices from stray radiation. The difficulty in designing slits for these devices also decreases the signal to noise discrimination since scattering from sample containers or environmental chambers often contaminate the pattern. Subtracting a pattern for the sample container alone, from that obtained from the sample plus container (Chen et al. 1998a, Chen et al. 1997, Parise and Chen 1997) can eliminate this. Another approach, when powder averaging is not a problem and angular resolution can be relaxed, to approximately $\Gamma = 0.03°$, is to use an energy discriminating PSD fitted with a slit (Fig. 2). While most commercial PSDs operate in the so-called "streaming mode," proportional counting and energy discrimination are possible with these devices. One such PSD is used at the X7A line of the NSLS (Smith 1991) and is particularly useful for small samples and for collection of data suitable for resonant scattering studies.

Detectors, data and data glut. Another concern with the use of area detectors is read out time. When transient processes occur on the sub-second scale, the time required to read detectors after exposure is a severe limitation. Recent improvements in online-IP technology (Thoms et al. 1998) and CCD capabilities will improve this situation (Larsson et al. 1998). One simple solution to this problem is to record the pattern continuously behind a slit and then process the data after the reaction is complete (Norby 1996, Norby 1997). The Translating Imaging Plate (TIP) detection system (Figs. 4, 5 and 6) allows acquisition of high quality monochromatic powder diffraction data through the phase transition, allowing the study of time resolved phenomena. This has allowed not only the elucidation of phase transitions, but also refinement of crystal structure based upon these data, thereby allowing the transition to be followed on an atomistic level (Chen et al. 1997, Parise and Chen 1997, Parise et al. 1998). However, a TIP has a limited translation range and requires real-time feedback to provide a trigger to activate the plate at a time appropriate to capture the transition. Ideally this would be done without resorting to guesswork, or to several loadings of the sample cell. Coupling a CCD detector with the TIP can provide this feedback. Such a device has been installed and tested on the large volume DIA apparatus (SAM-85) at the X17 beamline at the National Synchrotron Light Source (Fig. 4).

High time-resolution is afforded with energy dispersive diffraction (see above). While systematic errors are problematic, reliable structural refinements are possible for a limited class of experiment. The software and method to enable structure refinement using the Rietveld method and ED powder diffraction data are now well-established (Chen and Weidner 1997, Larson and von Dreele 1986). In a recent study of the partitioning of iron between the olivine and ringwoodite polymorphs of $(Mg,Fe)_2SiO_4$, energy dispersive data were sufficiently accurate to allow derivation of unit cell volumes

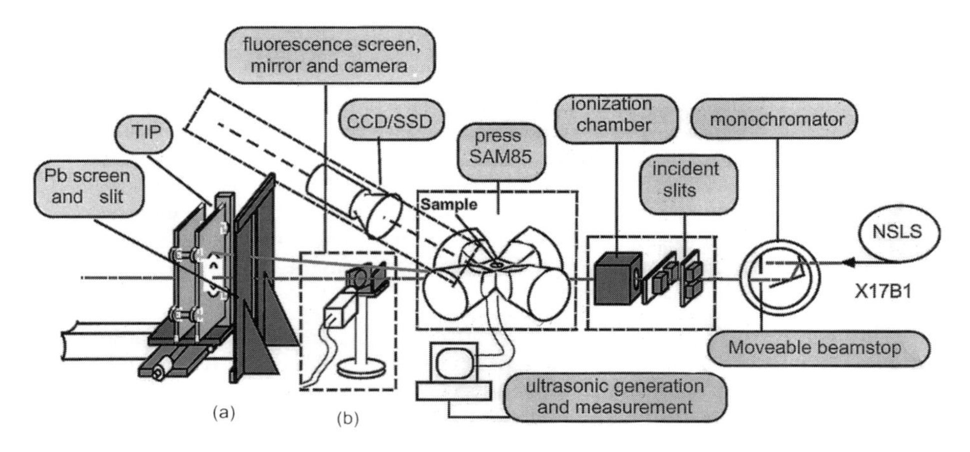

Figure 4. A typical experiment, using the imaging plate detector at the X17B1 beamline of the NSLS (http://www.nsls.bnl.gov). This beamline can operate in monochromatic mode, typically with E > 35 keV; or is configured for polychromatic beam. A movable beamstop allows easy conversion from monochromatic to polychromatic beam in less than 60 sec. For monochromatic X-ray scattering studies a Laue-Bragg monochromator is used. A double-plate IP allows the determination of the IP-to-sample distance with a single exposure and continuous monitoring of this distance throughout the experiment. A lead screen, with an adjustable slit, is installed before the plates, which can be translated to give a complete record of a sample undergoing a phase transition. The coupling of a variety of detectors, and the facile interchange between monochromatic and polychromatic modes, allows operation in a number of modes with simple adjustments to the X-ray beam position using a moveable beamstop after a Laue-Bragg monochromator. This view shows the beamline configured for *in-situ* high-pressure studies, being carried out in the large volume high-pressure device (LVHPD) known as a DIA. The capabilities include energy-dispersive X-ray diffraction with a solid-state detector (SSD), angle-dispersive X-ray diffraction with a combination of a Translating Imaging-Plate (TIP) and charge-coupled device (CCD), sample imaging by X-ray fluorescence, and ultrasonic measurement (Figs. 5-7). With appropriate sample cell design, diffraction and physical property measurements can be carried out simultaneously (Fig. 7).

(Chen and Weidner 1997). Without Rietveld refinement and the constraints provided by the calculated intensities, based on the known structures of these materials, peak overlap is too severe to allow derivation of volumes from the multi-component powder diffraction patterns. Since this partitioning is occurring at high pressure and temperatures (Chen and Weidner 1997), ED diffraction provides the required time resolution and the straightforward collimation making it easier to discriminate against parasitic scattering from sample containers.

When accurate determination of structural parameters is the objective, monochromatic data are preferred. Some beamlines are capable of changing between monochromatic and polychromatic modes of operation, and a description of such a set-up is given in two recent reviews (Parise and Chen 1997, Parise et al. 1998). The particular beamline configuration, including the availability of accessory equipment such as furnaces, cryo-equipment and pressure cells, needs to be ascertained *before* applying for beamtime. The beamline responsible scientist (BRS) can let the user know many of these details, and most facilities (Table 1) insist on the potential user contacting the BRS to discuss the feasibility of a particular experimental program.

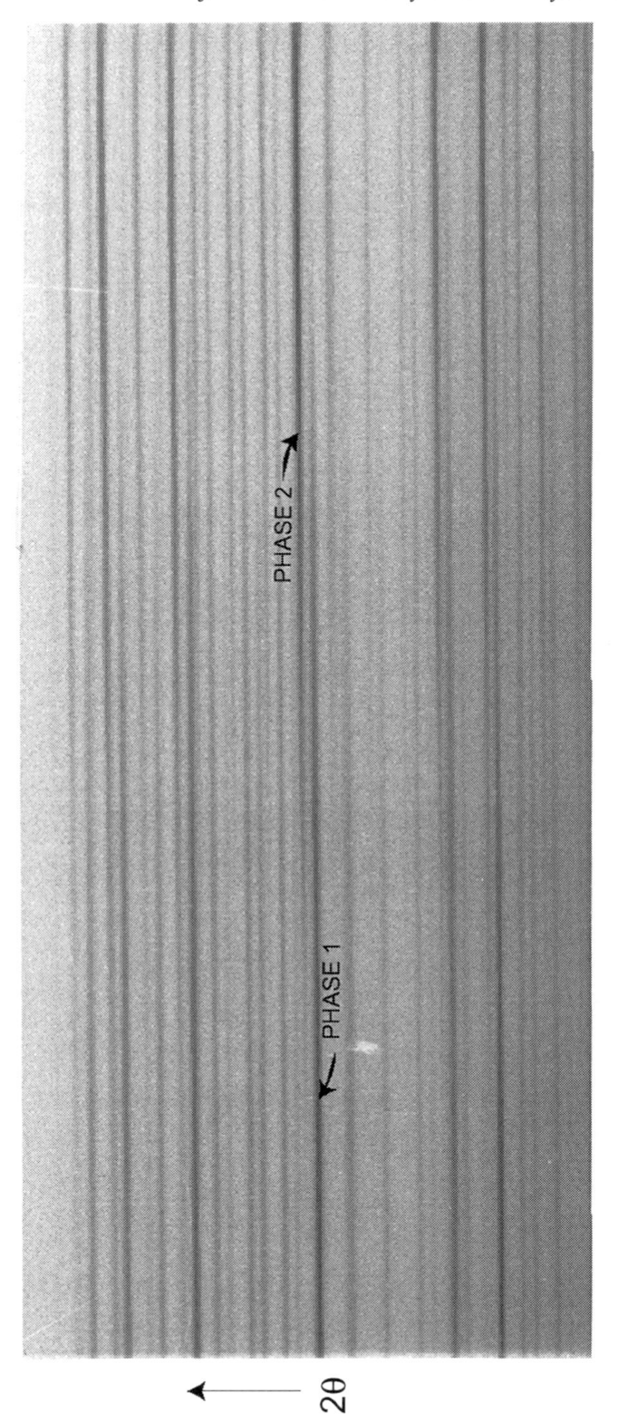

Figure 5. Example of an exposed imaging plate taken in a time resolved experiment to study the ion-exchange of K^+ into the Na^+-containing faujasitic zeolite X (Lee et al. 1998, 1999). The plate is exposed behind a 15 mm slit (Fig. 4) and translated in the direction shown. The transition is clearly evident.

As the use of brighter sources of X-rays and sensitive detectors become wide spread, the amount of data requiring evaluation increases. While visual inspection of time resolved data (Figs 5 and 6) is sufficient in some studies, it is desirable to use an unbiased and automated method to provide some real-time feedback during the course of a reaction. Such a method might provide information on systematic errors as well as the appearance of new phases and the disappearance of others. It should also provide visual queues to allow the choice of a manageable number (3 to 5) of diffraction patterns, out of the hundreds or thousands collected, to provide details of the crystal from Rietveld refinement (Larson and von Dreele 1986, Rietveld 1969). Iterative Target Transform Factor Analysis (Liang et al. 1996) is an unbiased mathematical treatment of the diffraction data that looks for changes as a function of time (Lee et al. 1999). From this processing, kinetic information and clues as to which patterns to use first for Rietveld refinement are obtained. This type of analysis will become important as the rate of data collection increases and as the number of users requiring such data expands (Lee et al. 1999). It will become especially important in the "structure imaging" studies alluded to in the introduction to this chapter.

An example of the "imaging" approach to phase transitions is the study of the structural changes accompanying ion exchange, dehydration and gas loading in zeolites (Cruciani et al. 1997, Lee et al. 1999, Parise 1995, Parise et al. 1995, Stahl and Hanson 1994). Using a translating imaging plate (TIP) detector (Norby 1997) or CCD (Lee et al. 1999) and appropriate sample cells (Norby et al. 1998) high quality diffraction data can be obtained with sufficient time resolution to allow full Rietveld structure refinement. By selecting which pattern to analyze, the complete path of the ion-exchange reaction can be followed for a small and manageable subset of the data collected. When K^+ exchanges for Na^+ in faujasitic zeolites, the path is found to have a structural dependence with Na^+ coordinated in a double 6-ring (D6R) the last to be replaced. Prior to its replacement the increasing amount of K^+ at other sites in the structures causes the restricted D6R, occupied by Na^+, to flex. This causes a transition to a structure with a larger unit cell with K^+ occupying more of the D6R-site as exchange proceeds to completion (Lee et al. 1998, 1999). Similar studies have been carried out for Sr^{2+} exchange and for the transitions involved upon zeolite dehydration (Cruciani et al. 1997, Stahl and Hanson 1994). Order/disorder reactions as a function of pressure and temperatures (Chen et al. 1996b, Chen et al. 1998a, Chen and Weidner 1997, Martinez-Garcia et al. 1999, Mezouar et al. 1999, Parise and Chen 1997, Parise et al. 1998, Thoms et al. 1998) have also been followed in real-time using these detectors and techniques. As detector technology continues apace, CCDs (Parise et al. 2000, Parise et al. 1998) and the in-hutch IP (Thoms et al. 1998) are replacing the TIP. This trend will continue as readout times for IP- (Thoms et al. 1998) and CCD-detectors decrease.

Multiple simultaneous techniques—a more complete picture of the phase transition.

Determination of crystal structure or unit cell volume in isolation of other physical property measurements is the routine practice in much of solid state research under both ambient and non-ambient conditions. This is often necessitated because the cell assemblies required for property measurements are not compatible with X-ray beams typically available in the laboratory. Centralized facilities, such as are available at the synchrotron, provide a cost-effective environment and opportunity to do more definitive experiments. One recent example from the Stony Brook laboratories will suffice to demonstrate what will become, I believe, the normal mode of operation for the study of important phase transitions in the future. For the study of mantle mineralogy, simultaneous measurements of elastic properties, structure and pressure is now established in large volume devices, (Chen et al. 1999) and being established in DACs as well

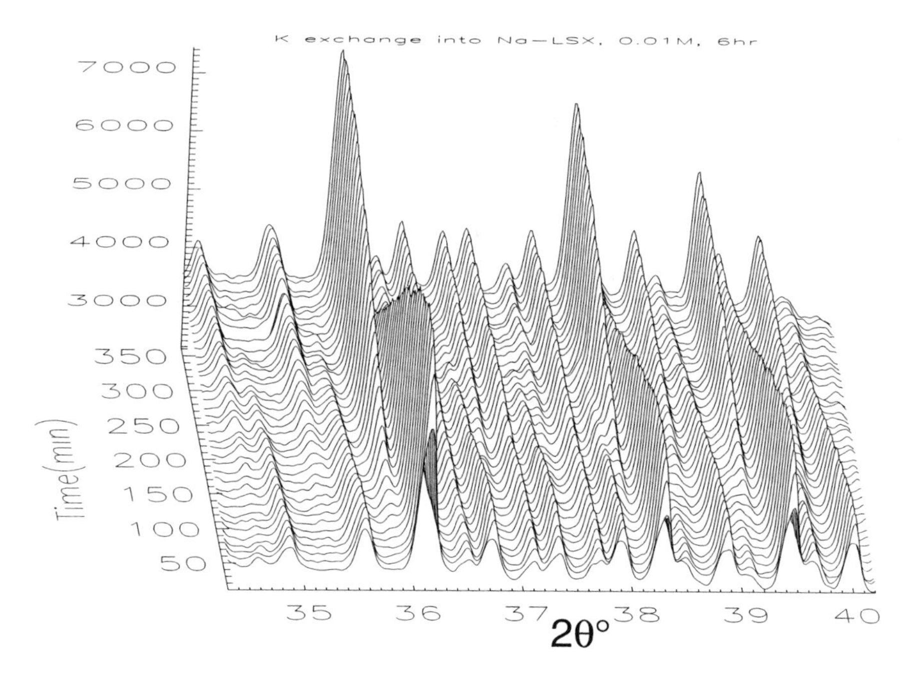

Figure 6. Integrated pattern obtained from the plate shown in Figure 5 and showing the time evolution as a K-rich phase, with peaks shifted to lower 2θ, replaces the Na^+-faujasitic phase. Selected powder diffraction patterns obtained from these data are suitable for Rietveld analysis and the derived structural models allow determination of individual site-occupancy factors within the structure as a function of time. These reveal the critical role of a site with restricted geometry, the last replaced in the Na-containing phase (Lee et al. 1998, 1999).

(Bassett et al. 2000). Measurements at the pressures *and* temperatures relevant to the earth are possible (Li et al. 1998, Liebermann and Li 1998) in large volume devices and will soon become widespread and commonplace. Another property of interest to the geophysical community, rheology (Ando et al. 1996, Chen et al. 1996a, Weidner 1998, Weidner et al. 1994b, 1995), is now studied at high pressures and temperatures (Chen et al. 1998b, Parise et al. 1998). While the study of rheology, the time dependent release of stress as a function of temperature, is thus far the domain of large volume devices (Ando et al. 1996, Chen et al. 1996a, Weidner et al. 1995), the strength of materials is determined *in situ* using both large volume devices (Chen et al. 1998b, Wang et al. 1998, Weidner et al. 1994a) and diamond anvil cells (Duffy et al. 1995, 1999a,b; Hemley et al. 1997, Singh et al. 1998a,b).

Naturally, the crystal structure, strength, rheology and elastic properties are interrelated, as are the pressure and temperature. It is clearly advantageous to measure all these properties simultaneously (Ando et al. 1996, Chen et al. 1996a, Weidner 1998, Weidner et al. 1994b, 1995). It would be especially revealing to measure *all* these properties on the same sample simultaneously *as* isochemical compositions undergo phase transitions (Weidner et al. 1999). A number of technical difficulties need to be overcome; these include maintaining the integrity of the sample, measuring its length to obtain elastic properties from travel time curves, proper calibration of temperature and pressure and suitable cell designs which accommodate transducers, heaters and thermocouples. These difficulties have essentially been eliminated in a recent tour-de-force by Li and co-workers (Fig. 7) in an experiment to measure the elastic properties of $CaSiO_3$ in the perovskite structure. Starting with a sample of the pyroxenoid wollastonite, the sample was pressurized into the stability field for the perovskite phase and heated, transforming most of the sample to $CaSiO_3$-perovskite. This material can not be retained to room pressure conditions, as many other phases of interest can, and so measurement of its properties can only be carried out while pressure is maintained. Despite the considerable compression of the sample (Fig. 7), high quality diffraction and ultrasonic data were obtained at high P and T. And even though, in this preliminary experiment, not all the sample was transformed, sufficient data were obtained to demonstrate conclusively how comprehensive a tool is now available at X17 at the NSLS for *in situ* measurement of structure and physical properties at high P and T.

Today we are able to follow, *in situ* and in real time, changes in crystal structure, elasticity, rheology and strength, which accompany change in phase. We can do this with greater facility and with greater accuracy and precision at synchrotron sources. This alone does not produce the sort of cultural shift that would justify further investment and a shift from laboratory based programs. Ability to perform many simultaneous measurements in order to obtain a complete picture of the transition does have the potential to stimulate such a shift. Conceptual extensions to simultaneous measurements of resistivity, magnetism, dielectric constant and a host of other behaviors, is straightforward given recent advances. All that is required is an incentive as powerful as those that prompted the experiment in Figure 7 are. These results show clearly that measurement of the properties most critical to interpreting the Earth are no longer in the realm of the possible, but are in the process of passing into the routine and user friendly.

ACKNOWLEDGMENTS

I am grateful to colleagues at Stony Brook and CHiPR, particularly Jiuhua Chen, Baosheng Li, Donald Weidner, Mike Vaughan, Ken Baldwin, Robert Liebermann, William Huebsch, Yongjae Lee, Aaron Celestian, Christopher Cahill and Carey Koleda for many of the ideas and results presented in this chapter. Apart from the opportunity to do unique science, the synchrotron facilities are a source of good company and collegiality. I am grateful to David Cox, Jon Hanson, Richard Nelmes, Malcolm McMahon, Scott Belmonte, Lachlan Cranswick, Martin Kunz, Daniel Hüsermann and other members of the IUCr high pressure commission for much stimulation over beam and beer. The NSF's EAR and DMR divisions funded this work, through grants EAR-9909145 and DMR-9713375 and funding to the NSFs Center for High Pressure Research. The work depends on continuing funding to the Daresbury Laboratory (UK), and the US DoEs support of the beamlines at the NSLS and GSECARS at the APS. At various times data were collected at beamlines 9.1 (Daresbury), X7A, X7B, X17B1 (NSLS), and GSECARS (APS); if the data were not used here, many of the concepts and ideas initiated at them were.

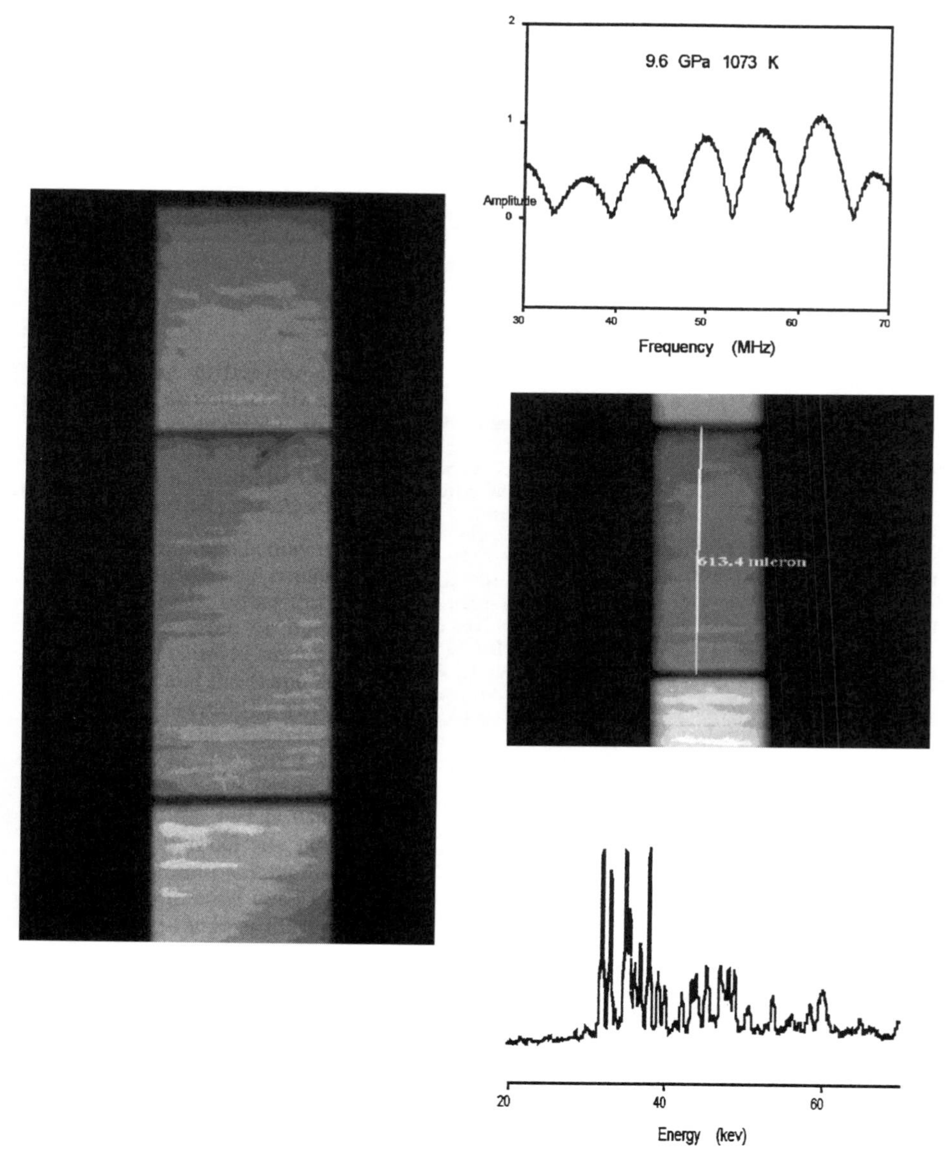

Figure 7. Composite of the various signals and images obtained from the same sample of $CaSiO_3$, at different conditions of temperature and pressure at beamline X17B1 (Fig. 4). Beginning with the wollastonite phase, imaged in the radiograph to the left, pressure is applied and the energy dispersive diffraction pattern, lower right and ultrasonic velocity, upper right can be followed continuously. Observations of deformation in the sample and accurate estimates of it length, shown on the image center right, can be made directly from this image. Thanks to Baosheng Lee for providing this unpublished diagram, which was part of a presentation at the 1999 Fall Geophysical Union Meeting, San Francisco (Weidner et al. 1999).

REFERENCES

Akagi R, Handa K, Ohtori N, Hannon AC, Tatsumisago M, Umesaki N (1999) High-temperature structure of K_2O-TeO_2 glasses. J Non-Cryst Solids 257:111-118

Ando J, Chen J, Weidner DJ, Wu Y, Wang Y (1996) *In situ* measurement of rheology in silicate garnets at high pressure and temperature. Trans Am Geophys Union EOS 77:715

Andrault D, Fiquet G, Kunz M, Visocekas F, Hausermann D (1997) The orthorhombic structure of iron: An *in situ* study at high-temperature and high-pressure. Science 278:831-834

Andreev YG, Bruce PG (1998) Solution of flexible structures from powder diffraction data using a simulated annealing technique. Mater Sci Forum 278:14-19

Andreev YG, Lightfoot P, Bruce PG (1997a) A general Monte Carlo approach to structure solution from powder-diffraction data: Application to poly(ethylene-oxide)$_3$:LiN(SO_2CF_3)$_2$. J Appl Crystallogr 30:294-305

Andreev YG, MacGlashan GS, Bruce PG (1997b) Ab initio solution of a complex crystal structure from powder-diffraction data using simulated-annealing method and a high degree of molecular flexibility. Phys Rev B-Condens Matter 55:12011-12017

Angel RJ, Ross NL, Wood IG, Woods PA (1992) Single-Crystal X-Ray-Diffraction At High-Pressures with Diamond-Anvil Cells. Phase Transit 39:13-32

Arfelli F, Bonvicini V, Bravin A, Cantatore G, Castelli E, Fabrizioli M, Longo R, Olivo A, Pani S, Pontoni D, Poropat P, Prest M, Rashevsky A, Rigon L, Tromba G, Vacchi A, Vallazza E (1999) A multilayer edge-on silicon microstrip single photon counting detector for digital mammography. Nucl Phys B-Proc Suppl 78:592-597

Aspelmeyer M, Klemradt U, Abe H, Moss SC, Peisl J (1999) Martensitic relief formation on an electropolished Ni-37 at % Al (001) surface by diffuse X-ray scattering under grazing angles. Mater Sci Eng A-Struct Mater Prop Microstruct Process 275:286-290

Atou T, Badding JV (1995) A high resolution laboratory-based high pressure X-ray diffraction system. Rev Sci Instrum 66:4496-4500

Badro J, Struzhkin VV, Shu JF, Hemley RJ, Mao HK, Kao CC, Rueff JP, Shen GY (1999) Magnetism in FeO at megabar pressures from X-ray emission spectroscopy. Phys Rev Lett 83:4101-4104

Bahr C (1994) Influence of dimensionality and surface ordering on phase-transitions—Studies of freely-suspended liquid-crystal films. Int J Mod Phys B 8:3051-3082

Bassett WA, Reichmann H-J, Angel RJ, Spetzler H, Smyth JR (2000) New diamond anvil cells for gigaheertz ultrasonic interferometry and X-ray diffraction. Am Mineral 85:288-295

Brenner S, McCusker LB, Baerlocher C (1997) Using a structure envelope to facilitate structure solution from powder diffraction data. J Appl Crystallogr 30:1167-1172

Burger K, Cox D, Papoular R, Prandl W (1998) The application of resonant scattering techniques to *ab initio* structure solution from powder data using $SrSO_4$ as a test case. J Appl Crystallogr 31:789-797

Burger K, Prandl W (1999) A new type of constraint in the maximum-entropy method using ambiguous phase information from anomalous-scattering powder data. Acta Crystallogr Sect A 55:719-728

Burns G, Glazer AM (1990) Space Groups for Solid State Scientists, 2nd Edition. Academic Press, Boston

Caliebe WA, Kao CC, Berman LE, Hastings JB, Krisch MH, Sette F, Hamalainen K (1996) Spin-resolved resonant Raman scattering. J Appl Phys 79:6509-6511

Caliebe WA, Kao CC, Hastings JB, Taguchi M, Uozumi T, de Groot FMF (1998) 1*s*2*p* resonant inelastic X-ray scattering in alpha-Fe_2O_3. Phys Rev B-Condens Matter 58:13452-13458

Chavanne J, Elleaume P, Van Vaerenbergh P (1998) The ESRF insertion devices. J Synchrot Radiat 5: 196-201

Chen GL, Cooke JA, Gwanmesia GD, Liebermann RC (1999) Elastic wave velocities of $Mg_3Al_2Si_3O_{12}$-pyrope garnet to 10 GPa. Am Mineral 84:384-388

Chen J, Inoue T, Wu Y, Weidner DJ, Vaughan MT (1996a) Rheology of dry and hydrous phases of the alpha and beta forms of $(Mg,Fe)_2SiO_4$. EOS 77:716

Chen J, Li R, Parise JB, Weidner DJ (1996b) Pressure-induced ordering in $(Ni,Mg)_2SiO_4$ olivine. Am Mineral 81:1519-1522

Chen J, Parise JB, Li R, Weidner DJ, Vaughan M (1998a) The imaging plate system interfaced to the large-volume press at beamline X17B1 of the National Synchrotron Light Source. *In* Manghnani MH, Yagi T (eds) High Pressure Research in Mineral Physics: Application to Earth and Planetary Sciences. Terra Scientific, Tokyo, p 139-144

Chen JH, Inoue T, Weidner DJ, Wu YJ, Vaughan MT (1998b) Strength and water weakening of mantle minerals, olivine, wadsleyite and ringwoodite. Geophys Res Lett 25:575-578

Chen JH, Kikegawa T, Shimomura O, Iwasaki H (1997) Application of an imaging plate to the large-volume press MAX80 at the photon factory. J Synchrot Radiat 4:21-27

Chen JH, Weidner DJ (1997) X-ray diffraction study of iron partitioning between alpha and gamma phases of the $(Mg,Fe)_2SiO_4$ system. Physica A 239:78-86
Clark SM, Evans JSO, O'Hare D, Nuttall CJ, Wong H-V (1994) Real-time *in situ* X-Ray Diffraction of Intercalation Reactions. J Chem Soc, Chem Commun 809
Coppens P (1992) Synchrotron Radiation Crystallography. Academic Press, London
Cowley JM (1992) Electron diffraction techniques. *In* Int'l Union Crystallogr Monogr Crystallogr 3:190ff, Oxford University Press, Oxford
Cox DE (1992) High resolution powder diffraction and structure determination. *In* Coppens P (ed) Synchrotron Radiation Crystallography. Academic Press, London, p 186-254
Cox DE, Papoular RJ (1996) Structure refinement with synchrotron data: R-factors, errors and significance tests. Mater Sci Forum 228:233-238
Cox DE, Toby BH, Eddy MM (1988) Acquisition of powder diffraction data with synchrotron radiation. Aust J Phys 41:117-131
Cox DE, Wilkinson AP (1994) Powder diffraction studies using anomalous dispersion. *In* Materlin G, Spurhs CJ, Fischer K (eds) Resonant Anomalous X-ray Scattering: Theory and Applications. North Holland, Amsterdam, p 13
Crozier ED (1997) A review of the current status of XAFS spectroscopy. Nucl Instrum Methods Phys Res Sect B-Beam Interact Mater Atoms 133:134-144
Cruciani G, Artioli G, Gualtieri A, Stahl K, Hanson JC (1997) Dehydration dynamics of stilbite using synchrotron X-ray powder diffraction. Am Mineral 82:729-739
Csoka T, David WIF, Shankland K (1998) Crystal structure determination from powder diffraction data by the application of a genetic algorithm. Mater Sci Forum 278:294-299
Dinnebier RE, Von Dreele R, Stephens PW, Jelonek S, Sieler J (1999) Structure of sodium para-hydroxybenzoate, $NaO_2C\text{-}C_6H_4OH$ by powder diffraction: application of a phenomenological model of anisotropic peak width. J Appl Crystallogr 32:761-769
Dmowski W, Akbas MK, Davies PK, Egami T (2000) Local structure of $Pb(Sc_{1/2}Ta_{1/2})O_3$ and related compounds. J Phys Chem Solids 61:229-237
Dorset D (1992) Electron crystallography. Trans Amer Crystallogr Assoc 28:1-182
Dubrovinsky LS, Lazor P, Saxena SK, Haggkvist P, Weber HP, Le Bihan T, Hausermann D (1999) Study of laser heated iron using third-generation synchrotron X-ray radiation facility with imaging plate at high pressures. Phys Chem Minerals 26:539-545
Duffy TS, Hemley RJ, Mao HK (1995) Equation of state and shear-strength at multimegabar pressures—magnesium-oxide to 227GPa. Phys Rev Lett 74:1371-1374
Duffy TS, Shen GY, Heinz DL, Shu JF, Ma YZ, Mao HK, Hemley RJ, Singh AK (1999a) Lattice strains in gold and rhenium under nonhydrostatic compression to 37 GPa. Phys Rev B-Condens Matter 60:15063-15073
Duffy TS, Shen GY, Shu JF, Mao HK, Hemley RJ, Singh AK (1999b) Elasticity, shear strength, and equation of state of molybdenum and gold from X-ray diffraction under nonhydrostatic compression to 24 GPa. J Appl Phys 86:6729-6736
Eng PJ, Rivers M, Yang BX, Schildkamp W (1995) Micro-focusing 4 keV to 65 keV X-rays with bent Kirkpatrick-Baez mirrors. *In* X-ray microbeam technology and applications Proc. S P I E 2516:41-51
Engel GE, Wilke S, Konig O, Harris KDM, Leusen FJJ (1999) PowderSolve—a complete package for crystal structure solution from powder diffraction patterns. J Appl Crystallogr 32:1169-1179
Etgens VH, Alves MCA, Tadjeddine A (1999) *In situ* surface X-ray diffraction studies of electrochemical interfaces at a high-energy third-generation synchrotron facility. Electrochim Acta 45:591-599
Evans JSO, Francis RJ, O'Hare D, Price SJ, Clark SM, Gordon J, Neild A, Tang CC (1994) An apparatus for the study of the kinetics and mechanism of hydrothermal reactons by *in situ* energy dispersive X-ray diffraction. Rev Sci Instrum 66:2442-2445
Eylem C, Hriljac JA, Ramamurthy V, Corbin DR, Parise JB (1996) Structure of a zeolite ZSM-5-bithiophene complex as determined by high resolution synchrotron X-ray powder diffraction. Chem Mater 8:844-849
Falcioni M, Deem MW (1999) A biased Monte Carlo scheme for zeolite structure solution. J Chem Phys 110:1754-1766
Fei YW, Frost DJ, Mao HK, Prewitt CT, Hausermann D (1999) *In situ* structure determination of the high-pressure phase of Fe_3O_4. Am Mineral 84:203-206
Finger LW (1989) Synchrotron powder diffraction. *In* Bish DL, Post JE (eds) Modern Powder Diffraction. Rev Mineral 20:309-331
Fitch AN (1996) The high resolution powder diffraction beam line at ESRF. Mater Sci Forum 228:219-221
Frenkel, AI, Cross, JO, Fanning, DM, Robinson, IK (1999) DAFS analysis of magnetite. J Synch Rad 6: 332-334

Fujihisa H, Fujii Y, Takemura K, Shimomura O, Nelmes RJ, McMahon MI (1996) Pressure dependence of the electron density in solid iodine by maximum-entropy method. High Press Res 14:335-340

Gasparik T, Parise JB, Eiben B, Hriljac JA (1995) Stability and structure of a new high pressure silicate $Na_{1.8}Ca_{1.1}Si_6O_{14}$. Am Mineral 80:1269-1276

Gasparik T, Parise JB, Reeder RJ, Young VG, Wilford WS (1999) Composition, stability, and structure of a new member of the aenigmatite group, $Na_2Mg_{4+x}Fe_{2-2x}{}^{3+}Si_{6+x}O_{20}$, synthesized at 13-14 GPa. Am Mineral 84:257-266

Gies H, Marler B, Vortmann S, Oberhagemann U, Bayat P, Krink K, Rius J, Wolf I, Fyfe C (1998) New structures—new insights: Progress in structure analysis of nanoporous materials. Microporous Mesoporous Mat 21:183-197

Glazer AM (1972) The classification of tilted octahedra in perovskites. Acta Crystallog B28:3384-3392

Gluskin E (1998) APS insertion devices: Recent developments and results. J Synchrot Radiat 5:189-195

Hamalainen K, Krisch M, Kao CC, Caliebe W, Hastings JB (1995) High-Resolution X-Ray Spectrometer Based On a Cylindrically Bent Crystal in Nondispersive Geometry. Rev Sci Instrum 66:1525-1527

Hamalainen K, Manninen S, Kao CC, Caliebe W, Hastings JB, Bansil A, Kaprzyk S, Platzman PM (1996) High resolution Compton scattering study of Be. Phys Rev B-Condens Matter 54:5453-5459

Hammersley AP, Svensson SO, Hanfland M, Fitch AN, Häusermann D (1996) Two-dimensional detector software: From real detector to idealized image or two-theta scan. High Press Res 14:235-248

Hanfland M, Hemley RJ, Mao HK (1994) Synchrotron infrared measurements of pressure-induced transformations in solid hydrogen. *In* Schmidt SC et al. (eds) High-Pressure Science and Technology—1993. Am Inst Physics, New York, p 877-880

Hanfland M, Hemley RJ, Mao HK, Williams GP (1992) Synchrotron infrared spectroscopy at megabar pressures: vibrational dynamics of hydrogen to 180 GPa. Phys Rev Lett 69:1129-1132

Hanfland M, Schwarz U, Syassen K, Takemura K (1999) Crystal structure of the high-pressure phase silicon VI. Phys Rev Lett 82:1197-1200

Harris KDM (1999) New approaches for solving crystal structures from powder diffraction data. J Chin Chem Soc 46:23-34

Harris KDM, Johnston RL, Kariuki BM (1998a) An evolving technique for powder structure solution—fundamentals and applications of the genetic algorithm. High Press Res 94:410-416

Harris KDM, Johnston RL, Kariuki BM (1998b) The genetic algorithm: Foundations and applications in structure solution from powder diffraction data. Acta Crystallogr Sect A 54:632-645

Harris KDM, Kariuki BM, Tremayne M (1998c) Crystal structure solution from powder diffraction data by the Monte Carlo method. Mater Sci Forum 278:32-37

Hemley RJ (1999) Mineralogy at a crossroads. Science 285:1026-1027

Hemley RJ, Goncharov AF, Lu R, Struzhkin VV, Li M, Mao HK (1998a) High-pressure synchrotron infrared spectroscopy at the National Synchrotron Light Source. Nuovo Cimento D20:539-551

Hemley RJ, Mao H, Cohen RE (1998b) High-pressure electronic and magnetic properties. *In* Hemley RJ (eds) Ultrahigh-Pressure Mineralogy. Rev Mineral 37:591-638

Hemley RJ, Mao HK, Goncharov AF, Hanfland M, Struzhkin VV (1996) Synchrotron infrared spectroscopy to 0.15 eV of H_2 and D_2 at megabar pressures. Phys Rev Lett 76:1667-1670

Hemley RJ, Mao HK, Shen GY, Badro J, Gillet P, Hanfland M, Hausermann D (1997) X-ray imaging of stress and strain of diamond, iron, and tungsten at megabar pressures. Science 276:1242-1245

Henderson CMB, Cressey G, Redfern SAT (1995) Geological Applications of Synchrotron-Radiation. Radiat Phys Chem 45:459-481

Herbst-Irmer R, Sheldrick GM (1998) Refinement of twinned structures with SHELXL97. Acta Crystallogr Sect B-Struct Sci 54:443-449

Hill JP, Kao CC, Caliebe WAC, Gibbs D, Hastings JB (1996) Inelastic X-ray scattering study of solid and liquid Li and Na. Phys Rev Lett 77:3665-3668

Hodeau J-L, Bordet P, Anne M, Prat A, Fitch AN, Dooryhee E, Vaughan G, Freund AK (1998) Nine-crystal multianalyzer stage for high-resolution powder diffraction between 6 keV and 40 keV. *In* X-ray Microbeam Technology and Applications. Proc SPIE 3448:353-361

Hu RZ, Egami T, Tsai AP, Inoue A, Masumoto T (1992) Atomic structure of quasi-crystalline $Al_{65}Ru_{15}Cu_{20}$. Phys Rev B-Condens Matter 46:6105-6114

Idziak SHJ, Safinya CR, Sirota EB, Bruinsma RF, Liang KS, Israelachvili JN (1994) Structure of complex fluids under flow and confinement—X-ray Couette-shear-cell and the X-ray surface forces apparatus. Am Chem Soc Symp Ser 578:288-299

Ikeda T, Kobayashi T, Takata M, Takayama T, Sakata M (1998) Charge density distributions of strontium titanate obtained by the maximum entropy method. Solid State Ion 108:151-157

Imai M, Mitamura T, Yaoita K, Tsuji K (1996) Pressure-induced phase transition of crystalline and amorphous silicon and germanium at low temperatures. High Pressure Res 15:167-189

Inui M, Tamura K, Oh'ishi Y, Nakaso I, Funakoshi K, Utsumi W (1999) X-ray diffraction measurements for expanded fluid-Se using synchrotron radiation. J Non-Cryst Solids 252:519-524

Ishii M, Yoshino Y, Takarabe K, Shimomura O (1999) Site-selective X-ray absorption fine structure: Selective observation of Ga local structure in DX center of $Al_{0.33}Ga_{0.67}As$:Se. Appl Phys Lett 74:2672-2674

Kahlenberg V (1999) Application and comparison of different tests on twinning by merohedry. Acta Crystallogr Sect B-Struct Sci 55:745-751

Kao CC, Caliebe WAL, Hastings JB, Gillet JM (1996) X-ray resonant Raman scattering in NiO: Resonant enhancement of the charge-transfer excitations. Phys Rev B-Condens Matter 54:16361-16364

Kariuki BM, Calcagno P, Harris KDM, Philp D, Johnston RL (1999) Evolving opportunities in structure solution from powder diffraction data—Crystal structure determination of a molecular system with twelve variable torsion angles. Angew Chem-Int'l Edition 38:831-835

Kariuki BM, Johnston RL, Harris KDM, Psallidas K, Ahn S, Serrano-Gonzalez H (1998) Application of a Genetic Algorithm in structure determination from powder diffraction data. Match-Commun Math Chem 123-135

Kendelewicz T, Liu P, Labiosa WB, Brown GE (1995) Surface EXAFS and X-ray standing-wave study of the cleaved CaO(100) surface. Physica B209:441-442

Kitamura H (1998) Present status of SPring-8 insertion devices. J Synchrot Radiat 5:184-188

Klug HP, Alexander LE (1974) X-ray Diffraction Procedures for Polycrystalline and Amorphous Materials: 2nd· edition. Wiley Interscience, New York

Knorr K, Krimmel A, Hanfland M, Wassilev-Reul C, Griewatsch C, Winkler B, Depmeier W (1999) High pressure behaviour of synthetic leucite $Rb_2[ZnSi_5O_{12}]$: a combined neutron, X-ray and synchrotron powder diffraction study. Z Kristallogr 214:346-350

Knorr K, Madler F (1999) The application of evolution strategies to disordered structures. J Appl Crystallogr 32:902-910

Koningsveld Hv, Jensen JC, Bekkum Hv (1987) The orthorhombic/monoclinic transition in single crystals of zeolite ZSM-5. Zeolites 7:564-568

Koningsveld Hv, Tuinstra F, Bekkum Hv, Jensen JC (1989a) The location of p-xylene in a single crystal of zeolite H-ZSM-5 with a new sorbate-induced, orthorhombic framework symmetry. Acta Crystallogr B45:423-431

Koningsveld Hv, Tuinstra F, Bekkum Hv, Jensen JC (1989b) On the location and disorder of the tetrapropylammonium (TPA) ion in zeolite ZSM-5 with improved framework accuracy. Acta Crystallogr B45:127-132

Krisch MH, Kao CC, Sette F, Caliebe WA, Hamalainen K, Hastings JB (1995) Evidence for a quadrupolar excitation channel at the L(III) edge of gadolinium by resonant inelastic X-ray-scattering. Phys Rev Lett 74:4931-4934

Kubota Y, Takata M, Sakata M, Ohba T, Kifune K, Tadaki T (1999) A charge density study of MgCu2 and MgZn2 by the maximum entropy method. Jpn J Appl Phys Part 1—Regul Pap Short Notes Rev Pap 38:456-459

Kunz C (1979) Synchrotron Radiation: Techniques and applications. *In* Topics in Current Physics. 10. Lotsch HKV (ed) Springer-Verlag, Berlin, p 442

Kunz M, Leinenweber K, Parise JB, Wu TC, Bassett WA, Brister K, Weidner DJ, Vaughan MT, Wang Y (1996) The baddeleyite-type high pressure phase of $Ca(OH)_2$. High Press Res 14:311-319

Langford JI, Louer D (1996) Powder diffraction. Rep Prog Phys 59:131-234

Larsson J, Heimann PA, Lindenberg AM, Schuck PJ, Bucksbaum PH, Lee RW, Padmore HA, Wark JS, Falcone RW (1998) Ultrafast structural changes measured by time-resolved X-ray diffraction. Appl Phys A-Mater Sci Process 66:587-591

Lasocha W, Schenk H (1997) A simplified, texture-based method for intensity determination of overlapping reflections in powder diffraction. J Appl Crystallogr 30:561-564

Lee Y, Cahill C, Hanson J, Parise JB, Carr S, Myrick ML, Preckwinkel UV, Phillips JC (1999) Characterization of K^+-ion exchange into Na-MAX using time resolved synchrotron X-ray powder diffraction and rietveld refinement. *In* 12th Int'l Zeolite Assoc Meeting, Baltimore, Maryland, IV: 2401-2408

Lee YJ, Carr SW, Parise JB (1998) Phase transition upon K^+ ion exchange into Na low silica X: Combined NMR and synchrotron X-ray powder diffraction study. Chem Mater 10:2561-2570

LeToullec R, Loubeyre P, Pinceaux JP, Mao HK, Hu J (1992) A system for doing low temperature-high pressure single-crystal X-ray diffraction with a synchrotron source. High Pressure Res 6:379-388

Li BS, Liebermann RC, Weidner DJ (1998) Elastic moduli of wadsleyite (beta-Mg_2SiO_4) to 7 gigapascals and 873 Kelvin. Science 281:675-677

Liang X, Andrews JE, de Haseth JA (1996) Resoltion of mixture components by target transformaton factor analysis and determinent analysis for the selection of targets. Analyt Chem 68:378-385

Liebermann RC, Li BS (1998) Elasticity at high pressures and temperatures. *In* Hemley RJ (ed) Ultrahigh-Pressure Mineralogy. Rev Mineral 37:459-492

Loubeyre P, LeToullec R, Hausermann D, Hanfland M, Hemley RJ, Mao HK, Finger LW (1996) X-ray diffraction and equation of state of hydrogen at megabar pressures. Nature 383:702-704

Lumpkin AH, Yang BX (1999) Potential diagnostics for the next-generation light sources. Nucl Instrum Methods Phys Res Sect A-Accel Spectrom Dect Assoc Equip 429:293-298

Manceau A, Drits VA, Lanson B, Chateigner D, Wu J, Huo D, Gates WP, Stucki JW (2000a) Oxidation-reduction mechanism of iron in dioctahedral smectites: II. Crystal chemistry of reduced Garfield nontronite. Am Mineral 85:153-172

Manceau A, Lanson B, Drits VA, Chateigner D, Gates WP, Wu J, Huo D, Stucki JW (2000b) Oxidation-reduction mechanism of iron in dioctahedral smectites: I. Crystal chemistry of oxidized reference nontronites. Am Mineral 85:133-152

Mao H, Hemley RJ (1998) New windows on the Earth's deep interior. *In* Hemley RJ (ed) Ultrahigh-Pressure Mineralogy. Rev Mineral 37:1-32

Martinez-Garcia D, Le Gòdec Y, Mezouar M, Syfosse G, Itie JP, Besson JM (1999) High pressure and temperature X-ray diffraction studies on CdTe. Phys Status Solidi B-Basic Res 211:461-467

Mayanovic RA, Anderson AJ, Bassett WA, Chou IM (1999) XAFS measurements on zinc chloride aqueous solutions from ambient to supercritical conditions using the diamond anvil cell. J Synchrot Radiat 6:195-197

McCusker LB, Von Dreele RB, Cox DE, Louer D, Scardi P (1999) Rietveld refinement guidelines. J Appl Crystallogr 32:36-50

McMahon MI, Nelmes RJ (1993) New High-Pressure Phase of Si. Phys Rev B-Condens Matter 47:8337-8340

McMahon MI, Nelmes RJ, Liu H, Belmonte SA (1996) Two-dimensional data collection in high pressure powder diffraction studies applications to semiconductors. High Pressure Res 14:277-286

Meden A (1998) Crystal structure solution from powder diffraction data—State of the art and perspectives. Croat Chem Acta 71:615-633

Meng Y, Weidner DJ, Fei Y (1993) Deviatoric stress in a quasi-hydrostatic diamond anvil cell: Effect on the volume-based pressure calibration. Geophys Res Lett 20:1147-1150

Mezouar M, Le Bihan T, Libotte H, Le Godec Y, Hausermann D (1999) Paris-Edinburgh large-volume cell coupled with a fast imaging-plate system for structural investigation at high pressure and high temperature. J Synchrot Radiat 6:1115-1119

Moller PJ, Li ZS, Egebjerg T, Sambi M, Granozzi G (1998) Synchrotron-radiation-induced photoemission study of VO_2 ultrathin films deposited on TiO_2(110). Surf Sci 404:719-723

Mora AJ, Fitch AN, Gates PN, Finch A (1996) Structural determination of $(CH_3)_2SBr_2$ using the Swiss-Norwegian beam line at ESRF. Mater Sci Forum 228:601-605

Mosselmans, JFW, Pattrick, RAD, Charnock, JM, Solé, VA (1999) EXAFS of copper in hydrosulfide solutions at very low concentratons: implications for speciation of copper in natural waters. Mineral Mag 63: 769-772.

Nagara H, Nagao K (1998) Vibrons and phonons in solid hydrogen at megabar pressures. J Low Temp Phys 111:483-488

Nanba T (1998) High-pressure solid-state spectroscopy at UVSOR by infrared synchrotron radiation. Nuovo Cimento Soc Ital Fis D-Condens Matter At Mol Chem Phys Fluids Plasmas Biophys 20:397-413

Neder RB, Burghammer M, Grasl T, Schulz H (1996a) Mounting an individual submicrometer sized single crystal. Z Kristallogr 211:365-367

Neder RB, Burghammer M, Grasl T, Schulz H (1996b) Single-crystal diffraction by submicrometer sized kaolinite; observation of Bragg reflections and diffuse scattering. Z Kristallogr 211:763-765

Neder RB, Burghammer M, Grasl T, Schulz H (1996c) A vacuum chamber for low background diffraction experiemnts. Z Kristallogr 211:591-593

Neder RB, Burghammer M, Grasl T, Schulz H, Bram A, Fiedler S (1999) Refinement of the kaolinite structure from single-crystal synchrotron data. Clays & Clay Minerals 47:487-494

Nelmes RJ, Allan DR, McMahon MI, Belmonte SA (1999a) Self-hosting incommensurate structure of barium IV. Phys Rev Lett 83:4081-4084

Nelmes RJ, Hatton PD, McMahon MI, Piltz RO, Crain J, Cernik RJ, Bushnellwye G (1992) Angle-dispersive powder-diffraction techniques for crystal-structure refinement at high-pressure. Rev Sci Instrum 63:1039-1042

Nelmes RJ, McMahon MI (1994) High-pressure powder diffraction on synchrotron sources. J Synchrotron Radiat 1:69-73

Nelmes RJ, McMahon MI, Belmonte SA, Parise JB (1999b) Structure of the high-pressure phase III of iron sulfide. Phys Rev B-Condens Matter 59:9048-9052

Nishibori E, Saso D, Takata M, Sakata M, Imai M (1999) Charge densities and chemical bonding of CaSi2. Jpn J Appl Phys Part 1—Regul Pap Short Notes Rev Pap 38:504-507
Norby P (1996) *In-situ* time resolved synchrotron powder diffraction studies of syntheses and chemical reactions. Mater Sci Forum 228-231:147-152
Norby P (1997) Synchrotron powder diffraction using imaging plates: crystal structure determination and Rietveld refinement. J Appl Crystallogr 30:21-30
Norby P, Cahill C, Koleda C, Parise JB (1998) A reaction cell for *in situ* studies of hydrothermal titration. J Appl Crystallogr 31:481-483
Nuhn H, Rossbach J (2000) LINAC-based short wavelength FELs: The challenges to be overcome to produce the ultimate X-ray source—the X-ray laser. Synchrot Radiat News 13:18-32
O'Hare D, Evans JSO, Francis R, Price S, O'Brien S (1998a) The use of *in situ* powder diffraction in the study of intercalation and hydrothermal reaction kinetics. Mater Sci Forum 278:367-378
O'Hare D, Evans JSO, Francis RJ, Halasyamani PS, Norby P, Hanson J (1998b) Time-resolved, *in situ* X-ray diffraction studies of the hydrothermal syntheses of microporous materials. Microporous Mesoporous Mat 21:253-262
Ohsumi K, Hagiya K, Ohmasu M (1991) Developemnt of a system to analyse the submicrometer-sixed single crystal by synchrotron X-ray diffraction. J Appl Crystallog 24:340-348
Ostbo NP, Goyal R, Jobic H, Fitch AN (1999) The location of pyridine in sodium-silver-Y zeolite by powder synchrotron X-ray diffraction. Microporous Mesoporous Mat 30:255-265
Parise JB (1995) Structural case studies of inclusion phenomena in zeolites: Xe in RHO and stilbene in ZSM-5. J Inclusion Phenom Molec Recog Chem 21:79-112
Parise JB, Cahill CL, Lee Y (2000) Dynamical X-ray diffraction studies at synchrotron sources. Can Mineral (in press):
Parise JB, Chen J (1997) Studies of crystalline solids at high pressure and temperature using the DIA multi-anvil apparatus. Eur J Solid State Inorg Chem 34:809-821
Parise JB, Corbin DR, Abrams L (1995) Structural changes upon sorption and desorption of Xe from Cd-exchanged zeolite rho: A real-time synchrotron X-ray powder diffraction study. Microporous Mater 4:99-110
Parise JB, Hriljac JA, Cox DE, Corbin DR, Ramamurthy V (1993) A high resolution synchrotron powder diffraction study of *trans*-stilbene in zeolite ZSM-5. J Chem Soc, Chem Comm 226-228
Parise JB, Weidner DJ, Chen J, Liebermann RC, Chen G (1998) *In situ* studies of the properties of materials under high-pressure and temperature conditions using multi-anvil apparatus and synchrotron X-rays. Ann Rev Mater Sci 28:349-374
Park J-H, Woodward PM, Parise JB (1998) Predictive modeling and high pressure-high temperature synthesis of perovskites containing Ag^+. Chem Mater 10:3092-3100
Pasternak MP, Rozenberg GK, Machavariani GY, Naaman O, Taylor RD, Jeanloz R (1999) Breakdown of the Mott-Hubbard state in Fe_2O_3: A first-order insulator-metal transition with collapse of magnetism at 50 GPa. Phys Rev Lett 82:4663-4666
Peng JB, Foran GJ, Barnes GT, Gentle IR (1997) Phase transitions in Langmuir-Blodgett films of cadmium stearate: Grazing incidence X-ray diffraction studies. Langmuir 13:1602-1606
Piltz RO, McMahon MI, Crain J, Hatton PD, Nelmes RJ, Cernik RJ, Bushnellwye G (1992) An Imaging Plate System For high-pressure powder diffraction—the data-processing side. Rev Sci Instrum 63:700-703
Prazeres R, Glotin F, Jaroszynski DA, Ortega JM, Rippon C (1999) Enhancement of harmonic generation using a two section undulator. Nucl Instrum Methods Phys Res Sect A-Accel Spectrom Dect Assoc Equip 429:131-135
Proffen T, Welberry TR (1997) Analysis of diffuse scattering via the reverse Monte Carlo technique: A systematic investigation. Acta Crystallogr Sect A 53:202-216
Proffen T, Welberry TR (1998) Analysis of diffuse scattering of single crystals using Monte Carlo methods. Phase Transit 67:373-397
Putz H, Schon JC, Jansen M (1999) Combined method for ab initio structure solution from powder diffraction data. J Appl Crystallogr 32:864-870
Qian YL, Sturchio NC, Chiarello RP, Lyman PF, Lee TL, Bedzyk MJ (1994) Lattice location of trace-elements within minerals and at their surface with X-ray standing waves. Science 265:1555-1557
Reeder RJ (1991) Crystals—Surfaces make a difference. Nature 353:797-798
Reeder RJ (1996) Interaction of divalent cobalt, zinc, cadmium, and barium with the calcite surface during layer growth. Geochim Cosmochim Acta 60:1543-1552
Reffner J, Carr GL, Sutton S, Hemley RJ, Williams GP (1994) Infrared microspectroscopy at the NSLS. Synch Rad News 7 (no. 2):30-37
Renaud G, Belakhovsky M, Lefebvre S, Bessiere M (1995) Correlated chemical and positional fluctuations in the gold-nickel system—a synchrotron X-ray diffuse-scattering study. J Phys III 5:1391-1405

Rietveld HM (1969) A profile refinement method for nuclear and magnetic structures. J Appl Crystallogr 2:65-71

Rius J, Miravitlles C, Gies H, Amigo JM (1999) A tangent formula derived from Patterson-function arguments. VI. Structure solution from powder patterns with systematic overlap. J Appl Crystallogr 32:89-97

Robinson IK, Eng PJ, Schuster R (1994) Origin of the surface sensitivity in surface X-ray diffraction. Acta Physica Polonica A 86:513-520

Robinson IK, Smilgies D-M, Eng PJ (1992) Cluster formation in the adsorbate-induced reconstruction of the O/Mo(100) surface. J Phys: Condens Matter 4:5845-5854

Rudolf PR (1993) Techniques For *Ab-initio* structure determination from X-ray-powder diffraction data. Mater Chem Phys 35:267-272

Rueff JP, Kao CC, Struzhkin VV, Badro J, Shu J, Hemley RJ, Hao HK (1999) Pressure-induced high-spin to low-spin transition in FeS evidenced by X-ray emission spectroscopy. Phys Rev Lett 82:3284-3287

Sakabe N, Nakagawa A, Sasaki K, Sakabe K, Watanabe N, Kondo H, Shimomura M (1989) Synchrotron radiation protein data-collection system using the newly developed Weissenberg camera and imaging plate for crystal-structure analysis. Rev Sci Instrum 60:2440-2441

Sarvestani A, Amenitsch H, Bernstorff S, Besch HJ, Menk RH, Orthen A, Pavel N, Rappolt M, Sauer N, Walenta AH (1999) Biological X-ray diffraction measurements with a novel two-dimensional gaseous pixel detector. J Synchrot Radiat 6:985-994

Sawers LJ, Carter VJ, Armstrong AR, Bruce PG, Wright PA, Gore BE (1996) *Ab initio* structure solution of a novel aluminium methylphosphonate from laboratory X-ray powder diffraction data. J Chem Soc–Dalton Trans 3159-3161

Schlegel ML, Manceau A, Chateigner D, Charlet L (1999) Sorption of metal ions on clay minerals I. Polarized EXAFS evidence for the adsorption of Co on the edges of hectorite particles. J Colloid Interface Sci 215:140-158

Schwarz U, Syassen K, Grzechnik A, Hanfland M (1999) The crystal structure of rubidium-VI near 50 GPa. Solid State Commun 112:319-322

Schwarz U, Takemura K, Hanfland M, Syassen K (1998) Crystal structure of cesium-V. Phys Rev Lett 81:2711-2714

Seward, TM, Henderson, CMB, Charnock, JM (2000) Indium (III) chloride complexing and solvation in hydrothermal solutions to 350°C, an EXAFS study. Chem Geol 167: 117-127.

Shankland K, David WIF, Sivia DS (1997) Routine *ab initio* structure determination of chlorothiazide by X-ray powder diffraction using optimised data collection and analysis strategies. J Mater Chem 7:569-572

Shaw S, Clark SM, Henderson CMB (2000) Hydrothermal studies of the calcium silicate hydrates, tobermorite and xonotlite. Chem Geol 167:129-140

Shimomura O, Takemura K, Ohishi Y, Kikegawa T, Fujii Y, Matsushita T, Amemiya Y (1989) X-ray-diffraction study under pressure using an imaging plate. Rev Sci Instrum 60:2437-2437

Shimomura O, Yamaoka S, Yagi T, Wakatsuki M, Tsuji K, Kawamura H, Hamaya N, Aoki K, Akimoto S (1985) Multi-anvil type X-ray system for synchrotron radiation. *In* Minomura S (ed) Solid State Physics Under Pressure: Recent Advances with Anvil Devices. KTK Science Publishers, Tokyo, p 351-356

Shiraishi Y, Nakai I, Tsubata T, Himeda T, Nishikawa F (1999) Effect of the elevated temperature on the local structure of lithium manganese oxide studied by *in situ* XAFS analysis. J Power Sources 82:571-574

Sicron N, Yacoby Y (1999) XAFS study of rhombohedral ferroelectric $PbHf_{0.9}Ti_{0.1}O_3$. J Synchrot Radiat 6:503-505

Sicron N, Yacoby Y, Stern EA, Dogan F (1997) XAFS study of the antiferroelectric phase transition in $PbZrO_3$. J Phys IV 7:1047-1049

Siddons DP, Furenlid L, Pietraski P, Yin Z, Li Z, Smith G, Yu B, Harlow R (1999) Instrumentation developmants for X-ray powder diffraction at Brookhaven. Synchrot Radiat News 12:21-26

Siddons DP, Yin Z, Furenlid L, Pietraski P, Li Z, Harlow R (1998) Multichannel analyzer/detector system for high-speed high-resolution powder diffraction. 3448:120-131

Singh AK, Balasingh C, Mao HK, Hemley RJ, Shu JF (1998a) Analysis of lattice strains measured under nonhydrostatic pressure. J Appl Phys 83:7567-7575

Singh AK, Mao HK, Shu JF, Hemley RJ (1998b) Estimation of single-crystal elastic moduli from polycrystalline X-ray diffraction at high pressure: Application to FeO and iron. Phys Rev Lett 80:2157-2160

Smith GC (1991) X-ray imaging with gas proportional detectors. Synch Radiat News 4:24-30

Somayazulu MS, Finger LW, Hemley RJ, Mao HK (1996) High-pressure compounds in methane-hydrogen mixtures. Science 271:1400-1402

Stahl K, Hanson J (1994) Real-time X-ray synchrotron powder diffraction studies of the dehydration processes in scolecite and mesolite. J Appl Crystallogr 27:543-550

Stefan PM, Krinsky S, Rakowsky G, Solomon L, Lynch D, Tanabe T, Kitamura H (1998) Small-gap undulator research at the NSLS: concepts and results. Nucl Instrum Methods Phys Res Sect A-Accel Spectrom Dect Assoc Equip 412:161-173

Takele S, Hearne GR (1999) Electrical transport, magnetism, and spin-state configurations of high-pressure phases of FeS. Phys Rev B-Condens Matter 60:4401-4403

Tamura K, Inui M, Nakaso I, Ohishi Y, Funakoshi K, Utsumi W (1999) Structural studies of expanded fluid mercury using synchrotron radiation. J Non-Cryst Solids 250:148-153

Tan K, Darovsky A, Parise JB (1995) The synthesis of a novel open framework sulfide, $CuGe_2S_5 \cdot (C_2H_5)_4N$ and its structure solution using synchrotron imaging plate data. J Am Chem Soc 117:7039-7040

Tatchyn R (1996) Image tuning techniques for enhancing the performance of pure permanent magnet undulators with small gap period ratios. Nucl Instrum Methods Phys Res Sect A-Accel Spectrom Dect Assoc Equip 375:500-503

Thoms M, Bauchau S, Hausermann D, Kunz M, Le Bihan T, Mezouar M, Strawbridge D (1998) An improved X-ray detector for use at synchrotrons. Nucl Instrum Methods Phys Res Sect A-Accel Spectrom Dect Assoc Equip 413:175-184

Toby BH, Egami T (1992) Accuracy of Pair Distribution Function-Analysis Applied to Crystalline and Noncrystalline Materials. Acta Crystallogr A48:336-346

Tomilin MG (1997) Interaction of liquid crystals with a solid surface. J Opt Technol 64:452-475

Toraya H, Hibino H, Ohsumi K (1996) A New Powder Diffractometer for Synchrotron Radiation with a Multiple-Detector System. J Synchrot Radiat 3:75-83

Tribaudino M, Benna P, Bruno E, Hanfland M (1999) High pressure behaviour of lead feldspar ($PbAl_2Si_2O_8$). Phys Chem Minerals 26:367-374

Vollhardt D (1999) Phase transition in adsorption layers at the air-water interface. Adv Colloid Interface Sci 79:19-57

Walker RP, Diviacco B (2000) Insertion devices—recent developments and future trends. Synchrot Radiat News 13:3342

Wang YB, Weidner DJ, Zhang JZ, Gwanrnesia GD, Liebermann RC (1998) Thermal equation of state of garnets along the pyrope-majorite join. Phys Earth Planet Inter 105:59-71

Warren BE (1969) X-ray Diffraction. Addison-Wesley, Reading, Massachusetts

Weidner DJ (1994) Workshop on synchrotron radiation at high pressure. Synchrot Radiat News 7:8

Weidner DJ (1998) Rheological studies at high pressure. *In* Hemley RJ (eds) Ultrahigh-Pressure Mineralogy. Rev Mineral 37:493-524

Weidner DJ, Li B, Wang Y (1999) Mineral physics constraints on transition zone structure. Trans Am Geophys Soc EOS 80:F25

Weidner DJ, Mao HK (1993) Photons at high pressure. NSLS Newsletter March, p 1-15

Weidner DJ, Wang Y, Chen G, Ando J (1995) Rheology measurements at high pressure and temperature. Trans Am Geophys Soc EOS 76:584

Weidner DJ, Wang Y, Vaughan MT (1994a) Strength of diamond. Science 266:419-422

Weidner DJ, Wang Y, Vaughan MT (1994b) Yield strength at high pressure and temperature. Geophys Res Lett 21:753-756

Welberry TR, Butler BD (1994) Interpretation of diffuse-X-ray scattering via models of disorder. J Appl Crystallogr 27:205-231

Welberry TR, Butler BD (1995) Diffuse X-ray scattering from disordered crystals. Chem Rev 95:2369-2403

Welberry TR, Mayo SC (1996) Diffuse X-ray scattering and Monte-Carlo study of guest-host interactions in urea inclusion compounds. J Appl Crystallogr 29:353-364

Welberry TR, Proffen T (1998) Analysis of diffuse scattering from single crystals via the reverse Monte Carlo technique. I. Comparison with direct Monte Carlo. J Appl Crystallogr 31:309-317

Woodward PM (1997a) Octahedral Tilting in Perovskites I : Geometrical Considerations. Acta Crystallogr B53:32-43

Woodward PM (1997b) Octahedral tilting in perovskites II: Stucuture-stabilizing forces. Acta Crystallogr B53:44-66

Woodward PM (1997c) POTATO—A program for generating perovskite structures distorted by tilting of rigid octahedra. J Appl Crystallogr 30:206

Yacoby, Y (1999) Local structure determination on the atomic scale. Current Opinion Solid State Mater Sci 4: 337-341

Yamanaka T, Ogata K (1991) Structure refinement of GeO_2 polymorphs at high-pressures and temperatures by energy-dispersive spectra of powder diffraction. J Appl Crystallogr 24:111-118

Yaoita K, Katayama Y, Tsuji K, Kikegawa T, Shimomura O (1997) Angle-dispersive diffraction measurement system for high-pressure experiments using a multichannel collimator. Rev Sci Instrum 68:2106-2110

Yoo CS, Cynn H, Gygi F, Galli G, Iota V, Nicol M, Carlson S, Hausermann D, Mailhiot C (1999) Crystal structure of carbon dioxide at high pressure: "Superhard" polymeric carbon dioxide. Phys Rev Lett 83:5527-5530

Young NA, Dent AJ (1999) Maintaining and improving the quality of published XAFS data: a view from the UK XAFS user group. J Synchrot Radiat 6:799-799

Young RA (1995) The Rietveld Method. International Union of Crtystallography, Oxford University Press, New York, 298 p

Zha CS, Duffy TS, Mao HK, Hemley RJ (1993a) Elasticity of hydrogen to 24 GPa from single-crystal Brillouin scattering and synchrotron X-ray diffraction. Phys Rev B-Condens Matter 48:9246-9255

Zha CS, Duffy TS, Mao HK, Hemley RJ (1993b) Elasticity of hydrogen to 24-GPa from single-crystal Brillouin-scattering and synchrotron X-ray-diffraction. Phys Rev B-Condens Matter 48:9246-9255

Zhang L, Stanek J, Hafner SS, Ahsbahs H, Grunsteudel HF, Metge J, Ruffer R (1999) ^{57}Fe nuclear forward scattering of synchrotron radiation in hedenbergite $CaFeSi_2O_6$ hydrostatic pressures up to 68 GPa. Am Mineral 84:447-453

12 Radiation-Induced Amorphization

Rodney C. Ewing[a,b], Alkiviathes Meldrum[c], LuMin Wang[a] and ShiXin Wang[a]

[a]Department of Nuclear Engineering & Radiological Sciences
[b]Department of Geological Sciences
University of Michigan
Ann Arbor, Michigan 48109

[c]Department of Physics
University of Alberta
Edmonton, Alberta T6G 2J1, Canada

INTRODUCTION

History and applications

The earliest description of a change in mineral property caused by radiation-damage was by Jöns Jacob Berzelius in 1814. Berzelius, a Swedish physician and mineral chemist, was certainly one of the most eminent scientists of his century, discovering the elements cerium, selenium and thorium. He discovered that some U- and Th-bearing minerals glowed on moderate heating, releasing, sometimes violently, large amounts of energy. This "pyrognomic" behavior was first observed in gadolinite, $(REE)_2FeBe_2Si_2O_{10}$, as it released stored energy. Today, this same type of catastrophic energy release is a significant concern in reactor safety where rapid energy release from neutron-irradiated graphite can lead to a rise in temperature and the rupture of fuel elements.

In 1893, Brøgger defined "metamikte" in an encyclopedia entry for amorphous materials. Metamict minerals were believed to have been previously crystalline, as judged by well-formed crystal faces, but had a characteristic glass-like, conchoidal fracture and were optically isotropic. Prior to the discovery of radioactivity by Becquerel in 1896, the metamict state was not recognized as a radiation-induced transformation. Hamberg (1914) was the first to suggest that metamictization is a radiation-induced, periodic-to-aperiodic transformation caused by α-particles that originate from the constituent radionuclides in the uranium and thorium decay-series. A detailed summary of the history of studies of radiation effects in minerals can be found in Ewing (1994).

In the 1950s, interest in metamict minerals was revived. Adolf Pabst rescued the term "metamict" from obscurity in his presidential address to the Mineralogical Society of America in 1951 (Pabst 1952). However, mineralogists paid limited attention to nuclear effects in minerals or the general importance of radiation effects in modifying the properties of materials. By the early 1940s, E.P. Wigner already had anticipated that the intense neutron flux in the Hanford Production reactors would alter the properties of the graphite moderator and the reactor fuel. This damage was called "Wigner's disease," and the early history of this work has been summarized by Seitz (1952). The first "cross-over" paper was by Holland and Gottfried (1955) in which they applied the understanding of radiation damage mechanisms to natural zircons of great age that were partially to completely metamict. Holland and Gottfried were the first to develop a model of damage accumulation for a mineral and to discuss the nature of this radiation-induced transition. The purpose of their study was to use damage accumulation as an age-dating technique.

During the past twenty-years, interest in radiation effects in minerals and "complex"

1529-6466/00/0039-0012$05.00

ceramics has increased because of the importance of radiation damage effects on the long-term behavior of solids used for the immobilization of high-level nuclear waste (Ewing 1975, Ewing and Haaker 1980, Lutze and Ewing 1988, Ewing et al. 1987, Ewing et al. 1988, Ewing et al. 1995a, Weber et al. 1998). Ion-beam irradiations were first systematically used to study radiation effects in minerals in the early 1990s (Wang and Ewing 1992a, Eby et al. 1992). The effects of radiation, such as volume and leach rate increases, are particularly important in the development of waste forms for actinides. Transuranium nuclides, such as ^{239}Pu from the dismantlement of nuclear weapons, generally transmute by α-decay events that can result in considerable atomic-scale damage (approximately 700-1000 atomic displacements per decay event), and the material becomes amorphous at doses equivalent to a fraction of a displacement per atom (Ewing 1999). The fundamentals of ion-beam damage to ceramics have assumed great importance, particularly because some materials are highly resistant to radiation damage, such as spinel (Yu et al. 1996, Wang et al. 1999), zirconia (Sickafus et al. 1999, Wang et al. 2000a) and Gd-zirconate (SX Wang et al. 1999c).

The use of ion-beam irradiations to simulate radiation effects in minerals has drawn heavily on the rapidly growing field of ion beam modification of materials. Ion beams are the favored technology for precisely controlling the surface and near-surface regions of materials. Many integrated circuit production lines rely on ion implantation systems. Ion beams are used to modify the mechanical, chemical and tribological properties of materials. Thus, considerable effort has been devoted to experimental studies and the development of a theoretical basis for understanding ion-beam amorphization of materials (Motta 1997, Okamoto et al. 1999). Fundamental questions, such as whether radiation-induced amorphization of ceramics is a phase transition are the subject of current debate (Salje et al. 1999). Of particular relevance, are studies of the ion-beam modification of insulators (Brown 1988, Hobbs 1979, Hobbs et al. 1994, Wang et al. 1998b). Present ion beam technology allows the modification of materials at the nano-scale (Wang et al. 2000b, Meldrum et al. 2000b). Thus, the fundamental issue of particle-solid interactions and radiation-induced transformations from the periodic-to-aperiodic state remain an active and important area of research.

MECHANISMS OF RADIATION DAMAGE

The general features of particle-solid interactions have been well described since the early work by Bohr (1913) and continuously refined throughout the century (Lindhard et al. 1963). The early history and basic principles are summarized by Ziegler et al. (1985). This brief review of principles draws heavily from Robinson (1994), Tesmer and Nastasi (1995) and Nastasi et al. (1996).

The initial step in the radiation damage process is the production of a primary recoil or "knock on" atom (PKA). This may be initiated by nuclear reactions, radioactive decay, scattering of incident radiation, or interactions with an ion beam or plasma. The PKA dissipates its energy by exciting electrons in the material along its trajectory (i.e. ionization effects) or by elastic interactions with target atoms (the formation of collision cascades). The kinetic energy of particles involved in the elastic interactions is nearly conserved. With decreasing velocity elastic interactions increasingly dominate (Fig. 1). If the PKA has enough kinetic energy, it imparts sufficient energy to the target atoms and causes atomic displacements ($E_{kinetic} > E_{displacement}$) from lattice sites. A sequence of elastic interactions results in a branching cascade of displaced atoms (Kinchin and Pease 1955), *the displacement cascade*. The dimensions of the cascade depend not only on E_d but also on the energy and mass of the incident particle. For a heavy ion the energy dissipation is high (>1 eV/atom), and the assumption of purely binary interactions breaks down, and the

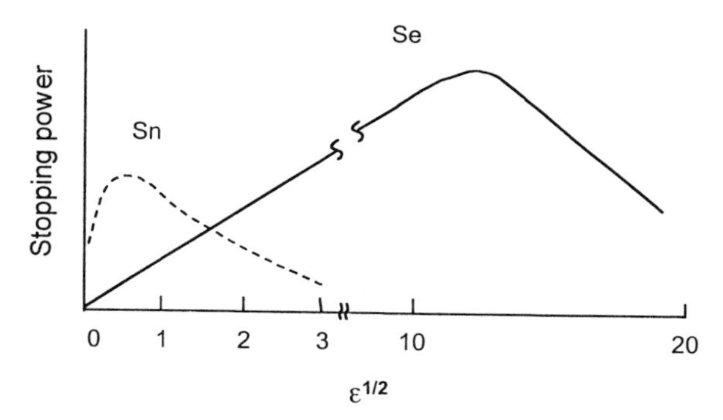

Figure 1. The Lindhard-Scharff-Schiøtt (LSS) nuclear and electronic stopping powers as a function of $\varepsilon^{1/2}$, a dimensionless, "reduced" variable proportional to energy (after Robinson 1994).

cascade may be described as a "hot zone" or thermal spike. Energy dissipation in the thermal spike can be sufficiently high so as to cause "melting" of the cascade volume in a very short time frame, 10^{-12} s (Davies 1992), as demonstrated by molecular dynamic computer simulations (Averback 1994, Spaczer et al. 1995). Due to the small volume of the thermal spike, a quench rate as high as 10^{15} Ks^{-1} may be attained.

The fate of the displaced atoms depends critically on their energy and range. The classic picture of a cascade in a metal consists of ejected atoms occupying interstitial sites around the periphery and vacancies concentrated in the interior of the cascade volume (Brinkman 1954). Particles may be channeled through the structure, in which case the range is increased because elastic interactions are less. Temporarily displaced atoms can return to their lattice sites or combine to form stable defect clusters. The energy from the initial PKA is dissipated within femtoseconds resulting in atomic displacements, electron excitation and heat. Many of the defects are unstable and rapidly rearrange themselves. The isolated defects can form Frenkel defect pairs, defect aggregates, isolated collision cascades or overlapping cascades that finally culminate in bulk amorphization. The relaxation stage lasts from 10^{-12} to 1 second and results in the final microstructure responsible for macroscopic radiation effects. Of particular importance in geologic materials are longer time-scale thermal annealing events. The final damage state is a result of the initial cascade or defect formation process (femtoseconds), the nearly instantaneous relaxation of the structure (picoseconds) and thermal annealing (seconds to billions of years). Other effects accompany these atomic scale interactions, such as chemical changes due to trans-mutation, radiolytic disruption of bonds (particularly of hydrated phases) and bubble formation due to the accumulation of He.

Systematic studies of particle-solid interactions began with Bohr (1948) and culminated in the general description of atomic collisions based on the Lindhard-Scharff-Schiøtt (LSS) theory (Lindhard et al. 1963, 1968). The universal treatment is described by a set of dimensionless or reduced variables that are proportional to the energy, distance and time functions.

$$\text{energy:} \quad \varepsilon = E/E_L \quad \text{with } E_L = \frac{Z_1 Z_2 e^2}{a_{12}} \frac{1+A}{A}, \tag{1}$$

$$\text{distance:} \quad \rho = x/R_L \quad \text{with } R_L = \frac{(1+A)^2}{4A} \frac{1}{n\pi a^2_{12}}, \tag{2}$$

time: $\tau = t/T_L$ with $T_L = \left(\frac{M_1}{2E_L}\right)^{1/2} R_L$, (3)

where Z_i and M_i are the atomic number and mass of the atoms, $i = 1$ for the projectile and 2 for the target atoms, $A=M_2/M_1$, e is the charge on the electron, n is the atomic density of the target, and a_{12} is a screening length used in the Thomas-Fermi description of the atoms. Using the dimensionless variables, the electronic stopping power, $(d\varepsilon/d\rho)_e$, and the nuclear stopping power, $(d\varepsilon/d\rho)_n$, can be plotted as a function of energy, ε (Fig. 1).

In minerals, the dominant damage mechanism is a result of the α-decay event associated with the decay of ^{238}U, ^{235}U and ^{232}Th and radionuclides in their decay chains. In nuclear waste forms the α-decay damage is the result of the decay of actinides, mainly ^{239}Pu. The "self-damage" from actinides is caused by two types of particle-irradiations: The α-particle (helium ion) has an energy of 4.5 to 5.8 MeV and a range of 10-30 μm; whereas, the recoil nucleus has an energy of 70-100 keV and a range of 10-20 nm. The α-particle carries approximately 98% of the energy of the decay-event. In estimating the consequences of irradiation of a solid by light and heavy particles of varying energies, it is important to estimate whether a given particle will deposit its energy primarily by displacive (elastic) or by electronic (inelastic) interactions (Fig. 1). This is determined by the relative velocity of the bombarding particle and that of the orbital electrons of the target ion. If the particle velocity is below that of the orbital electrons, the likelihood of electronic excitations is small and most of the energy will be transferred to the nucleus of the ion. However, if the particle velocity is higher than that of the orbital electrons, excitation will dominate. As a rough rule, inelastic processes are important if the energy of the bombarding particle, expressed in keV, is greater than its atomic weight. Thus an α-particle of atomic weight 4 and an energy of 5,000 keV will predominantly deposit its energy by ionization, while an α-recoil ion of mass 240 and an energy of 100 keV will lose most of its energy by elastic collisions. An α-particle will thus dissipate most of its energy by electronic interactions along its path of 10 to 30 μm with a limited number (e.g. 100) of elastic interactions occurring at the end of its trajectory. The α-recoil will cause a much larger number of atomic displacements (e.g. 700-1000). The types of defects created by these two irradiations, isolated defects vs. collision cascades, will affect the defect recombination and annealing processes.

EXPERIMENTAL METHODS

The crystalline-to-amorphous (c-a) transition in minerals has been investigated primarily by three experimental techniques for which different units and terminology are used in the calculation of dose.

Minerals containing U and Th

The earliest studies (Holland and Gottfried 1955) used natural minerals, mainly zircon, of known age and U- and Th-contents. The dose (α-decay events/g) is given by:

$$D = 8N_{238}[exp(t/\tau_{238})-1] + 6N_{232}[exp(t/\tau_{232})-1] \quad (4)$$

where D = α–recoil dose, N_{238} and N_{232} are the present number of atoms/g of ^{238}U and ^{232}Th, τ_{238} and τ_{232} are the half lives of ^{238}U and ^{232}Th, and t is the geologic age. For suites of zircons or even a single zoned crystal of zircon, the α-decay event dose can be correlated with changes in properties using a wide variety of techniques, such as X-ray diffraction, transmission electron microscopy, Raman spectroscopy, infrared spectroscopy and calorimetry (Holland and Gottfried 1955, Murakami et al. 1991, Ellsworth et al. 1994, Salje et al. 1999, Ríos et al. 2000, Zhang et al. 2000a,b,c). From such suites of samples estimates can be made of the α-decay event dose required for amorphization (Table 1). Detailed analytical electron microscopy studies have been completed on zoned zirconolites to determine radiation-damage effects as a function of dose. The entire range of dose for the crystalline-to-amorphous transition can be obtained from just a few zoned crystals (Lumpkin et al. 1994, 1997). The advantage of studying minerals is that they provide data on long-term damage accumulation (hundreds of millions of years) at very low dose rates ($<10^{-17}$ dpa/s). Most importantly, the bulk material is not highly radioactive and is amenable to study by a wide variety of analytical techniques. The main limitation is that minerals generally have a poorly documented thermal and alteration history, and subsequent thermal and alteration events can substantially anneal the damage; however, studies of suites of specimens allow one to correlate changes in properties with dose and "anomalous" samples are usually easily identified. Recently, considerable effort has been devoted to determining the thermal histories of metamict minerals and the effect on the crystalline-amorphous transition (Lumpkin et al. 1998, 1999).

Table 1. Alpha-decay-event dose for amorphization of selected minerals. The estimated dose can be highly variable because of the effect of thermal annealing over geologic time.

mineral	α-decay event dose ($\times 10^{18}$/g)	reference
aeschynite	>12	Gong et al. 1997
allanite	3.2 - 10	Janaczek and Eby 1993
britholite	>12	Gong et al. 1997
gadolinite	3.9 - 6.2	Janaczek and Eby 1993
perovskite	>12	Smith et al. 1999
pyrochlore (natural)	1-100	Lumpkin and Ewing 1988
titanite	5.0	Vance and Metson 1985
zircon	10	Holland and Gottfried 1955
	10	Headley et al. 1982 b
	10	Weber 1990
	8.0	Murakami et al. 1991
	4.5	Woodhead et al. 1991
	3.5	Salje et al. 1999
zirconolite	10 - 80	Lumpkin et al. 1998

Actinide-doping

A well-controlled (and highly accelerated) technique for studying the effects of α–decay damage is to dope synthetic crystals with short half-life actinides, such as ^{238}Pu ($\tau_{1/2}$ = 87.7 years) or ^{244}Cm ($\tau_{1/2}$ = 18.1 years). At typical concentrations of 10 atomic percent, specimens become amorphous in 5 to 10 years, and the dose rate is 10^{-10} to 10^{-8} dpa/s (Weber 1990). This represents an increase in the dose rate by a factor of $\sim 10^8$ over that of U- and Th-bearing minerals. The radiation fluence is generally measured in α–decay-events/g. The main advantage of actinide-doping experiments is that specimens spanning a complete range of damage can be obtained over extended experimental periods and the thermal history of the specimens is well controlled. However, the radioactivity of actinide-doped samples limits the types of analyses that can be done to simple determinations of density, unit cell refinements from X-ray diffraction data, and limited examination utilizing electron microscopy.

Charged-particle irradiation

Irradiation by a wide variety of ions of different energies is now a widely used means of investigating the crystalline-to-amorphous transition in metals (Johnson 1988), intermetallics (Motta 1997, Motta and Olander 1990), simple oxides (Gong et al. 1996a,b, 1998; Degueldre and Paratte 1998), and complex ceramics and minerals (Wang and Ewing 1992a, Ewing et al. 1987, 1995; Weber et al. 1998). A beam of energetic ions is focused onto a specimen, and the effects of radiation damage can be monitored using either *in situ* or *ex-situ* techniques. The ion flux, energy, total fluence, and specimen temperature are experimentally controlled. The *in situ* experiments are performed in a transmission electron microscope (TEM) interfaced to an ion accelerator. The ion beam penetrates the specimen that is simultaneously observed in an electron microscope. In the United States, this work is done at the IVEM- and HVEM-Tandem Facility at Argonne National Laboratory (Allen and Ryan 1997). The main advantage of the *in situ* irradiation technique is that the damage can be continuously monitored, mainly by selected-area electron diffraction, without the need to remove the specimen. The critical fluence for amorphization (Table 2) is measured directly, and fully damaged samples are produced in less than 30 minutes (dose rate = 10^{-5}

Table 2. Ion beam amorphization data for selected minerals.
The conversion equation between fluence and dose is given in the text.
Critical temperatures given for those phases for which there are T-dependent amorphization data.

Phase	Ion type and energy	Amorphization fluence ($\times 10^{14}$ ions/cm^2)	E_d (eV)	Amorphization dose (dpa)[1]	T_c[2] (°C)	Reference
acmite	1.5 MeV Kr	2.1				Eby et al. 1992
albite	1.5 MeV Kr	1.6				Eby et al. 1992
almandine	1.5 MeV Kr	2.9				Eby et al. 1992
analcime	1.5 meV Kr	1.0	25	0.10		Wang SX et al. 2000a
anorthite	1.5 MeV Kr	2.7				Eby et al. 1992
andalusite	1.5 MeV Xe	1.9		0.34	710	Wang SX et al. 1998a
andradite	1.5 MeV Kr	2.2				Eby et al. 1992
apatite	1.5 MeV Kr	2.0	25	0.20	200	Wang et al. 1994
berlinite	1.5 MeV Kr	0.9				Bordes and Ewing 1995
biotite	1.5 MeV Kr	1.7			750	Wang et al. 1998b
britholite	1.5 MeV Kr	4.3	25	0.49	437	Wang and Weber 1999
	1.5 MeV Xe	1.9	25	0.35	535	Wang and Weber 1999
chromite	1.5 MeV Kr	**				Wang et al. 1999
coesite	1.5 MeV Kr	1			>>600	Gong et al. 1996b
cordierite	1.5 MeV Xe	0.8		0.09	710	Wang SX et al. 1998a
	1.5 MeV Kr	3.2				Eby et al. 1992
corundum	1.5 MeV Xe	**			-150	Wang SX et al. 1998a
diopside	1.5 MeV Kr	2.6				Eby et al. 1992
enstatite	1.5 MeV Xe	1.2		0.20	520	Wang SX et al. 1998a
fayalite	1.5 MeV Kr	1.6				Eby et al. 1992
	1.5 MeV Kr	1.2	25	0.15	>400	Wang et al. 1999
forsterite	1.5 MeV Kr	5.4				Eby et al. 1992
	1.5 MeV Kr	5.3	25	0.50	200	Wang et al. 1999
gadolinite	1.5 MeV Kr	2.3				Wang et al. 1991
grossular	1.5 MeV Kr	2.5				Eby et al. 1992
huttonite	0.8 MeV Kr	1.4	25	0.20	720	Meldrum et al. 1999
ilmenite	0.2 MeV Ar	**				Mitchell et al. 1997
kyanite	1.5 MeV Xe	2.1		0.37	1000	Wang SX et al. 1998
lepidolite	1.5 MeV Kr	1.5			1030	Wang et al. 1998b
leucite	1.5 MeV Kr	1.5				Eby et al. 1992
microlite	1.0 MeV Kr	1.5	25	0.30	570	Wang SX et al. 2000d
monazite	0.8 MeV Kr	1.5	25	0.44	180	Meldrum et al. 1997
	1.5 MeV Kr	2.3	25	0.30	160	Meldrum et al. 1996
$LaPO_4$	0.8 MeV Kr	11.9	25		60	Meldrum et al. 1997
mullite	1.5 MeV Xe	2.6		0.45	320	Wang SX et al. 1998a
muscovite	1.5 MeV Kr	1.9				Eby et al. 1992
	1.5 MeV Kr	1.5			830	Wang et al. 1998b
neptunite	1.5 MeV Kr	1.7				Wang et al. 1991
perovskite	1.5 MeV Kr	18.0				Smith et al. 1997
	0.8 MeV Kr	8.5	25	1.75	170	Meldrum et al. 1998
phenakite	1.5 MeV Kr	17.0	25	1.50	100	Wang et al. 1999
phlogopite	1.5 MeV Kr	2.6				Eby et al. 1992
	1.5 MeV Kr	1.7			730	Wang et al. 1998b

to 10^{-2} dpa/s). Dual beam irradiations allow one to simulate both the α-particle and α-recoil damage processes. Because most of the ions pass through the thin TEM foil, the effects of implanted impurities are minimal. However, the use of thin foils is also a disadvantage, as the large surface area to volume ratio of the sample can affect damage accumulation due to defect migration to the surface (Wang 1998). Still, there are limitations to establishing the exact fluence for the onset of amorphization and determining the proportion of amorphous material using TEM, as it is difficult to interpret images of mixed periodic and aperiodic domains through the thin foil (Miller and Ewing 1992). Irradiation experiments using bulk single crystals have also been performed (Wang 1998). Energetic ions, usually an inert gas, are injected into the crystal creating a damaged or amorphous near-surface layer, usually no more than 300 nm thick. The specimen can be examined *ex-situ* using Rutherford Backscattering (RBS) in the channeling mode, X-ray diffraction, optical, and cross-sectional TEM techniques. Potential effects from the extended free surfaces of the TEM foils are avoided; however, the experiments are more laborious and results may be affected by the implanted impurity ions.

The specimen temperature during irradiation can be controlled between -250°C and 800°C. This allows the damage accumulation and simultaneous thermally-assisted recrystallization processes to be studied. At low temperatures (-250°C), no thermally-assisted annealing can occur in the specimen, and the measured amorphization dose depends only on the cascade structure and evolution. At higher temperatures, thermally-assisted annealing at cascade boundaries occurs, and the amorphization dose increases.

Table 2. (continued)

Phase	Ion type and energy	Amorphization fluence ($\times 10^{14}$ ions/cm^2)	E_d (eV)	Amorphization dose (dpa)[1]	T_c[2] (°C)	Reference
pyrochlore	1.0 MeV Kr	1.6	25	0.30	430	Wang SX et al. 2000c
$Cd_2Nb_2O_7$	1.2 MeV Xe	0.7	25	0.22	700	Meldrum et al. 2000
$Gd_2Ti_2O_7$	1.5 MeV Xe	1.3	25	0.63	1030	Wang SX et al. 1999b
	1.0 MeV Kr	2.6	25	0.50	840	Wang SX et al. 1999b
	0.6 MeV Ar	7.6	25	0.60	680	Wang SX et al. 1999b
$Gd_2Zr_2O_7$	1.0 MeV Kr	**				Wang SX et al. 1999a
$Y_2Ti_2O_7$	1.0 MeV Xe	4.0	25	0.80	510	Wang SX et al. 2000b
pyrope	1.5 MeV Xe	1.9		0.23	600	Wang SX et al. 1998a
quartz	1.5 MeV Xe	0.5		0.21	1130	Wang SX et al. 1998a
	1.5 MeV Kr	1.4				Eby et al. 1992
rhodonite	1.5 MeV Kr	1.7				Eby et al. 1992
sillimanite	1.5 MeV Xe	1.9		0.27	1100	Wang SX et al. 1998a
spessartine	1.5 MeV Kr	2.8				Eby et al. 1992
spinel	1.5 MeV Kr	**				Bordes et al. 1995
spodumene	1.5 MeV Kr	2.4				Eby et al. 1992
thorite	0.8 MeV Kr	1.6	25	0.21	830	Meldrum et al. 1999
titanite	1.5 MeV Kr	2.0	15	0.30		Wang et al. 1991
	3.0 MeV Ar	4.0				Vance et al. 1984
tremolite	1.5 MeV Kr	1.9				Eby et al. 1992
wollastonite	1.5 MeV Kr	1.8				Eby et al. 1992
zircon	1.5 MeV Kr	4.7				Eby et al. 1992
	1.5 MeV Kr	4.8	25	0.51	830	Weber et al. 1994
	1.0 MeV Ne	5.0	Zr(80),Si(20),O(45)	2.50	770	Devanathan et al. 1998
	0.8 MeV Xe	2.7	Zr(79),Si(23),O(47)	0.30	1340	Meldrum et al. 1999
	0.8 MeV Kr	1.6	Zr(79),Si(23),O(47)	0.37	860	Meldrum et al. 1999
	0.6 MeV Bi	not given	Zr(80),Si(20),O(45)	0.50		Weber et al. 1999
	1.5 MeV Xe	2.0	15	0.88		Wang and Ewing 1992
	0.7 MeV Kr	2.8	15	0.86		Wang and Ewing 1992
	1.5 MeV Kr	4.8	15	0.90		Wang and Ewing 1992
zirconolite	1.5 MeV Kr	4.0				Smith et al. 1997
	1.0 MeV Kr	4.0			380	Wang SX et al. 1999b
	1.5 MeV Xe	2.0			440	Wang SX et al. 1999b

[1]Determined by *in situ* TEM at room temperature (RT) with ion beam penetrating through the specimen (Wang and Ewing 1992, Wang 1998).
** indicates that the specimen showed no sign of amorphization at fluences >1 x 10exp16 ions/cm2 and probably with Tc < RT.
[2]Critical temperatures calculated using the model of Weber et al. (1994).

A temperature is eventually reached at which the mineral cannot be amorphized. This is the critical temperature (T_c), above which recrystallization is faster than damage accumulation, and amorphization does not occur. T_c varies from near absolute zero for MgO (periclase) to >1000°C for minerals such as quartz.

In the ion beam experiments, the beam-current density is measured and converted to a flux, ion/cm^{2}s. The fluence (in ions/cm^2) experienced by the specimen is the flux multiplied by time. In many papers, the measured fluence is converted to an ion dose in units of displacements per atom (dpa). This is essentially a measure of the actual amount of damage (i.e. number of atomic displacements) that results from cascade formation, but it does not consider subsequent relaxation or recrystallization events. The fluence-to-dose conversion is:

$$D = CF/n \tag{5}$$

where D is dose in dpa, C is the number of atomic displacements caused by each ion per unit depth, F is the fluence, and n is the atomic density (e.g. $n = 9.08 \times 10^{22}$ atoms/cm^3 for zircon). The number of displacements caused by each ion is a statistical quantity that depends on the crystal structure, incident ion mass and energy, and the displacement energy (E_d) for atoms of the target (i.e. the energy required to permanently displace an atom from its lattice site). Monte Carlo computer simulations (e.g. Ziegler 1999) are usually used to estimate C. C varies inversely with E_d; since E_d is not known for most minerals, it is estimated. Zinkle and Kinoshita (1997) have provided a list of experimentally determined E_d values for several oxides. A reasonable E_d value for minerals can be estimated based on a weighted average of the oxides (Wang et al. 1998b).

In order to compare the amount of damage in natural minerals with the ion irradiated specimens, it is necessary to convert between ppm U (or equivalent uranium), α–decay events per mg, and the dose in dpa:

$$Dose(\alpha - decays/mg) = \left[\frac{[eU(ppm)] \cdot (\exp \lambda t - 1) \cdot x}{10^6 \cdot MW} \right] \cdot A, \tag{6}$$

$$Dose(dpa) = \frac{Dose(\alpha - decays/mg) \cdot MW \cdot n}{a \cdot A}, \tag{7}$$

where λ is the decay constant for U, t is time, x is the number of α–decays in the decay chain, A is Avogadro's number, MW is the molecular weight of the mineral, n is the average number of displacements per α–decay in the decay chain, and a is the number of atoms per formula unit. The n term is analogous to C in Equation (5) and is obtained using computer simulation. For most minerals, n is between 700 and 1000 atomic displacements per α-decay event.

RADIATION DAMAGE IN MINERALS

In this section, we summarize radiation damage data for minerals; however, we focus only on those minerals for which substantial amounts of data are available or which are of considerable geologic interest. We do not include a discussion of phases that contain actinides but are not metamict (i.e. AO_2: A = Zr: baddeleyite; Th: thorianite; U: uraninite).

Zircon

Zircon ($ZrSiO_4$; $I4_1/amd$, Z = 4) is a widely occurring accessory mineral that typically contains several hundred to several thousand ppm U and Th. The zircon structure is composed of cross-linked chains of alternating, edge-sharing SiO_4 and ZrO_8 polyhedra (Hazen and Finger 1979). Uranium and thorium may substitute in relatively small quantities for zirconium (>1 wt % UO_2 is rare). Despite the relatively low concentrations of actinides in zircon, it is often reported to be partially or fully metamict, reaching doses of 10^{19} α-decay events/g. A high-uranium (10 wt %) zircon has been identified in the Chernobyl "lavas," formed from melted nuclear fuel, zircaloy cladding and concrete (Anderson et al. 1993). Radiation effects in zircon have been investigated more widely than for any other natural phase. Recent interest is focused on the use of zircon as an actinide waste form (Ewing et al. 1995b, Ewing 1999).

Alpha-decay-event damage. The effects of α–decay damage on the structure of zircon were first systematically investigated by Holland and Gottfried (1955). Based on density, refractive index, and XRD measurements, the accumulation of radiation damage in zircon was described as occurring in three stages: (1) the accumulation of isolated defects causing an expansion of the unit cell; (2) the formation of an intermediate, polyphase assemblage that may include undamaged zircon; (3) final conversion of the bulk zircon into an X-ray diffraction amorphous state. The dose for amorphization of the Sri Lankan zircon was 1×10^{19} α–decays/g. However, Vance and Anderson (1972) found no evidence of Stage 2 in their study of a Sri Lankan zircon. Based on optical absorption measurements, they reported the occurrence of cubic or tetragonal ZrO_2 precipitates in radiation-damaged zircon, in both untreated and annealed grains. Subsequent studies confirmed that the amorphization dose for Sri Lankan zircon is between $8\text{-}10 \times 10^{18}$ α-decay events/g (Headley et al. 1982b, 1992; Murakami et al. 1991). Optical absorption results (Woodhead et al. 1991) suggested an amorphization dose as low as 4.8×10^{18} α–decays/g. XRD data clearly showed residual crystallinity at this dose (Murakami et al. 1991) and the sensitivity of absorption measurements to residual crystallinity was uncertain (Biagini et al. 1997, Salje et al. 1999). No evidence for the formation of ZrO_2 was found in these or later studies of the Sri Lankan zircon. Single crystals of zircon also can become locally amorphous within discrete high uranium zones (Chakoumakos et al. 1987, Nasdala et al. 1996).

The seminal study by Holland and Gottfried (1955) identified important unanswered questions, especially concerning the fine-scale microstructure and the mechanism of radiation damage accumulation in zircon. Murakami et al. (1991) used a combination of experimental techniques, including electron microprobe analysis, X-ray diffraction, density measurements, and transmission electron microscopy to systematically characterize the Sri Lankan zircons as a function of increasing α–decay event dose. They proposed three stages of damage accumulation: (1) the formation of isolated defects caused by the α-particle resulting in an increase in unit cell parameters, but with large domains of undamaged zircon remaining; (2) formation of amorphous zones caused by elastic interactions of the α-recoil nucleus resulting in "islands" of mainly distorted but still crystalline zircon; (3) complete overlap of the amorphous domains leading to the metamict state. High resolution TEM was used to confirm the microstructure of these three stages. Murakami et al. (1991) found that the density of the specimens decreased sigmoidally as a function of dose (i.e. slow decrease at first, followed by a rapid decrease in density over the range of $2\text{-}8 \times 10^{18}$ α–decays/g, followed by an asymptotic approach to the minimum density value). The corresponding volume change as a function of dose is shown in Figure 2. These data fit a "double overlap model," first proposed for zircon by Weber (1991), in which zircon becomes metamict as a result of defect accumulation until there is structural collapse, requiring the overlap of at least three collision cascades to achieve complete amorphization. Recent high-resolution X-ray diffraction analysis has shown that even highly damaged zircon still has a substantial degree of crystallinity, and the accumulation of the amorphous fraction is consistent with the direct-impact model (Ríos et al. 2000). The results of this new study imply that zircon becomes amorphous directly within the highly-damaged cores of individual displacement cascades. Most recently, the accumulation of

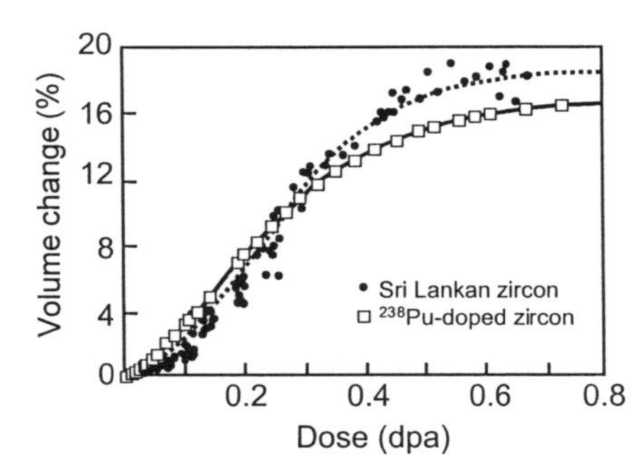

Figure 2. Dose dependence of the amorphous fraction for Pu-doped (squares) and natural Sri Lankan zircon (triangles), modified after Weber et al. (1994). The lines are fit using the double-overlap model for zircon. For low doses, the amorphous fraction is higher for the Pu-doped specimens than for the Sri Lankan zircon. This is due to an incubation dose of 0.05 dpa in the natural specimens.

damage has been described by a percolation model in which the amorphitization process is gradual (Trachenko et al. 2000). Raman spectroscopy has also been used to study the transition to the metamict state (Zhang et al. 2000a), and no evidence was found for the breakdown of zircon into ZrO_2 or SiO_2. However, on annealing of zircon that has experienced intermediate levels of damage, there is evidence in the Raman spectra (Zhang et al. 2000b) and infrared spectra (Zhang et al. 2000c) for the formation of an intermediate phase that disappears at higher temperatures.

The atomic structure of metamict or radiation-damaged natural zircon has been investigated using several techniques. The transformation to an aperiodic state has been confirmed by high-resolution electron microscopy of natural and ion-irradiated specimens (Headley et al. 1982b, Murakami et al. 1991, Weber et al. 1994). Despite the aperiodic nature of the specimens as viewed by TEM, short- or medium-range structure still exists. Farges and Calas (1991) and Farges (1994) used EXAFS to determine the shape and orientation of the cation polyhedra in metamict zircon. The coordination number of zirconium decreases from 8 in the crystalline state to 7 in metamict zircon, and the $\langle$Zr-O$\rangle$ distances decrease by 0.06Å. Heating results in the formation of Zr-rich "nanodomains" that could seed nuclei for the formation of the tetragonal ZrO_2 as described by Vance and Anderson (1972). Based on calorimetric measurements, Ellsworth et al. (1994) have also proposed the presence of ZrO_2-rich domains in highly damaged zircon. Nasdala et al. (1995, 1996) used a Raman microprobe technique to investigate the structure of metamict zircon on a scale of 1 to 5 microns and found that the environment around the Zr cation becomes increasingly disordered and that the SiO_4 units remain intact but are distorted and tilted. The overall structural picture of metamict zircon is one in which the long-range order is lost, and the medium and short range ordering is significantly modified. At least one of the main structural units of crystalline zircon (the ZrO_8 polyhedron) does not occur in the metamict state.

The energetics of radiation damage in the Sri Lankan zircon were investigated by Ellsworth et al. (1994) using high temperature drop calorimetry. The enthalpy of annealing at room temperature varies sigmoidally as a function of increasing radiation dose, and the large magnitude of the enthalpy of annealing plateau, -59 $\pm$ 3 kJ/mole, suggests that the damage is pervasive on the scale of Ångstroms, consistent with the loss of mid-range order. Similar to the early results of Vance and Anderson (1972), tetragonal zirconia was reported to crystallize after heating of the most radiation-damaged specimens. A schematic energy-level diagram, in which metamict zircon occupies the highest energy state, followed by tetragonal zirconia plus silica glass, baddeleyite plus glass, and finally crystalline zircon, was proposed to fit the experimental results. At higher temperatures, the energy barrier is overcome, and the amorphous or metamict zircon may recrystallize to ZrO_2 + SiO_2, or at higher temperatures, directly back to crystalline zircon.

Actinide-doping. Keller (1963) demonstrated that the actinide orthosilicates $PuSiO_4$, $AmSiO_4$, and $NpSiO_4$ are isostructural with zircon. Exharos (1984) presented the first results on radiation damage of zircon containing 10 wt % ^{238}Pu. Despite the large difference in the dose rate (8 orders of magnitude) between the ^{238}Pu-doped zircon and the Sri Lankan zircon (6.5 years vs. 570 Ma), the damage accumulation was nearly identical, as measured by variations in density and unit cell parameters with increasing dose (Weber 1990). There is, however, an incubation dose for the natural specimens as compared with the results for the ^{238}Pu-doped zircon, that is, for a given relative XRD intensity, the dose is only slightly higher for natural zircons. This difference decreases with increasing dose, and the dose for complete amorphization is the same in both cases. The differences at low dose were attributed to long-term annealing under the geologic conditions. Evidence for annealing of damage with increasing age was clearly shown by Lumpkin and Ewing (1988), who estimated the mean life of an α-recoil cascade in zircon is 400 Ma. Meldrum et al. (1998a) also reported clear evidence for an increasing α-decay event dose for amorphization of zircon with increasing age.

Volume swelling in actinide-doped zircon can reach values as high as 16% (Fig. 2). This swelling is attributed to the combined effects of unit-cell expansion at low dose and to the lower density of the amorphous phase at high dose. Weber (1990) found no evidence in the swelling data for recovery of the Sri Lankan zircon over geologic time. Based on the data for the Pu-doped samples, Weber (1990) proposed a double-overlap model to account for the accumulation of the amorphous phase in radiation-damaged zircon. Using advanced XRD techniques (7-circle diffractometer for measurement of diffuse diffraction maxima), Salje et al. (2000) have suggested, however, that the direct-impact model provides a better fit to the XRD data. One of the most remarkable data sets are the results of X-ray absorption and X-ray diffraction studies of ^{238}Pu and ^{239}Pu-doped (10 wt %) zircons that have reached a dose of 2.8×10^{19} α-decay events/g (Begg et al. 2000). Although the zircon was devoid of long-range order, it still retained a distorted zircon structure and composition over a length scale of 0.5 nm.

Ion-beam irradiation. Using ion beam irradiations, the mass and energy of the incident particles can be selected so as to investigate particular energy regimes during the damage process. The experiments can be completed at specific temperatures, thus allowing the study of kinetic effects on the accumulation and annealing of radiation damage.

In one of the first *in situ* irradiations of minerals, Wang and Ewing (1992b) irradiated zircon with 700 and 1500 keV Kr^+, 1500 keV Xe^+ and 400 keV He^+ at room temperature. These high energies were selected so that most of the ions pass through the thin TEM foil, significantly reducing the effects of implanted ions. The He^+ irradiation provided no indication of amorphization, implying that electronic energy loss processes are not important in the amorphization of zircon. Zircon was, however, amorphized at room temperature by Kr^+ and Xe^+ ions at ion fluences lower than 5×10^{14} ions/cm^2. This dose was reported to be equivalent to 0.9 dpa, assuming a displacement energy of 15 eV (see Eqns. 5-7 for the conversions between ion fluence, displacement dose, and α–decays/g). The actual displacement energy values for zircon are probably much higher. Recent computer simulations for zircon suggest displacement energy values of 79, 23, and 47 eV for Zr, Si, and O, respectively (Williford et al. 1999, Weber et al. 1998). The amorphization dose of 0.9 dpa estimated by Wang and Ewing (1992b) for zircon is probably high by a factor of three.

Weber et al. (1994) irradiated zircon with 1500 keV Kr^+ from -250 to +600°C. The amorphization dose increased as a function of temperature in two stages (Fig. 3). The increase in the amorphization dose with temperature was attributed to radiation-enhanced annealing concurrent with ion irradiation. The first stage (below room temperature) was tentatively attributed to the recombination of interstitials with nearby vacancies (close-pair recombination), and the second stage was attributed to irradiation-enhanced epitaxial recrystallization at the cascade boundaries. The critical temperature was estimated to be ~800°C. High-resolution imaging of the ion-irradiated zircon (Fig. 4) revealed micro-structures that are similar to those of α–decay damage in Sri Lankan zircon, supporting the three stage amorphization process of Murakami et al. (1991). Devanathan et al. (1998) found that zircon could not be amorphized by 800 keV electrons at -250°C, consistent with the earlier work of Wang and Ewing (1992b) that showed that electronic energy loss processes have no observable effect on the amorphization of zircon. Zircon can, however, be amorphized by 1 MeV Ne^+, and the critical temperature was estimated to be 480°C (Weber et al. 1999).

Meldrum et al. (1998b, 1999a) have irradiated synthetic zircon single crystals with 800 keV Kr^+ and Xe^+ ions. A two-stage process was observed. However, at elevated temperatures, zircon decomposed during irradiation into randomly oriented nanoparticles of cubic or tetragonal ZrO_2 embedded in amorphous silica. At temperatures between 600 and 750°C, the zircon initially becomes amorphous, but with increasing radiation dose, the

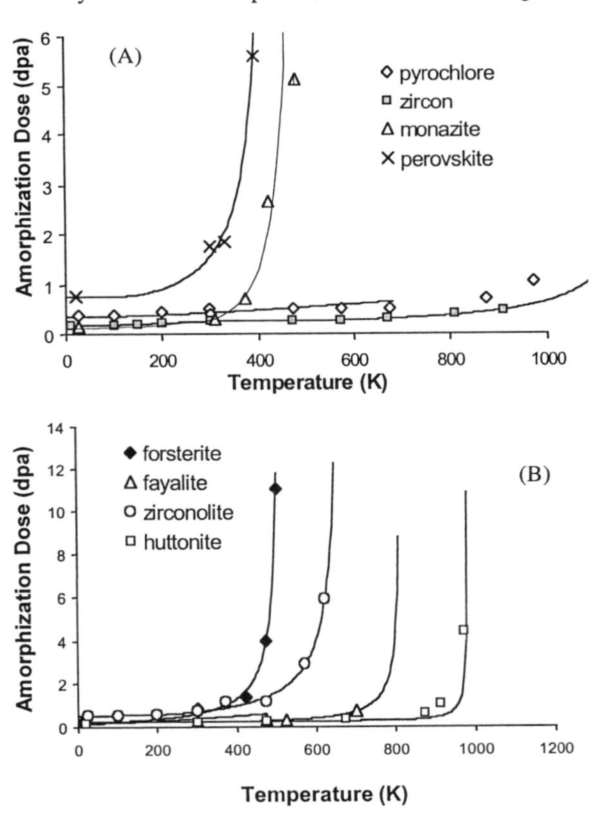

Figure 3. Temperature dependence of the amorphization dose for (A) pyrochlore, zircon, monazite and perovskite and for (B) fayalite, forsterite, zirconolite, and huttonite. The lines represent a least-squares fit to Equation (15).

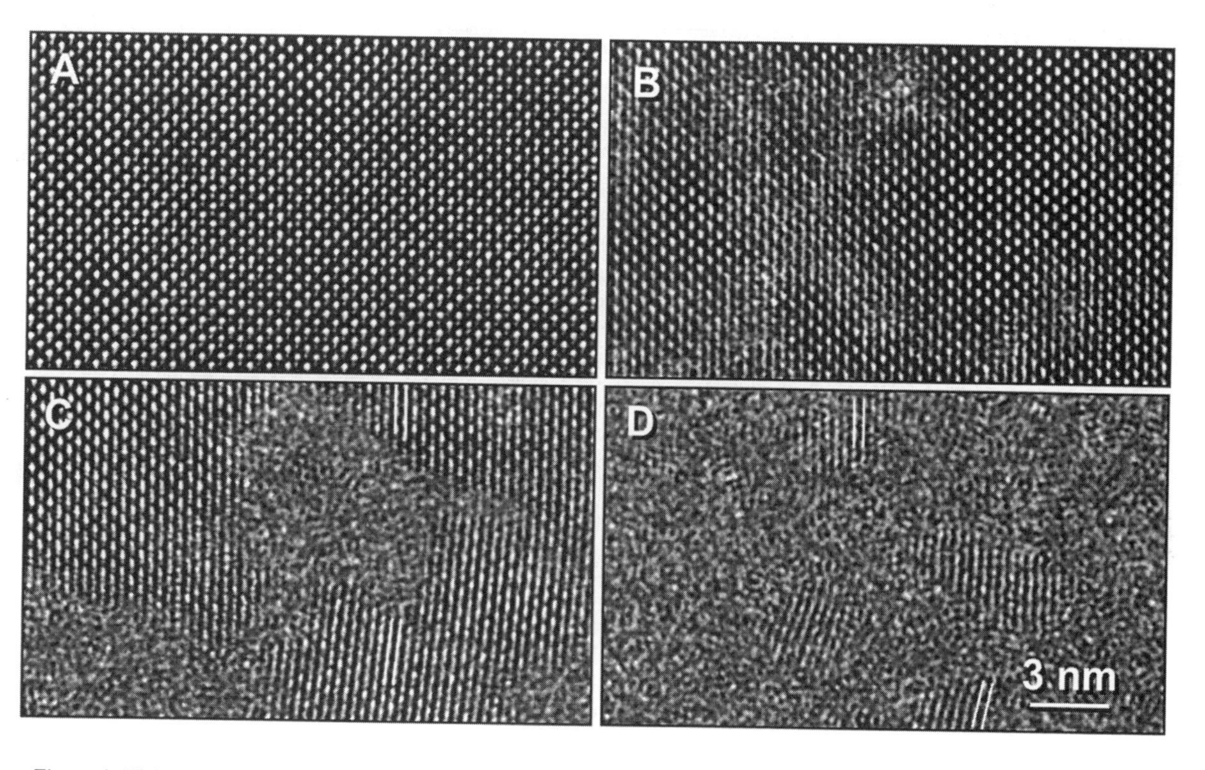

Figure 4. High-resolution transmission electron micrographs of zircon irradiated with 1500 keV Kr^+ ions at room temperature: (a) unirradiated, (b): 0.03 dpa, (c): 0.09 dpa, (d): 0.17 dpa.

amorphous zircon decomposes into component oxides. During irradiation at temperatures between 750 and 850°C, zircon decomposes directly to the oxides without first passing through the amorphous phase. The zircon phase diagram (Butterman and Foster 1967) shows that zircon melts incongruently to the component oxides above 1690°C. One interpretation of the data is that, at elevated temperatures, collision cascades in zircon "quench" slowly enough that crystalline ZrO_2 has sufficient time to nucleate in the "molten" cores of the collision cascades. Another interpretation is that the enhanced diffusion and defect mobility during irradiation provides the kinetic impetus to crystallize the ZrO_2. To date, however, this effect has only been observed for heavy ions, supporting the first interpretation. Nevertheless, further investigations must be carried out in order to understand the mechanism responsible for the phase decomposition. The results of Meldrum et al. (1999a) suggest that tetragonal ZrO_2 could form in natural zircon during periods of elevated temperature.

Weber et al. (1999) summarized the temperature-dependent *in situ* ion irradiation data obtained to date [1500 keV Kr^+ (Weber et al. 1994), 540 keV Pb^+ (one data point at -250K: Oliver et al. 1994), 1000 keV Ne^+ (Devanathan et al. 1998), 800 keV Kr^+ (Meldrum et al. 1999), 800 keV Xe^+ (Meldrum et al. 1999a)] and presented an additional curve for zircon irradiated with 600 keV Bi^+. The amorphization dose increased with increasing energy density (i.e. with increasing ion mass). This observation is consistent with a cascade overlap, defect accumulation process. The ion irradiation data are summarized in Table 2.

Monazite

Monazite (APO_4; A = lanthanides, Th, U, Ca) is a commonly occurring accessory mineral that contains relatively high concentrations of uranium and thorium (10 to 20 wt %). The monazite structure is composed of chains of alternating, edge-sharing PO_4 and AO_9 polyhedra parallel to the *c*-axis (Boatner and Sales 1988, references therein). The large, irregular AO_9 polyhedra can incorporate a large number of cations. Monazite often contains >5 wt % ThO_2 and has been reported with as much as 30 wt % ThO_2 (Della Ventura et al. 1996, Lumpkin 1998). UO_2 concentrations tend to be lower, typically around 1 wt %, with a maximum reported value of 16 wt % (Grammacioli and Segalstad 1978). Despite incorporating such high concentrations of radionuclides, monazite is almost never found in the metamict state. Frondel (1958), Ueda (1957) and Mitchell (1973) have reported "metamict" monazite, although in the latter two occurrences the degree of damage was minor. Clear evidence for radiation damage in natural monazite is limited to isolated domains within single crystals that contain high concentrations of Th, Si, and other impurities (Meldrum et al. 1998a). The difficulty of finding natural monazite grains with observable radiation damage precludes studies of naturally occurring α–recoil damage in monazite; thus, virtually everything known about radiation effects in monazite has been determined from ion-beam experiments.

Synthetic single crystals of $CePO_4$ have been irradiated at room temperature with 3 MeV Ar^+ ions (Karioris et al. 1981, 1982; Cartz et al. 1981). X-ray diffraction intensities were measured before and after a range of ion fluences from 0.2 to 3×10^{14} ions/cm^2. The decrease in the diffraction intensity with increasing fluence followed an exponential curve according to the equation $I/I_0 = \exp(-DMF)$, where F is the fluence and DM is the "damage cross section," a parameter related to the damaged area per ion. Monazite has a relatively high damage cross-section of 0.56 nm^2 (as compared with values of 0.33 and 0.27 nm^2 for zircon and huttonite, respectively). The experiments were performed at room temperature; hence, the effects of ion-beam damage also include simultaneous thermal recrystallization. The amorphization dose was 4×10^{14} ions/cm^2. The rarity of metamict monazite may be attributed to the low critical temperatures (~300°C) at which the radiation-damaged monazite completely recovered its atomic-scale periodicity (Ehlert et al. 1983, Karioris et al. 1982).

Radiation effects in monazite have been investigated using *in situ* ion irradiation monitored by transmission electron microscopy. The amorphization dose was measured as a function of temperature for monazite (Meldrum et al. 1996) and for a suite of synthetic rare-earth orthophosphates (A = La, Sm, Nd, Eu, Gd) (Meldrum et al. 1997b). The amorphization dose increased exponentially as a function of temperature (e.g. see Fig. 3 for natural monazite). The temperature at which the amorphization dose approached infinity (the critical amorphization temperature) can be estimated from the application of any of several models that fit the amorphization data. For the monazite-structure orthophosphates, the dose for amorphization is relatively constant below room temperature. The critical temperature increased regularly with increasing atomic number through the light rare-earths. For $LaPO_4$, the critical temperature is 60°C; for $GdPO_4$ the critical temperature reaches 210°C.

The critical temperatures for synthetic zircon-structure orthophosphates: xenotime (YPO_4), pretulite ($ScPO_4$), and $LnPO_4$ (Ln = Tb-Lu) are slightly higher than for the corresponding monazite-structure phases (Meldrum et al. 1997b). Critical temperatures increase with the mass of the A-site cation, from 200°C ($ScPO_4$) to 300°C ($LuPO_4$). At cryogenic temperatures, the amorphization dose is nearly independent of crystal structure for the orthophosphates. The relative amount of damage created in each displacement cascade must be similar for the two structures; however, the monazite structure is more readily recrystallized than the higher symmetry zircon structure.

Natural monazite has a higher critical temperature than expected on the basis of the average atomic number of the A-site cation. This suggests that the recrystallization process (but not the amount of damage produced) is slowed by high concentrations of impurities such as Th, Si, and Ca. The incorporation of impurities, including actinides, requires charge balance through a coupled substitution mechanism that may slow the recrystallization process. Additionally, the incorporation of large amounts of silicon may favor the zircon structure. Monazite- and zircon-structure silicates are much more readily amorphized at elevated temperatures (e.g. above room temperature) than their phosphate analogues (Meldrum et al. 2000a).

A computer simulation based on a binary collision approximation (BCA) code MARLOWE was used to investigate the evolution of collision cascades in monazite (Robinson 1983). As with the TRIM code (Ziegler et al. 1985), the trajectories of the ions are divided into distinct binary or two-body encounters. The TRIM code, however, does not account for crystal structure and the atoms are placed randomly; whereas, MARLOWE is designed to handle atomic-scale periodicity. Using MARLOWE, Robinson (1983) found evidence for amorphization via overlap of amorphous domains that formed within the cascade.

The *in situ* studies confirm the earlier investigations (Karioris et al. 1981, 1982) in that monazite is readily amorphized at room temperature. In contrast to these previous results; however, the dose required for amorphization at 20K was the same for the monazite- and zircon-structure orthophosphates, suggesting that the damage cross-sections are the same for the monazite and zircon structures (Meldrum et al. 2000a). The results in Figure 3 also show that the monazite-structure orthophosphates are impossible to amorphize at temperatures equal to or higher than 210°C. This increase in the amorphization dose is a result of recrystallization that occurs during ion irradiation. Thus, even at relatively low temperatures, radiation damage does not accumulate in monazite because of annealing that occurs during ion irradiation. Thus, metamict monazite is not rare because of its "resistance" to radiation damage, but instead because of rapid recrystallization of damage domains at relatively low temperatures.

Thorite and huttonite

The two polymorphs of $ThSiO_4$, thorite and huttonite, have the zircon and monazite structure types, respectively (Pabst and Hutton 1951, Taylor and Ewing 1978). Thorite commonly occurs in the metamict state (Pabst 1952); however, there are many examples of partially crystalline thorite (e.g. Speer 1982, Lumpkin and Chakoumakos 1988). Huttonite, the high-temperature, high-pressure polymorph of $ThSiO_4$ (Dachille and Roy 1964) is rare but invariably crystalline (Speer 1982). Pabst (1952) was the first to suggest that the structure-type of these polymorphs controls susceptibility to radiation damage.

Meldrum et al. (1999a) completed *in situ* irradiation experiments on synthetic thorite and huttonite single crystals. Specimens were irradiated with 800 keV Kr^+ ions over a range of temperature: –250°C to 800°C. Both phases amorphized at a low dose (0.16 and 0.18 dpa extrapolated to 0 K, respectively) and are comparable to zircon (0.17 dpa at 0 K) up to 400°C. At room temperature, the difference in amorphization dose was within experimental error. In the absence of high-temperature annealing, both phases are equally susceptible to irradiation-induced amorphization. However, at elevated temperatures the behaviors are different. Above 400°C, huttonite requires a higher dose to become amorphous. The critical temperature for huttonite is ~650°C; whereas, for thorite, 800°C. Recrystallization of metamict $ThSiO_4$ at temperatures > 850°C usually results in the huttonite structure (Pabst 1952). Because huttonite appears to be the primary phase observed after recrystallization, huttonite may also be preferentially formed as a result of irradiation-enhanced diffusion-driven crystallization that occurs during irradiation, resulting in a lower critical temperature.

Pyrochlore and zirconolite

Radiation effects in cubic pyrochlore and derivative structures, such as monoclinic zirconolite, have been widely investigated because they are actinide-bearing phases in Synroc, a polyphase, Ti-based, nuclear waste form (Ringwood et al. 1979, 1988). The general formula is $A_{1\text{-}2}B_2X_6Y_{0\text{-}1}$ ($Fd3m$, Z=8). Pyrochlore is an anion-deficient derivative of the MX_2 cubic fluorite structure in which the A- and B-site cations are ordered on the M-site and 1/8 of the X anions are absent. The BO_6 octahedra form a continuous corner-sharing network (McCauley 1980, Chakoumakos 1984). The (111) layers of pyrochlore have the (001) hexagonal tungsten bronze topology, consisting of three- and six-membered rings of octahedra. The large, 8-fold coordinated A-site cation and the seventh oxygen occupy channels in the (B_2O_6) network (Aleshin and Roy 1962). The crystal structure of pyrochlore can accommodate a wide variety of cations on either site, including actinides (Chakoumakos and Ewing 1985). Zirconolite, ideally $CaZrTi_2O_7$, is a derivative of the pyrochlore structure. Zirconolite has three polytypes. $CaZrTi_2O_7$ is monoclinic (2M); however, doping with actinides or other cations can result in trigonal (3T) or monoclinic (4M) polytypes. In nature, U and Th substitute into the A-site in pyrochlore and replace Ca in zirconolite. $UO_2 + ThO_2$ concentrations typically range from 1 to 30 wt % (Lumpkin and Ewing 1992, 1995, 1996; Lumpkin et al. 1999).

Alpha-decay event damage. Pyrochlore and zirconolite are often metamict (Lumpkin and Ewing 1988). Sinclair and Ringwood (1981) investigated natural zirconolite from several localities. They found that zirconolite transforms to a defect fluorite structure after a cumulative dose of 8×10^{18} α–decays/g. Specimens with more than 1.1×10^{19} α–decays/g were generally metamict, although some crystalline domains were identified by

TEM. Ewing and Headley (1983) showed that the crystallinity at high doses was probably due to electron beam-heating of the specimen in the TEM. In a detailed study using a variety of analytical techniques, Lumpkin et al. (1986) confirmed that the fully-damaged zirconolite has a random, three-dimensional network with no atomic periodicity extending beyond the first coordination sphere.

Lumpkin and Ewing (1988) studied a suite of pyrochlores with a wide variation in U- and Th-contents and a range of geologic ages. The α–decay dose for amorphization varied from $\sim 10^{18}$ to 10^{20} α–decay events/g, depending on the age of the specimen. Based on the broadening of the X-ray diffraction maxima, strain dominates the first half of the crystalline-to-metamict transition, reaching a maximum of 0.003, then decreasing to <0.001. Line broadening due to decreasing crystallite size dominates the latter half of the transition. For fully metamict samples crystallite size was <15 nm. The amorphization dose increased exponentially as a function of age and was described by an equation that includes a consideration of the annealing of the damaged areas: $D = D_0 exp(Bt)$, where D_0 is the extrapolated amorphization dose at zero time, t is time, and B is a recrystallization rate constant. Lumpkin and Ewing (1988) calculated a mean life for α–recoil tracks in pyrochlore of 100 Ma. The rate constant B has been estimated to be $\sim 1.7 \times 10^{-9}$/yr (Lumpkin et al. 1994). A similar method of analysis was used for a suite of zirconolites, and the amorphization dose increased from 10^{19} α–decays/g for specimens younger than 100 Ma to almost 8×10^{19} α-decays/g for samples older than 2 Ga. The recrystallization rate constant B was estimated to be lower by nearly a factor of two (1.0×10^{-9}/yr) for zirconolite as compared with pyrochlore (Lumpkin et al. 1994).

The microstructure and bonding geometry of amorphous pyrochlore and zirconolite have been extensively investigated using X-ray absorption spectroscopy. For uranium-bearing betafite, Greegor et al. (1984, 1985a,b; 1989) found that the <U-O> bond lengths decrease, and there is a disruption in the second nearest neighbor periodicity. The <Ti-O> bond lengths also decrease in metamict betafite, and the Ti-site has a reduced symmetry and lower average coordination number. In microlite, the nearest neighbor coordination sphere around Ta is also distorted, the mean second nearest neighbor <Ta–X> distances increase, and there is a wider distribution of second nearest-neighbor distances (Greegor et al. 1987). In general, these observations show a decrease in cation-oxygen bond length and an increase in the metal-metal distances. The offsetting effects of the smaller size of the cation coordination polyhedra together with the volume increase associated with the larger interpolyhedral distances may account for the relatively limited swelling observed in metamict pyrochlore. In metamict zirconolite, Farges et al. (1993) found evidence for: (1) 7-fold coordination for Zr; whereas, impurity Th remained in 8-fold coordination; (2) broadening of the distribution of <(Zr,Th)-O> distances; (3) a wider distribution of M-O-M bond angles (i.e. tilting of the cation polyhedra across bridging oxygens). Like pyrochlore, the increase in unit cell volume, as measured by X-ray diffraction in radiation-damaged zirconolite, is small (2%), but the expansion, as determined by density measurements, reaches values of 7 to 10% (Wald and Offermann 1982). Farges et al. (1993) modeled the change in volume and determined that most of the expansion can be accounted for by the changes in bond lengths and slight tilting of coordination polyhedra across bridging oxygens.

Actinide-doping. $Gd_2Ti_2O_7$ pyrochlore has been doped with ^{244}Cm and became amorphous after a cumulative α–decay dose of 2.98×10^{18} α–decays/g (Wald and Offermann 1982,Weber et al. 1985, Weber et al. 1986). Zirconolite doped with ^{244}Cm or ^{238}Pu required a cumulative dose of approximately 4.59×10^{18} α–decays/g to become X-ray diffraction amorphous (Wald and Offermann 1982, Clinard et al. 1984, Foltyn et al. 1985, Clinard 1986, Weber et al. 1986). X-ray diffraction results suggested cation disordering prior to amorphization (Clinard et al. 1984). The maximum irradiation-induced swelling was 5-6 vol.% (as compare with 18 vol.% for zircon). The dose for amorphization in the actinide-doped pyrochlores was lower than that calculated for natural pyrochlores older than 100 Ma, probably due to annealing of the natural samples over geologic time.

Ion-beam irradiation. Karioris et al. (1982) irradiated zirconolite with 3 MeV Ar^+ and found that the damage cross section was 4-7 times lower than that of monazite and zircon, implying that zirconolite actually retains less damage for each α–decay event. *In situ* ion irradiations were first employed by Ewing and Wang (1992) to investigate the crystalline-to-amorphous transition, mostly in pure synthetic powders of zirconolite or $Gd_2Ti_2O_7$ pyrochlore. Smith et al. (1997) compared the effects of 1.5 MeV Kr^+ irradiation on zirconolite and pyrochlore at room temperature. The transformation to the amorphous state was monitored by electron diffraction, and five stages were identified: (1) decrease in the intensity of zirconolite or pyrochlore superlattice maxima; (2) appearance of diffuse diffraction rings; (3) disappearance of zirconolite or pyrochlore superlattice maxima; (4) disappearance of some fluorite maxima; (5) disappearance of all remaining fluorite maxima and intensification of the diffuse rings. Meldrum et al. (2000c) showed that steps 2 and 3 are reversed for light-ion irradiation of pyrochlore.

The ion fluence for amorphization of both zirconolite and pyrochlore was 4×10^{14} ions/cm^2 (Smith et al. 1997). This is consistent with previous results for α–decay damaged natural crystals, for which the same amorphization dose (~10^{19} α–decays/g; corrected for geologic annealing) was obtained for both phases.

SX Wang et al. (1999b) found that the critical temperature of pyrochlore and zirconolite depends on the mass and energy of the incident ions. T_c for pyrochlore varies from 1027°C (1.5 MeV Xe^+) to 837°C (1 MeV Kr^+) to 677°C (0.6 MeV Ar^+). T_c for zirconolite decreases from ~ 437°C to 381°C for 1.5 MeV Xe^+ to 1.0 MeV Kr^+ irradiations, respectively (Fig. 9, below). At room temperature, the measured amorphization dose for zirconolite was approximately 1.5 times the measured dose for pyrochlore (SX Wang et al. 1999b), although this difference (which was not observed by Smith et al. 1997) may be close to the experimental error in this temperature range. As discussed above, the recrystallization rate constant B is higher (more rapid annealing) for pyrochlore than for zirconolite, nevertheless, zirconolite has lower T_c values in the ion beam experiments.

There have been a number of systematic studies of the effect of composition on the amorphization dose and the critical temperature of pyrochlore (SX Wang et al. 2000b, 2000d) and zirconolite (SX Wang et al. 2000c), as well as of the relationship between the order-disorder structural transformations (SX Wang et al. 1999b). Prior to amorphization, pyrochlore is transformed to the disordered fluorite structure; zirconolite is first transformed to a partially disordered cubic pyrochlore, followed by the disordered fluorite structure prior to amorphization. For both structures the critical temperature for amorphization increased with increasing ion mass due to the increasing size of the cascade (SX Wang et al. 1999b). The same relationship was observed for ion-beam irradiation studies of a series of pyrochlores in which the A-site cations were Gd, Sm, Eu and Y. The difference in amorphization dose and critical temperature is attributed to the different cascade sizes resulting from the different cation masses of the target. Based on a model of cascade quenching, the larger cascade is related to a lower amorphization dose and higher critical temperature (SX Wang et al. 1997, 1998a,b; 2000d). Compositional variations in the A-site can also affect the tendency to form stable amorphous domains, due to the glass-forming ability of particular compositions. As an example, Ca, as a network modifier, reduces the glass-forming ability of zirconolite, thus decreasing the critical temperature (SX

Wang et al. 2000c). The most dramatic effect of composition is seen in the solid-solution series between $Gd_2Ti_2O_7$ and $Gd_2Zr_2O_7$. The critical temperature decreases from ~1000°C for $Gd_2Ti_2O_7$ to below room temperature as the Ti is replaced by Zr in increasing concentrations. $Gd_2Zr_2O_7$ was not amorphized by 1.5 MeV Kr^+ irradiation at –250°C at doses as high as 10 dpa (SX Wang et al. 1999a, SX Wang et al. 1999c).

Meldrum et al. (2000c) recently irradiated single crystals of the pyrochlore $Cd_2Nb_2O_7$ with 320 keV Xe^{2+} and 70 keV Ne^+. The damage accumulation was measured as a function of fluence using RBS in the channeling mode. *In situ* irradiation experiments of the same phase were also completed using 1200 keV Xe^{2+} and 280 keV Ne^+. The amorphization dose at room temperature was approximately 40% higher for the bulk irradiation experiment than for the thin TEM specimens (all data were normalized to dpa). The results are consistent with the previous observations by SX Wang et al. (1999a,b,c) that the amorphization mechanism in pyrochlore depends on the mass of the incident ion.

Perovskite

Perovskite (nominally $CaTiO_3$) has an orthorhombic structure that is a cubic close-packed array of oxygen anions with one fourth of the oxygens regularly replaced by an A-site cation (e.g. Ca, Sr, Ba, K). The B-site cations (e.g. Ti, Ta, Nb) occupy octahedral sites that are defined by oxygen atoms. The orthorhombic deviation in $CaTiO_3$ is caused by an offset of Ti from the center of the octahedra. At temperatures above ~1200°C, $CaTiO_3$ is isometric. Depending on the A- and B-site cations, perovskite structure-types may undergo as many as four displacive phase transitions with increasing temperature: rhombohedral, orthorhombic, tetragonal, cubic. Some perovskite-structures (e.g. $KTaO_3$) are isometric at all temperatures. In natural perovskite, U and Th may substitute for Ca on the A-site in concentrations typically lower than 1 wt % UO_2+ThO_2. Highly altered specimens with concentrations as high as 6 wt % ThO_2 have been reported (Lumpkin et al. 1998).

Studies of radiation damage in natural perovskite are few. An early investigation of several natural perovskite crystals suggested that D_c is in the range of 0.3 to 2.6×10^{19} α–decays/g (Van Konynenburg and Guinan 1983). Smith et al. (1997; 1998) report that a perovskite from the Lovozero intrusion, Russia, was fully crystalline after an accumulated dose of 0.5×10^{19} α–decays/g. Lumpkin et al. (1998) reported on perovskite from localities ranging in age from ~300 to ~500 Ma with doses up to 1.2×10^{19} α–decays/g. Most specimens were highly crystalline, and none were metamict. A specimen from Bratthagen, Norway, (dose = 1.2×10^{19} α–decays/g) was "moderately damaged" based on the appearance of a single weak diffuse halo in the electron-diffraction pattern. Another highly altered and moderately damaged Bratthagen specimen had a calculated dose of 1.7×10^{19} α-decays/g. On the basis of these data, Lumpkin et al. (1998) estimated an amorphization dose of 3-6 × 10^{19} α-decays/g for natural perovskite.

Systematic temperature-dependent amorphization results on perovskite structure compounds were obtained by Meldrum et al. (1998c). Synthetic specimens of $CaTiO_3$, $SrTiO_3$ (tausonite), and several other perovskite-structure compounds not found in nature ($BaTiO_3$, $KNbO_3$, and $KTaO_3$) were irradiated with 800 keV Kr^+ ions. $CaTiO_3$ and $SrTiO_3$ required the highest ion dose to become amorphous at all irradiation temperatures, as compared with the other perovskite-structure compositions. The critical temperature for both $CaTiO_3$ and $SrTiO_3$ was approximately 150°C, similar to that of monazite (Fig. 3). These two perovskite compositions could not be amorphized by 280 keV Ne^+ irradiation at temperatures as low as –250°C (Meldrum et al. 2000c). Monte Carlo computer simulations (Ziegler 2000) suggest that Ne^+ ions are too light to produce dense collision cascades, resulting instead in the production of isolated point defects. The point defects may be sufficiently mobile in titanate perovskites that amorphization through a purely defect accumulation process does not occur. Electronic energy loss processes may also inhibit the amorphization process because of ionization-induced diffusion (Meldrum et al. 1998c).

Bulk implantation of Pb^+ ions to high fluences (~10^{15} ions/cm^2) were used to amorphize surface layers of $CaTiO_3$ and $SrTiO_3$ synthetic single crystals (White et al. 1989). The epitaxial recrystallization rate was faster for low-symmetry crystallographic directions at low temperatures, but at high temperatures, the regrowth rate was faster for the high-symmetry directions. Synthetic single crystals of $SrTiO_3$ were irradiated with 1.0 MeV Au^+ at room temperature, and the amorphization dose was measured by RBS-channeling (Thevuthasan et al. 1999; Weber et al. 1999). The results showed a strongly sigmoidal dependence of disorder with ion dose. These data were interpreted to suggest that heavy-ion, irradiation-induced amorphization of $SrTiO_3$ is achieved mostly by a defect accumulation mechanism, rather than by direct impact amorphization. The room-temperature amorphization dose, in units of dpa, was identical in the bulk single crystal irradiations of Weber et al. (1999) and the *in situ* experiments of Meldrum et al. (1998c).

Titanite

Titanite ($CaTiSiO_4O$) is composed of parallel chains of TiO_6 octahedra cross-linked by SiO_4 tetrahedra. Large cavities in the network of corner-sharing octahedra are filled with Ca that may be coordinated with 7 to 9 oxygen anions (Higgins and Ribbe 1976). U and Th can substitute for Ca, generally at a level of a few hundred to a few thousand ppm. Titanite is susceptible to metamictization (e.g. Cerny and Povondra 1972; Cerny and Sanseverino 1972). Higgins and Ribbe (1976) reported two metamict titanite grains that contain "significant" U and Th, but they did not give the age of the specimens. Based on their analytical totals, the U+Th concentration could not have exceeded a few thousand ppm.

Vance and Metson (1985) performed X-ray diffraction measurements on a suite of specimens from Canada (Cardiff mine and Bear Lake) and Norway (various localities). Although the age of the specimens was not given, the uranium content was reported to be between 0 and 1100 ppm. Damage ranged from "slight" to "moderate" in specimens with a measurable actinide content. The XRD results were interpreted to imply two separate types of damaged material, one rich in point defects and the other rich in "quasi-amorphous material." Fleet and Henderson (1985) also concluded that radiation damage in natural titanite consists of isolated point defects rather than discrete amorphous zones. Based on their results and the earlier work of Cerny and Povondra (1972), the α–decay dose for metamictization of titanite was estimated to be ~5×10^{18} α–decays/g (Vance and Metson 1985), nearly a factor of two lower than for Sri Lankan zircon. This difference was taken as evidence for the greater susceptibility of titanite to α-decay damage as compared with zircon. However, Lumpkin et al. (1991) observed that the mean diameter of α-recoil tracks is nearly the same for the titanite and zircon (~4 nm). An annealing study (Lumpkin et al. 1991) of titanite (300 to 700°C, N_2) showed a two-stage recovery process. Track diameter decreased at 400°C. An intermediate phase developed at 500°C, and nearly all tracks were epitaxially recrystallized. At 700°C, all tracks and the intermediate phase were gone.

Hawthorne et al. (1991) used a wide variety of analytical techniques (EMPA, TGA, powder XRD and IR, single crystal XRD, Mössbauer spectroscopy, ^{29}Si MAS-NMR, EXAFS/XANES spectroscopies and HRTEM) to determine the structure of α-decay damaged titanite. The metamictization process was proposed to begin with the formation of isolated defects caused by α-particles and amorphous domains caused by α-recoil tracks which, with increasing dose, overlapped to produce the metamict state. Fe^{3+} was reduced to Fe^{2+} with increasing α-decay dose. In a set of annealing experiments of radiation-damaged titanite at temperatures up to 1090°C, the "moderately damaged" titanites were completely recrystallized, but the most highly damaged titanite was not completely recrystallized. Based on the variety of analytical techniques used in this study, it is clear that the measurement of the residual crystallinity depends critically on the analytical method used.

Chrosch et al. (1998) found additional evidence for the formation of two separate phases in a titanite from Cardiff. Peak splitting was clearly observed in the X-ray diffraction patterns of highly damaged titanite. This was attributed to the presence of two phases, one consisting of highly strained titanite with larger d-spacings, the other consisting of apparently "normal," unstrained titanite. The first may be attributed to partial recrystallization in the cascade cores, or possibly to the presence of completely undamaged titanite in local, low-U regions.

Ion-beam-irradiation investigations of titanite are limited. Vance et al. (1984) irradiated synthetic titanite with 3 MeV Ar^+ at room temperature. Consistent with the work on the natural specimens, the damage cross section for titanite was determined to be 0.6 ± 0.2 nm^2, as compared to 0.33 ± 0.07 and 0.56 ± 0.13 nm^2 for zircon and monazite, respectively. This observation is consistent with the relatively low concentration of actinides (> 3000 ppm) in the highly damaged natural specimens. However, the data of Vance et al. (1984) suggests an amorphization dose of 4×10^{14} ions/cm^2 for 3 MeV Ar^+ irradiation of titanite powder at room temperature, comparable to the amorphization dose for zircon. Using RBS, Cherniak (1993) concluded that a room temperature ion fluence of 1×10^{15} ions/cm^2 of 100 kV Pb^+ amorphized the near-surface region of titanite single crystals. Clearly, the measured amorphization dose depends on the mass and energy of the incident ions, as well as on the measurement technique.

Apatite

Apatite, ideally $Ca_5(PO_4)_3(OH,F)$, is a widely occurring accessory mineral that incorporates minor quantities of U and Th that substitute for Ca. Silicate apatite (i.e. britholite) has the general formula $(Ce,Ca)_5(SiO_4,PO_4)_3(O,OH,F)$. Like monazite, natural apatite is only rarely reported to sustain radiation damage from α-decay events, although fission tracks are common (Fleischer 1975). The durability of natural apatite is well demonstrated at the Oklo natural reactors in the Republic of Gabon, Africa; there the apatite has retained its crystalline structure for nearly two billion years despite a significant ^{235}U enrichment and high fission-product content (Bros et al. 1996). Only natural apatites containing silica appear to be susceptible to amorphization over geologic time. Linberg and Ingram (1964) have reported a radiation-damaged SiO_2-rich apatite. The X-ray diffraction pattern of this specimen sharpened considerably on annealing. Gong et al. (1997) have described a 270 Ma natural britholite that has become metamict after a cumulative α-decay dose of 1.2×10^{19} α-decay events/g.

Apatite has been proposed as an actinide waste form (Bart et al. 1997); thus, extensive studies of radiation damage have been conducted using actinide-doping and ion beam irradiation (Weber et al. 1997). Radiation-induced amorphization has been observed in synthetic rare earth-containing silicate apatite, i.e. ^{244}Cm-doped $Ca_2Nd_8(SiO_4)_6O_2$ (Weber 1983) and ion-beam irradiated $Ca_2La_8(SiO_4)_6O_2$ (Wang et al. 1994, Weber and Wang 1994). Both conventional TEM results for the ^{244}Cm-doped (1.2 wt%) $Ca_2Nd_8(SiO_4)_6O_2$ (Weber and Matzke 1986) and HRTEM results for ion-beam-irradiated $Ca_2La_8(SiO_4)_6O_2$ (Wang et al. 1994, Wang and Weber 1999) have confirmed that amorphization occurred directly in the displacement cascades (Figs. 5 and 6). This is consistent with the direct impact model (Gibbons 1972) used to describe macroscopic swelling with increasing dose

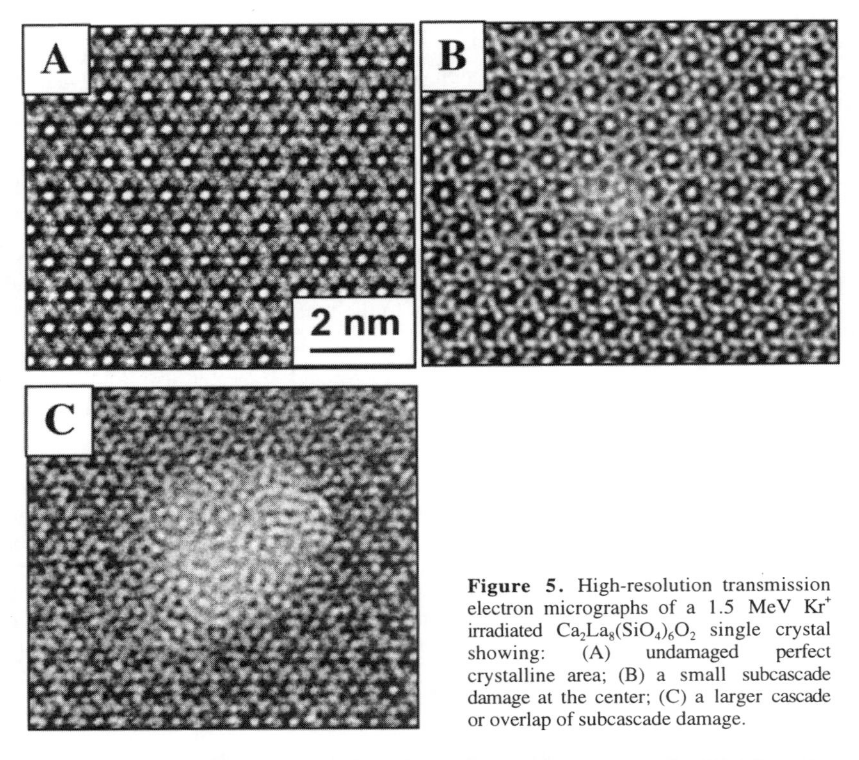

Figure 5. High-resolution transmission electron micrographs of a 1.5 MeV Kr^{+} irradiated $Ca_2La_8(SiO_4)_6O_2$ single crystal showing: (A) undamaged perfect crystalline area; (B) a small subcascade damage at the center; (C) a larger cascade or overlap of subcascade damage.

Figure 6. Cross-sectional high resolution transmission electron micrograph of an 800 keV Kr^{+}-irradiated $Ca_2La_8(SiO_4)_6O_2$ single crystal showing an amorphous track of subcascades created by the passage of a single ion.

in ^{244}Cm-doped apatite (Weber 1993). The critical temperatures of $Ca_2La_8(SiO_4)_6O_2$ under 0.8 MeV Ne^+ and 1.5 MeV Kr^+ irradiations have been determined to be ~90°C and ~440°C, respectively (Weber and Wang 1994). The Ne^+ ions created much smaller cascade volumes than those created by the Kr^+ due to their lower mass. The smaller amorphous volumes are more susceptible to epitaxial recrystallization due to their higher interface to volume ratio. This explains the lower critical amorphization temperature for the lighter-ion irradiations. Indeed, the typical size of amorphous domains observed by HRTEM for a fixed ion dose decreased with the decreasing ion mass (Wang and Weber 1999). HRTEM results have also shown a reduction in the size of amorphous domains with increasing temperature and electron dose during TEM observation. These results demonstrate the role of thermal and ionization energy deposited by the electrons in the epitaxial recrystallization of silicate apatite. Based on a model that considers the kinetics of the amorphization and the dose rate effect, Weber et al. (1997) have predicted that the amorphization rate in rare-earth silicate apatite containing 10 wt % ^{239}Pu will decrease rapidly at temperatures above 10°C, and complete amorphization will not occur for temperatures 28°C , even for periods >10^6 yr.

Natural F-apatite has been irradiated with 1.5 MeV Kr^+ (Wang et al. 1994) and 0.8 MeV Kr^+ (A. Meldrum, unpublished data). In both cases, amorphization was only observed below a critical temperature of ~200°C. This relatively low critical temperature is consistent with the absence of metamict apatite in nature and has been attributed to the low activation energy for recovery associated with fluorine mobility. Increasing the silica concentration toward the britholite end-member increases the critical temperature (i.e. silica–rich apatite may be amorphized at temperatures up to 437°C). This is similar to the case of monazite, for which an increase in silica content reduces the tendency of the material to recrystallize, resulting in higher critical temperatures (Meldrum et al. 1998a, 2000a). Amorphization was not observed in natural F-apatite under 200 keV electron irradiation (mainly ionizing radiation) despite the formation of a complex labyrinth of cavities and CaO precipitates (Cameron et al. 1992). As with the silicate apatite, ionization energy loss during electron irradiation of amorphous apatite causes rapid recrystallization (Meldrum et al. 1997a).

Olivine and spinel

Although olivine and spinel do not contain U or Th, these structure types have many technological applications in high radiation environments (e.g. Pells 1991, Ibarra 1996, Yu et al. 1994), such as the use of spinel as an inert matrix nuclear fuel (Boczar et al. 1997). Additionally, because there are so many data available on the structure and properties of olivine and spinel structure-types and compositions, this system is ideal for systematic studies of radiation effects.

$MgAl_2O_4$ spinel is not easily amorphized by ion irradiation. Using *in situ* ion irradiation, Bordes et al. (1995) found that, with increasing dose of 1.5 MeV Kr^+ at –250°C, the lattice parameter of spinel is reduced to one-half of its initial value. Continued irradiation to 5.4 dpa (E_d = 40 eV) produced no further structural changes and amorphization was not observed. The decrease in the lattice parameter was caused by disordering of the A- and B-site cations over octahedral and tetrahedral sites, while the ccp oxygen sublattice remained unchanged. If the spinel is pre-implanted with Ne^+ ions at –170°C, isolated amorphous domains were produced by a subsequent *in situ* Kr^+ irradiation. This has been attributed to the formation of "stable defects" produced during the Ne^+ pre-implantation. Devanathan et al. (1996a,b) subsequently reported that single crystals of $MgAl_2O_4$ become amorphous after implantation with 400 keV Xe^+ ions to a dose of 1×10^{16} ions/cm^2 at 100 K. Sickafus et al. (1995), Devanathan et al. (1996a), Zinkle (1995) and Zinkle and Snead (1996) have also investigated defect formation and disordering in spinel. The nature of the defects formed (e.g. size of dislocations loops) and their spatial density and location depend strongly on the irradiation conditions. Ionizing radiation dramatically decreases the rate of damage accumulation in spinel (Zinkle 1995). The extreme radiation resistance of $MgAl_2O_4$ has been attributed to the near-perfect ccp packing of the anion sublattice for which only minor structural relaxations are required to restore the oxygen anion periodicity (Wang et al. 1999).

The effect of chemical composition on the relative radiation resistance of several spinel phases ($MgAl_2O_4$, $FeCr_2O_4$, and γ-$SiFe_2O_4$) were investigated by Wang et al. (1995) using *in situ* ion irradiation. $FeCr_2O_4$ was more easily damaged than $MgAl_2O_4$. γ–$SiFe_2O_4$ was amorphized at a relatively low dose at all temperatures investigated. The differences in susceptibility to radiation damage were attributed to structural and bonding parameters, such as variations in Pauling ionicity and bond lengths, deviation from ideal packing and the Gibbs free energies. The γ-$SiFe_2O_4$ phase forms at high pressure and is metastable at ambient temperature and pressure. Consequently, recrystallization (thermal or in-cascade) may not occur because there is no thermodynamic driving force to recrystallize γ-$SiFe_2O_4$ at ambient pressure. In general, bonding and structural parameters can often be used successfully to predict the relative ranking of radiation resistance for isostructural compounds; however, many exceptions to the expected order occur for compounds that have different crystal structures.

Compounds with the olivine structure tend to be more readily amorphized than their thermodynamically stable spinel-structure analogues. For phases with the olivine structure, T_c increases from ~200°C for Mg_2SiO_4 to 300°C for Mg_2GeO_4, and to >400°C for Fe_2SiO_4. Be_2SiO_4 (which does not have the olivine structure) has the lowest value of ~50°C. These differences can be generally explained by variations in the structure and bonding parameters (Wang and Ewing 1992c); however, anomalous behavior has been noted for the fayalite-fosterite solid solution series (Wang et al. 1999). The generally lower radiation resistance of the olivine structure was attributed to its distorted hcp oxygen sublattice (Wang et al. 1999). The olivine structure does not form a perfect hcp anion array upon relaxation; whereas, for spinel, perfect ccp packing restores the anion periodicity.

MODELS OF RADIATION DAMAGE MECHANISMS

The mechanisms of radiation-induced amorphization have been investigated for more than three decades, and a variety of models have been developed to describe the processes that lead to radiation-induced amorphization of different types of materials. Early work was mainly based on the results for irradiated intermetallic compounds and semiconductors. In general, these models have been applied to ceramics with only minor modifications (Weber 2000). The models are of two types: the direct impact or cascade quenching models (Morehead and Crowder 1970, Gibbons 1972) and defect accumulation models (Gibbons 1972, Jackson 1988, Ziemann 1985, Ziemann et al. 1993, Titov and Carter 1996, Motta and Olander 1990, Okamoto et al. 1999).

Direct impact vs. defect accumulation models

The direct impact or cascade quenching model assumes that an amorphous domain forms directly within the core of a displacement cascade in a manner similar to the result of rapid quenching of a melt. Complete amorphization of a bulk specimen is achieved when the amorphous domains increase in abundance with increasing ion dose, eventually occupying the entire sample volume. In contrast, the point defect accumulation model assumes that the incoming particle produces isolated point defects, and amorphization occurs spontaneously when the local defect concentration reaches a critical value. A model based on thermodynamic melting for solid-state amorphization (Okamoto et al. 1999) is mainly based on the concept of defect accumulation followed by structural collapse to an aperiodic state. The direct impact model is based on a heterogeneous amorphization process, while the defect accumulation models are based on the assumption of a homogeneous process of defect accumulation followed by a phase transformation. Few ceramic materials are considered to undergo purely homogeneous radiation-induced amorphization due to defect accumulation, except under electron irradiation (Motta 1997).

In addition to these two conceptual models, there are several modified or composite models that have been developed in an attempt to better explain experimental observations. The more recent models are usually based on the concept of defect accumulation, but address the heterogeneous nature of the amorphization process, as has been observed in many semiconductors and ceramic materials by electron spin resonance (Dennis and Hale 1976) and HRTEM (Headley et al. 1981, Miller and Ewing 1992, Wang 1998). These models include the cascade-overlap model (Gibbons 1972, Weber 2000), defect complex overlap model (Pedraza 1986), nucleation and growth model (Campisano et al. 1993, Bolse 1998) or models that involve a combination of these mechanisms.

Most models of radiation-induced amorphization have developed characteristic mathematical descriptions of the accumulation of the amorphous fraction (f_a) as a function of dose (D) based on a set of differential equations (see Weber 2000, for a recent review).

The rate of amorphization by direct impact can be expressed as:

$$df_a/dt = \sigma_a \phi (1 - f_a) \tag{8}$$

where σ_a is the amorphization cross-section, ϕ is the ion flux and $(1 - f_a)$ is the probability of impact in an undamaged region. Integrating Equation (8), the change in amorphous fraction (f_a) with ion dose (D) is a simple exponential function:

$$f_a = 1 - exp(-\sigma_a D) \tag{9}$$

In contrast, the defect accumulation model or cascade overlap model predicts sigmoidal functions of f_a versus D, but with various curvatures. For example, the double-overlap model, which assumes that the formation of an amorphous domain requires the overlap of three cascades, is represented by the equation:

$$f_a = 1 - [(1 + \sigma_t D + \sigma_t^2 D^2/2) exp(-\sigma_t D)] \tag{10}$$

where σ_t is the total damage cross-section.

The f_a vs. D curves generated from the various models have been compared with the limited amount of experimental data to infer which type of model, e.g. direct impact vs. double overlap, is most applicable. As shown in Figure 7, the measured dependence of f_a on D for Cm-doped $Ca_2Nd_8(SiO_4)_6O_2$ is consistent with the predicted curve for the direct impact model; for ion-irradiated zircon, the measured dependence of f_a on D is consistent with the double-overlap model; $SrTiO_3$, the defect accumulation model (Weber 2000). It has been generally accepted that for heavy ion irradiations, especially at low temperatures, the amorphization process in ceramic materials is the result of amorphization within the cascade. For lighter ions (or electrons and neutrons) or at high temperatures, the dominant mechanism is considered to be cascade-overlap or defect accumulation.

There are, however, a number of issues that must be addressed prior to comparing experimental data to any of these models. First, the sensitivity and the accuracy of the techniques used in determining the amorphous fraction in a damaged material affects the measurement of the amorphous fraction (Webb and Carter 1979, Salje et al. 1999).

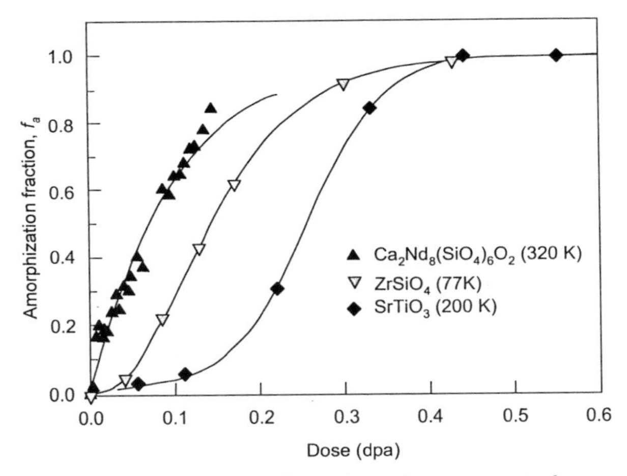

Figure 7. Comparison of experimental measurement of amorphous fraction accumulation as a function of dose with curves fit to various models for irradiation induced amorphization (modified after Weber 2000).

Second, the simple models used for predicting the f_a vs. D relationship have not considered the kinetics of defect recovery for recrystallization of the amorphous domains. Experiments have shown that defect recovery or cascade annealing can be substantial at room temperature, or even lower temperatures (Wang and Weber 1999, Bench et al. 2000). As an example, two recent studies of zircon can be used to support either the double-overlap (Weber 1990) or direct impact (Ríos et al. 2000) models.

Another major problem in applying the various models is that in order to match experimental results, some investigators have had to assume abrupt changes in mechanism (thereby applying a completely different set of equations) for slight changes in irradiation temperature, ion energy or target composition (Dennis and Hale 1978). This is illustrated by the results of studies of ion beam-induced amorphization of $Al_xGa_{1-x}As$. Using *in situ* on-axis bright-field TEM, Bench et al. (1991, 2000) have demonstrated direct impact amorphization in heavy ion-implanted GaAs at 30 K. They have reported that the probability of an individual ion impact forming an amorphous zone and the average dimensions of the zone increased with increasing ion mass, but decreased with increasing temperature. Researchers from the same research group at the University of Illinois (Turkot et al. 1996, Lagow et al. 1998), as well at the Australian National University (Tan et al. 1996), have reported that the ion dose needed to amorphize $Al_xGa_{1-x}As$ increases with the Al-concentration. However, the proposed amorphization mechanism had to be changed from a direct impact model at x 0.2, to a combination of a direct impact and point defect accumulation model for $0.2 < x < 0.83$ (or cascade overlap at $x = 0.6$), and changed again to a model of nucleation of amorphous domains on planar defects for $x \geq 0.83$. Similarly, because the amorphization efficiency per ion decreases with the increasing irradiation temperature for most ceramic materials, one must change to different amorphization mechanisms as the temperature is increased in order to fit the experimental data. The basis for not considering the direct impact model as the main cause of amorphization has mainly been due to the difficulty of identifying discrete amorphous volumes. This is, of course, a limitation of the experimental technique used in the analysis. Further, the dimensions of these domains decrease with increased dynamic recovery (recrystallization). In summary, the atomic-scale understanding of radiation-induced amorphization would be greatly improved by a generally applicable model of damage accumulation that considers amorphous domains of a range of dimensions, from isolated defects to the size of a collision cascade.

A modified model of direct impact amorphization

Recently, SX Wang et al. (1997, 2000e) have developed a model based on the direct-impact amorphization mechanism. This model can also be used to analyze the effects of variations in temperature, dose rate and ion mass. The basic assumptions of the model are: (1) each incident ion creates one or more displacement cascades or subcascades that may be quenched into an amorphous phase during the "quenching" stage of the cascade; (2) partial or complete recrystallization of the amorphous volume may occur depending on the atomic mobility and the rate of energy dissipation under irradiation; (3) because of the presence of a crystalline matrix surrounding the aperiodic cascade, only epitaxial recrystallization is considered; (4) the rate of epitaxial recovery is related to the amorphous fraction in the material such that an increasing amorphous fraction in the material reduces the probability for recrystallization. The initial equation for the description of the process is based on the lifetime of a single cascade. The formation of the individual cascade with a volume, V_0, removes all of the crystallinity within its boundary. A parameter, "crystallization efficiency," A, is defined to represent the volume fraction of the recrystallized shell within a single cascade in a crystalline matrix. If there are amorphous "cores" remaining in the displacement cascades after "quenching" the cascade, the amorphous fraction will accumulate with the increasing ion dose. This process is expressed by the following differential equation:

$$\frac{dV_c}{dN} = -mV_0 \cdot \frac{V_c}{V_T} + A \cdot mV_0 \cdot \frac{V_c}{V_T} + A(1-A) \cdot mV_0 \left(\frac{V_c}{V_T} \right)^3 \qquad (11)$$

where N is the number of ions; m is the number of individual subcascades created by one incident ion; V_T is the total volume of the damaged layer; V_c is the crystalline volume within V_T; V_0 is the volume of a subcascade. The first term on the right side of Equation (11) is the crystalline fraction lost due to cascade formation by a single incident ion. The second and the third terms are the volumes that have recrystallized during "quenching." Integrating Equation (11) from zero to N ions, and using the initial condition that $V_c/V_T = 1$ when $N = 0$, the amorphous fraction, f_a, is:

$$f_a = 1 - \frac{V_c}{V_T} = 1 - \frac{1}{\sqrt{A + (1-A) \cdot exp\left(\frac{mV_0}{V_T} \cdot 2(1-A)N \right)}} \qquad (12)$$

Using D for fluence (ions/cm^2), the amorphous volume fraction is:

$$f_a = 1 - \frac{1}{\sqrt{A + (1-A) \cdot exp\left(\frac{mV_0}{h} \cdot 2(1-A) \cdot D \right)}} \qquad (13)$$

where h is the sample thickness (assuming $h <$ ion range). The term mV_0/h is the cross-sectional area of an ion track. Equation (13) can be simplified by combining $[D \cdot (mV_0/h)]$ as a normalized radiation dose, D_n, which represents the total number of atoms in all damaged regions (or cascades) divided by the total number of atoms in the irradiated volume. The resulting expression for the amorphous fraction as a function of ion dose is:

$$f_a = 1 - \frac{1}{\sqrt{A + (1-A) \cdot exp[2 \cdot (1-A) \cdot D_n]}} \tag{14}$$

The amorphous fraction as a function of D_n is plotted for different crystallization efficiencies in Figure 8. For small A, f_a is a simple exponential function. In fact, f_a = 1 - $exp(-D_n)$ when A = 0 in Equation (14). This is the Gibbons' direct impact model (Gibbons 1972) which does not consider epitaxial recrystallization around the shell of the cascade. For large A, the f_a vs. D curve is sigmoidal, and the sigmoidal character of the curve becomes more pronounced when A approaches 1. When A = 1, f_a is zero; the amorphous domains are fully recrystallized after the cooling phase of the cascade, and the material cannot be amorphized. The f_a-D curves in Gibbons' overlap model with various numbers of overlaps can be closely matched by Equation (14) using different A values. Because A represents the efficiency of recrystallization, it varies with temperature, the properties of target materials, as well as the cascade size which is related to the mass and energy of the incident particle. The relationship of the crystallization efficiency, A, to material properties is not an explicit part of this model, but the relationship must be determined before the model can be used to predict the amorphization behavior of specific materials. In contrast to the discontinuous changes required in the multiple-overlap model that change the number of terms in the differential equation, the A value can be changed continuously with the variations in the composition of a particular structure type. The different values of A incorporate the temperature dependence of the amorphization dose. Many of the experimental results previously considered illustrative of multiple-overlap models or defect-accumulation models can now be described by this direct-impact model.

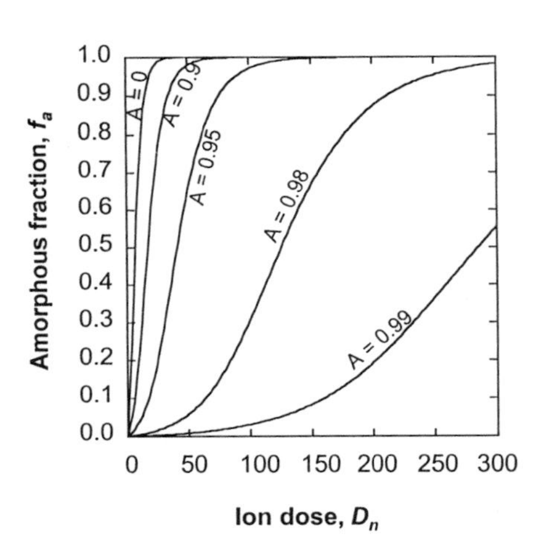

Figure 8. Amorphous volume fraction, f_a, as a function of normalized ion dose for different values of crystallization efficiency, A..

Temperature-dependence of the amorphization dose

The dose required for amorphization is a function of the kinetics of simultaneous dynamic recovery processes. The recovery process is accelerated at elevated temperatures and, in many cases, is greatly increased by radiation-enhanced defect migration. These simultaneous recovery processes may be associated with defect recombination or annihilation, epitaxial recrystallization at crystalline-amorphous interfaces (Carter and Nobes 1991), or nucleation and growth recrystallization in the bulk of the amorphous state. For any crystalline solid, there is a critical temperature, T_c, above which the rate of amorphization is less than the rate of recovery, thus amorphization cannot occur. However, T_c also depends on the energy and mass of the incident beam, as well as the dose rate.

The amorphization kinetics associated with previous models have recently been reviewed by Weber (2000). The relationship between temperature and dose to achieve a similar amorphous state can be expressed by the equation:

$$ln[1 - (D_0/D)^{1/m}] = C - E/nkT \tag{15}$$

where D_0 is the dose required to achieve a specific amorphous fraction at 0 K, E is an activation energy for the recovery process, C is a constant dependent on ion flux and damage cross section, and both m and n are model dependent order parameters. Although expressions in the form of Equation (15) often provide excellent fits to the observed temperature dependence of the critical amorphization dose for both semiconductors and ceramics (Fig. 3), the activation energies determined from such fits to experimental data are often low, which may reflect the almost athermal nature of the irradiation-assisted processes (Weber 2000).

Meldrum et al. (1999) recently derived an expression to describe the relationship between amorphization dose and temperature based on a composite model involving both direct impact and close-pair recombination processes. This model provides more realistic values for the activation energies of thermal recovery processes, but the curves generated from the expression predict a sharp curvature near T_c.

The temperature dependence of amorphization dose can also be derived by including the kinetics for thermal recovery in the crystallization efficiency parameter, A, which can be written as (SX Wang 1997, SX Wang et al. 2000e):

$$A = A_0 \cdot \exp[-E_a/(kT)]. \qquad (16)$$

where A_0 is a pre-exponential constant; E_a is the activation energy for dynamic annealing of a cascade; k is the Boltzmann's constant; T is the sample temperature. Based on a model of cascade quenching and recrystallization, an approximate relation between T_c and other parameters may be derived:

$$T_c \cong T_m - (T_m - T_g) \cdot R_{cryst}/(r_0 \cdot B) \qquad (17)$$

where T_g is the glass-transition temperature; R_{cryst} is the crystallization rate; r_0 is the subcascade radius; B is a constant related to thermal diffusivity. The expression shows several important relationships: (1) for materials that are good glass-formers (larger T_g, lower crystallization rate), T_c is larger; (2) smaller cascade size results in a lower T_c; (3) T_c is higher for materials with a high crystallization rate.

The temperature dependence of the amorphization dose can be easily obtained by solving for D in Equation (13). The simplified solution is:

$$D_c = \frac{D_0}{1 - exp\left[\frac{E_a}{k}\left(\frac{1}{T_c} - \frac{1}{T}\right)\right]}. \qquad (18)$$

Equation (18) is identical to that derived by Weber et al. (1994).

The effects of ion mass and energy

Because the nuclear stopping power is larger for heavier incident ions, the cascade size generally increases with ion mass. A heavier-ion irradiation leads to a higher T_c for a fixed target material due to a smaller value of A associated with a larger cascade size (i.e. the surface area available for epitaxial recrystallization decreases relative to the volume of the damaged domain). This relationship is clearly shown in Equation (17) and also demonstrated in Figure 9A by both experimental data and curves fit to experimental data (SX Wang et al. 1999b). In Figure 9B, the reciprocal of cascade radius, r_0, has an approximately linear relationship to T_c, as predicted by Equation (17). The relative cascade radii are derived from the square root of the displacement cross sections, which were generated using the Monte Carlo code TRIM96 (Ziegler et al. 1985).

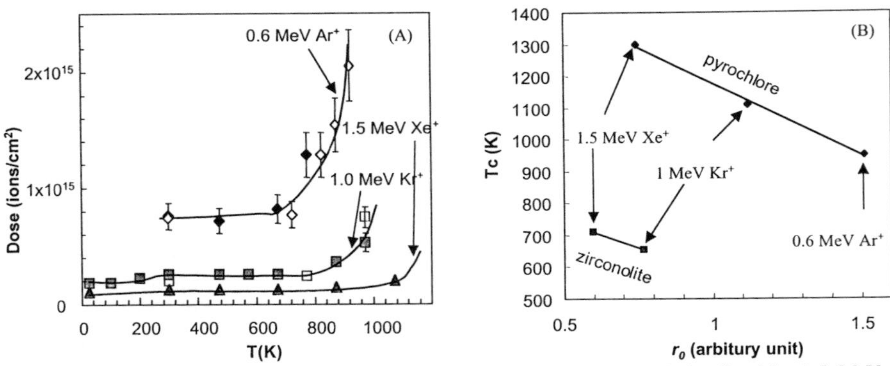

Figure 9. (A) Temperature dependence of amorphization dose of $Gd_2Ti_2O_7$ irradiated by 1.5 MeV Xe^+, 1 MeV Kr^+ and 0.6 MeV Ar^+, (B) The trend of T_c for $Gd_2Ti_2O_7$ changes with the cascade size, r_0, for different ions.

The effect of ion energy is reflected both in cascade size and activation energy. Ions of higher energy usually lose more energy by ionization (Fig. 1), and this may enhance the annealing kinetics. Thus, one possible effect of increasing energy is to decrease E_a. This has the effect of increasing the value of A. Also, when the ion energy is increased, ion range increases. Individual cascades tend to split into multiple smaller subcascades. This will increase the crystallization efficiency due to the greater surface area between the cascade and crystalline matrix. Thus, in most cases, higher energy particles will result in a larger A-value. This leads to a higher amorphization dose and a more distinctly sigmoidal shape for the f_a-D curve.

SUSCEPTIBILITY TO RADIATION-INDUCED AMORPHIZATION

A number of criteria have been proposed for the evaluation of the susceptibility of a material to ion irradiation-induced amorphization: ionicity (Naguib and Kelly 1975); crystallization temperature (Naguib and Kelly 1975); structural topology (Hobbs 1994); bonding and structure (SX Wang et al. 1998b).

Naguib and Kelly (1975) were among the first to attempt a systematic evaluation of the susceptibility of phases to ion beam-induced amorphization. They proposed two empirical criteria: bond-type (ionicity) and a temperature ratio, T_{cryst}/T_m (crystallization temperature of fission tracks: melting temperature). Although this approach provided a qualitative grouping of materials according to their susceptibility to radiation-induced amorphization, it failed when extended to more complex ceramics, such as minerals.

Hobbs (1994) evaluated the susceptibility of materials to irradiation-induced amorphization based on a consideration of structural topology. This approach is derived from the theory of glass formation that emphasizes the structural geometry of the crystalline equivalent to the glass composition (Cooper 1978, Gupta 1993). The degree of structural freedom of a phase is determined by its topology. The crystal structure can be described by the connectivity of rigid identical structural units, called polytopes. The term "polytope" generalizes the concept of the geometric representation of a coordination unit, such as a polygon or polyhedron to a space of arbitrary dimension (Gupta 1993). The ease of amorphization was related to the structural freedom, f, calculated by (Gupta 1993):

$$f = d - C\cdot[\delta - \delta\cdot(\delta + 1)/2V] - (d-1)\cdot(Y/2) - [(p-1)\cdot d - (2p-3)]\cdot(Z/p), \tag{19}$$

where d is the number of degrees of freedom (= 3), C is the connectivity, which is the number of polytopes common to a vertex, δ is the dimensionality of the structural polytope, V is the number of vertices of the structural polytope, Y is the fraction of edge-sharing vertices, and Z is the fraction of vertices sharing p-sided faces.

Highly constrained structures tend to remain crystalline, and marginally- or under-constrained structures tend to adopt aperiodic configurations. One limitation of this approach is that it does not consider chemical effects, such as bond-type and bond-strength. A number of studies have shown that isostructural phases exhibit large differences in their susceptibility to amorphization. Olivine and spinel structure-types (Wang et al. 1999) and monazite and zircon (Meldrum et al. 1997b) show significant differences in their susceptibility to amorphization even among structure-types that are the same or closely related. Nevertheless, the topologic criterion works for many simple compounds, as it is indicative of the tendency toward glass-formation.

SX Wang et al. (1998) have developed a parameter, S, to address the susceptibility to amorphization for oxides based on the analysis of glass-forming ability. The S parameter combines three factors: the structural connectivity, bond strength, and thermodynamic stability. The use of the three factors is based on a consideration of their affect on the tendency of the damaged regions to recrystallize. The S parameter is defined as:

$$S = f(T_s) \cdot \sum_i x_i \cdot G_i \cdot F_i^c , \tag{20}$$

in which,

$$G_i = \frac{1}{C_i} \cdot \left(\frac{n_{sh}}{n_T} \right)^b , \tag{21}$$

$$F_i = \frac{\left(z_{cation} / N_{cation} \right) \cdot \left(z_{anion} / N_{anion} \right)}{a_i^2} , \tag{22}$$

$$f(T_s) = \frac{1}{T_s^d} , \tag{23}$$

where T_s is the phase transition temperature (K), x_i is the relative abundance of polytope i in the structure; C_i is connectivity, which is the number of polytopes shared at one anion and equals the coordination number of the anion; n_{sh} is the number of shared polytopic vertices; n_T is the total number of vertices of a polytope; z_{cation} and z_{anion} are electrostatic charges of cation and anion; N is the coordination number; a_i is the cation-anion distance (Å), and b, c and d are constants (b = 2; c = 0.3; d = 0.1). These equations can be rearranged into a condensed formula (SX Wang et al. 1998b):

$$S = 100 \times \frac{1}{2} \times \frac{1}{T_s^{0.1}} \cdot \sum_i \left(x_i \cdot \frac{\left(\frac{Z_i}{N_i} \right)^{1.6}}{a_i^{0.6}} \right) , \tag{24}$$

in which i is the number of different polyhedra, Z_i is the charge of the cation; N_i is the coordination number of the cation; a_i is the cation-oxygen distance (Å); T_s is the phase transition temperature (K).

The S parameter provides a significant improvement over criteria that only consider the topology of a structure. However, this approach still, in some cases, fails to predict which materials are susceptible to irradiation-induced amorphization (SX Wang et al. 1999a). The

S value, however, can be improved by additional experimental data. For phases in MgO-Al_2O_3-SiO_2 system, the S value was found to be proportional to $1/(T_m - T_c)$ (SX Wang et al. 1998b). This suggests that the S value is related to the crystallization efficiency, A.

GEOLOGIC APPLICATIONS

Kulp et al. (1952) proposed that the amount of radiation damage in zircon, as well as other minerals, could be used for geologic age-dating. This motivated the study of radiation effects in zircon by Holland and Gottfried (1955), as well as the later studies by Deliens et al. (1977) who used infrared absorption and Makayev et al. (1981) who used X-ray diffraction to estimate the geologic ages of zircon. The major difficulty with this approach has been the effect of long term thermal annealing on the final damage state.

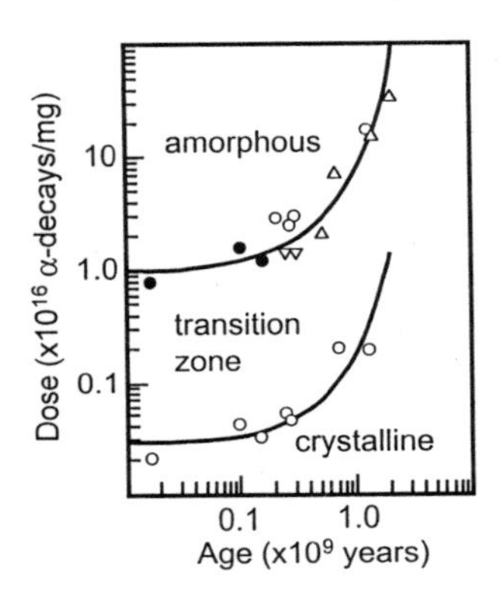

Figure 10. Variation of α-decay dose as a function of age for natural pyrochlore (modified after Lumpkin and Ewing 1988). The lower boundary marks the onset of observable radiation damage, as determined by transmission electron microscopy. The upper curve represents complete amorphization. Open circles: specimens from a single locality which accurately bracket the α-decay dose for amorphization, closed circles: localities for which the amorphization dose was estimated from low-dose specimens, upward triangles: localities for which complete amorphization was not observed, downward triangles: localities which had only amorphous specimens.

Clear evidence for thermal annealing under ambient conditions has been demonstrated for natural pyrochlores (Lumpkin and Ewing 1988). The α-decay dose for metamictization increased exponentially with the age of the specimen (Fig. 10). The upper curve represents the minimum α-decay dose for metamictization of pyrochlore, as measured by TEM. The lower curve is the minimum dose for the onset of amorphization. For specimens of greater age, metamictization required an apparently higher α-decay dose because of thermal annealing under ambient conditions. Based on a model of α-recoil-track fading, the mean life of an α-recoil track has been estimated: pyrochlore (100 Ma); zircon (400 Ma) and zirconolite (400 Ma) (Lumpkin and Ewing 1988, Lumpkin et al. 1998). Eyal and Olander (1990) estimated a maximum time for complete annealing of α-recoil tracks in monazite of only 1 Ma, further evidence supporting rapid recrystallization kinetics of monazite.

For ion-beam data that are collected as a function of temperature, damage accumulation can be modeled as a function of increasing dose, and the critical temperature determined (Fig. 3). At any given temperature, the amorphization dose depends on cascade evolution and on time. If a given dose is obtained over longer times, the residual damage can be annealed due to thermal diffusion. At temperatures well below T_c, thermal annealing is too slow to be observed in the laboratory; but over geologic time, this process may become important. The data from ion beam irradiation experiments can be applied to natural samples by modeling the effect of long-term thermal annealing on the critical temperature (Meldrum et al. 1998a, 1999a). Effectively, T_c will be lower if there is more time for thermal recrystallization to occur.

Meldrum et al. (1998a) have modeled the crystalline-to-metamict transition using the following equation:

$$N_c(ppm) = \left[\frac{6x10^6 \cdot D_0(dpa)}{\left[1 - e^{C(1-T_c/T)}\right] \cdot (e^{\lambda \cdot t} - 1) \cdot n \cdot x} \right] \quad (25)$$

N_c is the radionuclide content required to cause the metamict state, in ppm uranium, D_0 is the extrapolated ion-beam amorphization dose at 0 K, C is E_a/kT_c from the model of Weber et al. (1994), n is the average number of α-decay events in the decay chain, and x is the number of atomic displacements per α-decay event. For a given temperature, a specimen containing a uranium concentration above N_c should be metamict.

The effects of temperature are incorporated into the left-hand exponent in the denominator. The equation uses the model of Weber et al. (1994) to account for the increase in amorphization dose with increasing temperature. No explicit meaning is attached to the model, rather it simply represents a function fitted to the ion beam data. Long-term, diffusion-driven recrystallization is accounted for by considering the effect on T_c (Meldrum et al. 1999b):

$$T_C = \frac{E_a}{k \cdot ln\left[\frac{A \cdot f_a}{P(1-\varepsilon)\cdot(1-f_a)}\right]} \quad (26)$$

where A is the lattice vibration frequency (~10^{13} Hz), E_a is the activation energy for the epitaxial thermal recrystallization of amorphous zones, f_a is the amorphous fraction (f_a = ~0.95 for electron diffraction: Miller and Ewing 1992), ε is the experimentally determined ratio of displacement damage that recovers during Stage-I and Stage- II recrystallization processes [$\varepsilon = (D_{O(II)}-D_{O(I)})/D_{O(II)}$]; it is zero if there is only a single stage of recrystallization in the amorphization curves (Table 3), and k is the Boltzmann constant. With decreasing dose rate P, T_c decreases because more time is needed to achieve a given α-decay dose (Fig. 11). The decrease in the effective critical temperature is caused by gradual recrystallization of α-recoil damage over geologic time. In natural minerals, the dose rate will depend on the equivalent uranium concentration (approximately equal to [U + 0.3 Th], in ppm).

The dose rate for natural specimens gradually decreases due to radioactive decay. This has a limited affect on the calculated value of T_c. For example, a zircon grain with 1,000 ppm uranium initially will experience an initial dose rate of 5.2×10^{-18} dpa/s. After 500

Table 3. Constants used in Equations 25-27.

Mineral	D_0 (dpa)	C [1]	T_c [1] (°C)	x	e [2]	B (g) [3]
zircon	0.16-0.17	0.66-1.10 (1.10)	230-360 (230)	700-900 (750)	0.68	5.8×10^{-19}
thorite	0.08-0.19 (0.15)	4.2	260	725	0.4-0.57 (0.4)	na
huttonite	0.18	3.4	220	730	0.47	na
monazite	0.3 - 0.4 (0.3)	1.1-2.0 (2.0)	-40	800-900 (850)	0	na
pyrochlore	0.37	0.19	3204	1350	0	na

[1] C is calculated using least squares to determine for E_a and ln(1/ϕστ) in the model of Weber et al. (1994). C is E_a/kT_c.

[2] The constant ε is calculated using the model of Meldrum et al. (1999). This term models the discontinuity in the the amorphization dose for materials that show more than one stage in the ion-beam amorphization data.

[3] The mass of damaged material per cascade, B, has been estimated only for zircon. Values for other minerals are probably similar within a factor of two.

[4] Assumes 20,000 ppm U.

Note: in many cases a range of values is given, owing to variations in the ion beam data, mineral composition, or to different model fits. The number in parantheses is the recommended value based on a comparison with α-decay- damaged minerals or for a composition that is closest to those compositions found in nature.

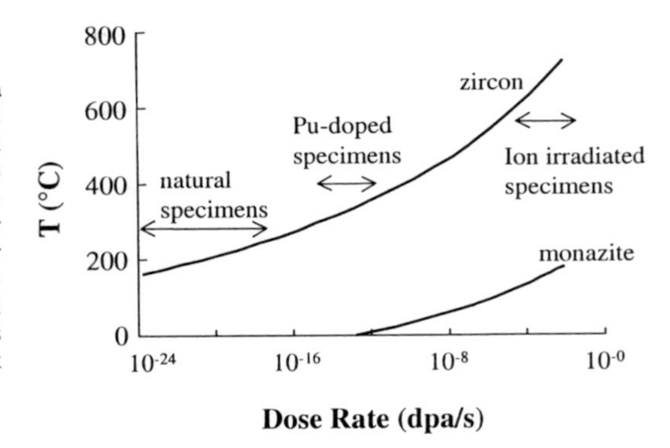

Figure 11. Variation of T_c as a function of dose rate for zircon and monazite, modified after Meldrum et al. (1999a). Reducing the dose rate effectively lowers the critical temperature since, for a given dose, a lower dose rate provides more time for thermal recrystallization. The calculated critical temperature of natural monazite is so low that amorphous specimens of pure monazite are not expected to occur in nature.

Ma, the uranium concentration will be 925 ppm, and the dose rate will have decreased to 4.8×10^{-18} dpa/s. The corresponding decrease in T_c is only 0.5°C and is ignored in the calculation. For most natural zircons, T_c is essentially constant at ~250°C. For minerals that incorporate higher concentrations of actinides (e.g. pyrochlore or monazite), T_c will be somewhat higher (e.g. 20,000 ppm eU results in a critical temperature of 380°C).

Following the analysis of Lumpkin and Ewing (1988), the minimum dose for the observable onset of amorphization can be calculated. According to Miller and Ewing (1992), the minimum amount of amorphous material that can be detected by electron diffraction is approximately 35% (this number can vary greatly as a function of specimen thickness). If a direct impact amorphization model is assumed, then the following equation relates the amorphous fraction, f_a, to the amorphization dose (Gibbons 1972):

$$f_a = 1 - exp(-BD_c) \tag{27}$$

where B is a constant ($B = 5.8 \times 10^{-19}$g for zircon; Weber et al. 1997). This is essentially Equation (9). According to Equation (27), if f_a decreases from 95% to 35%, $D_{c(\text{minimum detectable})}/D_{c(\text{complete amorphization})} = 0.15$. Since N_c is linearly related to D_0 (Eqn. 25), the minimum detectable amorphous fraction occurs at a value of only $0.15N_c$. This value is in good agreement with the earlier results of Lumpkin and Ewing (1988) who found by TEM analysis that the minimum observable α-decay dose for the onset of amorphization of pyrochlore is ~20% of the total amorphization dose.

This model for the crystalline-to-metamict transformation has been evaluated using literature data for zircon (Fig. 12). Crystalline and metamict specimens of zircon of known age and uranium concentration are plotted and compared to the curve obtained from Equations (25) and (26). The lines are plotted for a temperature of 100°C, and the ages of the metamict specimens were obtained by dating cogenetic crystalline zircon. The curves delineate the crystalline-metamict boundary fairly well, despite the uncertain thermal histories of the various specimens. Equations (26) and (27) can probably be generalized to any mineral for which sufficient ion beam irradiation data exist (Table 3).

Equation (25) demonstrates that metamictization can be employed as a qualitative geothermometer. A minimum temperature can be calculated for which a given mineral is expected to remain crystalline for a given uranium concentration. For example, according to Equation (25), one of the 4.2 Ga Mt. Narryer zircon specimens with 589 ppm U (Froude et al. 1983) must have experienced a minimum average temperature close to T_c (250°C) in order for it to have remained crystalline. Equation (25) can, therefore, be used to estimate a minimum average temperature. Equation (25) can also be used to calculate the crystalline-

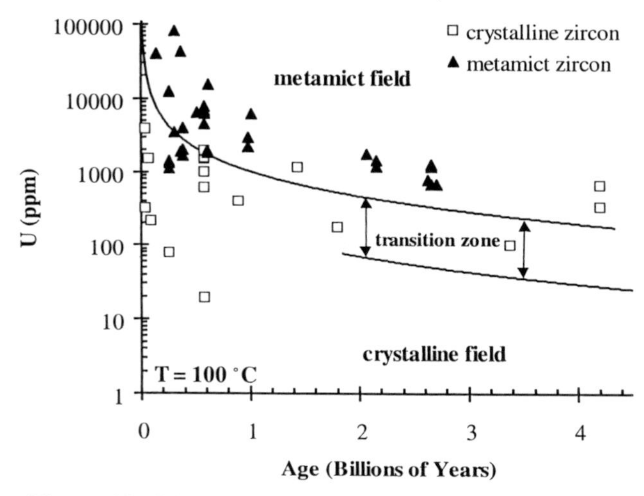

Figure 12. Calculation of the critical radionuclide content for zircon at a temperature of 100°C, modified after Meldrum et al. (1998a). The lower boundary marks the calculated onset for amorphization and the upper curve represents complete amorphization. The data points give the equivalent uranium concentration and age for zircon from a variety of localities. The open squares are for selected crystalline zircons, and the closed triangles are for metamict zircon whose age was assumed to be the same as that of cogenetic crystalline samples. The calculated lines separate the data fairly well despite the uncertain thermal histories of the various specimens. Specimen localities and references are given in Meldrum et al. (1998a).

to-amorphous transition for nuclear waste forms as a function of actinide content and temperature (Weber et al. 1997, Meldrum et al. 1998a).

This discussion emphasizes the use of ion irradiation experiments to model the crystalline-to-metamict transformation in minerals. The final calculation for N_c involves the use of constants obtained from several models relating the amorphization dose to temperature. Several variables must be obtained from these models (i.e. C, ε, T_c). As these models are further developed and refined, the constants may be improved. Despite the uncertainties in the models, Equation (25) is in reasonable agreement with the available data for zircon and pyrochlore. This approach can be generalized to any mineral for which temperature-dependent ion-beam data exist for the crystalline-to-amorphous transition.

ACKNOWLEDGMENTS

The research program by RCE on radiation effects in ceramics and minerals has been sustained by funding from the Office of Basic Energy Sciences of the U.S. Department of Energy. Without this long-term support, it would not have been possible to explore new areas of application or develop new techniques for studying radiation effects. The BES program has supported three Ph.D. students: M.L. Miller, G.R. Lumpkin, A. Meldrum and SX Wang, who have each made their own important contributions to the understanding of radiation effects in minerals. We also acknowledge Professor Takashi Murakami and Bryan Chakoumakos, who were the first in our group to begin the systematic study of radiation damage in zircon.

We also acknowledge the important role that W.J. Weber has played in our work. When we were not collaborating, we were competing, and in both modes we made progress. Lynn Boatner provided considerable inspiration, a steady stream of good crystals and did much to educate us in the proper use of ion-beam techniques.

Finally, we are grateful to the staff (L.L. Funk, E.A. Ryan and S.T. Ockers) at the HVEM-Tandem Facility at Argonne National Laboratory where much of the ion irradiation and *in situ* TEM was completed. Without their skilled assistance, we could not have made much progress in our research program.

REFERENCES

Aleshin E, Roy R (1962) Crystal chemistry of pyrochlore. J Am Ceram Soc 45:18-25

Allen CW, Ryan EA (1997) *In situ* ion beam research in Argonne's intermediate voltage microscope. Mater Res Soc Symp Proc 439:277-288

Anderson EB, Burakov BE, Pazukhin EM (1993) High-uranium zircon from "Chernobyl lavas." Radiochimica Acta 60:149-151

Averback RS (1994) Fundamental aspects of defect production in solids. Nucl Instr Meth B127/128:1-11

Bart F, L'Hermite V, Houpert S, Fillet C, Pacaud F, Jacquet-Francillon N (1997) Development of apatite-based glass ceramics for actinide immobilization. Proc 99th Annual Meeting Am Ceramic Soc, CD

Begg BD, Hess NJ, Weber WJ, Conradson SD, Schweiger MJ, Ewing RC (2000) XAS and XRD study of annealed ^{238}Pu- and ^{239}Pu-substituted zircons ($Zr_{0.92}Pu_{0.08}SiO_4$). J Nucl Mater 278:212-224

Bench MW, Robertson IM, Kirk MA (1991) Transmission electron microscopy investigation of the damage produced in individual displacement cascades in GaAs and GaP. Nucl Instr Meth B59/60:372-376

Bench MW, Robertson IM, Kirk MA, Jencic I (2000) Production of amorphous zones in GaAs by the direct impact of energetic heavy ions. J Appl Phys 87:49-56

Berzelius J (1814) An attempt to establish a pure scientific system of mineralogy by the application of the electro-chemical theory and the chemical properties. Afhandl Fys Kem Min 4: 217-353

Biagini R, Memmi I, Olmi F (1997) Radiation damage in zircons. Neues Jahrb Mineral Monatsh 6:257-270

Boatner LA, Sales BC (1988) Monazite. *In* Radioactive Waste Forms For the Future. W Lutze, RC Ewing (eds) Elsevier, Amsterdam, p 495-564

Bohr N (1913) On the theory of the decrease of velocity of moving electrified particles on passing through matter. Phil Mag and J Science 25:10-31

Bohr N (1948) The penetration of atomic particles through matter. Dansk Vidensk Selsk Mat Fys Medd 18:1-144

Bolse W (1998) Formation and development of disordered networks in Si-based ceramics under ion bombardment. Nucl Instr Meth B141:133-139

Bordes N, Ewing RC (1995) Ion-beam and electron beam induced amorphization of berlinite ($AlPO_4$). Mater Res Soc Symp Proc 373:395-400

Bordes N, Wang LM, Ewing RC, Sickafus KE (1995) Ion-beam induced disordering and onset of amorphization in spinel by defect accumulation. J Mater Res 10:981-985

Boczar PG, Gagnon MJN, Chan PSW, Ellis RJ, Verrall RA, Dastur AR (1997) Advanced CANDU systems for plutonium destruction. Can Nucl Soc Bull 18:2-10

Brinkman JA (1954) On the nature of radiation damage in metals. J Appl Phys 25:961-970

Brøgger WC (1893) Amorf Salmonsens store illustrerede. Konversationslexikon 1:742-743

Bros R, Carpena J, Sere V, Beltritti A (1996) Occurrence of Pu and fissiogenic REE in hydrothermal apatites from the fossil nuclear reactor 16 at Oklo (Gabon). Radiochim Acta 74:277-282

Brown WL (1988) The many faces of ion beam interactions with insulators. Nucl Instr Meth B32:1-10

Butterman WC and Foster WR (1967) Zircon stability and the ZrO_2-SiO_2 phase diagram. Am Mineral 52:880-885

Cameron M, Wang LM, Crowley KD, Ewing RC (1992) HRTEM observations on electron irradiation damage in F-apatite. Proc 15th annual meeting Electron Microscopy Society of America. GW Bailey, J Bentley, JA Small (eds) San Francisco Press, San Francisco, p 378-379

Campisano SU, Coffa S, Raineri V, Priolo F, Rimini E (1993) Mechanisms of amorphization in ion-implanted crystalline silicon. Nucl Instr Meth B80:514-518

Carter G, Nobes MJ (1991) A phenomenological model of ion-induced crystallization and amorphization. J Mater Res 10:2103-2107

Cartz L, Karioris FG, Fournelle RA, Appaji Gowda K, Ramasami K, Sarkar G, Billy M. (1981) Metamictization by heavy ion bombardment of α-quartz, zircon, monazite, and nitride structures. *In* JG Moore (ed) Scientific Basis for Nuclear Waste Management. Plenum Press, New York, p 421-427
Cerny P and Povandra P (1972) An Al, F-rich metamict titanite from Czechoslovakia. Neues Jahrb Mineral Monatsh 400-406
Cerny P and Sanseverino LR (1972) Comments on the crystal chemistry of titanite. Neues Jahrb Mineral. Monatsh 97-103
Chakoumakos BC (1984) Systematics of the pyrochlore structure type, ideal $A_2B_2X_6Y$. J Solid State Chem 53:120-129
Chakoumakos BC, Ewing RC (1985) Crystal chemical constraints on the formation of actinide pyrochlores. Mater Res Soc Symp Proc 44:641-646
Chakoumakos BC, Murakami T, Lumpkin GR, and Ewing RC (1987) Alpha-decay-induced fracturing in zircon: The transition from the crystalline to the metamict state. Science 236:1556-1559
Cherniak DJ (1993) Lead diffusion in titanite and preliminary results on the effects of radiation damage on Pb transport. Chem Geology 110:177-194
Chrosch J, Colombo M, Malcherek T, Salje EKH, Groat LA, Bismayer U (1998) Thermal annealing of radiation damaged titanite. Am Mineral 83:1083-1091
Clinard FW Jr (1986) Review of self-irradiation effects in Pu-substituted zirconolite. Ceram Bull 65:1181-1187
Clinard FW, Rohr DL Jr., Roof RB (1984) Structural damage in a self-irradiated zirconolite-based ceramic. Nucl Instr Meth B 1:581-586
Cooper AR (1978) Zachariasen's rules, Madelung constant, and network topology. Phys Chem Glasses 19:60-68
Davies JA (1992) Fundamental concepts of ion-solid interactions: Single ions, 10^{-12} seconds. Mater Res Soc Bull 17:26-29
Dachille F, Roy R (1964) Effectiveness of shearing stresses in accelerating solid phase reactions at low temperatures and high pressures. J Geol 72:243-247
Degueldre C, Paratte JM (1998) Basic properties of a zirconia-based fuel material for light water reactors. Nucl Tech 123:21-29
Deliens M, Delhal J, Tarte P (1977) Metamictization and U-Pb systematics–a study by infrared absorption spectrometry of Precambrian zircons. Earth Planet Sci Lett 33:331-344
Della Ventura G, Mottana A, Parodi GC, Raudsepp M, Bellatreccia F, Caprilli E, Rossi P, Fiori S (1996) Monazite-huttonite solid-solutions from the Vico volcanic complex, Latium, Italy. Mineral Mag 60:751-758
Dennis JR, Hale EB (1976) Amorphization of silicon by ion implantation: Homogeneous or heterogeneous nucleation? Radiation Effects 30:219-225
Dennis JR, Hale EB (1978) Crystalline to amorphous transformation in ion-implanted silicon: a composite model. J Appl Phys 49:1119-1127
Devanathan R, Yu N, Sickafus KE, Nastasi M (1996a) Structure and properties of irradiated magnesium aluminate spinel. J Nucl Mater 232:59-64
Devanathan R, Yu N, Sickafus KE, Nastasi M (1996b) Structure and property changes in spinel irradiated with heavy ions. Nucl Instr Meth B127/128:608-611
Devanathan R, Weber WJ, Boatner LA (1998) Response of zircon to electron and Ne^+ irradiation. Mater Res Soc Symp Proc 481:419-424
Eby RK, Ewing RC, Birtcher RC (1992) The amorphization of complex silicates by ion-beam irradiation. J Mater Res 7:3080-3102
Eby RK, Janeczek J, Ewing RC, Ercit TS, Groat LA, Chakoumakos BC, Hawthorne FC, Rossman GR (1993) Metamict and chemically altered vesuvianite. Can Mineral 31:357-369
Ehlert TC, Gowda KA, Karioris FG, Cartz L (1983) Differential scanning calorimetry of heavy ion bombarded synthetic monazite. Radiation Effects 70:173-181
Ellsworth S, Navrotsky A, Ewing RC (1994) Energetics of radiation damage in natural zircon ($ZrSiO_4$). Phys Chem Mineral 21:140-149
Ewing RC (1975) The crystal chemistry of complex niobium and tantalum oxides IV. The metamict state: Discussion. Am Mineral 60:728-733
Ewing RC (1994) The metamict state: 1993—the centennial. Nucl Instr Meth B91:22-29.
Ewing RC (1999) Nuclear waste forms for actinides. Proc National Academy Science, USA 96:3432-3439
Ewing RC, Haaker RF (1980) The metamict state: Implications for radiation damage in crystalline waste forms. Nuclear Chem Waste Management 1:51-57
Ewing RC, Headley TJ (1983) Alpha-recoil damage in natural zirconolite ($CaZrTi_2O_7$). J Nucl Mater 119:102-109

Ewing RC, Wang LM (1992) Amorphization of zirconolite: Alpha-decay event damage versus krypton ion irradiations. Nucl Instr Meth B65:319-323

Ewing RC, Chakoumakos BC, Lumpkin GR, Murakami T (1987) The metamict state. Mater Res Soc Bulletin 12:58-66

Ewing RC, Chakoumakos BC, Lumpkin GR, Murakami T, Greegor RB, Lytle FW (1988) Metamict minerals: natural analogues for radiation damage effects in ceramic nuclear waste forms. Nucl Instr Meth B32:487-497

Ewing RC, Weber WJ, Clinard, FW Jr. (1995a) Radiation effects in nuclear waste forms. Progress Nucl Energy 29:63-127

Ewing RC, Lutze W, Weber WJ (1995b) Zircon: A host-phase for the disposal of weapons plutonium. J Mater Res 10:243-246

Exarhos GJ (1984) Induced swelling in radiation damaged $ZrSiO_4$. Nucl Instr Meth B1:538-541

Eyal Y, Olander DR (1990) Leaching of uranium and thorium from monazite: I. Initial leaching. Geochim Cosmochim Acta 54:1867-1877

Farges F (1994) The structure of metamict zircon: A temperature-dependent EXAFS study. Phys Chem Minerals 20:504-514

Farges F, Calas G (1991) Structural analysis of radiation damage in zircon and thorite: An X-ray absorption spectroscopic study. Am Mineral 76:60-73

Farges F, Ewing RC, Brown GE Jr. (1993) The structure of aperiodic, metamict $(Ca,Th)ZrTi_2O_7$ (zirconolite): An EXAFS study of the Zr, Th, and U sites. J Mater Res 8:1983-1995

Fleet ME, Henderson GS (1985) Radiation damage in natural titanite by crystal structure analysis. Mater Res Soc Symp Proc 50:363-370

Fleischer RL (1975) Nuclear Tracks in Solids. Univ. Calif. Press, 605 p

Foltyn EM, Clinard FW Jr, Rankin J, Peterson DE (1985) Self-irradiation effects in ^{238}Pu-substituted zirconolite: II. Effect of damage microstructure on recovery. J Nucl Mater 136:97-103

Frondel C (1958) Hydroxyl substitution in thorite and zircon. Am Mineral 38:1007-1018

Froude DO, Ireland TR, Kinney PD, Williams IS, Compston W, Williams IR, Myers JS (1983) Ion microprobe identification of 4,100-4,200 Myr.-old terrestrial zircons. Nature 304:616-618

Gibbons JG (1972) Ion implantation in semiconductors—Part II: Damage production and annealing. Proc IEEE 60:1062-1096

Gong WL, Wang LM, Ewing RC, Fei Y (1996a) Surface and grain-boundary amorphization: Thermodynamic melting of coesite below the glass transition temperature. Phys Rev B53:2155-2158

Gong WL, Wang LM, Ewing RC, Zhang J (1996b) Electron-irradiation- and ion-beam-induced amorphization of coesite. Phys Rev B54:3800-3808

Gong WL, Wang LM, Ewing RC, Chen LF, Lutze W (1997) Transmission electron microscopy study of α-decay damage in aeschynite and britholite. Mater Res Soc Symp Proc 465:649-656

Gong WL, Wang LM, Ewing RC (1998) Cross-sectional transmission electron microscopy study of 1.5 MeV Kr^+ irradiation-induced amorphization in α-quartz. J Appl Phys 84:4204-4208

Grammacioli CM, Segalstad TV (1978) A uranium- and thorium-rich monazite from a south-alpine pegmatite at Piona, Italy. Am Mineral 63:757-761

Greegor RB, Lytle FW, Ewing RC, Haaker RF (1984) Ti-site geometry in metamict, annealed and synthetic complex Ti-Nb-oxides by X-ray absorption spectroscopy. Nucl Instr Meth B1:587-594

Greegor RB, Lytle FW, Chakoumakos BC, Lumpkin GR, Ewing RC (1985a) An investigation of metamict and annealed natural pyrochlore by X-ray absorption spectroscopy. Mater Res Soc Symp Proc 44:655-662

Greegor RB, Lytle FW, Chakoumakos BC, Lumpkin GR, Ewing RC (1985b) An investigation of uranium L-edges of metamict and annealed betafite. Mater Res Soc Symp Proc 50:387-392

Greegor RB, Lytle FW, Chakoumakos BC, Lumpkin GR, Ewing RC, Spiro CL, Wong J (1987) An X-ray absorption spectroscopy investigation of the Ta site in alpha-recoil damaged natural pyrochlore. Mater Res Soc Symp Proc 84:645-658

Greegror RB, Lytle FW, Chakoumakos BC, Lumpkin GR, Warner JK, Ewing RC (1989) Characterization of radiation damage at the Nb site in natural pyrochlores and samarskites by X-ray absorption spectroscopy. Mater Res Soc Symp Proc 127:261-268

Gupta PK (1993) Rigidity, connectivity, and glass-forming ability. J Am Ceram Soc 76:1088-1095

Hamberg A (1914) Die radioaktiven Substanzen und die geologische Forschung. Geol För Förh 36:31-96

Hawthorne FC, Groat LA, Raudsepp M, Ball NA, Kimata M, Spike FD, Gaba R, Halden NM, Lumpkin GR, Ewing RC, Greegor RB, Lytle FW, Ercit TS, Rossman GR, Wicks FJ, Ramik RA, Sherriff BL, Fleet ME, McCammon C (1991) Alpha-decay damage in titanite. Am Mineral 76:370-396

Hazen RM, Finger LW (1979) Crystal structure and compressibility of zircon at high pressure. Am Mineral 64:196-201

Headley TJ, Ewing RC, Haaker RF (1981) High resolution study of the metamict state in zircon. Proc 39th Annual Meeting Electron Microscopy Society of America, p 112-113

Headley TJ, Arnold GW, Northrup CJM (1982a) Dose-dependence of Pb-ion implantation damage in zirconolite, hollandite, and zircon. Mater Res Soc Symp Proc 11:379-387

Headley TJ, Ewing RC, Haaker RF (1982b) TEM study of the metamict state. Physics of Minerals and Ore Micrscopy—IMA 1982, p 281-289

Higgins JB, Ribbe PH (1976) The crystal chemistry and space groups of natural and synthetic titanites. Am Mineral 61:878-888

Hobbs LW (1979) Application of transmission electron microscopy to radiation damage in ceramics. J Am Ceram Soc 62:267-278

Hobbs LW (1994) Topology and geometry in the irradiation-induced amorphization of insulators. Nucl Instr Meth B91:30-42

Hobbs LW, Clinard FW Jr., Zinkle SJ, Ewing RC (1994) Radiation effects in ceramics. J Nucl Mater 216:291-321

Holland HD, Gottfried D (1955) The effect of nuclear radiation on the structure of zircon. Acta Crystallogr 8:291-300

Ibarra A, Vila R, Garner FA (1996) Optical and dielectric properties of neutron-irradiated $MgAl_2O_4$ spinel. J Nucl Mater 237:1336-1339

Jackson KA (1988) A defect model for ion-induced crystallization and amorphization. J Mater Res 3:1218-1226

Janeczek J, Eby RK (1993) Annealing of radiation damage in allanite and gadolinite. Phys Chem Minerals 19:343-356

Johnson WL (1988) Crystal-to-glass transformation in metallic materials. Mater Sci Eng 97:1-13

Karioris FG, Appaji Gowda K, Cartz L (1981) Heavy ion bombardment of monoclinic $ThSiO_4$, ThO_2 and monazite. Radiation Effects Lett 58:1-3

Karioris, FG, Appaji Gowda K, Cartz L, Labbe JC (1982) Damage cross- sections of heavy ions in crystal structures. J Nucl Mater 108/109:748-750

Keller VC (1963) Untersuchungen über die Germanate und Silikate des typs ABO_4 der vierwertigen Elemente Thorium bis Americium. Nukleonik 5:41-48

Kinchin GH, Pease RS (1955) The displacement of atoms in solids by radiation. Rep Prog Phys 18:1-51

Kulp JL, Volchok HL, Holland HD (1952) Age from metamict minerals. Am Mineral 37:709-718

Lagow BW, Turkot BA, Robertson IM, Coleman JJ, Roh SD, Forbs DV, Rehn LE, Baldo PM (1998) Ion implantation in $Al_xGa_{1-x}As$: Damage structures and amorphization mechanisms. IEEE J Selected Topics Quantum Electronics 4:606-618

Linberg ML, Ingram B (1964) Rare-earth silicatian apatite from the Adirondack mountains, New York. U S Geol Surv Prof Paper 501-B:B64-B65

Lindhard J, Nielsen V, Scharff M, Thomsen PV (1963) Integral equations governing radiation effects (Notes on atomic collisions (III). K Dan Vidensk Matematisk-fysiske Meddelelser 33:1-42

Lindhard J, Nielsen V, Scharff M (1968) Approximation method in classical scattering by screened coulomb fields (Notes on atomic collisions (I). K Dan Vidensk Selsk Matematisk-fysiske Meddelelser 36(10):1-32

Lumpkin GR (1998) Rare-element mineralogy and internal evolution of the Rutherford #2 pegmatite, Amelia County, Virginia: A classic locality revisited. Can Mineral 36:339-353

Lumpkin GR, Chakoumakos BC (1988) Chemistry and radiation effects of thorite-group minerals from the Harding pegmatite, Taos County, New Mexico. Am Mineral 73:1405-1419

Lumpkin GR, Ewing RC (1988) Alpha-decay damage in minerals of the pyrochlore group. Phys Chem Mineral 16:2-20

Lumpkin GR, Ewing RC (1992) Geochemical alteration of pyrochlore group minerals: Microlite subgroup. Am Mineral 77:179-188

Lumpkin GR Ewing RC (1995) Geochemical alteration of pyrochlore group minerals: Pyrochlore subgroup. Am Mineral 80:732-743

Lumpkin GR, Ewing RC (1996) Geochemical alteration of pyrochlore group minerals: Betafite subgroup. Am Mineral 81:1237-1248

Lumpkin GR, Ewing RC, Chakoumakos BC, Greegor RB, Lytle FW, Foltyn EM, Clinard FW Jr, Boatner LA, Abraham MM (1986) Alpha-recoil damage in zirconolite ($CaZrTi_2O_7$). J Mater Res 1:564-576

Lumpkin GR, Eby RK, Ewing RC (1991) Alpha-recoil damage in titanite ($CaTiSiO_5$): Direct observation and annealing study using high resolution transmission electron microscopy. J Mater Res 6:560-564

Lumpkin GR, Smith KL, Blackford MG, Gieré R, Williams CT (1994) Determination of 25 elements in the complex oxide mineral zirconolite by analytical electron-microscopy. Micron 25:581-587

Lumpkin GR, Smith KL, Gieré R (1997) Application of analytical electron microscopy to the study of radiation damage in the complex oxide mineral zirconolite. Micron 28:57-68

Lumpkin GR, Smith KL, Blackford MG, Gieré R, Williams CT (1998) The crystalline-amorphous transformation in natural zirconolite: evidence for long-term annealing. Mater Res Soc Symp Proc 506:215-222

Lumpkin GR, Day RA, McGlinn PJ, Payne TE, Gieré R, Williams CT (1999) Investigation of the long-term performance of betafite and zirconolite in hydrothermal veins from Adamello, Italy. Mater Res Soc Symp Proc 556:793-800

Lutze W, Ewing RC (1988) Radioactive Waste Forms for the Future. North-Holland, Amsterdam, 778 p

Makayev AI, Levchenkov OA, Bubnova RS (1981) Radiation damage as an age measure for natural zircons. Geokhimiya 2:175-182 (in Russian)

McCauley RA (1980) Structural characteristics of pyrochlore formation. J Appl Phys 51:290-294

Meldrum A, Wang LM, Ewing RC (1996) Ion beam-induced amorphization of monazite. Nucl Instr Meth B116:220-224

Meldrum A, Wang LM, Ewing RC (1997a) Electron-irradiation-induced phase segregation in crystalline an amorphous apatite: A TEM study. Am Mineral 82:858-869

Meldrum A, Boatner LA, Wang LM, Ewing RC (1997b) Displacive irradiation effects in the monazite- and zircon-structure orthophosphates. Phys Rev B56:13805-13814

Meldrum A, Boatner LA, Weber WJ, Ewing RC (1998a) Radiation damage in zircon and monazite. Geochim Cosmochim Acta 62:2509-2520

Meldrum A, Zinkle SJ, Boatner LA, Ewing RC (1998b) A transient liquid-like state in the displacement cascades of zircon, hafnon, and thorite. Nature 395:56-58

Meldrum A, Boatner LA, Ewing RC (1998c) Effects of ionizing and displacive irradiation on several perovskite-structure oxides. Nucl Instr Meth B141:347-352

Meldrum A, Zinkle SJ, Boatner LA, Wang SX, Wang LM, Ewing RC (1999a) Effects of dose rate and temperature on the crystalline-to-metamict transformation in the ABO_4 orthosilicates. Can Mineral 37:207-221

Meldrum A, Zinkle SJ, Boatner LA, Ewing RC (1999b) Heavy-ion irradiation effects in the ABO_4 orthosilicates: decomposition, amorphization, and recrystallization. Phys Rev B59:3981-3992

Meldrum A, Boatner LA, Ewing RC (2000a) A comparison of radiation effects in crystalline ABO_4-type phosphates and silicates. Mineral Mag 64:183-192

Meldrum A, Boatner LA, White CW, Ewing RC (2000b) Ion irradiation effects in nonmetals: Formation of nanocrystals and novel microstructures. Mater Res Innovations 3:190-204

Meldrum A, White CW, Keppens V, Boatner LA, Ewing RC (2000c) Irradiation-induced amorphization of $Cd_2Nd_2O_7$ pyrochlore. Can J Phys (submitted)

Miller ML, Ewing RC (1992) Image simulation of partially amorphous materials. Ultramicroscopy 48:203-237

Mitchell RS (1973) Metamict minerals: A review. Mineralogical Record 4:214-223

Morehead FF Jr., Crowder BL (1970) A model for the formation of amorphous Si by ion bombardment. Radiat Eff 6:27-32

Motta AT (1997) Amorphization of intermetallic compounds under irradiation—a review. J Nucl Mater 244:227-250

Motta AT, Olander DR (1990) Theory of electron-irradiation-induced amorphization. Acta Materialia 38:2175-2185

Murakami T, Chakoumakos BC, Ewing RC, Lumpkin GR, Weber WJ (1991) Alpha-decay event damage in zircon. Am Mineral 76:1510-1532

Naguib HM, Kelly R (1975) Criteria for bombardment-induced structural changes in non-metallic solids. Radiation Effects 25:1-12

Nasdala L, Irmer G, Wolf D (1995) The degree of metamictization in zircon: a Raman spectroscopic study. Neues Jahrb Mineral 7:471-478

Nasdala L, Pidgeon RT, Wolf D (1996) Heterogeneous metamictization of zircon on a microscale. Geochim Cosmochim Acta 60:1091-1097

Nastasi M, Mayer JW, Hirvonen JK (1996) Ion-Solid Interactions. Cambridge University Press, Cambridge, UK, 540 p

Okamoto PR, Lam NQ, Rehn LE (1999) Physics of crystal-to-glass transformations. Solid State Phys 52:1-135

Oliver WC, McCallum JC, Chakoumakos BC, Boatner LA (1994) Hardness and elastic modulus of zircon as a function of heavy-particle irradiation dose: II. Pb-ion implantation damage. Radiation Effects Defects in Solids 132:131-141

Pabst A (1952) The metamict state. Am Mineral 37, 137-157

Pabst A, Hutton CO (1951) Huttonite: a new monoclinic thorium silicate. Am Mineral 36:60-69

Pedraza DF (1986) Mechanisms of the electron irradiation-induced amorphous transition in intermetallic compounds. J Mater Res 1:425-441

Pells GP (1991) Radiation-induced electrical conductivity in $MgAl_2O_4$ spinel. J Nucl Mater 184:183-190
Ringwood AE, Kesson SE, Ware NG, Hibberson W, Major A (1979) Immobilisation of high level nuclear reactor wastes in SYNROC. Nature 278:219-223
Ringwood AE, Kesson SE, Reeve KD, Levins DM, Ramm EJ (1988) *In* Radioactive Waste Forms for the Future. W Lutze, RC Ewing (eds) North-Holland, Amsterdam, 233-334
Ríos S, Salje EKH, Zhang M, Ewing RC (2000) Amorphization in zircon: evidence for direct impact damage. J Phys: Condens Mater 12:2401-2412
Robinson MT (1983) Computer simulation of collision cascades in monazite. Physical Review B 27:5347-5359
Robinson MT (1993) Computer simulation of atomic collision processes in solids. Mater Res Soc Symp Proc 279:3-16
Robinson MT (1994) Basic physics of radiation damage production. J Nucl Mater 216:1-28
Salje EKH, Chrosch J, Ewing RC (1999) Is "metamictization" of zircon a phase transition? Am Mineral 84:1107-1116
Seitz F (1952) Radiation effects in solids. Physics Today 5:6-9
Sickafus KE, Larson AG, Yu N, Nastasi M, Hollenburg GW, Garner FA, Bradt RC (1995) Cation disordering in high-dose neutron-irradiated spinel. J Nucl Mater 219:128-134
Sickafus KE, Matzke Hj, Hartmann Th, Yasuda K, Valdez JA, Chodak P III, Nastasi M, Verrall RA (1999) Radiation damage effects in zirconia. J Nucl Mater 274:66-77.
Sinclair W, Ringwood AE (1981) Alpha-recoil damage in natural zirconolite and perovskite. Geochemical Journal 15:229-243
Smith KL, Zaluzec NJ, Lumpkin GR (1997) In situ studies of ion irradiated zirconolite, pyrochlore and perovskite. J Nucl Mater 250:36-52
Smith KL, Zaluzec NJ, Lumpkin GR (1998) The relative radiation resistance of zirconolite, pyrochlore, and perovskite to 1.5 MeV Kr^+ ions. Mater Res Soc Symp Proc 506:931-932
Spaczer M, Caro A, Victoria M (1995) Evidence of amorphization in molecular-dynamics simulations on irradiated intermetallic NiAl. Phys Rev B 52:7171-7178
Speer JA (1982) The actinide orthosilicates. Rev Mineral 5:113-135
Tan HH, Jagadish C, Williams JS, Zou J, Cockayne DJH (1996) Ion damage buildup and amorphization processes in GaAs-AlGaAs multilayers. J Appl Phys 80:2691-2701
Taylor M, Ewing RC (1978) The crystal structures of the $ThSiO_4$ polymorphs: Huttonite and thorite. Acta Crystallogr B34:1074-1079
Tesmer JR and Nastasi M (editors) Handbook of Modern Ion Beam Materials Analysis. Materials Research Society, Pittsburgh, 704 p
Thevuthasan S, Jiang W, Weber WJ, McCready DE (1999) Damage accumulation and thermal recovery in $SrTiO_3$ implanted with Au^{2+} ions. Mat Res Soc Symp Proc 540:373-378.
Titov AI, Carter G (1996) Defect accumulation during room temperature N^+ irradiation of silicon. Nucl Instr Meth B119:491-500
Trachenko K, Dove MT, Salje EKH (2000) Modelling the percolation-type transition in radiation damage. J Appl Phys 87:7702-7707
Turkot BA, Robertson IM, Rehn LE, Baldo PM, Forbes DV, Coleman JJ (1998) On the amorphization processes in $Al_{0.6}Ga_{0.4}As$/GaAs heterostructures. J Appl Phys 83:2539-2547
Ueda T (1957) Studies on the metamictization of radioactive minerals. Memoirs Faculty of Engineering, Kyoto University 24:81-120
Van Konynenburg RA, Guinan MW (1983) Radiation effects in Synroc-D. Nuclear Technology 60:206-217
Vance ER, Anderson BW (1972) Study of metamict Ceylon zircons. Mineral Mag 38:605-613
Vance ER, Metson JB (1985) Radiation damage in natural titanites. Phys Chem Minerals 12:255-260
Vance ER, Karioris FG, Cartz L, Wong MS (1984) Radiation effects on sphene and sphene-based glass-ceramics. *In* Advances in Ceramics Vol 8, GG Wicks, WA Ross (eds) American Ceramic Society, Westerville, OH, p 62-70
Wald JW, Offermann P (1982) A study of radiation effects in curium-doped $Gd_2Ti_2O_7$ (pyrochlore) and $CaZrSi_2O_7$ (zirconolite). Mater Res Soc Sym Proc 11:368-378
Wang LM (1998) Application of advanced electron microscopy techniques to the studies of radiation effects in ceramic materials. Nucl Instr Meth B141:312-325
Wang LM, Ewing RC (1992a) Ion beam induced amorphization of complex ceramic materials—minerals. Mater Res Soc Bull XVII:38-44
Wang LM, Ewing RC (1992b) Detailed in situ study of ion beam-induced amorphization of zircon. Nucl Instr Meth B65:324-329
Wang LM, Ewing RC (1992c) Ion beam-induced amorphization of $(Mg,Fe)_2SiO_4$ olivine series: An *in situ* transmission electron microscopy study. Mater Res Soc Symp Proc 235:333-338

Wang LM, Weber WJ (1999) Transmission electron microscopy study of ion-beam-induced amorphization of $Ca_2La_8(SiO_4)_6O_2$. Phil Mag A79:237-253

Wang LM, Eby RK, Janeczek J, Ewing RC (1991) In situ TEM study of ion-beam-induced amorphization of complex silicate structures. Nucl Instr Meth B59/B60:395-400

Wang LM, Cameron M, Weber WJ, Crowley KD, Ewing RC (1994) *In situ* TEM observation of radiation induced amorphization of crystals with apatite structure. *In* Hydroxyapatite and Related Materials. P.W. Brown and B. Constantz (eds) CRC Press, London, p 243-249

Wang LM, Gong WL, Bordes N, Ewing RC, Fei Y (1995) Effects of ion dose and irradiation temperature on the microstructure of three spinel compositions. Mater Res Soc Symp Proc 373:407-412

Wang LM, Wang SX, Gong WL, Ewing RC (1998a) Temperature dependence of Kr ion-induced amorphization of mica minerals. Nucl Instr Meth B 141:501-508

Wang LM, Wang SX, Gong WL, Ewing RC, Weber WJ (1998b) Amorphization of ceramic materials by ion beam irradiation. Mater Sci Eng A253:106-113

Wang LM, Gong WL, Wang SX, Ewing RC (1999) Comparison of ion-beam irradiation effects in X_2YO_4 compounds. J Am Ceram Soc 82:3321-3329

Wang LM, Wang SX, Ewing RC (2000a) Amorphization of cubic zirconia by cesium ion implantation. Phil Mag Lett 80:341-347.

Wang LM, Wang SX, Ewing RC, Meldrum A, Birtcher RC, Newcomer Provencio P, Weber WJ, Matzke Hj (2000b) Irradiation-induced nanostructures. Mater Sci Eng A286:72-80

Wang SX (1997) Ion beam irradiation-induced amorphization: nano-scale glass formation by cascade quenching. PhD Thesis, University of New Mexico

Wang SX, Wang LM, Ewing RC (1997) Ion irradiation-induced amorphization in the $MgO-Al_2O_3-SiO_2$ system: a cascade quenching model. Mater Res Soc Symp Proc 439:619-624

Wang SX, Wang LM, Ewing RC, Doremus RH (1998a) Ion beam-induced amorphization and glass formation in $MgO-Al_2O_3-SiO_2$. I. Experimental and theoretical basis. J Non-Cryst Solids 238:198-213

Wang SX, Wang LM, Ewing RC, Doremus RH (1998b) Ion beam-induced amorphization and glass formation in $MgO-Al_2O_3-SiO_2$. II. Empirical model. J Non-Cryst Solids 238:214-224

Wang SX, Wang LM, Ewing RC, Kutty KVG (1999a) Ion irradiation effects for two pyrochlore compositions: $Gd_2Ti_2O_7$ and $Gd_2Zr_2O_7$. Mater Res Soc Symp Proc 540:355-360

Wang SX, Wang LM, Ewing RC, Was GS, Lumpkin GR (1999b) Ion irradiation-induced phase transformation of pyrochlore and zirconolite. Nucl Instr Meth B148:704-709

Wang SX, Begg BD, Wang LM, Ewing RC, Weber WJ, Kutty KVG (1999c) Radiation stability of gadolinium zirconate: a waste form for plutonium disposition. J Mater Res 14:4470-4473

Wang SX. Wang LM, Ewing RC (2000a) Electron and ion irradiation of zeolites. J Nucl Mater 278:233-241

Wang SX, Wang LM, Ewing RC, Kutty KVG (2000b) Ion irradiation of rare-earth- and yttrium-titanate-pyrochlores. Nucl Instr Meth B (in press)

Wang SX, Lumpkin GR, Wang LM, Ewing RC (2000c) Ion irradiation-induced amorphization of six zirconolite compositions. Nucl Instr Meth B (in press)

Wang SX, Wang LM, Ewing RC (2000d) Nano-scale glass formation in pyrochlore by heavy ion irradiation. J Non-Cryst Solids (in press)

Wang SX, Wang LM, Ewing RC (2000e) Irradiation-induced-amorphization: the effects of temperature, ion species, dose rate and materials properties. Phys Rev B (submitted)

Webb R and Carter G (1979) Difficulties in deducing disordering mechanisms from experimental studies of disorder – ion fluence functions in ion irradiation of semiconductors. Radiation Effects 42:159-168

Weber WJ (1983) Radiation-induced swelling and amorphization in $Ca_2Nd_8(SiO_4)_6O_2$. Radiation Effects 77:295-308

Weber WJ (1990) Radiation-induced defects and amorphization in zircon. J Mater Res 5:2687-2697

Weber WJ (1991) Self-radiation damage and recovery in Pu-doped zircon. Radiation Effects Defects in Solids 115:341-349

Weber WJ (1993) Alpha-decay-induced amorphization in complex silicate structures. J Am Ceram Soc 76:1729-1738

Weber WJ (2000) Models and mechanisms of irradiation-induced amorphization in ceramics. Nucl Instr Meth (in press)

Weber WJ, Matzke Hj (1986) Effects of radiation on microstructure and fracture properties in $Ca_2Nd_8(SiO_4)_6O_2$. Mater Lett 5:9-16

Weber WJ, Wang LM (1994) Effects of temperature and recoil-energy spectra on irradiation-induced amorphization in $Ca_2La_8(SiO_4)_6O_2$. Nucl Instr Meth B91:63-66

Weber WJ, Wald JW, Matzke Hj (1985) Self-radiation damage in $Gd_2Ti_2O_7$. Mater Lett 3:173-180

Weber WJ, Wald JW, Matzke Hj (1986) Effects of self-radiation damage in Cm-doped $Gd_2Ti_2O_7$ and $CaZrTi_2O_7$. J Nucl Mater 138:196-209

Weber WJ, Ewing RC, Wang LM (1994) The radiation-induced crystalline-to-amorphous transition in zircon. J Mater Res 9:688-698
Weber WJ, Ewing RC, Lutze W (1996) Performance assessment of zircon as a waste form for excess weapons plutonium under deep borehole burial conditions. Mater Res Soc Symp Proc 412:25-32
Weber WJ, Ewing RC, Meldrum A (1997) The kinetics of alpha-decay-induced amorphization in zircon and apatite containing weapons-grade plutonium or other actinides. J Nucl Mater 250:147-155
Weber WJ, Ewing RC, Catlow CRA, Diaz de la Rubia T, Hobbs LW, Kinoshita C, Matzke Hj, Motta AT, Nastasi M, Salje EHK, Vance ER, Zinkle SJ (1998) Radiation effects in crystalline ceramics for the immobilization of high-level nuclear waste and plutonium. J Mater Res 13:1434-1484
Weber WJ, Devanathan R, Meldrum A, Boatner LA, Ewing RC, Wang LM (1999) The effect of temperature and damage energy on amorphization of zircon. Mater Res Soc Symp Proc 540:367-372
Weber WJ, Jiang W, Thevuthasan S, Williford RE, Meldrum A, Boatner LA (2000) Ion-beam-induced defects and defect interactions in perovskite-structure titanates. *In* Defects and Surface-Induced Effects in Advanced Perovskites. Kluwer Academic Publishers, Dordrecht, The Netherlands (in press)
White CW, McHargue CJ, Sklad PS, Boatner LA, Farlow GC (1989) Ion implantation and annealing of crystalline oxides. Mater Sci Rep 4:41-146.
Williford RE, Weber WJ, Devanathan R, Cormack AN (1999) Native vacancy migrations in zircon. J Nucl Mater 273:164-170
Woodhead JA, Rossman GR, Silver LT (1991) The metamictization of zircon: radiation dose-dependent structural characteristics. Am Mineral 76:74-82
Yu N, Sickafus KE, Nastasi M (1994) First observation of amorphization in single crystal $MgAl_2O_4$ spinel. Phil Mag Lett 70:235-240
Yu N, Maggiore CJ, Sickafus KE, Nastasi M, Garner FA, Hollenberg GW, Bradt RC (1996) Ion beam analysis of $MgAl_2O_4$ spinel irradiated with fast neutrons to 50-250 dpa. J Nucl Mater 239:284-290
Zhang M, Salje EKH, Farnan I, Graeme-Barber A, Daniel P, Ewing RC, Clark AM, Leroux H (2000a) Metamictization of zircon: Raman spectroscopic study. J Phys: Condens Matter 12:1915-1925
Zhang M, Salje EKH, Capitani GC, Leroux H, Clark AM, Schlüter J, Ewing RC (2000b) Annealing of α-decay damage in zircon: a Raman spectroscopic study. J Phys: Condens Matter 12:3131-314
Zhang M, Salje EKH, Ewing RC, Farnan I, Ríos S, Schlüter J, Leggo P (2000c) α-decay damage and recrystallization in zircon: Evidence for an intermediate state from infrared spectroscopy. J Phys: Condens Matter (in press)
Ziegler JF (2000) The stopping and range of ions in matter (SRIM)-2000. http://www.research.ibm.com/ionbeams
Ziegler JF, Biersack JP, Littmark U (1985) The Stopping and Range of Ions in Solids. Pergamon Press, New York, 321 p
Ziemann P (1985) Amorphization of metallic systems by ion beams. Mater Sci Eng 69:95-103
Ziemann P, Miehle W, Plewnia A (1993) Amorphization of metallic alloys by ion bombardment. Nucl Instr Meth B80/81:370-378
Zinkle SJ (1995) Effect of irradiation spectrum on the microstructural evolution in ceramic insulators. J Nucl Mater 219:113-127
Zinkle SJ, Snead LL (1996) Influence of irradiation spectrum and implanted ions on the amorphization of ceramics. Nucl Instr Meth B116:92-101.
Zinkle SJ, Kinoshita C (1997) Defect production in ceramics. J Nucl Mater 252:200-217